# Atomic Force Microscopy for Energy Research

# Emerging Materials and Technologies

*Series Editor*
Boris I. Kharissov

Drug Delivery Using Nanomaterials
*Yasser Shahzad, Syed A.A. Rizvi, Abid Mehmood Yousaf, and Talib Hussain*

Nanomaterials for Environmental Applications
*Mohamed Abou El-Fetouh Barakat and Rajeev Kumar*

Nanotechnology for Smart Concrete
*Ghasan Fahim Huseien, Nur Hafizah A. Khalid, and Jahangir Mirza*

Nanomaterials in the Battle against Pathogens and Disease Vectors
*Kaushik Pal and Tean Zaheer*

MXene-Based Photocatalysts
Fabrication and Applications
*Zuzeng Qin, Tongming Su, and Hongbing Ji*

Advanced Electrochemical Materials in Energy Conversion and Storage
*Junbo Hou*

Emerging Technologies for Textile Coloration
*Mohd Yusuf and Shahid Mohammad*

Emerging Pollutant Treatment in Wastewater
*S.K. Nataraj*

Heterogeneous Catalysis in Organic Transformations
*Varun Rawat, Anirban Das, and Chandra Mohan Srivastava*

2D Monoelemental Materials (Xenes) and Related Technologies
Beyond Graphene
*Zongyu Huang, Xiang Qi, and Jianxin Zhong*

Atomic Force Microscopy for Energy Research
*Cai Shen*

Self-Healing Cementitious Materials
Technologies, Evaluation Methods, and Applications
*Ghasan Fahim Huseien, Iman Faridmehr, and Mohammad Hajmohammadian Baghban*

For more information about this series, please visit: https://www.routledge.com/Emerging-Materials-and-Technologies/book-series/CRCEMT

# Atomic Force Microscopy for Energy Research

Edited by

Cai Shen

CRC Press
Taylor & Francis Group
Boca Raton London New York

CRC Press is an imprint of the
Taylor & Francis Group, an **informa** business

First edition published 2022
by CRC Press
6000 Broken Sound Parkway NW, Suite 300, Boca Raton, FL 33487-2742

and by CRC Press
4 Park Square, Milton Park, Abingdon, Oxon, OX14 4RN



CRC Press is an imprint of Taylor & Francis Group, LLC



ISBN: 978-1-032-00407-5 (hbk)
ISBN: 978-1-032-00411-2 (pbk)
ISBN: 978-1-003-17404-2 (ebk)

DOI: 10.1201/9781003174042

Typeset in Times
by codeMantra

# Contents

# Preface

The invention of scanning tunneling microscopy (STM) in 1981 and atomic force microscopy (AFM) in 1986 completely changed human's ability to understand the world. Since then, a variety of STM/AFM-based scanning probe microscopy technologies have emerged with exciting innovations. Over the past 10 years, AFM technology has not been as "exciting" as it was when it was first invented. As a researcher who has used both STM and AFM technologies extensively during the past 15 years, I once fell into deep thinking and dilemma, and even doubted whether AFM technology is still an advanced science and technology. I am glad that after a period of careful observation, I found that there are many scientists in the world working hard in their respective fields and who give up. Just as I was soliciting for writing this book, I received warm response and support from scientists. It is a great inspiration and an overwhelming experience for me to see that those young scientists propel the diversity of AFM.

Compared to books that describe the application of AFM for the study of biology, little information can be found regarding books that describe the application of AFM for energy research. This is because biological systems are more about imaging; in the first place, AFM is an ideal tool like confocal microscopy for people to use it to image samples, whether in the atmosphere or in the liquid environment. The growing application of AFM for energy materials study has been witnessed mostly in the last decade. Nowadays, we are experiencing more severe global climate changes, making it necessary to more actively explore new energy materials. In the process of exploring new energy materials, humans find that only after fully understanding the surface/interface characteristics of materials, especially in the nanometer scale, they can design materials with better performance. In addition to simple imaging of samples, AFM has advanced models that can measure the microscopic electrical and mechanical characteristics of samples. Combined with the spectroscopy technology developed in recent years, AFM can achieve comprehensive characterization of sample morphology and physical and chemical characteristics that provide powerful real-time, direct evidence, revealing details and phenomena that were previously unobserved by other techniques. It should be noted that the application of AFM in the field of energy has made AFM leap from the characterization of the basic morphology of samples to the characterization of material physicochemical properties that can form complementary evidence with the characterization results of macro scale and strongly support the research in modern materials science.

The aim of this book is to describe the basic principles of AFM as well as the advanced modes that are powerful tools for the study of energy materials. The main contents include Chapter 1 Principles and Basic Modes of Atomic Force Microscopy; Chapter 2 Advanced Modes of Electrostatic and Kelvin Probe Force Microscopy for Energy Applications; Chapter 3 Piezoresponse Force Microscopy and Electrochemical Strain Microscopy; Chapter 4 Hybrid AFM Technique: Atomic Force Microscopy – Scanning Electrochemical Microscopy; Chapter 5 Scanning Microwave Impedance Microscopy; Chapter 6 Atomic Force Microscopy-based

Infrared Microscopy for Chemical Nano-imaging and Spectroscopy; Chapter 7 Application of AFM in Lithium Batteries Research; Chapter 8 Application of AFM in Solar Cell Research; Chapter 9 Application of AFM for Analyzing the Microstructure of Ferroelectric Polymer as an Energy Material; Chapter 10 Application of AFM in Microbial Energy Systems; and Chapter 11 Practical Guidance of AFM Operations for Energy Research.

These topics cover the most recent advances in AFM as well as its applications. It would be impossible to include every aspect of energy materials in this book, for example the study of catalysis by AFM is an important and emerging research field; application of AFM in oil recovery is another interesting topic that is worth to be highlighted. However, due to the limited time and space, these interesting topics are not discussed in this book. However, I am sure these topics will become the spotlights in the near future.

I would like to thank Prof. Manfred Buck and Prof. Flemming Besenbacher for introducing me to the fantastic world of scanning probe microscopy. I would like to thank all contributors for their support to publish this book. Many colleagues and friends provided me support during my career, which are much appreciated. Finally, I would like to express my gratitude to my family especially my wife Ling-Zhi for her tremendous understanding and encouragement in the past decade.

**Cai Shen**

# Editor

**Cai Shen** earned a PhD in chemistry from the University of St Andrews, UK, in 2008. Sequentially, he continued his research at the University of Maryland, Heidelberg University and Aarhus University before he joined Ningbo Institute of Materials Technology and Engineering, Chinese Academy of Sciences as an Associate Professor in 2013. He was promoted to Professor in 2021. He joined the University of Nottingham Ningbo China in 2022. He has published 100 papers in peer-reviewed journals. He is on the editorial boards of a number of international journals, including *Journal of Microscopy*. His current research interests are lithium-ion batteries and applications of atomic force microscopy for the study of energy materials and biology.

# Contributors

**Nicholas Antoniou**
PrimeNano Inc.
Santa Clara, California

**Kai Cai**
Beijing Center for Physical and Chemical Analysis
Beijing, China

**Shuang Cao**
College of Chemistry and Chemical Engineering
Qingdao University
Qingdao, China

**Martí Checa**
Center for Nanophase Materials Sciences
Oak Ridge National Laboratory
Oak Ridge, Tennessee

**Ravi Chintala**
PrimeNano Inc.
Santa Clara, California

**Liam Collins**
Center for Nanophase Materials Sciences
Oak Ridge National Laboratory
Oak Ridge, Tennessee

**Anyang Cui**
Technical Center for Multifunctional Magneto-Optical Spectroscopy
Department of Materials, School of Physics and Electronic Science
East China Normal University
Shanghai, China

**Menghan Deng**
Technical Center for Multifunctional Magneto-Optical Spectroscopy
Department of Materials, School of Physics and Electronic Science
East China Normal University
Shanghai, China

**Youjie Fan**
Bruker (Beijing) Scientific Technology Co., Ltd.
Beijing, China

**Xin Guo**
Bruker (Beijing) Scientific Technology Co., Ltd.
Beijing, China

**Dong Guo**
School of Medical Instrument and Food Engineering
University of Shanghai for Science and Technology
Shanghai, China

**Zhigao Hu**
Technical Center for Multifunctional Magneto-Optical Spectroscopy
Department of Materials, School of Physics and Electronic Science
East China Normal University
Shanghai, China

**Shuang-Yan Lang**
Institute of Chemistry, Chinese Academy of Sciences

**Chen Liu**
Bruker (Beijing) Scientific Technology Co., Ltd.
Beijing, China

**Yang Liu**
Bruker (Beijing) Scientific Technology Co., Ltd.
Beijing, China

**Yaolun Liu**
Bruker (Beijing) Scientific Technology Co., Ltd.
Beijing, China

**Sabine M. Neumayer**
Center for Nanophase Materials Sciences
Oak Ridge National Laboratory
Oak Ridge, Tennessee

**Wenhui Pang**
Bruker (Beijing) Scientific Technology Co., Ltd.
Beijing, China

**Fei Peng**
Bruker (Beijing) Scientific Technology Co., Ltd.
Beijing, China

**Zhen Zhen Shen**
Key Laboratory of Molecular Nanostructure and Nanotechnology
Institute of Chemistry
Chinese Academy of Sciences
and
Research/Education Center for Excellence in Molecular Sciences
and
University of Chinese Academy of Sciences
Chinese Academy of Sciences
and
Beijing National Laboratory for Molecular Sciences
Beijing, People's Republic of China

**Hao Sun**
Bruker (Beijing) Scientific Technology Co., Ltd.
Beijing, China

**Tong Sun**
College of Chemistry and Chemical Engineering
Qingdao University
Qingdao, China

**Xiaochun Tian**
Institute of Urban Environment
Chinese Academy of Sciences
Xiamen, China

**Ahmed Touhami**
Department of Physics and Astronomy, College of Sciences
University of Texas Rio Grande Valley
Brownsville, Texas

**Wan-Yu Tsai**
Chemical Science Division
Oak Ridge National Laboratory
Oak Ridge, Tennessee

**Jing Wan**
Institute of Chemistry, Chinese Academy of Sciences

**Shurui Wang**
Bruker (Beijing) Scientific Technology Co., Ltd.
Beijing, China

**Xiang Wang**
Technical Center for Multifunctional Magneto-Optical Spectroscopy
Department of Materials, School of Physics and Electronic Science
East China Normal University
Shanghai, China

**Xin Wang**
Bruker (Beijing) Scientific Technology Co., Ltd.
Beijing, China

**Rui Wen**
Institute of Chemistry, Chinese Academy of Sciences

**Jingshu Xu**
School of Materials Science and Engineering
Beihang University
Beijing, China

**Xiaoji G. Xu**
Department of Chemistry
Lehigh University
Bethlehem, Pennsylvania

**Yongliang Yang**
PrimeNano Inc.
Santa Clara, California

**Yan Ye**
Technical Center for Multifunctional Magneto-Optical Spectroscopy, Department of Materials, School of Physics and Electronic Science
East China Normal University
Shanghai, China

**Kaiyang Zeng**
Department of Mechanical Engineering
National University of Singapore
Singapore, Singapore

**Qibin Zeng**
Department of Mechanical Engineering
National University of Singapore
Singapore, Singapore

# 1 Principles and Basic Modes of Atomic Force Microscopy

*Anyang Cui, Menghan Deng, Yan Ye,
Xiang Wang, and Zhigao Hu*
East China Normal University

**CONTENTS**



Scanning probe microscopy (SPM) is able to measure the surface morphology and the physical/chemical functionalities of materials with high spatial resolution, on the basis of the interaction between sample surface and probe. In SPM family, scanning tunneling microscopy (STM) and atomic force microscopy (AFM) were invented successively in 1980s.[1,2] For STM, the topographic imaging depends on the quantum tunneling effect. It obtains the topography on atomic scale by collecting the tunneling current in a separation between the conductive tip with a sharp apex and the measured surface. STM is only used to measure the topography for conductor and semiconductor materials. Fortunately, AFM overcomes this restriction and is competent for imaging morphology of insulator materials, with nanometric resolution in lateral *x*/*y* direction and vertical (*z*) resolution of beyond 0.1 nm in some cases.

As one of the near-field techniques, AFM presents many strengths: (1) ultrahigh spatial resolution of reaching the sub-nanometric scale; (2) better force control in the

DOI: 10.1201/9781003174042-1

range of nano-to-micro newton between AFM tip and the sample; and (3) easy specimen preparation and flexible imaging environments such as vacuum, air, and liquid. In addition, other than the surface morphology, AFM allows synchronous quantified characterizations for chemical, mechanical, electrical, and magnetic properties of functional materials. An AFM probe could be regarded as a nano "operator" to manipulate chemical or physical properties such as chemical etching, electric polarization, and atom/molecule migration.

In this chapter, the fundamental working principles and classic imaging modes of AFM will be first reviewed, followed by the discussion of force and quantitative nanoscale mechanical property imaging. Finally, recent developments in high-resolution imaging, AFM test in different environments (air, liquid, and ultrahigh vacuum), and AFM for electrical conductivity measurement will be demonstrated.

## 1.1 WORKING PRINCIPLES OF AFM

Different from the "seeing" method of the optical and electron microscopy techniques, a nanoscale probe of AFM is used to "touch" and visualize the surface morphology. An AFM system is primarily comprised of the tip-sample motion unit, the feedback system, as well as the data processing and display system, where the first two parts constitute the mainframe of AFM. An AFM probe with a micro-cantilever of about 100–500 μm in length and less than 60 μm in width is used to detect the surface topography of the sample, where one end of the cantilever probe is fixed, while the other end has a sharp tip. In the tip-sample motion unit, a piezo scanner controls the scan motion of the probe or sample with a spatial ($x$, $y$, $z$) accuracy in sub-nanometer level. Supposing that the scanner controls the motion of probe and the sample is immobile, a basic AFM framework is roughly sketched in Figure 1.1. Sub-nanoscale ($x$, $y$, $z$) positioning is driven by a piezoelectric ceramic component assembled into the scanner under high-voltage simulation.

A laser beam is emitted from a laser diode in a laser and coupling optics module, and incident to the end of cantilever. To achieve the mechanical behavior of the probe, a quadrant photodiode is considered as a force sensor to measure the reflected laser beam. As shown in Figure 1.2, the vertical movement of laser beam on a photodetector demonstrates the vertical bending of the cantilever. Simultaneously, the lateral shift of laser spot indicates the cantilever twisting. Then, the collected light signal of the laser spot will be transferred into the electrical data and processed in an internal controller, according to a set of algorithms, which briefly follow that the value of (A+B)−(C+D) is proportional to vertical position, and (A+C)−(B+D) is proportional to lateral deflection. The measured light signals are normalized by the sum response of (A+B+C+D), so as to eliminate the effect of cantilever reflectivity. Therefore, the optical lever method could efficiently obtain the angle and the vertical displacement of cantilever deflection.

As seen in Figure 1.1, a parameter $\boldsymbol{F}(z)$ in internal controllers is defined to describe the probe-sample interaction, and input into the feedback system for controlling the separation between the probe and the measured surface. When the interaction $\boldsymbol{F}(z)$ deviates from the pre-customized setpoint ($F_0$), the feedback system will produce a differential signal, or called an error signal $E(t)$, to examine and correct the constant

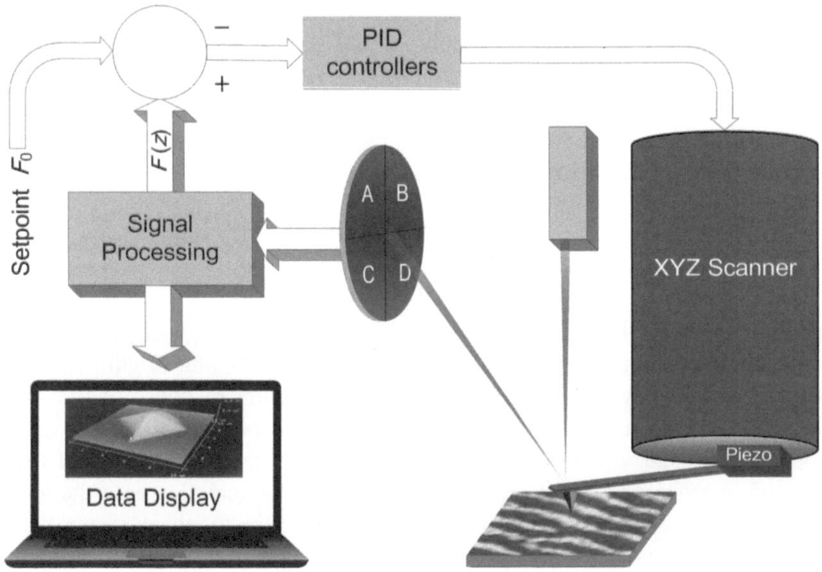


**FIGURE 1.1** A schematic diagram of an AFM system.

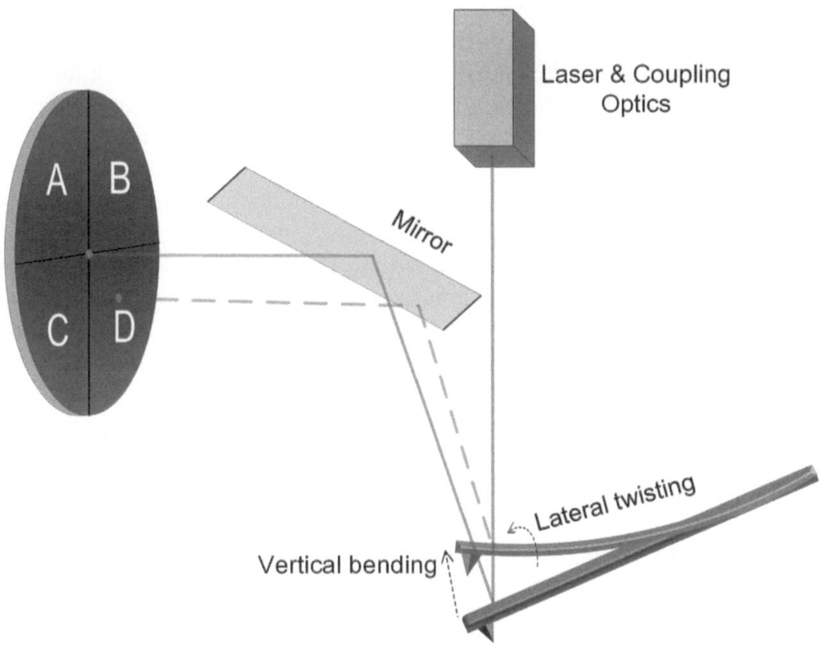


**FIGURE 1.2** Scheme of the optical detection system. The solid lines indicate the laser beams for the natural cantilever state without deflection, while the dashed lines mark the reflected beams by the cantilever with vertical-bending and lateral-twisting deflection.

tip-sample interaction. This differential signal in the feedback loop will be first processed by proportional-integral-derivative (PID) controllers that provide the three terms of proportional gain (*P*), integral gain (*I*), and differential gain (*D*) to compensate the error signal and optimize the probe-sample interaction. The output of PID controllers is fed into the piezo scanner to adjust Z voltage $\Delta V_z$ for leading the probe to move up or down, following[3]

$$\Delta V_z = PE(t) + I\int_t^0 E(t)\,dt + D\frac{dE(t)}{dt} \tag{1.1}$$

where *P* as an overall gain is proportional to the error, *I* is used to decrease the steady-state error signals by low-frequency compensation, and *D* gain is responsible for high-frequency compensation to modify transient response of the system.[4] During topographic imaging, these PID gains should be optimized for achieving a good AFM image. For example, too high proportional gain (*P*) would lead the scanner-probe system to oscillate, while too low *P* value may decrease the response of the scanner-probe system. Therefore, on the basis of the feedback system, scanner can control the *z*-direction motion of the probe for minimizing the error signal and keep a constant probe-sample interaction to measure a real surface morphology.

In addition, owing to the nonlinear response of piezoelectric ceramic, the linear external voltages will result in the nonlinear motion of the scanner. In the available commercial AFMs, the position accuracy can be corrected by two approaches.[3] (1) The feedback close loop employs XYZ position transducers to examine the actual motion of the scanner, and then modifies the added voltages for making sure that the scanner reaches the targeted positions. (2) In the open loop method, the scanner keeps the linear motion by applying the nonlinear voltage functions. It follows the nonlinear models fitted by a dataset including AFM images collected at different scanning conditions.

After the basic demonstration of the tip-sample motion unit and feedback system in AFM, the physical behaviors in the tip-sample junction will be introduced. For observing the surface morphology of the sample, the tip is controlled to approach the sample surface. During the approaching or separating process, atomic interaction between the tip apex and sample surface can be considered as a repulsive force or an attractive force. The widely used working modes of AFM have been categorized based on the tip-sample interaction in the range of repulsion and attraction.

In fact, there are various types of interactions between atoms at the tip and the sample surface such as interatomic repulsion, van der Waals attraction, friction force, electrostatic force, and chemical force, among which van der Waals attraction and interatomic repulsion are the main ones. The tip-sample distance determines the short-range repulsive force and the long-range attractive force. The van der Waals force (***F***) as a function of the tip-sample distance (***r***) is shown in Figure 1.3. When two atoms get close to each other, they will first attract each other. With the tip-sample distance decreasing, the repulsive force between the atoms begins to cancel out the attractive force until an equilibrium is reached at $\mathbf{r}_0$ position. As the distance continues to decrease, the van der Waals force changes from attractive to repulsive range, and the repulsive force increases sharply. AFM operating mode could be realized

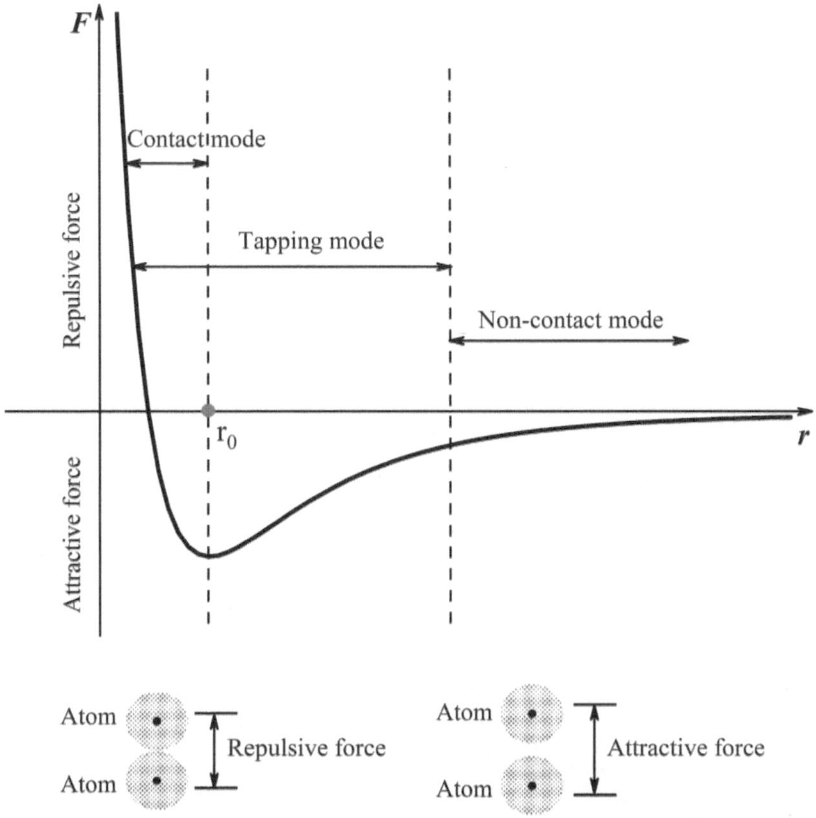


**FIGURE 1.3** Lennard-Jones potential diagram of the interaction force vs. distance between the tip apex and the sample surface.

by controlling the mechanical interaction and the distance between the tip and the sample. At the same time, the achieved surface property may vary significantly for different operating modes.

## 1.2 CONTACT MODE

As shown in Figure 1.3, the contact mode operates in the force region of the interatomic repulsion, where the interaction between the tip and the sample keeps a constant repulsive force of about $10^{-8} \sim 10^{-11}$ N between the atoms in contact with each other. The tip is in close with the sample and slides over the sample surface. During the scan, the cantilever maintains a vertical deflection, which is recorded to image the topography, as shown in Figure 1.4. Based on the detection of the cantilever deflection, the AFM system could operate in the constant height mode or the constant force mode to obtain topographic images or graphic files. In the constant height mode, the height of the scanner remains unchanged, and the morphology image of the sample is directly obtained from the deflection information of the cantilever in vertical space, as shown in Figure 1.5a. The constant height mode is sensitive to the

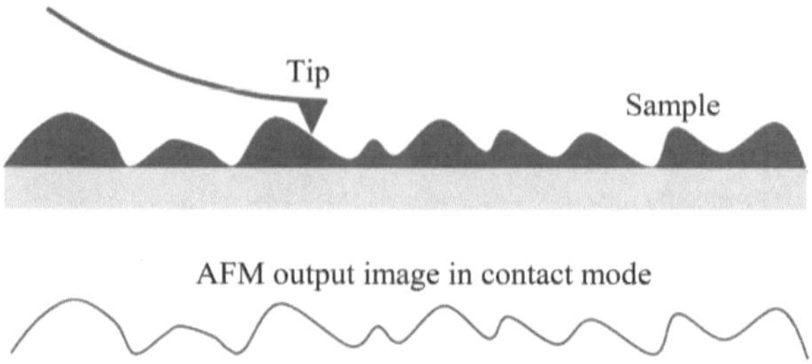


**FIGURE 1.4** AFM contact mode used to scan the morphology of the measure sample.

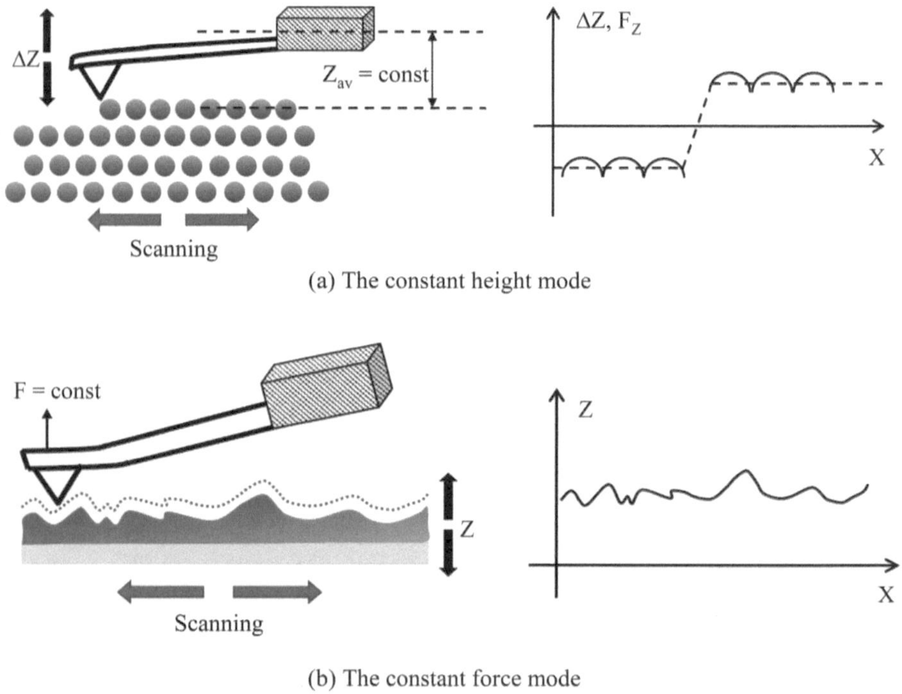


**FIGURE 1.5** Schematic diagrams of (a) the constant height mode and (b) the constant force mode in the contact mode of AFM scanning.

change in surface height, could realize rapid scanning, and is suitable for the observation of molecules and atoms.[5,6] However, the constant height mode is not suitable for the sample surface with large fluctuation, whereas it is usually used for testing a flat surface. At the same time, the constant height mode is competent to measure dynamic changes on the surface in real time and fast scanning speed.[7]

In the constant force mode, the deflection degree of the cantilever is controlled by the feedback loop to keep the constant tip-sample interaction. The scanner movement in the *z* direction is recorded to obtain the surface topography image, as illustrated in Figure 1.5b. In other words, the constant force mode adjusts the tip-sample separation according to the undulate height of the sample. Compared with the constant height mode, the scanning speed of the constant force mode is limited by the response time of the feedback loop. Additionally, the constant force model has wider applications for obtaining rougher surface morphology.[8]

It is noted that the contact mode would possibly damage or deform the measured surface morphology due to the tight contact between the tip and the sample. For imaging the topography of a relatively hard sample, the magnitude of the force generally has a little effect on the obtained topography. However, for easily deformable soft surface, the imaging force has a great impact on the measured surface. When the probe contacts the sample surface, the imaging force leads the sample to be deformed in a micro area. If the micro-deformation of the sample surface induced by the tip is relatively large, the obtained morphology information will deviate from the actual morphology of the sample. Therefore, using the contact mode, we suggest the operator to choose an AFM probe with a relatively small elastic coefficient. A soft cantilever could effectively reduce the deviation of the morphology information caused by the deformation of materials under the mechanical interaction of the tip. In addition, an AFM probe with small elastic constant could improve the detection sensitivity and collect the tiny interaction in the tip-sample junction.

Figure 1.6 shows the height images achieved by different imaging forces, where below each image is the height profile of the cross section at the central position.[9] The sample is a multilayer polyethylene (PE) material, which is composed of PE layers with a density of 0.92 g/cm (high density) and 0.86 g/cm (low density). Figure 1.6d shows a schematic diagram of component distribution and the relative position of the sample and the tip, where the narrow band corresponds to the high density PE. As can be seen from the scan height map, the narrow band is elevated under the large tip force. With the variation of the imaging force, the contrast of the height image also changes correspondingly. It is ascribed to the deformation on the surface component (low density PE) induced by the contact force. Operator should pay attention to judge the deformation to be elastic or plastic, where the latter would damage the surface. Therefore, for observing the soft surface by using the contact mode, it should be concerned that the obtained height image reflects the real topography of the sample.

Therefore, the contact mode may be not suitable for studying biological cell and soft macromolecule samples with low elastic modulus. In addition, most of the samples have a covering layer adsorbed on the surface with the thickness of several nanometers in the atmospheric environment such as the condensed water evaporation or other organic pollutants. When the tip contacts the adsorption layer, there is an enhanced adhesion force between the tip and the sample due to the presence of capillary action. The existence of adhesion force would increase the contact area between the tip and the sample, and reduce the imaging resolution. Except for the adhesion effect, the shear force may also affect the topographic imaging and lead to the deformation of the sample surface.

The AFM probe detects not only the force in the perpendicular ($z$) direction between the tip and the sample, but also the lateral frictional force. When the scanning direction is perpendicular to the length of the cantilever, the friction force between the tip and the sample can be measured. The friction force twists the cantilever with a lateral deflection behavior, which can be used to study the friction

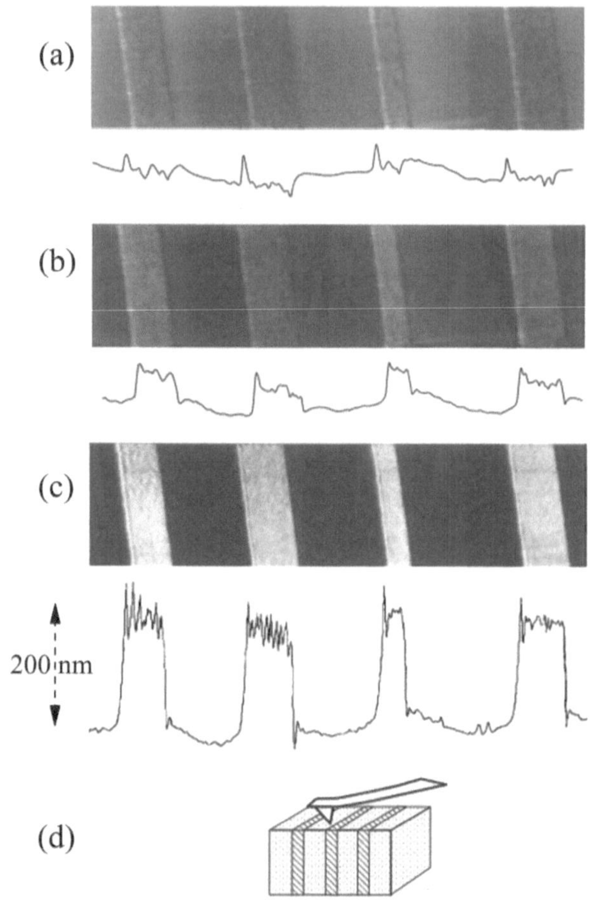


**FIGURE 1.6** The topography images and height profiles of multilayer PE samples achieved at the tip forces of (a) 10 nN, (b) 30 nN, and (c) 80 nN, respectively. (d) A three-dimensional schematic of the measured multilayer PE. (Reprinted with permission from Magonov,[9] Copyright (2000) John Wiley and Sons.)

properties of the sample surface. Based on this principle, a lateral force microscopy (LFM) has been developed.

## 1.3 TAPPING MODE

The tapping mode (also called the intermittent contact mode) works in the wide force range from the repulsive to attractive force between the tip and the sample. The tapping mode uses two methods to drive the vibration of the micro-cantilever: One is the indirect vibration method, in which the high-frequency sound waves generated by the piezoelectric ceramics in the scanner drive the vibration of the cantilever, also known as the acoustic-driven mode; the other is the direct vibration method, in which the magnetic cantilever is directly excited by an alternating magnetic field generated by a magnetic coil installed nearby the cantilever, also known as the magnetic drive mode. The magnetic drive mode has the advantage of achieving a simpler cantilever response without triggering the background noise, which is usually inherent in the acoustic-driven mode, especially for imaging in liquids. In the tapping mode, the excitation frequency of the tip is generally set at or near the free resonance frequency of the tip. Before scanning, it is necessary to determine the free resonance frequency of the tip through the frequency sweep test.

During the scanning process, the tip contacts the sample surface intermittently. When the tip is not in contact with the sample surface, the cantilever vibrates "freely" with a higher amplitude, as shown in Figure 1.7a. When the tip moves toward the sample surface in vibration and contacts with the surface, the vibration amplitude of the micro-cantilever will be reduced due to the limited space for continuing movement, as presented in Figure 1.7b. Subsequently, the tip moves away from the sample surface. The vibration amplitude increases with enough space for the vibration of the cantilever. The vertical amplitude of the tip is generally used as a reference for the force in the tapping mode. The amplitude reference point during the scanning process is set to be smaller than the free vibration amplitude of the tip. The smaller the set reference amplitude value, the greater the tip-sample force. The Feedback loop would examine the amplitude signal measured by the photodetector, and then modify the distance between the tip and the sample to control the amplitude of the cantilever. Along with the topography imaging, the phase change of the cantilever oscillation is imaged as phase contrast signal. Besides, the interaction between the tip and the sample in the tapping mode is in the range of 0.1–1 nN, which is much smaller than that in the contact mode. The tapping dynamics overcomes the tip-sample adhesion

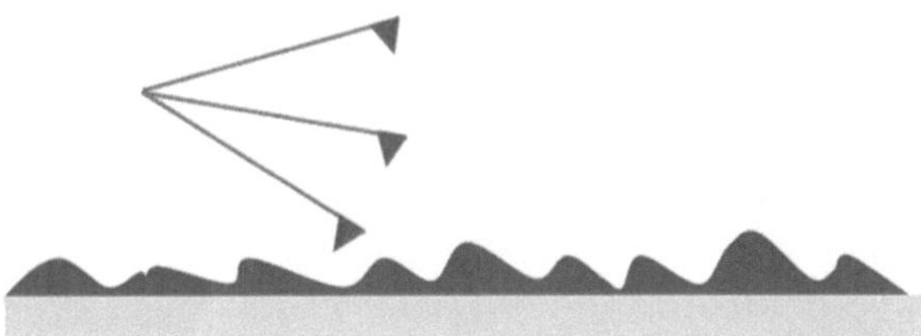

(a)

(b)

**FIGURE 1.7** Working diagrams of the cantilever vibrational behavior in the tapping mode. (a) The cantilever vibrates freely with a bigger amplitude, when the tip is not in contact with the sample surface. (b) Reduced amplitude of the cantilever vibration, when tip moves toward the sample surface with the limited tapping space.

effect and could avoid the damage of the surface.[10] The tapping mode could not only obtain a higher spatial resolution but also better eliminate the lateral force of the tip on the sample during scanning. Therefore, the tapping mode is suitable for the observation of soft, easy-deformed and viscous samples.[11–13]

The amplitude, phase, and resonance frequency are regarded as three key parameters of the tapping mode. It should be noted that in general the tapping mode refers to the amplitude modulation mode, while the tip-sample force is determined by measuring the resonance frequency variation of the probe based on the frequency modulation mode. In tapping mode imaging, the ratio of the reference amplitude to the free vibration amplitude would affect the measured morphology. If this ratio is too small, corresponding to a large applied force, the induced deformation on the sample surface with different mechanical properties will be greatly different. It will cause the deviation between the measured morphology and the actual one, which is attributed to the deviation between the measured amplitude and the reference one. Therefore, controlling a suitable tip-sample force is one of the most important issues for getting a better AFM image. The small imaging force results in a small contact area around tip-sample junction and benefits high-resolution imaging.

In the tapping mode, phase signal is another significant data for studying the nanomechanical properties of materials, according to the phase difference between the driving voltage signal of the cantilever and the actual response of the tip. The change of phase difference is derived from a comprehensive effect of the surface elasticity, viscoelasticity, friction, and surface topography of the sample. Compared with the friction microscopy mode, the phase imaging has a wider application scope for studying the nanomechanical properties of the sample surface with large adhesion or a soft component. In vacuum environment, the tapping mode can obtain atomic images of samples with atomic-scale resolution.[14,15] At present, the dynamic tapping mode has also shown a promising development momentum on quantitative nanomechanical characterization.[16–18] Similar to the contact mode, the tapping mode could be applied to probe the topography in air, vacuum, and liquid environments.[19–21]

Here, we give a brief summary to demonstrate the comparison of the contact mode and the tapping mode of AFM, as shown in Table 1.1.

## 1.4 PEAKFORCE TAPPING MODE

Although the tapping mode makes the force between the probe and the sample less than that of the contact mode, the tapping mode still cannot accurately control the force between the probe and the sample through controlling the amplitude of probe vibration. It is attributed that the force loaded on different degrees of soft and hard samples by probe with the same tapping amplitude is different. Recently, a newborn imaging mode known as PeakForce tapping has attracted an increasing interest, due to its nondestructive characterization, high-resolution imaging, and simultaneous quantitative sample mechanical properties.[22,23] The PeakForce tapping mode combines the characteristics of the contact mode and the tapping mode. The force between the probe and the sample is controlled by scanner vibration rather than probe-cantilever vibration. The most important characteristic of this imaging mode is that it can accurately control the probe-sample force, so it can not only make a

**TABLE 1.1**
**The Comparison of Two Working Modes of AFM**

| Operating Mode | Tip-sample Force | Resolution | Effect on the Sample | Advantage | Disadvantage |
|---|---|---|---|---|---|
| Contact mode | Constant | Highest | May destroy the sample | ① Fast scanning speed;<br>② Feedback control is relatively simple;<br>③ The only operating mode capable of obtaining "atomic resolution" images;<br>④ More suitable for observing hard samples with obvious changes in the vertical direction of AFM | ① Shear forces affect image quality;<br>② The combination of lateral force and adhesive force would result in a damage in patial resolution;<br>③ (The tip scraping) Large lateral force and normal force could damage soft samples |
| Tapping mode | Changing | Higher | No damage | ① The influence of lateral force is well eliminated;<br>② The force caused by the adsorption liquid layer is reduced;<br>③ The image horizontal resolution is higher;<br>④ Suitable for observing soft, fragile or sticky samples | ① Slower scanning speed than contact mode;<br>② Feedback control is more complicated;<br>③ Difficult to operate in a vacuum environment;<br>④ Submerged operation is more difficult |

good surface morphology image of soft sample but also obtain the mechanical information of the sample surface quantitatively.[24]

Figure 1.8 illustrates the working schematic of the PeakForce tapping AFM mode. In the PeakForce tapping mode, the surface morphology is measured by controlling the probe-sample force near the maximum value (peak force) on each pixel. First, the PeakForce tapping mode uses a sinusoidal wave as the driving signal based on the synchronization algorithm, which is used to set the synchronization window to about half a period to extract the actual vertical peak force. The vertical peak force signal is then distinguished from the interference signal, so that the value can be set small enough. Second, in the PeakForce tapping mode, the subtracting background difference algorithm is adopted to eliminate the influence of additional deflection signals caused by environmental damping or rough sample surface.[25] Finally, the synchronization algorithm effectively separates the contact and non-contact ranges. The data in the non-contact range are averaged to get a horizontal line as the baseline, which effectively overcomes the fluctuation of interference signals on the baseline. It allows the PeakForce tapping mode to begin the next cycle without waiting for the interference to be completely attenuated, consequently ushering in the era of high-speed scanning for quantitative nanomechanical imaging of AFM.

Because the PeakForce tapping mode can effectively eliminate the influence of background signals and has good force control ability, it has unique advantages in imaging biological and soft macro-molecule samples in near physiological environment, compared with contact and tapping modes of AFM. Most importantly, the

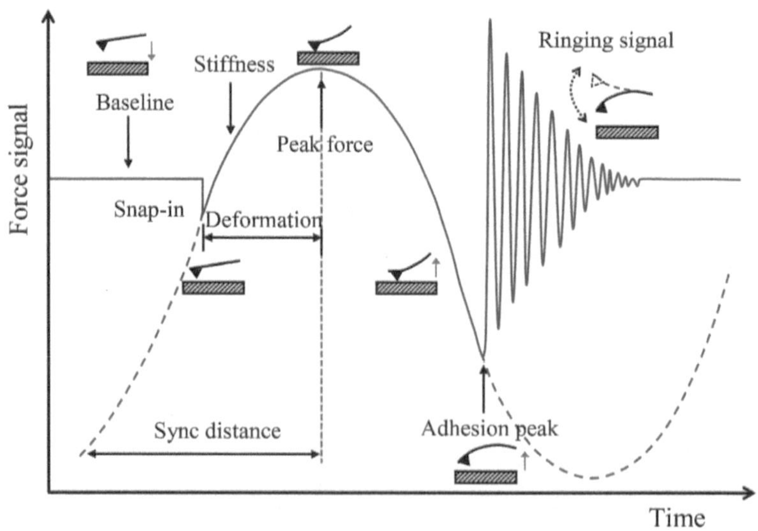


**FIGURE 1.8** Tip-sample interaction profile in the PeakForce tapping mode. Under a driving sinusoidal wave, the scanner sets a synchronization window. The peak force point is extracted in about half a period. The tip gradually approaches the measured surface with a stable zero-force baseline before the snap-in state of tip-sample contact. After reaching the peak force, the tip retracts and is off the surface at the adhesion peak state.

**TABLE 1.2**
**Different Feedback Physical Parameter for the Corresponding Imaging Mode**

| Feedback Parameter | The Imaging Mode |
|---|---|
| Tunneling current | Scanning Tunneling Microscopy |
| Vibrational amplitude of cantilever | Tapping Mode AFM |
| Deflection of cantilever | Contact Mode AFM |
| Torsional angle of the cantilever | Torsional Resonance Mode AFM |
| Tip-sample interaction | PeakForce Tapping AFM |

PeakForce tapping mode can obtain force-distance curves in real time at high speed, which reflect a variety of nanomechanical properties of the specimen. Many superior operating modes have been developed based on the PeakForce tapping mode. PeakForce Scanning Electrochemical Microscopy (PeakForce SECM)[26] for high-resolution imaging of soft materials in liquid environments, PeakForce Tunneling AFM (PeakForce TUNA)[27] for electrical/electrochemical performance characterization, and PeakForce Quantitative Nanoscale Mechanical characterization (PeakForce QNM)[28] for quantitative analysis of nanomechanical properties are three typical applications based on PeakForce tapping technology.[29]

To be concluded, no matter which mode is chosen for AFM imaging, a good control of the tip-sample force is considered as the key factor for getting a real surface morphology. The imaging force becomes the "soul" of AFM operation. The mechanics of micro-cantilever and its deflection behavior reflect the tip-sample interaction during the scan, and are fed into the feedback loop for improving the probe movement. Table 1.2 briefly summarizes the feedback parameter in a variety of AFM imaging modes.

## 1.5 FORCE MEASUREMENT AND QUANTITATIVE NANOSCALE MECHANICAL MEASUREMENT

On the basis of AFM topography measurement, the sample surface properties, i.e. the nano structure, height difference, roughness, and grain size distribution, can be obtained with high resolution. Force measurement is one of the most fundamental but important functions of AFM for studying the mechanical properties of materials on micro-to-nano scale, and has been widely applied in surface science research. In this subsection, we will carefully introduce the force-distance curve test of AFM and the principle of PeakForce QNM characterization.

### 1.5.1 Force-Distance Curve

Force-distance curves measured by AFM can be used to characterize the mechanical properties such as elasticity, hardness, Hamaker constant, and surface mechanical information including peeling force, adhesion force, and friction force.[30] The force-distance curve can sufficiently describe the interaction between the tip and the sample, or the function of the cantilever deflection with respect to the tip-sample distance. The force-distance curve is widely used, especially for studying polymers and biomaterials. When the probe stiffness is close to the contact stiffness between the tip and the sample, the sensitivity of the force-distance curve is relatively high.

In the force-distance curve mode, the probe is driven by retracting/extending the piezoelectric ceramic of scanner in the $z$ direction and the cantilever deflection is recorded against the $z$ position, producing a force-distance curve. Figure 1.9 is a diagram of the force-distance curve, showing the general process of probe-sample interaction. The probe approaches the sample surface slowly from a position far away from the sample surface without the deflection of the cantilever. When the distance between the probe and the sample decreases to a certain value, the attractive force gradient will be greater than the spring constant of the cantilever. At this point, the probe will suddenly jump-in and contact with the sample surface instantaneously. Then, as the probe continues to descend, the repulsive force increases gradually, the total force changes from attraction to repulsive interaction, and the cantilever turns to bend up. The probe cantilever continues to move downward until the preset force or deflection is reached and then slowly retracts from the sample surface. During the retracting process, an adhesive interaction between the tip and the sample surface occurs and prevents the probe from leaving the sample surface. If the upward force generated by the deflection of cantilever exceeds the maximum adhesion force, the probe will suddenly be pulled out from the sample surface and reach a free state again. Therefore, the adhesive force between the tip and the sample can be measured from the abrupt change of force on the force-distance curve when the probe is pulled out.

During the test of the force-distance curve, the force is actually recorded as a curve against the displacement of scanner rather than the deformation of the sample. When the tip presses into the sample surface, the displacement of the scanner ($z_p$), the

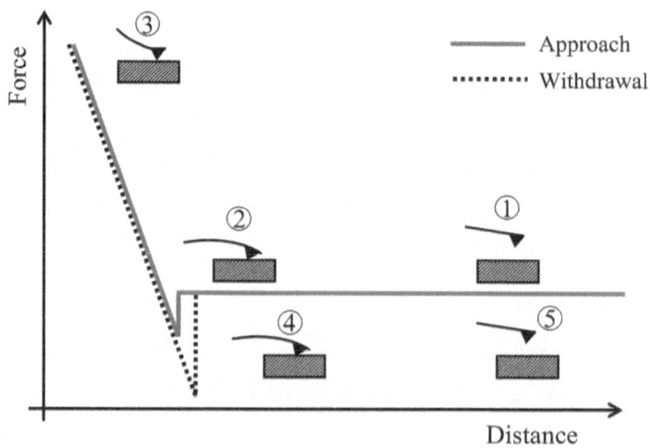


**FIGURE 1.9** Force-distance curve of the contact process between the tip and the sample. It shows a cycle of the tip-sample contact process: ① the approaching, ② the jump-in, ③ the contact at the maximum tip-sample interaction, ④ the adhesion state during the retracting process, and finally ⑤ pull-off the sample surface.

deflection of the probe cantilever ($z_c$), and the deformation ($\delta$) of the sample satisfy the following relationship:

$$z_p = z_c + \delta \tag{1.2}$$

The scanner displacement and the deflection of the probe cantilever can generally be directly measured by the instrument, and thus the deformation of the sample can be obtained. The mechanical properties of the sample surface can be obtained by fitting the force-distance curves according to an appropriate contact model. The accuracy of force measurement determines the veracity of mechanical property analysis of samples. To obtain accurate force, it needs to calibrate two parameters, deflection sensitivity and spring constant of the probe cantilever. The deflection of cantilever measured by the photosensitive detector is the change of the voltage signal. When the probe deformation is small, the relationship between the change of the voltage signal and the deflection of the cantilever is approximately linear. Deflection sensitivity (unit: nm/V) measures how many nanometer the deflection of cantilever is per unit voltage change. Deflection sensitivity can be obtained by measuring the force-distance curve on a very hard sample, where the sample deformation is negligible and the displacement of scanner is approximately the same as the deflection of the cantilever. In this case, by measuring the displacement of the scanner and the change in the reflected voltage of the cantilever, the value of deflection sensitivity can be obtained. This process has been automated in today's AFM.

After calibrating the spring constant of the cantilever, the force loaded on the sample can be determined by cantilever deflection. For a rectangular cantilever with homogeneous materials and a uniform cross section, its spring constant can be expressed as

$$k = \frac{Et^3 w}{4L^3} \tag{1.3}$$

where $E$ is Young's modulus of the cantilever, and $t$, $w$, and $L$ are the thickness, width, and length of the cantilever, respectively. The calibration of spring constant of the cantilever is important for quantitative mechanical testing. Several common methods are introduced below:

1. Sader method

   The Sader method is a simple method to determine the spring constant of the cantilever by measuring the resonant frequency of the cantilever. For a rectangular cantilever, the spring constant can be calculated as[31]

$$k = 4\pi^2 M_e \rho wtL f_{vac}^2 \tag{1.4}$$

   where $M_e$ is the Sader constant, and its value is 0.2427 when the ratio of $L$ to $w$ is greater than 5.0, $\rho$ is the density of the cantilever. $f_{vac}$ is the free resonant frequency of the cantilever in vacuum. $t$, $w$, and $L$ are the thickness, width, and length of the cantilever, respectively. In practical application, the length and width of the cantilever can be easily determined with the help of optical microscope, but it is difficult to determine the thickness and density of the cantilever accurately. In addition, the measurement of the free resonance frequency of the cantilever is usually carried out under atmospheric conditions. Due to the damping characteristics of air medium, the actual measured frequency is lower than that measured in vacuum, so there may be an error in the above spring constant measurement method.

In order to optimize this method, Sader et al. considered the influence of fluid on the cantilever resonance frequency, and obtained the relationship between the resonant frequency of the cantilever in vacuum and in fluid, which can be expressed as[32]

$$\omega_{vac} = \omega'\left(1 + \frac{\pi\rho_f w}{4\rho_c t}\Gamma_r\left(\omega_f\right)\right)^{1/2} \tag{1.5}$$

where $\omega_{vac}$ and $\omega'$ are the resonant angular frequencies of the cantilever in vacuum and fluid, respectively, $\rho_f$ is the fluid density, and $\rho_f t$ is the specific surface mass density, which can be calculated by

$$\rho_c t = \frac{\pi\rho_f w}{4}\left[Q_f\Gamma_i\left(\omega_f\right) - \Gamma_r\left(\omega_f\right)\right] \tag{1.6}$$

where $\Gamma_i$ and $\Gamma_r$ are the imaginary part and real part of the hydrodynamic function, respectively, $Q_f$ is the quality factor of the first-order vibration mode of the cantilever in the fluid medium, $\Gamma(\omega)$ only depends on the Reynolds number, but independent of the thickness and density of the cantilever. Combining the above formulas, the spring constant of the cantilever considering the viscosity of the surrounding fluid medium can be calculated by

$$k = 0.1906\rho_f w^2 L Q_f \Gamma_i\left(\omega_f\right)\omega_f^2 \tag{1.7}$$

2. Thermal noise method[33]

When the probe is far away from the sample surface, the cantilever movement is mainly from thermal oscillation. The spring constant of the probe can be determined by measuring the thermal oscillating movement of the cantilever at the resonant frequency. In the thermal calibration method, the probe cantilever is equivalent to a spring oscillator, according to Equipartition Theorem,

$$\frac{1}{2}m(\omega_0 z_n)^2 = \frac{1}{2}k_B T \tag{1.8}$$

where $m$ is the effective mass of the cantilever, $\omega_0$ and $z_n$ are the angular frequency and noise amplitude of the cantilever free end, $k_B$ and $T$ are the Boltzmann constant and absolute temperature in Kelvin, respectively. For the spring oscillator model, according to $\frac{1}{2}kz_n^2 = \frac{1}{2}k_BT$, the spring constant of the probe cantilever can be expressed as

$$k = \frac{k_BT}{z_n^2} \tag{1.9}$$

where average of the square of the free end amplitude $\langle z_n^2 \rangle$ is the area enclosed under the power spectra density curve. The thermal noise method is widely used because of its simple operation and relatively accurate results. Therefore, most of AFM analysis software has integrated this method. In addition, other methods, such as the microsphere adhesion method[34] and the reference cantilever method,[35] can also calibrate the spring constant of the probe, which will not be introduced in detail here.

Young's modulus of the sample can be determined by fitting the force-distance curve with the suitable contact mechanics model. Therefore, the contact mechanics models between the tip and the sample are very important and have been widely investigated. Here, we will introduce several commonly used models. According to the shape of the tip, the models can be divided into Hertz model and Sneddon model, respectively. While considering the magnitude of adhesion force, the models between can be divided into Hertz model, Derjaguin-Muller-Toporov (DMT) model, and Johnson-Kendall-Roberts (JKR) model, respectively.

1. Hertz model

   The Hertz model is mainly used to describe the contact between the ball tip and the rigid surface without considering the adhesion force.[36] For Hertz model, the magnitude of the force is related to the tip radius, indentation depth, and the effective Young's modulus. It is expressed as follows:

$$F = \frac{4}{3}E^* \sqrt{R}d^{3/2} \tag{1.10}$$

$$\frac{1}{E^*} = \frac{1-v_t^2}{E_t} + \frac{1-v_s^2}{E_s} \tag{1.11}$$

   where $F$ is the force between the tip and the sample, $R$ is the tip radius, $d$ is the sample deformation, $E^*$ is the effective Young's modulus, $E_t$ and $E_s$ are the Young's modulus of the tip and the sample, $V_t$ and $V_s$ are the Poisson's ratio of the tip and the sample, respectively.

2. Sneddon model

   The Sneddon model[37] is suitable for a wider range of AFM probe types such as flat, cylindrical, parabolic, conical, spherical, and other arbitrary axisymmetric shapes of the tip. The magnitude of the force is mainly related to the shape of the tip, the depth of the indentation, and the effective Young's modulus. The Sneddon model can be expressed as follows:

$$d = \int_0^1 \frac{f'(x)}{\sqrt{1-x^2}} dx \tag{1.12}$$

$$F = \frac{3}{2} E^* a \int_0^1 \frac{x^2 f'(x)}{\sqrt{1-x^2}} dx \tag{1.13}$$

where $f(x)$ is the function representing the shape of the tip, and $a$ is the contact radius, respectively.

3. DMT model

Both the Hertz model and Sneddon model do not consider the influence of adhesion force. If it is necessary to consider the influence of adhesion force, the DMT model and JKR model can be used. The DMT model is generally used in the case that the adhesion force between the tip and the sample is small and mainly acts outside the contact surface.[38] It is more suitable for small tip radius and large elastic modulus. The DMT model is mainly related to the tip radius, indentation depth, effective Young's modulus, and the surface energy of the sample, and can be expressed as

$$F = \frac{4}{3} E^* \sqrt{R} d^{3/2} - 4\pi R\gamma \tag{1.14}$$

where the last item $-4\pi R\gamma$ is the magnitude of the adhesion force between the tip and the sample, and $\gamma$ is the surface energy, respectively.

4. JKR model

The JKR model is generally used to describe the situation that the adhesion force between the tip and the sample is large, mainly acts within the contact surface, and is suitable for large tip radius and low elastic modulus.[39] The mechanical model is mainly related to the tip radius, sample surface energy, contact radius, effective Young's modulus, and indentation depth. The mechanical model is expressed as

$$F = \frac{4E^* d^3}{3R} - \sqrt{8\pi E^* a^3 \gamma} \tag{1.15}$$

### 1.5.2 PeakForce Quantitative Nanoscale Mechanical Method

Force-distance curve measurement is usually used to analyze the mechanical properties of a single point on a sample surface. The array mode based on the force-distance curve, also known as the force volume mode, can be used for the imaging analysis of the mechanical properties of the sample surface, but it has a low resolution and generally takes a very long time to test. In the force volume mode, the scanner needs to maintain a small vibration frequency, usually less than 100 Hz, in order to suppress the resonance of the piezoelectric scanner.[40] As a result, its scanning speed is slow, usually taking several hours per frame.[41] In addition, the force-distance curve mode usually requires the force load of several nN on the sample surface,[30] which is still not enough for particularly soft samples.

The PeakForce QNM mode is a new quantitative nanomechanical measurement method based on the PeakForce tapping imaging mode, which can simultaneously perform high-resolution imaging and quantitative mechanical properties measurement of the test area.[42] This mode can accurately measure and control the maximum force (peak force) of the interaction between the tip and the sample at each pixel point, and simultaneously obtain the force-distance curve of each pixel point, thus achieving high-resolution mapping of the mechanical information of the sample. Moreover, the force-distance curves can be fitted and analyzed in real time using the synchronization algorithm. A high scanning rate can be achieved. The PeakForce QNM mode can control the normal peak force in the range of less than 100 pN, so it is widely used in soft specimens and biomaterials.[43,44]

The basic principle of the tip-sample force for the imaging mode of PeakForce QNM is shown in Figure 1.10. AFM uses the peak force as the feedback signal to adjust the rise and fall of the scanner so as to draw the surface topography of the sample. Figure 1.10a shows the force cycle curve captured by the PeakForce tapping mode, with the blue line of approaching process and the red line of retracting process. In this mode, the probe is oscillated at a low frequency (1 ~ 2 kHz) and the force curve is recorded for each contact with the sample surface. Under the direct control of the interaction between the probe and the sample, the PeakForce tapping mode can image the sample surface under weak repulsive force or even attraction state, so as to avoid the damage of the sample to the greatest extent.[45] The marked A, B, C, D, and E spot of the curve in Figure 1.10 completely records the interaction process of the probe from slowly approaching the surface to reaching the surface and finally away from the surface. At point A, the probe is in the non-contact state, keeping a certain distance above the sample surface. Then, the scanner drives the probe down until enough near the surface, reaching point B. A small attraction generates between the probe and the sample, and pulls the probe into the surface. The sample surface deforms as the probe presses on the sample, reaching peak force point C. Next, the

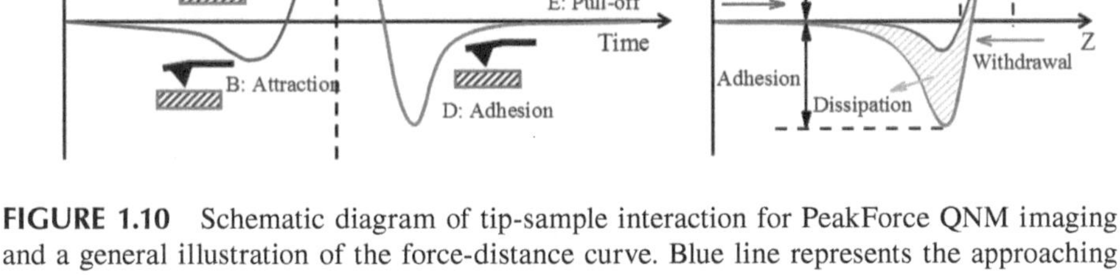


**FIGURE 1.10** Schematic diagram of tip-sample interaction for PeakForce QNM imaging and a general illustration of the force-distance curve. Blue line represents the approaching process of the probe, while the red line represents the retracting one.

scanner begins to lift up to pull the probe away from the surface at point D, i.e. the point of maximum adhesion force. Finally, after getting rid of the adhesion force, the probe returns to the non-contact position, corresponding to the point E. This cycle is repeated at each pixel until the entire area is scanned pixel by pixel.

Under the great control of the tip position and the feedback loop, the curves of the interaction and the separation between the tip and the sample can be obtained at each pixel point. As shown in Figure 1.10b, the blue line represents the approaching process of the probe, while the red line represents the retracting one. The mechanical properties at each pixel of the sample, such as deformation, Young's modulus, adhesion force, and energy dissipation, can be achieved by the force-distance curves.

**Deformation**: That the abscissa of the peak force point of the force-distance curve minus the abscissa of the position where the interaction force is zero represents the maximum deformation of the sample, which reflects the deformability of the sample. With increasing the load of the tip on the sample, the sample deformation increases and reaches the maximum value at the peak force. The measured deformation includes plastic deformation and elastic deformation. There may be some experimental deviations, because the contact point between the tip and the sample surface is usually not the zero intersection of the curve.

**Adhesion:** The adhesion corresponds to the difference between the bottom of the red line and the baseline, and represents the maximum attraction when the tip was pulled out from the sample. Adhesion interaction includes a range of attractive force, such as van der Waals force, capillary force, and electrostatic force. For instance, if the sample or probe surface is hydrophilic, the capillary force is strong, resulting in a strong adhesion. Adhesion force usually increases with the bigger radius of the tip. In addition, adhesion effect becomes even more important for the functionalized imaging by AFM. In this case, the adhesion force reflects the chemical interaction between the tip and the specific molecule on the sample, and provides more chemical information.

**Energy dissipation:** The energy dissipation is calculated by the area of the hysteresis loop between the approaching process (blue line) and the retracting process (red line), which represents the energy loss of one cycle. The magnitude of the energy dissipation indicates the viscoelastic properties of the sample.

**Young's modulus:** To measure Young's modulus, contact mechanics models are used to fit the force-distance curves. The Hertz model, JKR model, and DMT model have already been introduced above and will not be explored in detail here. The Hertz model is a nonadhesive contact between two elastic spheres, while the DMT and JKR models consider the external and internal adhesion forces of the contact surface. The difference between the two models is that the JKR model is suitable for soft samples with high adhesion and large tip radius, such as living cells, while the DMT model can be used for small tip radius and rigid samples, such as polymers or composites with small adhesion. We can also select the appropriate model according to the ratio of the adhesion to elasticity.[46]

In addition, for quantitative nanomechanical measurement using PeakForce QNM mode, several factors dramatically affect the measurement such as the shape of the tip, the spring constant of the cantilever, and the choice of contact mechanics model. When choosing the probe, the cantilever whose spring constant is close to the sample should be selected. This is because if the cantilever is too hard relative to the sample, it will be difficult for the cantilever to deflect sufficiently to provide an accurate force measurement. However, if the cantilever is too soft relative to the sample, the deformation of the sample will be small, so the deformation signal detected is not accurate enough. Second, when measuring the elastic modulus of the soft sample, the result of the blunt tip will be more accurate than that of the sharp tip. Finally, the influence of adhesion force should be considered in the selection of the contact mechanics model.

## 1.6 HIGH-RESOLUTION IMAGING OF AFM

Compared with STM, AFM has the ability to measure a series of properties of sample surface with high resolution, even down to the atomic scale. As one of the basic concepts for AFM imaging, the resolution can be described with lateral and vertical resolution. The factors that determine the resolution are mainly considered from electronic and mechanical noise, tip radius, sample compliance, the interaction between tip and surface, and nonlinearity of the AFM imaging process.

### 1.6.1 Vertical Resolution

In general, the vertical resolution of AFM is described as the minimum change of step height that can be measured on the surface. It is primarily restricted by noise floor of the AFM system, including vertical cantilever deflection noise, $z$ scanner noise, seismic noise, and environmental acoustic. Being in a quiet environment and implementing appropriate seismic isolation can be helpful for reducing the environmental acoustic and seismic noise. As for the vertical deflection noise, it consists of the detector noise and the thermal noise of the deflection detection system. The source of these noises is statistically independent of each other, as shown below:

$$\delta A \approx \delta z = \sqrt{\delta z_{th}^2 + \delta z_{\text{det}}^2} \tag{1.16}$$

The detector noise is primarily from the optical detector noise. The thermal noise is usually much less than detector noise. Hence, the detector noise determines the ultimate limit of the maximum resolution of AFM. In dynamic AFM, thermal noise ($\delta z_{th}$) is calculated by:

$$\delta z_{th} = \sqrt{\frac{4k_B TQB}{\pi f_0 k}} \tag{1.17}$$

where $k_B$ is the Boltzmann constant, $k$ is the spring constant of the cantilever, $T$ is the absolute temperature, and $Q$ is the quality factor, respectively.[47,48] The thermal noise depends on the geometry and environment of the cantilever. It can be reduced by using the cantilever with higher resonance frequency and larger spring constant. Ideally, most commercial AFM instruments can reach a vertical resolution as low as 0.01 nm for rigid cantilevers. The $z$ scanner noise is another factor for the vertical resolution of AFM imaging. The electronic noise in the high voltage applied to the scanner and the bit resolution of $z$-drive digital-to-analog converter results in the $z$ scanner noise. This noise can be reduced by choosing narrow bandwidth and better electronic design.

### 1.6.2 Lateral Resolution

It is assumed that the tip and the surface are non-deformable objects; the lateral resolution of AFM is defined as the minimum detectable distance where the dimple depth of the two sample features is larger than the noise, as shown in Figure 1.11.[49] The dimple depth is derived from the image intersection of two sample features. The lateral resolution can be given as follows:

$$d = \sqrt{2R}\left(\sqrt{\Delta z} + \sqrt{\Delta z + \Delta h}\right) \tag{1.18}$$

where the minimum detectable distance ($d$) is related to the curvature radius of the tip ($R$), the vertical resolution ($\Delta z$), and a small relative height ($\Delta h$) of two spikes, respectively.[49,50] Therefore, the lateral resolution of AFM is greatly limited by the radius of the tip. Under ideal conditions, if the deformation of the tip and the sample is neglected, a sharp enough tip enables AFM to image with atomic and sub-nanometer resolution. Owing to the limited sharpness of AFM tips, the apparent width of surface feature is larger than the actual width, as shown in Figure 1.11. The lateral resolution is also affected by the data recording process, where the key parameter is the relationship between the number of pixels used for imaging and the scanning speed. For images with a lateral size $\Delta x$, the spatial resolution is not better than $\Delta x/N_p$, in which $N_p$ represents the number of scan pixels.[51]

However, the force exerted by the tip always leads to the deformation of the sample surface. Considering that situation, the lateral resolution cannot be better than the actual or effective tip-surface contact area.[52] The Hertz model proposes the lateral resolution ($l$) as a function of the tip radius ($R$), applied force ($F$), and the effective Young's modulus of the tip-surface system ($E_{eff}$) as follows:

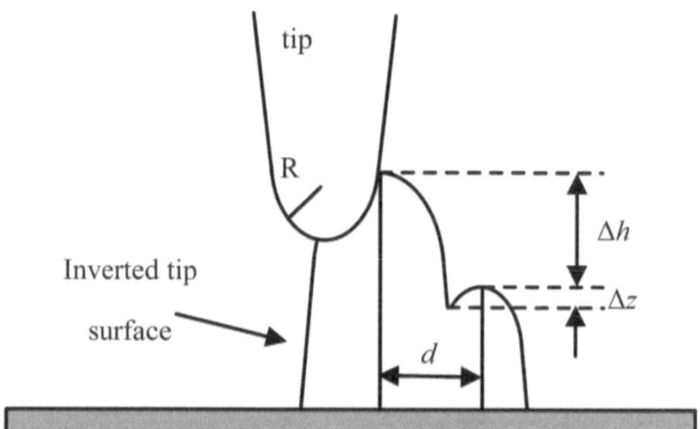


**FIGURE 1.11** Diagram of AFM lateral resolution and convolution effect,[49] where $R$ is the curvature radius of the tip, $d$ is the detectable minimum lateral distance, $\Delta z$ is the vertical resolution, and $\Delta h$ is a small relative height of two spikes.

$$l = 2\left(\frac{3RF}{4E_{eff}}\right)^{1/3} \tag{1.19}$$

Equation (1.19) gives the explanation about why atomic and sub-nanometer resolution is easier to achieve on stiff surfaces than on soft ones. As the tip apex is made of silicon or silicon dioxide with high Young's modulus (>50 GPa), the deformation of the tip is usually negligible. Equation (1.19) also shows that a proper imaging force is vital to high-resolution AFM imaging. The deformation of the sample would increase with the increase in the applied force, leading to the poor lateral resolution. The interactions between tip and sample surface are complicated including long-range van der Waals force, short-range forces, capillary force, and electric double layer (EDL) in liquid. Previous studies have shown that the repulsive force is more localized[53,54]; consequently using the lowest short-range tip-sample interaction is the first choice for high-resolution imaging. Pyne et al. observed the DNA helical structure using AFM with high resolution.[55] As shown in Figure 1.12a, the DNA double helical structure can be clearly recognized with double-banded corrugation and its diameter can be accurately measured as approximately 2 nm. In this work, the resolution of images critically relies on the force that the tip applied to the surface. The measured corrugation and the overall height of the DNA obviously reduce as the imaging force increases (Figure 1.12b–f). These results indicate that the low tip-sample force is significant to accurately measure the oligonucleotide diameter and obtain a high-resolution image of a DNA structure.

Additionally, when the aspect ratio of sample feature is larger than the tip apex, there will be significant distortion on AFM images caused by the tip. This phenomenon is called as dilation.[56,57] Figure 1.13 shows the nature of AFM dilation. The apparent imaging point of the tip is not consistent with the real imaging point. When

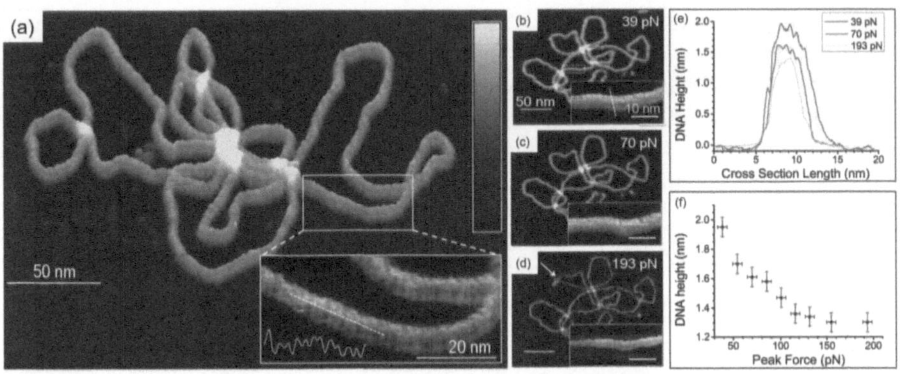


**FIGURE 1.12** DNA double-helix structure was obtained with high resolution. (a) DNA imaged at a peak force of 49 pN. Inset: high-resolution AFM topography of DNA and a height profile taken along the dashed line, Color scales: 3.5 nm; (b)–(d) DNA imaged at peak forces of 39, 70, 193 pN, respectively. (e) Height profiles measured across the DNA for different peak forces. (f) Measured height as a function of peak force. (Reprinted with permission from Pyne et al.,[55] Copyright (2014) John Wiley and Sons.)

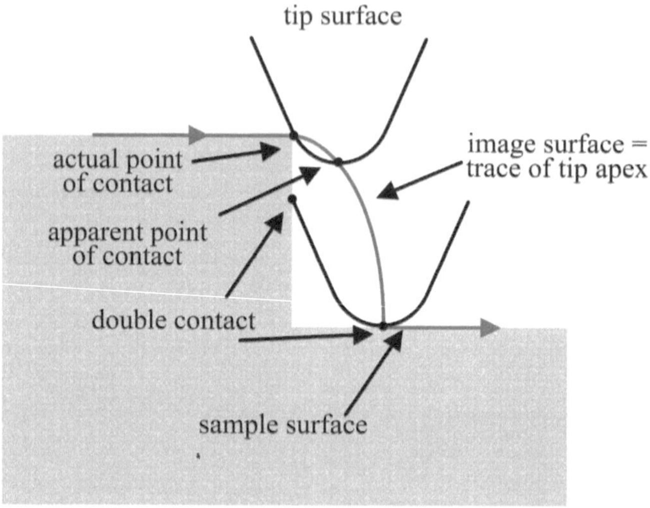


**FIGURE 1.13** The diagram of AFM dilation shown in the trace (red arrows) of the tip apex, due to the size mismatch between tip and the measured object.[57]

the curvature of sample surface feature is larger than the tip curvature, the tip will contact the two points simultaneously. As a consequence, the presence of double contact point indicates the undetectable areas during scanning. An ideal probe needs to have a sharp tip and its side walls need to be sharper and steeper than the sample features.

Visualization of the double helix of RNA has also been realized by AFM, whose sufficient resolution enough reveals the periodicity of A-form sub-helical pitch.[58] RNA was imaged in Figure 1.14a and b, where a periodic corrugation was observed clearly within the molecular scale. The profile taken along the molecule shows major and minor grooves with diameter in the range of 1.5–1.7 nm. Additionally, different imaging modes of AFM were applied to image RNA molecules such as the amplitude modulation AFM (AM-AFM), drive amplitude modulation (DAM-AFM), and Jumping Mode plus (JM+). With a proper interacting force and an enough sharp tip, a similar high resolution can be achieved in three imaging modes.

### 1.6.3 Atomic and Sub-nanometer Resolution

It is valuable to provide a practical definition of lateral resolution at atomic and subnanometer scales. Atomic resolution of AFM is defined as the ability to distinguish single or multiple atomic-scale features (spaced at atomic scales). Atomic-scale features include atoms, molecules, point defects, and grain boundaries.[59] It is also called "true atomic resolution".[47] Similarly, sub-nanometer resolution is defined as the ability to distinguish single or multiple sub-nanometer features. Sub-nanometer resolution images are particularly useful for biological studies. In order to obtain a given resolution, the contact size must be equal to or less than the size of the feature to be observed. AFM images with atomic resolution can be obtained on relatively hard

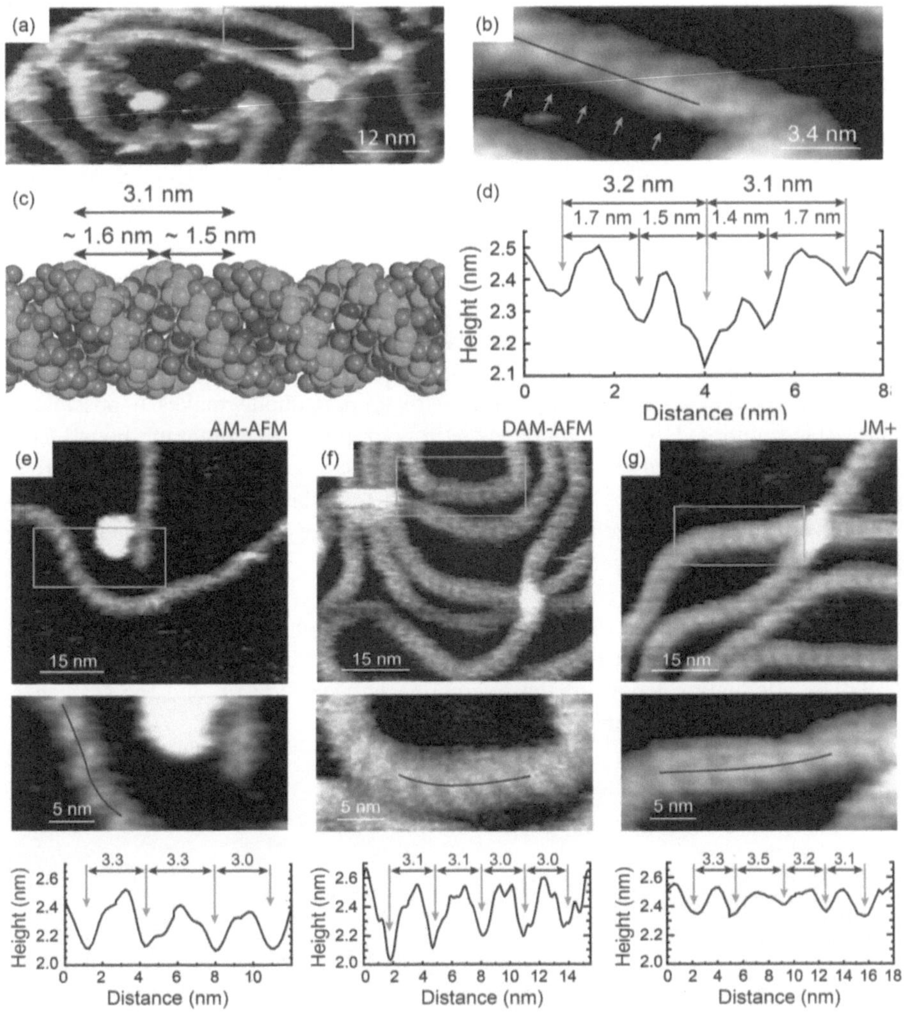


**FIGURE 1.14** RNA helical structure was imaged by AM-AFM with high-resolution. (a) A periodic corrugation and a helical structure along the RNA molecule are obtained. (b) Enlarged part of (a) with double-band corrugation corresponding to the major (green arrow) and minor (gray arrow) grooves. (c) Model of RNA with relevant dimensions. (d) Height profiles along the line in (b). High-resolution AFM images of RNA acquired in different imaging modes: (e) Amplitude Modulation (AM-AFM), (f) Drive Amplitude Modulation (DAM-AFM), and (g) Jumping Mode plus (JM+). (Adapted with permission from Ares et al.,[58] Copyright (2016) Royal Society of Chemistry.)

crystalline materials with atomically flat surfaces. For soft materials and samples with surface height variations of more than 1 nm, lateral resolution is somewhat limited. The first attempt to achieve AFM image with atomic resolution was carried out on semiconductor surfaces with chemical forces (about 1 nN).[60] Muller C. and co-workers obtained AFM images of purple membrane surfaces with lateral resolution of 1.1 nm.[61] AFM atomic resolution images of graphene show various contrast patterns depending on the tip apex and the tip-sample distance.[62] Sugimoto et al. gave several reports that showed that topographic AFM measurements with atomic resolution have the capability to get the access to the real structure.[63–66] In Figure 1.15, Sugimoto et al. employed dynamic AFM to identify single atoms Sn and Pb on Si (111) substrate and achieve atomic imaging resolution. The dynamic AFM allows the detection of the interaction forces between the tip apex and surface atoms being images.[63] The reduction of topographic contrast in Figure 1.15d–e illustrates a small concentration of substitutional Si defect. They also quantified attractive short-range force to identify individual atoms in a multi-element system, where Si, Sn, and Pb atoms mixed in equal proportions, as shown in Figure 1.15f–i.

In addition to atomic resolution, AFM is a powerful tool to achieve molecular and sub-molecular scale imaging. AFM with tapping mode makes it possible to image soft samples. High-resolution AFM topographies of membrane-bound LLO oligomers were recorded in buffer solutions.[67] The sub-nanometer scale structure of tubulins was also imaged by means of frequency modulation AFM in liquids.[68] It is relatively difficult to achieve an AFM image with high resolution. The conditions for high-resolution imaging can be simplified as follows:

1. The instrument noise should be less than the target resolution;
2. The lateral resolution cannot be less than the size of the tip apex;
3. The attenuation length of the interaction between the tip and the surface should be smaller than that of the tip apex;

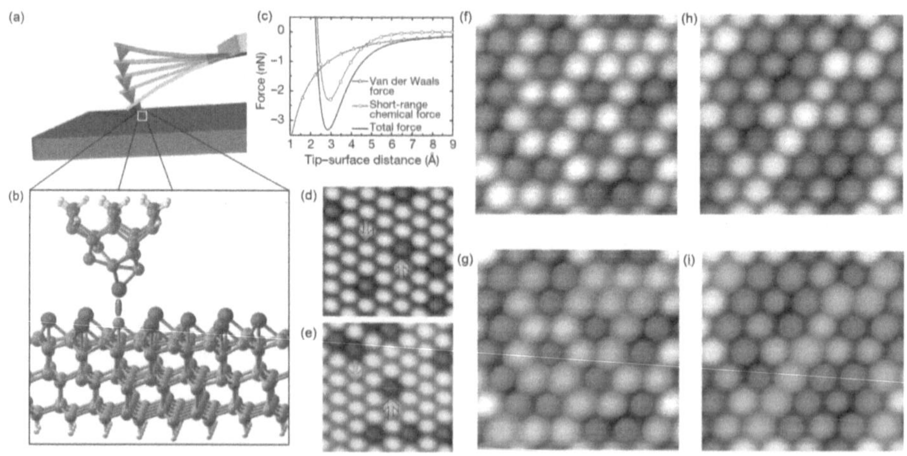


**FIGURE 1.15** A working schematic of the dynamic force microscopy with atomic-resolved resolution. (a) Diagram of AFM in dynamic mode (b) and the chemical bonding between the outmost tip atom and a surface atom. (c) The relationship between tip-surface distance and different forces. Topographic images of a single-atomic layer of (d) Sn and (e) Pb grown on a Si (111) substrate. (f),(h) Topographic image and (g),(i) chemical composition of a surface alloy composed by Si, Sn, and Pb atoms blended in equal proportions at different positions. (Reprinted with permission from Sugimoto et al.,[63] Copyright (2007) Springer Nature.)

4. The isolated 3D objects limit the lateral resolution;
5. Adhesion energy should be minimized, which can be realized with a sharp tip;
6. The imaging force should be as small as possible;
7. Under the same conditions (tip radius and imaging force), a harder sample would get a better lateral resolution.

Last but not least, the contaminant control is another important factor for high-resolution imaging, especially for atomic resolution imaging.

## 1.7 IMAGING IN AIR, LIQUID, UHV

AFM can obtain surface morphology information of the sample through the weak interaction force (atomic force) between the probe and the sample to be measured. Therefore, AFM can observe not only the surface structure of conductive samples but also the surface of nonconductive samples. For example, the polyvinyl alcohol (PVA) film/water interface morphology is stable as the annealing time changed from 1 to 120 minutes, as shown in Figure 1.16a–e.[69] The AFM height profile of the few-layer GaSe flake of the typical vertical GaSe/$VO_2$ mixed-dimensional heterostructure is about 11.96 nm in Figure 1.16f.[70] The triangular monolayer $MoS_2$ flake on the perforated substrate grown by chemical vapor deposition was imaged by AFM in Figure 1.16g. The thickness of monolayer $MoS_2$ was determined to be about 1.0 nm.[71] AFM also enables to distinguish the spatial distribution structures of $WSe_2$-$WS_2$ lateral heterojunction in Figure 1.16h, confirming that the monolayer heterojunction has a thickness of approximately 0.8 nm for the shell region and 0.6 nm for the core region, respectively.[72]

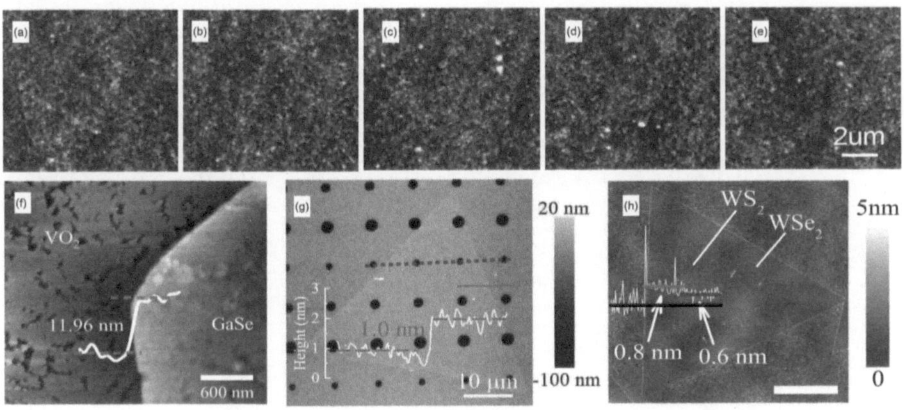


**FIGURE 1.16** AFM images of surface morphology for PVA films after annealing at 150°C with (a) 1 minute, (b) 30 minutes, (c) 60 minutes, (d) 90 minutes, and (e) 120 minutes. (Adapted with permission from Wu et al.,[69] Copyright (2019) American Association for the Advancement of Science.) (f) Topography of the GaSe/$VO_2$ heterostructure. (g) Topography of a single-layered triangular $MoS_2$ on a perforated substrate. (h) AFM image of a monolayer $WSe_2$-$WS_2$ heterostructure.[72]

On the other hand, AFM can obtain a real three-dimensional (3D) image of the sample surface and measure the 3D information of the sample. Therefore, the height image of AFM can be used to quantify the high-resolution roughness, pore size, particle distribution, and size of a given micro region on the sample surface. The resolution of the AFM images can reach the atomic level with a horizontal resolution of 0.1 nm and a vertical resolution of 0.01 nm. The average-surface roughness $R_a$, the maximum-height roughness $R_m$, and the distribution asymmetry $S_k$ (skewness) are the parameters commonly used to describe the surface roughness. The average surface roughness $R_a$ is the arithmetic mean value of the absolute height deviation measured relative to the central plane in the investigated area, which is defined as $R_a = \frac{1}{n}\sum_{i=1}^{n}|y_i|$, where $y_i$ is the height of surface contour based on the center line and $i$ is the number of contour offset. The root-mean-square roughness $R_q$ is another parameter to be used for quantifying the surface roughness, which is defined as $R_q = \sqrt{\frac{1}{n}\sum_{i=1}^{n}(y_i)^2}$. Figure 1.17 shows an example of surface roughness measurement using AFM. The vanadium oxide-based $V_{1-x}W_xO_2$ films with different tungsten concentrations and preparation conditions were grown on a single-crystal *c*-cut $Al_2O_3$ substrate using a pulse laser deposition technique.[73] The 3D AFM images show the surface morphologies in Figure 1.17. The measured surface roughness ($R_q$) of samples 1–4 are 1.65, 1.41, 2.05, and 2.07 nm, respectively. The measured roughness results

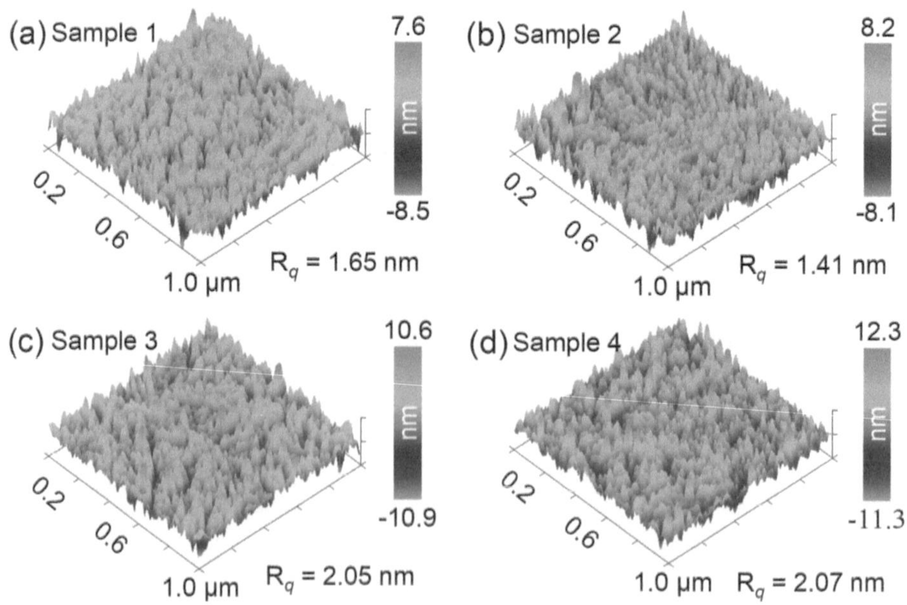


**FIGURE 1.17** Topographies of vanadium oxide-based $V_{1-x}W_xO_2$ thin films with different tungsten concentrations and preparation conditions: (a) Sample 1, (b) Sample 2, (c) Sample 3, and (d) Sample 4, respectively.[70]

by AFM illustrate that the surface roughness of $V_{1-x}W_xO_2$ films does not change much by the variation of deposition time, oxygen pressure, and energy density.[73]

In addition to inorganic materials, AFM is applied to image the surface morphology for a number of soft materials and biological samples such as DNA, virus, proteins, and cells.[74–79] Liu et al. investigated the heterogeneous feature of actin filaments (ACFs) associated with the cellular membrane in HeLa and HCT-116 cells at the nanoscale level via AFM coupled with the fluorescence microscopy as shown in Figure 1.18a. The topography and distribution of microvilli detected by AFM with fluorescence microscopy in Figure 1.18b–d confirmed that the observed structural features in AFM images belonged to the ACFs of the HCT-116 cells.[80] Belaidi et al. used AFM to observe the dynamic process of pore formation for living cells mediated by executioner molecules with high resolution and noninvasive capabilities.[74] As shown in Figure 1.18e–f, pore-like structures on the surface of the cell membrane in living OVA-B16 cells confirmed the morphological changes induced by the perforin treatment.[74] The application of AFM enables the visualization and understanding of its pore formation. It is critical to understand the basic process of immune clearance of pathogens and tumor cells, as well as pore-induced cell death.

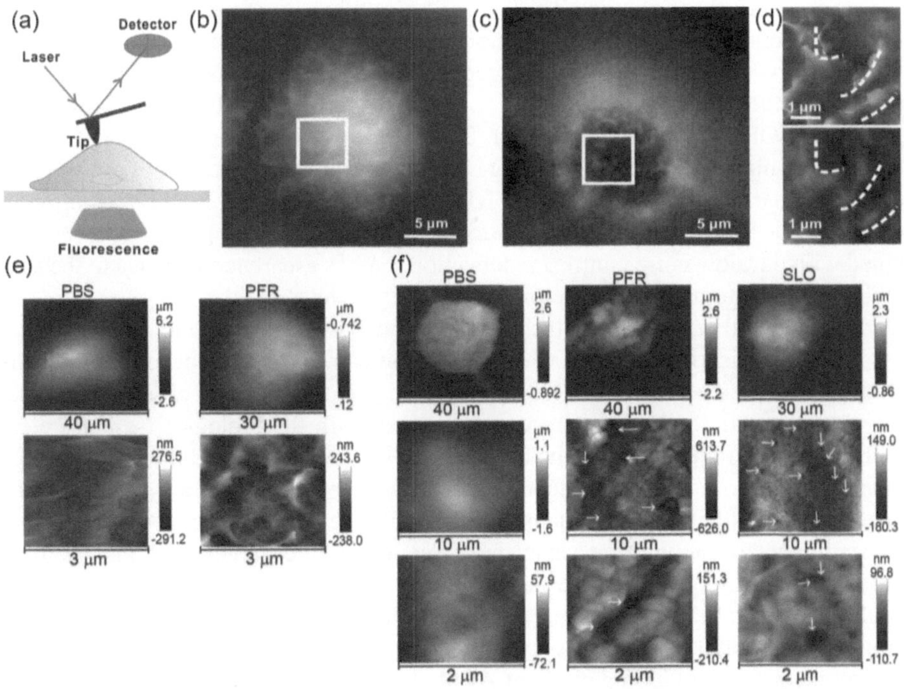


**FIGURE 1.18** (a) Schematic of AFM setup with the fluorescence microscopy (FM) to image cells. In situ (b) AFM and (c) FM images of the stained cell. (d) The enlarged AFM and FM images of the same cell region. (Adapted with permission from Liu et al.,[80] Copyright (2020) John Wiley and Sons.) (e) and (f) Visualization of SLO/perforin-induced pore formation using AFM. (Adapted with permission from Liu et al.,[74] Copyright (2019) Springer Nature.)

Voigt et al. employed the tapping mode to detect monomolecular chemical reactions on the surface of DNA-folding scaffolds under ambient environment, and proved the high yield and chemical selectivity of monomolecular chemical reactions, thus demonstrating the feasibility of chemical modification of DNA nanostructures after assembly.[81] AFM is also used to observe the self-assembly process of nanomaterials and its related influencing factors, providing a basis for in-depth understanding of the structure and function of nanomaterials. Ruan et al. designed a peptide structure containing nine residues and observed the morphology of the peptide samples by tapping mode of AFM in air. The results showed that the polypeptides can self-assemble into nanofiber networks, and the morphology of the fibers varies with the change in polypeptide concentration.[82]

When the sample is exposed to air, its surface is usually covered with water molecules in the form of clusters or thin films. When the AFM tip approaches the surface covered with a water film, a liquid neck forms around the approach contact. Compared with the case where there is no liquid neck, an increased force should be applied, due to the capillary effect. Therefore, the tip-surface junction would have effect from the capillary force in ambient environment, resulting in the reduction in imaging resolution. Hence, AFM imaging in liquid can eliminate the contribution from capillary force to improve the resolution.[83–85] In addition, AFM in liquid can reduce the long-range van der Waals force[86,87] and has a precise control of electrostatic interactions.[45] Fukuma et al. demonstrated true atomic resolution of the cleaved (001) surface of muscovite mica via frequency-modulation AFM in liquid.[88] Zhang et al. imaged a series of cylindrical cavities covered by highly oriented pyrolytic graphite (HOPG) in liquid and in air. They demonstrated that harmonic AFM in liquid has a higher subsurface imaging capability than its counterpart in air, because the higher-harmonic amplitude sensitivity to local stiffness in liquid was an order of magnitude larger that in air.[89] Additionally, the technique of dynamic AFM combined with ultralow noise cantilever deflection sensors greatly contributes to the submolecular and atomic resolution for a series of systems.[90]

An additional merit is that the technology of imaging in liquid would greatly expand the application of AFM, which has become one of the important tools for the study of biological samples. It allows AFM to be performed in situ observation under the physiological conditions, where one can maintain the biological samples alive with native functional and morphological state.[83,91] Yang et al. used the AFM tapping mode to in situ observe the nucleation and growth process of multilayer elastin-like peptide I (EPI) fibers in a liquid environment. The results showed that the EPI fiber nucleation needs enough transverse space to allow the fiber to expand, and the fiber expansion mainly occurs on the flat step rather than the edge of the defect.[92] Chen et al. investigated the in situ morphology and surface hydrophilic and hydrophobic properties of poly (N-isopropylacrylamide) materials at 20°C and 37°C in a liquid environment via the tapping mode of AFM. The results showed that the increase in temperature leads to the decrease in the nanoscale papillate structure and the change of surface from hydrophilic to hydrophobic.[93] El-Kirat-Chatel et al. used AFM and optical microscopy to image living cells at the close-to-native state in a liquid environment. Figure 1.19a shows the setup of AFM coupled with optical microscopy of imaging cells. When the fluorescence images showed the difference between fungal

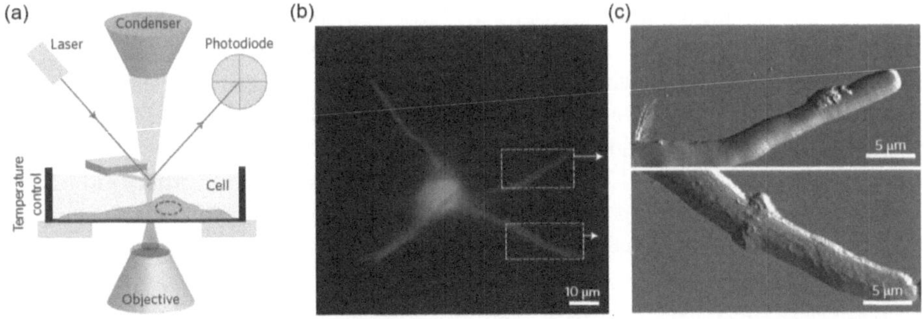


**FIGURE 1.19** (a) Schematic of AFM setup with optical microscopy for characterizing living cells in liquid. Fluorescence image (b) and correlative AFM images (c) of a macrophage incubated for 3 hours with cells from Candida albicans. Images in panel (c) are enlarged views of the dashed areas shown in the fluorescence image. Internalized (bottom) and externalized (top) hyphae exhibits major structural differences. (Reprinted with permission from El-Kirat-Chatel and Dufrêne,[94] Copyright (2012) American Chemical Society.)

cells and macrophages, AFM images exhibited the biologically related nanostructures of both cell types in Figure 1.19b–c.[94]

Additionally, it is valuable to measure electrical properties at the nanometer scale in liquids for many applications. Since the emergence of AFM technology, it has been widely used in the analysis and research of lithium-ion batteries (LIBs). AFM can detect lithium ion diffusion, electron migration, and electrode-electrolyte interface reaction in real time under electrochemical reaction conditions.[95–97] AFM also plays an important role in studying the formation, deformation, and rupture process of solid electrolyte interphase (SEI) films, which provides a reliable basis for the study of SEI films.[98,99] Wan et al. conducted in operando characterizations of the morphology evolution of monolayer $MoS_2$ during the live formation of SEI layer, where, during the lithiation process, the SEI film grew, propagated, and branch as wrinkles and eventually became a nanofold structure at the interface. In their work,[98] the thickness of SEI film was about 0.6 nm at the beginning, and then reached at 1.5 nm, measured by AFM. This study found the fact that the additive of fluoroethylene carbonate induced the formation of ultra-thin and dense films to protect the anode electrodes and also revealed the capacity degradation mechanism of $MoS_2$. Besides, Verde et al. studied the phase transition mechanism of $Li_4Ti_5O_{12}$ cathode in the process of lithium-intercalation/delithium by in situ conductive AFM. Through observing the change of electrode surface morphology and current intensity under different charge states, it demonstrates that the cathode could undergo reversible transformation between $Li_4Ti_5O_{12}$ and $Li_7Ti_5O_{12}$ phase.[100] The ability of AFM to detect morphology and performance evolution on nanometer scale with low destructivity is conductive to a deep understanding of the structure and related properties of SEI films, the anode and anode materials of LIBs. It has laid a solid foundation for the research works on LIBs, and further promoted the development of LIBs.

Since most of commercial AFMs can be carried out in liquid and high-resolution images, many advanced functional techniques on the basis of AFM have also been designed to probe piezoelectricity, electrochemistry, and electrical transport in aqueous environment. The traditional conductive AFM probe is not suitable to test the electric signal in a conductive liquid environment, owing to the electrochemical reactions and stray capacitance effect. Figure 1.20a shows a simple experimental schematic of a deriving conductive SPM mode in a liquid environment. For imaging electric properties better, a kind of the pre-mounted nanoelectrode probe (#PFSECM probe) with a nanoscale tip apex, *k* value of ~1.5 N/m, and resonant frequency of ~62 kHz, has been invented for realizing high-resolution test by PFM, conductive AFM, or Scanning Electrochemical Microscopy (SECM) modes. The optical photograph at right inset and the bottom scanning electron microscopic (SEM) images of Figure 1.20a shows the structure of the PFSECM probe, where the premounted probe

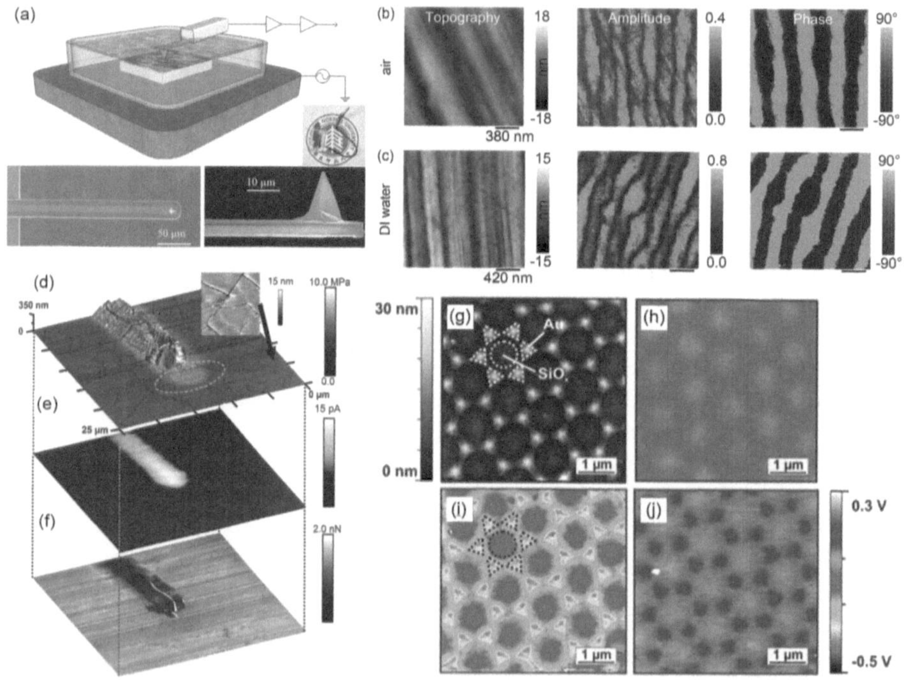


**FIGURE 1.20** (a) Experimental setup of electrical AFM mode in a liquid tank, the optical photograph (the inset at right), and the bottom SEM images of the pre-mounted nanoelectrode probe (#PFSECM). High-resolution images of topography, amplitude, and phase from left to right in (b) DI water and (c) electrolytes, respectively. (d) 3D topography with Young's modulus, (inset shows topography of the cracked PDMS), (e) faradaic current image, and (f) tip-sample adhesion image of a gold electrode patterned onto PDMS recorded in 2.5 mM $Fc(MeOH)_2$/0.1 M KCl. (Adapted with permission from Knittel et al.,[104] Copyright (2016) American Chemical Society.) (g) Topography image of a $SiO_x$/Au structure, KPFM images of the contact potential differences measured in (h) air and decane (i) before and (j) after exposure to hexadecanethiol for 90 minutes. (Adapted with permission from Domanski,[105] Copyright (2012) American Chemical Society.)

is encapsulated in glass sealed with a chemical-resistant epoxy compatible in a broad range of electrochemical environments. The encapsulation is designed in a large size. A Pt conductive path of 15-μm width runs in the core of the probe.

Recently, we have attempted to employ piezoresponse force microscopy (PFM) in the conductive liquid to obtain high-resolution PFM phase and amplitude images. Nanoscale ferroelectric domain boundaries and domain pattern in air and the deionized (DI) water have been achieved on a lead (Pb)-based relaxor ferroelectric single crystal, as shown in Figure 1.20b–c. Imaging in a liquid environment can reduce the long-range electrostatic contributions, van der Waals forces, and has precise control tip-sample interaction.[101–103] Knittel et al. have combined QNM and SECM modes (QNM-AFM-SECM), enabling to provide quantitative nanomechanical information along with electrochemical properties in electrolytes at the same time. QNM-AFM-SECM simultaneously obtained topographic data, Young's modulus, and the current image of the fabricated soft gold microelectrode on the PDMS substrate, as shown in Figure 1.20d–f.[104] Domanski et al. applied Kelvin probe force microscopy (KPFM) in electrically insulating nonpolar solvents to investigate the work function of a gold surface upon chemisorption of hexadecanethiol. The work function of SiO*x*/Au patterned substrates varied from each other as different adsorption processes were employed, as shown in Figure 1.20g–j.[105] However, when a charged surface and tip is exposed to a polar liquid, due to the ionization or ion adsorption, electric double layer (EDL) would form near the solid-liquid interface. A model referred to the Gouy-Chapman-Stern (GCS) model has been proposed to describe the behavior of EDL. When two surfaces are close to each other, the EDL force caused by the EDLs interaction would lead to a reduced resolution for electrical imaging by SPM.[85,106] Therefore, owing to the effect of EDL, an electric field screening effect has been proposed in the tip-sample junction, leading to a reduction of AFM resolution in conductive liquids.

Finally, along with the development of AFM technology, significant progress has also been made in the development of high-resolution AFM under ultra-high vacuum (UHV). AFM imaging in UHV condition is of many merits to eliminate the capillary force and reduce the long-range van der Waals force, as AFM imaging in liquid does. Besides, imaging in UHV can avoid the effect of EDL force, contaminant, and some unexpected chemical reactions for experiment. Gryzia and co-workers have applied UHV AFM/STM in the non-contact mode to perform AFM and KPFM micrograph of the absorbed $[Mn^{III}_6Cr^{III}](ClO_4)_3$ single-molecule magnets (SMM) on HOPG to characterize its topography and local contact potential differences.[107] Barletta and Wandelt studied the freshly fractured Baltic Amber samples with UHV-AFM in micrometer/nanometer resolution, where AFM images proved the completely amorphous structure of the sample by imaging the 2D pair-distance distributions. These results supported that it is possible to obtain AFM image of amorphous dielectric surface with atomic resolution.[108]

In summary, AFM can be used to detect surface properties of the conductor and semiconductor, insulator materials in gas, vacuum, and liquid environments. Therefore, AFM has great significance and broad application prospects in surface science, material science, electrochemistry, and life science.

## 1.8 AFM FOR ELECTRICAL CONDUCTIVITY IMAGING

Sections 1.1–1.7 have specifically introduced the fundamental principles and the commercially available imaging modes of AFM in detail. Based on the basic imaging modes, a great amount of the functional modes of SPM have been produced in the past decades such as piezoresponse force microscopy (PFM), KPFM, and conductive AFM (CAFM). In these deriving functional modes, the measured variable by the AFM probe is different for interpreting the functional characteristics of materials. Here, Table 1.3 lists some mainstream functional SPM modes and their related physical variables, which are collected into the AFM controller by probe for mapping the physical or chemical property of the measured materials. Some functional SPM modes and their recent significant results have been addressed carefully in the following chapters of this book.

As mentioned in Table 1.3, PFM is able to achieve the piezoelectric signal and ferroelectric polarization distribution by sensing the electric-field induced displacement of piezoelectric surface by a conductive PFM probe.[71,101–103] The CAFM mode takes the AFM probe as a nanoscale "ampere meter" for collecting the current-voltage (I-V) response at each pixel of a resistance-typed sample surface. Conventional CAFM depends on the contact imaging mode, forming a conductive circuit junction in tip-sample contact. Unfortunately, the contact-mode-based CAFM limits the application in soft conducting samples, such as conductive organics and polymers, due to the severe topographic feedback and large imaging force in vertical/lateral directions.

An appropriate imaging force plays an important role in accurately collecting the electrical current signal. Recently, TUNA based on PeakForce tapping mode has been realized for imaging conductive property of materials,[109] with better resolution and higher sensitivity than traditional CAFM, in which the aforementioned PeakForce tapping mode allows the precise control of imaging force around the tip-sample contact area. Hence, the TUNA mode enables the nanoscale electrical

**TABLE 1.3**
**Some Mainstream SPM Functional Imaging Modes and Their Core Detected Variables**

| The Detected Physical Variable | Functional Imaging Mode |
|---|---|
| Phase signal of cantilever vibration | Phase imaging |
| Tip-sample lateral force | Lateral force microscopy (LFM) |
| Phase or frequency of cantilever vibration | Electrostatic/Magnetic force microscopy (EFM/MFM) |
| Surface potential | Kelvin probe force microscopy (KPFM) |
| Electric induced surface displacement | Piezoresponse force microscopy (PFM) |
| Current between tip and sample | Conductive AFM/ Tunneling AFM (CAFM/TUNA) |
| Surface capacitance change with respect to the applied alternating voltage $dC/dV$ | Scanning capacitance microscopy (SCM) |

imaging to be suitable for more conductive materials, including the soft conducting samples. The CAFM or TUNA mode is equipped with three main elements, involving a basic AFM operating unit, a conductive probe, and a current sensor. Generally, a *dc* voltage is added to the tip or the sample, and then the resulting current crossing the tip-sample junction is collected by the current sensor and amplified by a low-noise high-bandwidth current amplifier. As an intermittent contact mode, the PeakForce tapping-based TUNA mode allows the conducting tip to stay on the surface with the tip-surface interaction around the peak force value. In this duration, the current characteristics are collected. To accurately examine the measured current response, PeakForce TUNA could measure three kinds of current signals: (1) Peak current, (2) the cycle-averaged current, and (3) the contact-averaged current. The peak current as an instantaneous response is collected at the moment of tip surface reaching the defined peak force, while the cycle-averaged one is an average current over one whole tapping cycle, and considers the measured signals when tip is in contact with the surface and off the surface, whereas the contact-averaged current only collects the averaged current signal when the tip-sample is in contact. These measured current could be simultaneously visualized with the topography and mechanical properties during the PeakForce TUNA scanning. Except for the imaging of current signal under a bias, the TUNA mode can measure nanoscale I-V curves on the sample, where the AFM feedback loop is set as the contact mode.

Recently, with the rapid development of nanoscale electrical AFM imaging in a conductive liquid environment, CAFM and TUNA techniques realize the high-resolution nanoscale observation of electrical conductivity in different liquid environments. Electrical imaging technique by AFM in a conducting liquid has greatly promoted the studies for real-time exploring energy storage, biofunctionality, and electroactive soft matter systems.[101–103] There is a classic experimental example to introduce the current mappings and I-V spectra by the TUNA mode in different imaging environments. Figures 1.21 and 1.22 show the topography and current images by PeakForce TUNA mode in argon, nonpolar dimethyl carbonate (DMC, $C_3H_6O_3$), and the polar KCl electrolyte, respectively, on a standard sample of a 50-nm-thick partially patterned silicon-nitride ($Si_3N_4$) layer deposited on a Pt substrate. The voltages of ±0.5 V are applied to the tip to collect current images. Current response exhibits strong saturation on the Pt substrate and almost no electrical signal on $Si_3N_4$. Simultaneously, good electric imaging is also closely related to achieve a background noise signal as low as possible. As shown in Figures 1.21d, e and 1.22d, e, very low noise currents of approximately less than 6 pA (peak-to-peak value, $V_{pp}$) are estimated at the modulating bias of ± 0.5 V. Note that the pre-mounted insulated nanoelectrode probe (#PFSECM) was utilized for conductive imaging in the conducting KCl electrolyte environment. It clearly confirms the efficient operation of PeakForce TUNA technique in gas and liquid environments.

For the domain of energy research, conductive AFM has become a powerful tool in investigating the underlying interplay between the topography and its electrical performance of organic solar cell materials and devices, LIBs, and nanostructures. Here, for readers who are interested in energy research using conductive AFM, we

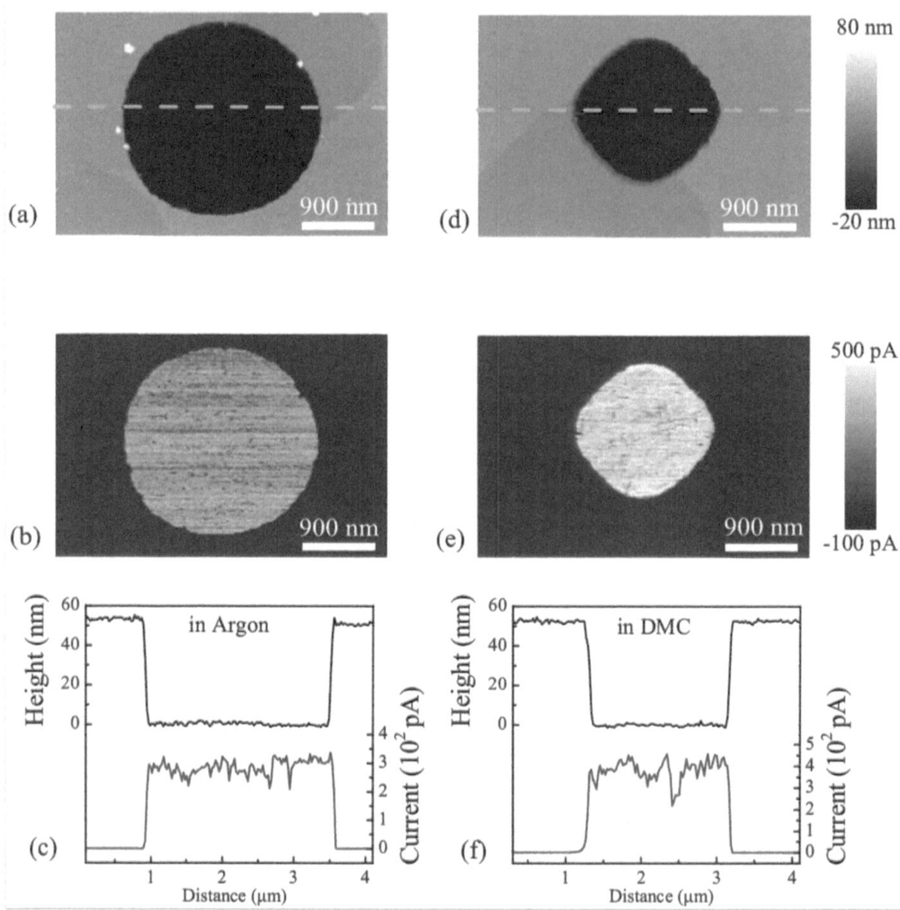


**FIGURE 1.21** Using PeakForce tapping mode-based TUNA method, conductive measurement was implemented in (a)–(c) argon and (d)–(f) DMC solution on a 50-nm-thick partially patterned silicon-nitride layer deposited on a Pt substrate. (a), (d) The topography and (b), (e) current mapping of $Si_3N_4$ on Pt under the adding voltage of $\pm 0.5$ V. (c), (f) Line profiles of topography at the upper, and the resulting current under $\pm 0.5$ V at the bottom.

would like to recommend the work of Li et al.,[109] in which readers can find more technical details and many representative conductive AFM example cases, including Thermal Annealing Effect on P3HT Thin Film; P3HT:PCBM Organic Solar Cell; Material Assignments in $Li[Ni_{1/3}Mn_{1/3}Co_{1/3}]O_2$ Composite Cathode; Optimizing PVDF+AB Content in $LiNi_{0.8}Co_{0.15}Al_{0.05}O_2$ Cathode; and Loosely Bound Carbon Nanotubes.

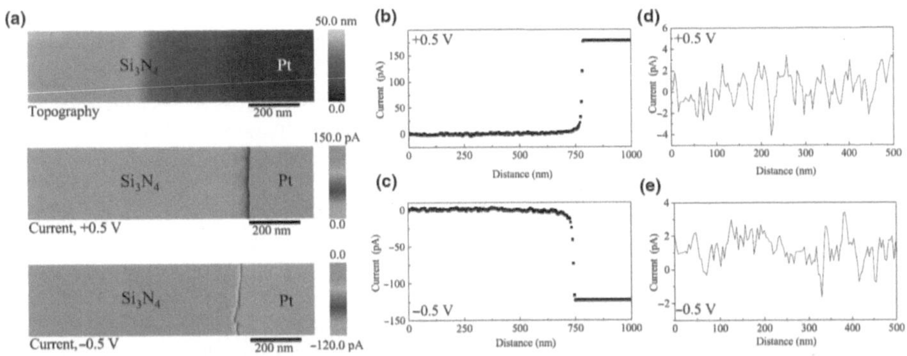


**FIGURE 1.22** Using PeakForce tapping mode-based TUNA method, conductive measurement was implemented in 1 M KCl solution on a 50-nm-thick partially patterned silicon-nitride layer deposited on a Pt substrate. (a) The topography and current mapping of $Si_3N_4$ on Pt with a tip voltage of $\pm 0.5$ V. (b)–(c) Line profiles of contact currents at $\pm 0.5$ V. (d)–(e) Background current ($\pm$ 0.5 V on $Si_3N_4$) manifesting good electric performances with low-noise signals of less than 6 pA ($V_{pp}$).[101]

## REFERENCES

1. Hansma Paul, K., and J. Tersoff. 1987. Scanning tunneling microscopy. *J. Appl. Phys.* 61 (2):R1–R24.
2. Binnig, G., Quate, C. F., and Gerber, C. 1986. Atomic force microscope. *Phys. Rev. Lett.* 56 (9): 930–933.
3. Cai, J. 2018. *Atomic Force Microscopy in Molecular and Cell Biology.* Springer Nature Pte Ltd. (Singapore).
4. Ang, K. H., Chong, G., and Li, Y. 2005. PID control system analysis, design, and technology. *IEEE Trans. Control. Syst. Technol.* 13(4):559–576.
5. Li, M., Zhou, W., Pang, H., and Lai, L. 2019. Improving the atomic-resolution AFM imaging of monolayer $MoS_2$ for worn tips: A molecular dynamics study. *Jpn. J. Appl. Phys.* 58 (5):055003.
6. Kolodziej, J. J., Goryl, M., Konior, J., Reichling, M., and Szymonski, M. 2007. Direct real-space imaging of the *c*(2×8)/(2×4) GaAs (001) surface structure. *Phys. Rev. B* 76(24):245314.
7. Diakowski, P. M., and Ding, Z. 2007. Interrogation of living cells using alternating current scanning electrochemical microscopy (AC-SECM). *Phys. Chem. Chem. Phys.* 9(45):5966–5974.
8. Ali, M., Hamzah, E. B., and Mohd Toff, M. R. H. J. 2007. Deposition and characterization of tin-coated steels at various $N_2$ gas flow rates with constant etching by using CAPVD technique. *Surf. Rev. Lett.* 14(1): 93–100.
9. Magonov, S. N. 2000. Atomic force microscopy in analysis of polymers. In *'Encyclopedia of Analytical Chemistry'* Ed by Meyers, R. A., Chichester, UK: John Wiley & Sons Ltd.
10. Saadi, M. A. S. R., Uluutku, B., Parvini, C. H., and Solares, S. D. 2020. Soft sample deformation, damage and induced electromechanical property changes in contact- and tapping-mode atomic force microscopy. *Surf. Topogr. Metrol. Prop.* 8(4): 045004.
11. Ruozi, B., Tosi, G., Tonelli, M., Bondioli, L., Mucci, A., Forni, F., and Vandelli, M. A. 2009. AFM phase imaging of soft-hydrated samples: A versatile tool to complete the chemical-physical study of liposomes. *J. Liposome Res.* 19(1): 59–67.

12. Akram, N., Zia, K. M., Saeed, M., Mansha, A., and Khan, W. G. 2018. Morphological studies of polyurethane based pressure sensitive adhesives by tapping mode atomic force microscopy. *J. Polym. Res.* 25(9): 194.
13. Kamruzzahan, A. S. M., Kienberger, F., Stroh, C. M., Berg, J., Huss, R., Ebner, A., Zhu, R., Rankl, C., Gruber, H. J., and Hinterdorfer, P. 2004. Imaging morphological details and pathological differences of red blood cells using tapping-mode AFM. *Biol. Chem.* 385(10): 955–960.
14. Dokou, E., Zhang, L., and Barteau, M. A. 2002. Comparison of atomic force microscopy imaging methods and roughness determinations for a highly polished quartz surface. *J. Vac. Sci. Tech.* 20(6): 2183–2186.
15. Yu, J., He, H., Zhang, Y., and Hu, H. 2017. Nanoscale mechanochemical wear of phosphate laser glass against a $CeO_2$ particle in humid air. *Appl. Surf. Sci.* 392:523–530.
16. Kocun, M., Labuda, A., Meinhold, W., Revenko, I., and Proksch, R. 2017. Fast, high resolution, and wide modulus range nanomechanical mapping with bimodal tapping mode. *ACS Nano* 11(10): 10097–10105.
17. Stan, C., and King, S. W. 2020. Atomic force microscopy for nanoscale mechanical property characterization. *J. Vac. Sci. Technol. B* 38(6): 060801.
18. Mousa, M., and Dong, Y. 2018. Novel three-dimensional interphase characterisation of polymer nanocomposites using nanoscaled topography. *Nanotechnology* 29(38):385701.
19. Wei, Q.-Q., Chen, S.-F., Cheng, X.-Y., Yu, X.-B., Hu, J., Li, M.-Q., and Zhu, P. H. 2000. Topography of skeletal muscle ryanodine receptors studied by atomic force microscopy. *J. Vac. Sci. Technol.* 18(2):636–638.
20. Micic, M., Hu, D., Suh, Y. D., Newton, G., Romine, M., and Lu, H. P. 2004. Correlated atomic force microscopy and fluorescence lifetime imaging of live bacterial cells. *Colloids Surf. B* 34(4): 205–212.
21. Anderson, M. S. 2019. Nanofluidic chromatography using a vibrating atomic force microscope tip. *Rev. Sci. Instrum.* 90(9):093701.
22. Hu, S., Mininni, L., Hu, Y., Erina, N., Kindt, J., and Su, C. 2012. High-speed atomic force microscopy and peak force tapping control. *Proc. SPIE Int. Soc. Opt. Eng.* 8324:832410.
23. Su, C., Shi, J., Hu, Y., Hu, S., and Ma, J. 2008. Method and apparatus of using peak force tapping mode to measure physical properties of a sample. *Bruker Nano Inc.* US9291640B2.
24. Pavel, T., Josef, K., and Udo, V. 2012. On the use of peak-force tapping atomic force microscopy for quantification of the local elastic modulus in hardened cement paste. *Cem. Concr. Res.* 42(1): 215–221.
25. Gazze, S. A., Hallin, I., Quinn, G., Dudley, E., Matthews, G. P., Rees, P., van Keulen, G., Doerr, S. H., and Francis, L. W. 2018. Organic matter identifies the nano-mechanical properties of native soil aggregates. *Nanoscale* 10(2): 520–525.
26. Nellist, M. R., Chen, Y., Mark, A., Goedrich, S., Stelling, C., Jiang, J., Poddar, P., Li, C., Kumar, R., Papastavrou, G., Retsch, M., Brunschwig, B., Huang, Z., Xiang, C., and Boettcher, S. W. 2017. Atomic force microscopy with nanoelectrode tips for high resolution electrochemical, nanoadhesion and nanoelectrical imaging. *Nanotechnology* 28(9): 095711.
27. Gutierrez, J., Mondragon, I., and Tercjak, A. 2014. Quantitative nanoelectrical and nanomechanical properties of nanostructured hybrid composites by PeakForce tunneling atomic force microscopy. *J. Phys. Chem. C* 118(2): 1206–1212.
28. Dokukin, M. E., and Sokolov, I. 2012. Quantitative mapping of the elastic modulus of soft materials with HarmoniX and Peak Force QNM AFM modes. *Langmuir* 28(46): 16060–16071.

29. Xu, K., Sun, W., Shao, Y., Wei, F., Zhang, X., Wang, W., and Li, P. 2018. Recent development of PeakForce tapping mode atomic force microscopy and its applications on nanoscience. *Nanotechnol. Rev.* 7(6): 605–621.
30. Cappella, B., and Dietler, G. 1999. Force-distance curves by atomic force microscopy. *Surf. Sci. Rep.* 34(1–3): 1–104.
31. Sader, J. E., Larson, I., Mulvaney, P., and White, L. R. 1995. Method for the calibration of atomic force microscope cantilevers. *Rev. Sci. Instrum.* 66(7): 3789–3798.
32. Sader, J. E., Chon, J. W. M., and Mulvaney, P. 1999. Calibration of rectangular atomic force microscope cantilevers. *Rev. Sci. Instrum.* 70(10): 3967–3969.
33. Hutter, J. L., and Bechhoefer, J. 1993. Calibration of atomic force microscope tips. *Rev. Sci. Instrum.* 64(7): 1868–1873.
34. Cleveland, J. P., Manne, S., Bocek, D., and Hansma, P. K. 1993. A nondestructive method for determining the spring constant of cantilevers for scanning force microscopy. *Rev. Sci. Instrum.* 64(2): 403–405.
35. Gibson, C. T., Watson, G. S., and Myhra, S. 1996. Determination of the spring constants of probes for force microscopy/spectroscopy. *Nanotechnology* 7(3): 259–262.
36. Hertz, H. 1882. Ueber die verdunstung der flüssigkeiten, insbesondere des quecksilbers, im luftleeren raume. *Annalen Der Physik* 253(10): 177–193.
37. Pharr, G. M., Oliver, W. C., and Brotzen, F. R. 1992. On the generality of the relationship among contact stiffness, contact area, and elastic modulus during indentation. *J. Mater. Res.* 7(3): 613–617
38. Pashley, M. D. 1984. Further consideration of the DMT model for elastic contact. *Colloids Surf.* 12(1–2): 69–77.
39. Johnson, K. L., Kendall, K., and Roberts, A. D. 1971. Surface energy and the contact of elastic solids. *Proc. R. Soc. Lond. A.* 324(1558): 301–313.
40. Sokolov, I., Zorn, G., and Nichols, J. M. 2016. A study of molecular adsorption of a cationic surfactant on complex surfaces with atomic force microscopy. *Analyst* 141(3): 1017–1026.
41. Chopinet, L., Formosa, C., Rols, M. P., Duval, R. E., and Dague, E. 2013. Imaging living cells surface and quantifying its properties at high resolution using AFM in QI™ mode. *Micron* 48: 26–33.
42. Pittenger, B., Erina, N., and Su, C. 2010. Quantitative Mechanical Property Mapping at the Nanoscale with PeakForce QNM. *Application Note Veeco Instruments Inc.* Doi: 10.13140/RG.2.1.4463.8246.
43. Beckwitt, E. C., Simon, N., Carnaval, I., Kisker, C., Carell, T., and Van Houten, B. 2018. Peakforce tapping AFM reveals that human XPA binds to DNA damage as a monomer producing a 60 degrees bend. *Biophys. J.* 114(3): 93a.
44. Faouri, A. R., Henry, R., Biris, A. S., Sleezer, R., and Salamo, G. J. 2017. Adhesive force between graphene nanoscale flakes and living biological cells. *J. Appl. Toxicol.* 37(11): 1346–1353.
45. Butt, H. J., Cappella, B., and Kappl, M. 2005. Force measurements with the atomic force microscope: technique, interpretation and applications. *Surf. Sci. Rep.* 59(1):1–152.
46. Maugis, D. 1992. Adhesion of spheres-the JKR-DMT transition using a dugdale model. *J. Colloid Interface Sci.* 150(1): 243–269.
47. Ohnesorge, F., and Binnig, G. 1993. True atomic-resolution by atomic force microscopy through repulsive and attractive forces. *Science* 260(5113): 1451–1456.
48. Ando, T. Uchihashi, T., and Fukuma, T. 2008. High speed atomic force microscopy for nano visualization of dynamic biomolecular processer. *Prog. Surf. Sci.* 83: 337.
49. Bustamante, C., and Keller, D. 1995. Scanning force microscopy in biology. *Phys. Today* 48(12): 32–38.

50. Garcia, R. 2010. Amplitude modulation atomic force microscopy. Chap. 8: Resolution, Noise, and Sensitivity. 103. Doi:10.1002/9783527632183
51. Shannon, C. E. 1949. Communication in the presence of noise. *Proc. Institute Radio Eng.* 31(1): 10–21.
52. Bustamante, C., Rivetti, C., and Keller, D. J. 1997. Scanning Force Microscopy under aqueous solution. *Curr. Opin. Struct. Biol.* 7: 709.
53. Gross, L., Mohn, F., Liljeroth, P., Repp, J., Giessibl, F. J., Meyer, G. 2009. Measuring the charge state of an Adatom with noncontact atomic force microscopy. *Science* 324(5933): 1428–1431.
54. Welker, J., and Giessibl, F. J. 2012. Revealing the angular symmetry of chemical bonds by atomic force microscopy. *Science* 336(6080): 444–449.
55. Pyne, A., Thompson, R., Leung, C., Roy, D., and Hoogenboom, B. W. 2014. Single-molecule reconstruction of oligonucleotide secondary structure by atomic force microscopy. *Small* 10(16): 3257–3261.
56. Keller, D. 1991. Reconstruction of STM and AFM imaged distorted by finite size tips. *Surf. Sci.* 253: 353.
57. Keller, D. J., and Franke, F. S. 1993. Envelope reconstruction of probe microscope images. *Surf. Sci.* 294: 409.
58. Ares, P., Fuentes-Perez, M. E., Herrero-Galán, E., Valpuesta, J. M., Gil, A., Gomez-Herrero, J., and Moreno-Herrero, F. 2016. High resolution atomic force microscopy of double-stranded RNA. *Nanoscale* 8: 11818–11826.
59. Gan, Y. 2009. Atomic and subnanometer resolution in ambient conditions by atomic force microscopy. *Surf Sci Rep.* 64(3): 99–121.
60. Giessibl, F. J. 1995. Atomic resolution of the silicon (111)-(7*7) surface by atomic force microscopy. *Science* 267(5194): 68–71.
61. Moller, C., Allen, M., Elings, V., Engel, A., and Muller, D. J. 1999. Tapping mode atomic force microscopy produces faithful high resolution images protein surfaces. *Biophys. J.* 77: 1150.
62. Boneschanscher, M. P., van der Lit, J., Sun, Z., Swart, I., Liljeroth, P., and Vanmaekelbergh, D. 2012. Quantitative atomic resolution force imaging on epitaxial graphene with reactive and nonreactive AFM probes. *ACS Nano* 6(11): 10216–10221.
63. Sugimoto, Y., Pou, P., Abe, M., Jelinek, P., Pérez, R., Morita, S., and Custance, Ó. 2007. Chemical identification of individual surface atoms by atomic force microscopy. *Nature* 446: 64–67.
64. Sugimoto, Y., Pou, P., Custance, Ó., Jelinek, P., Morita, S., Pérez, R., and Abe, M. 2006. Real topography, atomic relaxations, and short-range chemical interactions in atomic force microscopy: The case of the α-Sn/Si(111)-$\left(\sqrt{3}\times\sqrt{3}\right)$ R30° surface. *Phys. Rev. B* 73: 205329.
65. Onoda, J., Miyazaki, H., and Sugimoto, Y. 2020. Chemical Identification of the foremost tip atom in atomic force microscopy. *Nano Lett.* 20: 2000–2004.
66. Onoda, J., Niki, K., and Sugimoto, Y. 2015. Identification of Si and Ge atoms by atomic force microscopy. *Phys. Rev. B* 92: 155309.
67. Mulvihill, E., van Pee, K., Mari, S. A., Müller, D. J., and Yildiz, Ö. 2015. Directly observing the lipid-dependent self-assembly and pore-forming mechanism of the cytolytic toxin listeriolysin O. *Nano Lett.* 15: 6965–6973.
68. Asakawa, H., Ikegami, K., Setou, M., Watanabe, N., Tsukada, M., and Fukuma, T. 2011. Submolecular-scale imaging of a-helices and C-terminal domains of tubulins by frequency modulation atomic force microscopy in liquid. *Biophys. J.* 101(5): 1270–1276.
69. Wu, S., He, Z., Zang, J., Jin, S., Wang, Z., Wang, J., Yao, Y., and Wang, J. 2019. Heterogeneous ice nucleation correlates with bulk-like interfacial water. *Sci. Adv.* 5 (4): eaat9825.

70. Zhou, J., Xie, M., Ji, H., Cui, A., Ye, Y., Jiang, K., Shang, L., Zhang, J., Hu, Z., and Chu, J. 2020. Mixed-dimensional Van der Waals heterostructure photodetector, *ACS Appl. Mater. Interfaces* 12(16): 18674–18682.
71. Wang, X., Cui, A., Chen, F., Xu, L., Hu, Z., Jiang, K., Shang, L., and Chu, J. 2019. Probing effective out-of-plane piezoelectricity in van der Waals layered materials induced by flexoelectricity. *Small* 15: 1903106.
72. Jiang, T., Wang, F., Cui, A., Guo, S., Jiang, K., Shang, L., Hu, Z., and Chu, J. In situ exploration of the thermodynamic evolution properties in the type II interface from the $WSe_2$-$WS_2$ lateral heterojunction. 2018. *Nanotechnology* 29(43): 435703.
73. Zhou, J., Xie, M., Cui, A., Zhou, B., Jiang, K., Shang, L., Hu, Z., and Chu, J. 2018. Manipulating behaviors from heavy tungsten doping on interband electronic transition and orbital structure variation of vanadium dioxide films. *ACS Appl. Mater. Interfaces* 10 (36): 30548–30557.
74. Liu, Y., Zhang, T., Zhou, Y., Li, J., Liang, X., Zhou, N., Lv, J., Xie, J., Cheng, F., Fang, Y., Gao, Y., Wang, N., and Huang, B. 2019. Visualization of perforin/gasdermin/complement-formed pores in real cell membranes using atomic force microscopy. *Cell. Mol. Immunol.* 16: 611–620.
75. Hu, J., Zhang, Y., Gao, H. B., Li, M. Q., and Hartmann, U. 2002. Artificial DNA patterns by mechanical nanomanipulation. *Nano Lett.* 2: 55–57.
76. Kuznetsov, Y. G., Malkin, A. J., Lucas, R. W., Plomp, M., and McPherson, A. 2001. Imaging of viruses by atomic force microscopy. *J. Gen. Virol.* 82: 2025–2034.
77. Goncalves, R. P., and Scheuring, S. Manipulating and imaging individual membrane proteins by AFM. *Surf. Interface Anal.* 2006, 38, 1413–1418.
78. Scheuring, S., Casuso, I., and Rico, F. 2011. Biological AFM: Where we come from-where we are-where we may go. *J. Mol. Recognit.* 24: 406–413.
79. Engel, A., and Muller, D. J. 2000. Observing single biomolecules at work with the atomic force microscope. *Nat. Struct. Biol.* 7: 715–718.
80. Liu, L., Wei, Y., Liu, J., Wang, K., Zhang, J., Zhang, P., Zhou, Y., and Li, B. 2020. Spatial high resolution of actin filament organization by PeakForce atomic force microscopy. *Cell Proliferat.* 53: e12670.
81. Voigt, N. V., Torring, T., Rotaru, A., Jacobsen, M. F., Ravnsbaek, J. B., Subramani, R., Mamdouh, W., Kjems, J., Mokhir, A., Besenbacher, F., and Gothelf, K. V. 2010. Single-molecule chemical reactions on DNA origami. *Nat. Nano.* 5: 200–203.
82. Ruan, L., Zhang, H., Luo, H., Liu, J., Tang, F., Shi, Y.-K., and Zhao, X. 2009. Designed amphiphilic peptide forms stable nanoweb, slowly releases encapsulated hydrophobic drug, and accelerates animal hemostasis. *PNAS* 106(13): 5105–5110.
83. Drake, B., Prater, C. B., Weisenhorn, A. L., Gould, S. A. C., Albrecht, T. R., Quate, C. F., Cannell, D. S., Hansma, H. G., and Hansma, P. K. 1989. Imaging crystals, polymers, and processes in water with the atomic force microscope. *Science* 243(4898): 1586–1589.
84. Weisenhorn, A. L., Hansma, P. K., Albrecht, T. R., and Quate, C. F. 1989. Forces in atomic force microscopy in air and water. *Appl. Phys. Lett.* 54: 2651–2653.
85. Israelachvili, J. N. 2011. *Intermolecular and Surface Forces*, 3rd ed. Elsevier Pte Ltd. (Singapore)
86. Hartmann, U. 1991. Van der Waals interactions between sharp probes and flat sample surfaces. *Phys. Rev. B* 43: 2404.
87. Weisenhorn, A. L., Maivald, P., Butt, H. J., and Hansma, P. K. 1992. Measuring adhesion, attraction, and repulsion between surfaces in liquids with an atomic-force microscope. *Phys. Rev. B* 45: 11226.
88. Fukuma, T., Kobayashi, K., Matsushige, K., and Yamada, H. 2005. True atomic resolution in liquid by frequency-modulation atomic force microscopy. *Appl. Phys. Lett.* 87: 034101.

89. Zhang, W., Chen, Y., Liu, H., and Zheng, L. 2018. Subsurface imaging of cavities in liquid by higher harmonic atomic force microscopy. *Appl. Phys. Lett.* 113: 193105.
90. Fukuma, T., Higgins, M. J., and Jarvis, S. P. 2007. Direct imaging of individual intrinsic hydration layers on lipid bilayers at Angstrom resolution. *Biophys. J.* 92: 3603–3609.
91. Pfreundschuh, M., Martinez-Martin, D., Mulvihill, E., Wegmann, S., and Muller, D. J. 2014. Multiparametric high-resolution imaging of native proteins by force-distance curve–based AFM. *Nat. Protocols* 9: 1113–1130.
92. Yang, G., Wong, M. K., Lin, L. E., and Yip, C. M. 2011. Nucleation and growth of elastin-like peptide fibril multilayers: an in situ atomic force microscopy study. *Nanotechnology* 22: 494018.
93. Chen, L., Liu, M., Bai, H., Chen, P., Xia, F., Han, D., and Jiang, L. 2009. Antiplatelet and thermally responsive poly (N-isopropylacrylamide) surface with nanoscale topography, *J. Am. Chem. Soc.* 131(30): 10467–10472.
94. El-Kirat-Chatel, S., and Dufrêne, Y. F. 2012. Nanoscale imaging of the candida-macrophage interaction using correlated fluorescence-atomic force microscopy. *ACS Nano* 6(12): 10792–10799.
95. Zhang, J., Wang, R., Yang, X., Lu, W., Wu, X., Wang, X., Li, H., and Chen, L. 2012. Direct observation of inhomogeneous solid electrolyte interphase on MnO anode with atomic force microscopy and spectroscopy. *Nano Lett.* 12(4): 2153–2157.
96. Becker, C. R., Strawhecker, K. E., McAllister, Q. P., and Lundgren, C. A. 2013. In situ atomic force microscopy of lithiation and delithiation of silicon nanostructures for lithium ion batteries. *ACS Nano* 7(10): 9173–9182.
97. Becker, C. R., Prokes, S. M., and Love, C. T. 2016. Enhanced lithiation cycle stability of ALD-coated confined a Si microstructures determined using in situ AFM. *ACS Appl. Mater. Interfaces* 8(1): 530–537.
98. Wan, J., Hao, Y., Shi, Y., Song, Y.-X., Yan, H.-J., Zheng, J., Wen, R., and Wan, L.-J. 2019. Ultra-thin solid electrolyte interphase evolution and wrinkling processes in molybdenum disulfide-based lithium-ion batteries. *Nat. Commun.* 10: 3265.
99. Steinhauer, M., Stich, M., Kurniawan, M., Seidlhofe, B.-K., Trapp, M., Bund, A., Wagner, N., and Friedrich, K. A. 2017. In situ studies of solid electrolyte interphase (SEI) formation on crystalline carbon surfaces by neutron reflectometry and atomic force microscopy. *ACS Appl. Mater. Interfaces* 9: 35794–35801.
100. Verde, M. G., Baggetto, L., Balke, N., Veith, G. M., Seo, J. K., Wang, Z., and Meng, Y. S. 2016. Elucidating the phase transformation of $Li_4Ti_5O_{12}$ lithiation at the nanoscale. *ACS Nano* 10(4): 4312–4321.
101. Cui, A., Zhu, L., Jiang, K., Xu, L., Hu, Z., Xu, G., Sun, H., Huang, Z., Poddar, R., and Chu, J. 2019. Probing nanoscale electromechanical behaviors of relaxor ferroelectrics in highly conductive liquid environments. *Phys. Rev. Appl.* 11: 054037.
102. Ye, Y., Cui, A., Zhu, L., Hu, Z., Jiang, K., Shang, L., Li, Y., Xu, G., and Chu, J. 2019. Electric double layer oriented field screening effect on high-resolution electromechanical imaging in conductive solutions. *Phys. Rev. Appl.* 12: 034006.
103. Cui, A., de Wolf, P., Ye, Y., Hu, Z., Dujardin, A., Huang, Z., Jiang, K., Shang, L., Ye, M., Sun, H., and Chu, J. 2019. Probing electromechanical behaviors by datacube piezoresponse force microscopy in ambient and aqueous environments. *Nanotechnology* 30: 235701.
104. Knittel, P., Mizaikoff, B., and Kranz, C. 2016. Simultaneous nanomechanical and electrochemical mapping: combining peak force tapping atomic force microscopy with scanning electrochemical microscopy. *Anal. Chem.* 88: 6174–6178.
105. Domanski, A. L. 2012. Kelvin probe force microscopy in nonpolar liquids. *Langmuir* 28: 13892.

106. Helmholtz, H. 1853. Ueber einige gesetze der vertheilung elektrischer ströme in körperlichen leitern mit anwendung auf die thierisch-elektrischen versuche. *Ann. Phys. Chem.* 89: 211–233, 353–7.
107. Gryzia, A., Volkmann, T., Brechling, A., Hoeke, V., Schneider, L., Kuepper, K., Glaser, T., and Heinzmann, U. 2014. Crystallographic order and decomposition of $Mn^{III}{}_6Cr^{III}]^{3+}$ single-molecule magnets deposited in submonolayers and monolayers on HOPG studied by means of molecular resolved atomic force microscopy (AFM) and Kelvin probe force microscopy in UHV. *Nanoscale Res. Lett.* 9: 60.
108. Barletta, E., and Wandelt, K. 2011. High resolution UHV-AFM surface analysis on polymeric materials: Baltic Amber. *J. Non-Cryst. Solids* 357: 1473–1478.
109. Li, C., Minne, S., Pittenger, B., and Mednick, A. 2011. Simultaneous Electrical and mechanical property mapping at the nanoscale with PeakForce TUNA. Bruker Nano Surfaces Division, Application Note AN132, Rev. A0.

# 2 Advanced Modes of Electrostatic and Kelvin Probe Force Microscopy for Energy Applications

*Martí Checa, Sabine M. Neumayer, Wan-Yu Tsai, and Liam Collins*
Oak Ridge National Laboratory

**CONTENTS**



DOI: 10.1201/9781003174042-2

## 2.1 INTRODUCTION

The invention of the Atomic Force Microscope (AFM)[1] in the last quarter of the 20th century opened the door for the systematic exploration of the nanoscale world. Soon after its initial development, it was realized that the sensitivity of AFM to probe minute forces could be extended to applications for probing long-range magnetic and electrostatic forces. This began a continuous realization of functional modes of AFM operation which went beyond topographical imaging to allow simultaneous mapping of material properties with nanoscale spatial resolution. In the very early days, Electrostatic Force Microscopy (EFM)[2,3] and Kelvin Probe Force Microscopy (KPFM)[4] emerged as two prominent functional mapping approaches. The enormous potential of these modes is evidenced by their broad and continuous application to this day (see Figure 2.1).

Both EFM and KPFM are non-destructive surface-sensitive techniques that allow us to map a variety of electronic/electrochemical properties including work function distributions (for metals or semiconductors), surface charge or surface potential distributions (for insulators), and/or dielectric properties with high lateral resolution (tens of nanometers). Both EFM and KPFM techniques are based on a biased vibrating AFM probe that is scanned across the sample surface measuring electrostatic forces (i.e., amplitude modulated) or force gradients (i.e., frequency modulated) between probe and sample. The main difference between the two techniques is that a bias voltage feedback is involved in KPFM enabling it to directly obtain the absolute work function values of the sample. Therefore, EFM is considered a qualitative approach whereas KPFM is considered a quantitative one. Despite its qualitative nature, certain groups have successfully quantified local properties, including dielectric constants from the EFM signal (discussed in Section 2.2.3).

The principle and standard operation mode of EFM and KPFM will be discussed in Sections 2.2, and 2.3, respectively. It is noteworthy that KPFM can suffer from problems associated with crosstalk and artifacts, which will be dealt with in Section 2.3.5 along with the significant efforts to eliminate these issues. Technical details, such as advanced data analysis and interpretation,[5,6] quantitative EFM (Section 2.2.3), artifacts and crosstalk between tip and sample, and how to eliminate these issues (Section 2.3.5) are also reviewed.

Some of the earliest applications of EFM, KPFM, and related techniques dealt with applications involving semiconductor surfaces[7,8] and metallic nanostructures[9] and quickly expanded into more diverse areas of investigation including biological systems,[10,11] corrosion science,[12–14] and ferroelectrics.[15] In recent years, there has been a steady rise in applications of EFM/KPFM for energy materials and devices, which will be reviewed in Section 2.4.

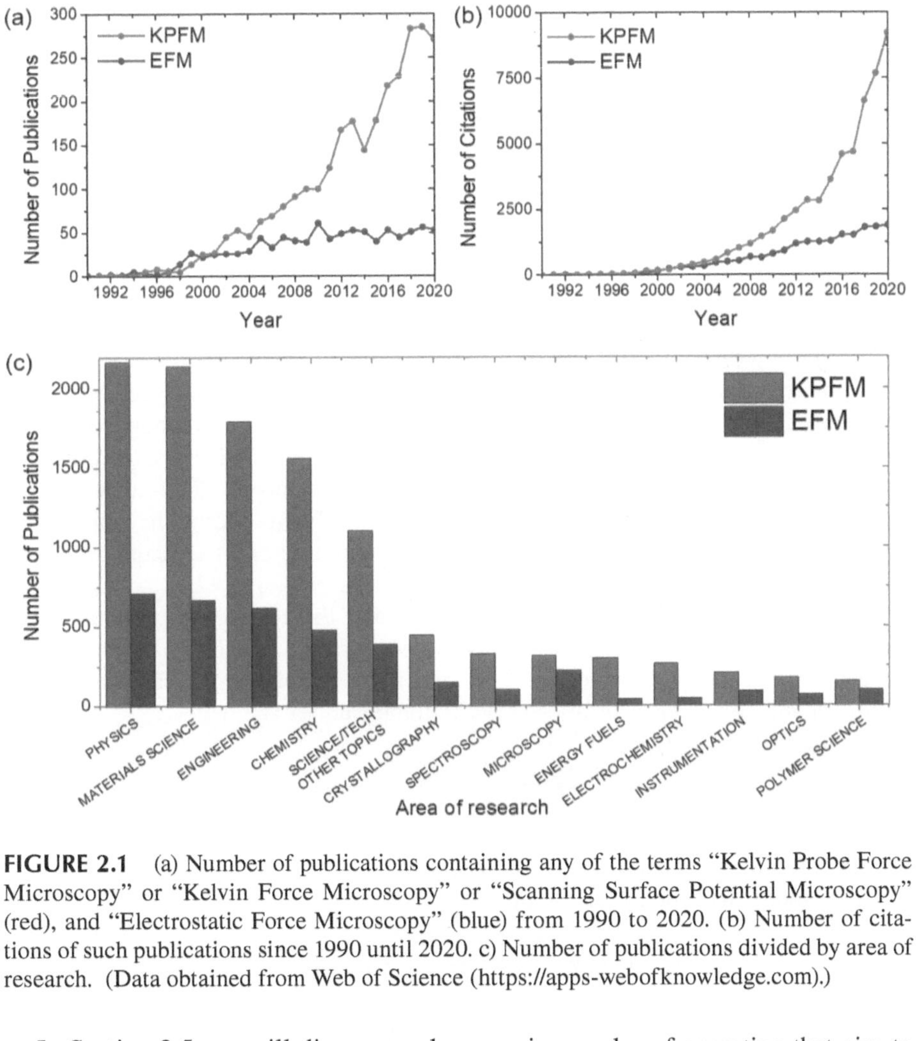


**FIGURE 2.1** (a) Number of publications containing any of the terms "Kelvin Probe Force Microscopy" or "Kelvin Force Microscopy" or "Scanning Surface Potential Microscopy" (red), and "Electrostatic Force Microscopy" (blue) from 1990 to 2020. (b) Number of citations of such publications since 1990 until 2020. c) Number of publications divided by area of research. (Data obtained from Web of Science (https://apps-webofknowledge.com).)

In Section 2.5, we will discuss newly emerging modes of operation that aim to improve the veracity, resolution, and sensitivity of the EFM/KPFM techniques. Pertinent to energy research, Section 2.5.5 will provide an overview of recent developments in time-resolved EFM/KPFM methods, opening the possibility to probe fast processes (picosecond-millisecond) in turn providing a nanoscale view of phenomena such as photon-induced charge accumulation and charge transfer, charge injection, ion dynamics, etc.[16,17]

Finally, the current advances in in situ/operando AFM characterization of energy devices such as batteries,[18–20] in addition to well-established current-based SPM modes such as Scanning Electrochemical Microscopy (SECM),[21,22] could be complemented with the further expansion of KPFM/EFM to operate in liquid environments. Consequently, looking forward, in Section 2.6 we will briefly outline efforts toward the application of KPFM and EFM at the solid-liquid interface, which would be a welcome development for understanding electrochemical processes underpinning important energy technologies from batteries to supercapacitors, electrolyzers, and fuels cells.

## 2.2 ELECTROSTATIC FORCE MICROSCOPY

### 2.2.1 Principles of EFM

In the late 1980s, Martin et al.,[3] along with Stern[23] and Terris et al.,[24] independently adapted the AFM with conductive probes for the detection of electrostatic forces. Their collective approaches would become known as EFM, although operated by fundamentally different detection methodologies (e.g., electrostatic force[3] vs. force gradient detection[23]). Unlike current sensitive potentiometric measurements (e.g., Scanning Tunneling Microscopy (STM)),[25] EFM presented the first opportunity for nanoscale electrical characterization of all materials (e.g., conductors, semiconductors, insulators, etc.). This is made possible as EFM detects the long-range electrostatic forces (as opposed to currents) acting on an AFM probe, while offering far superior spatial resolution compared to the traditional microscopic Scanning Kelvin Probe (SKP)[26,27] approach described in Section 2.3.1.

In the EFM method described by Martin et al.,[3] the conductive probe is driven electrically with an AC voltage by $V_{tip} = V_{DC} + V_{AC}\cos(\omega t)$, where $V_{AC}$ is the driving voltage and $V_{DC}$ is a potential offset. The tip-sample system forms a capacitor whose plates accumulate an electrostatic energy of the form:

$$U_{es}(z) = -\frac{1}{2} C_z \left(V_{tip} - V_{CPD}\right)^2 \tag{2.1}$$

The tip can then be used to measure locally the resultant electrostatic force, which has a parabolic dependence on the tip voltage:

$$F_{es}(z) = -\frac{1}{2} C_z' \left(V_{tip} - V_{CPD}\right)^2 \tag{2.2}$$

Here, $V_{CPD}$ is the Contact Potential Difference (CPD) difference between probe and sample (see Section 2.3.1) and $C_z'$ is the capacitance gradient of the probe-sample system, which is dependent on tip geometry, surface topography, and tip-surface separation, $z$. In certain configurations of EFM or KPFM, the probe can be simultaneously driven both electrically and mechanically. If this is the case, the electrostatic force, $F_{es}$, between the biased conductive tip and the surface results in a change of the cantilever resonance frequency (mechanically excited) that is proportional to the force gradient:

$$\Delta\omega = \frac{\omega_0}{2k} \frac{\partial F_{es}}{\partial z} \tag{2.3}$$

where $k$ is the spring constant and $\omega_0$ is the resonance frequency of the cantilever. Thus, in Frequency-Modulated EFM (FM-EFM), the resonant excitation is usually maintained by adjusting the driving frequency, $\omega_p$, and the frequency shift, $\Delta\omega$ (or phase), is collected as the EFM image. In a similar fashion to the electrostatic force, the force gradient acting between the probe apex (tip) and sample can be described by:

$$F'_{es}(z) = \frac{\partial F_{es}}{\partial z} = -\frac{1}{2}C''_z\left(V_{\text{tip}} - V_{\text{CPD}}\right)^2 \tag{2.4}$$

Where $C''_z$ is the derivative of the capacitance gradient.

Practically, the modulated tip voltages result in spectral components, ($F_{es} = F_{\text{dc}} + F_\omega + F_{2\omega}$), including a DC force component (static bending), a force at the modulation frequency ($\omega$, the first harmonic), and a force at double the frequency of the tip voltage ($2\omega$, the second harmonic), as described by equations (2.5)–(2.7):

$$F_{dc} = -C'_z\left(\frac{1}{2}(V_{\text{DC}} - V_{\text{CPD}})^2 + \frac{1}{4}V_{\text{AC}}^2\right) \tag{2.5}$$

$$F_\omega = -C'_z\left(V_{\text{DC}} - V_{\text{CPD}}\right)V_{\text{AC}}\sin(\omega t) \tag{2.6}$$

$$F_{2\omega} = C'_z\frac{1}{4}V_{\text{AC}}^2\cos(2\omega t) \tag{2.7}$$

Thus, the DC components can be determined by directly monitoring the raw static deflection of the cantilever and will contribute to the topography if not compensated for,[28,29] whereas the dynamic cantilever response at the harmonic frequencies $F_\omega$ and $F_{2\omega}$ are used to measure CPD and capacitance, respectively. The amplitude (and phase) of these dynamic channels is normally captured using traditional lock-in amplifier (LIA) detection, in the so-called Amplitude-Modulated EFM (AM-EFM) or by KPFM as will be shown in Section 2.3.2. As in most AFM approaches, the dynamic response of the cantilever is more easily detected than static forces, which are often concealed by large *1/f* noise components. Generally, the frequency bandwidth used in EFM ranges from ~kHz to ~MHz (typically limited by the tip mechanical response) and can be used on resonance (bringing a higher Signal-to-Noise Ratio (SNR)), or off-resonance, which allows easier quantification, but limits the sensitivity.

Alternatively, the latest cutting-edge data-driven technical implementations of voltage-modulated SPM measurements, consisting of the acquisition of the full cantilever response and further post-processing of the rich hyperspectral datasets to reconstruct the measured signal (such as General Acquisition Mode (G-Mode) AFM[30,31]), can also be applied to the measurement of the electrostatic force, as a modern and profitable alternative to Phase-Locked Loop (PLL) or LIA schemes.

A benefit of using force-based over current-based sensing techniques (like scanning capacitance microscopy[32]) is that it takes advantage of the direct physical link of the electrostatic force (and its gradient) with the gradient of capacitance (and the second derivative of the capacitance) of the system which overcomes the challenge of measuring the capacitance itself through impedimetric measures, which is

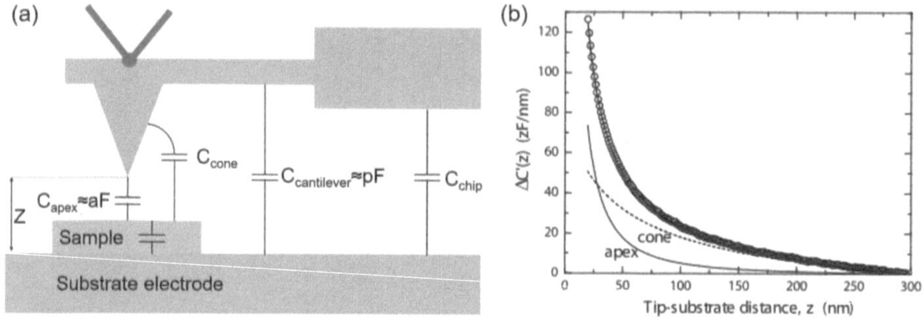


**FIGURE 2.2** (a) Schematics of tip-sample capacitive coupling in an EFM experiment. (b) Total (circles), apex (line), and cone (dashed line) contribution of the measured signal during an EFM force-distance curve. (Figure b is adapted from Fumagalli and Gomila.[34])

intrinsically complex due to the large parasitic non-local contributions along with the shortcomings of current amplifiers. This fact, coupled with the high sensitivity of AFM to measure forces (in the picoNewton range, corresponding to zeptoFarads/nm in capacitance gradient units), is essential to achieve locality (see Figure 2.2), as the local capacitance measured by the tip apex (on the order of aF) is hidden under the larger total capacitance of the cone, cantilever, and chip (on the order of pF). That is to say that the gradient of the capacitance with tip-sample distance is much more sensitive to apex variations (closer to the sample) than to cone/cantilever variations (further from the sample). Therefore, when the electrostatic force is measured (i.e., AM-EFM), the spatial resolution can reach the 10 nm range, whereas when the gradient of the force is measured (i.e., FM-EFM) it can increase to sub-10 nm resolution (mostly limited by the tip radius and cone angle). Thus, by monitoring the electrostatic contributions as the probe is raster scanned across a sample surface, it is possible to generate qualitative maps of surface charge density[33] and/or dielectric properties[34] depending on the harmonic response which is captured.

These attributes resulted in EFM being applied to mapping electric field distributions in devices,[35] imaging of Self-Assembled Monolayers (SAMs) on surfaces,[36,37] nanoparticles,[82] potential and polarization mapping on DNA,[38,39] and proteins,[40,41] dielectric constant mapping on cells,[42–44] surface potential variations in oxide bicrystals,[45,46] static and dynamic properties of ferroelectric materials,[47,48] piezoelectricity of 2D materials,[49] subsurface imaging of nanocomposites,[50–53] optoelectronic properties such as light-induced charge separation,[54,55] as well as observation of charge storage and leakage in various materials.[56–58] Although the nature of the interaction is always the same, several possible technical implementations of EFM exist today.[33,59]

### 2.2.2 EFM Scanning Modes

Most commonly, EFM is operated in lift mode, where in a first pass the tip is electrically grounded and used to track the sample's topography (either by contact, non-contact, or intermittent contact mode), and during the second pass the tip is lifted a certain height and the previously determined topography is retraced while the tip-sample electrostatic force is measured out of contact. Other probe-sample trajectories (i.e., constant height mode[60]) can be followed during the second pass if (instead of retracing the measured topography) the tip is scanned through a parallel path with respect to the sample substrate. This approach is of special interest for dielectric characterization, as it eliminates the topographic crosstalk in the electrostatic signal (induced for the z changes during the second scan), which can be sometimes difficult to eliminate by postprocessing.[61]

Single-pass mode,[33] where both topography and electrostatic force are acquired simultaneously by exciting two different frequencies (one mechanically and the other electrically), offers faster imaging rates and higher spatial resolution but is usually not chosen (especially in quantitative studies or for fragile samples) as other drawbacks like mechanical excitation of higher eigenmodes, topographically induced changes in the resonant frequency, or electrical damage of delicate specimens can complicate the quantification.

Moreover, force volumetric (3D) approaches have also been implemented[42,62] where a force-distance curve is acquired for every location across a grid of points, comprising the region of interest. Traditionally, force volumetric approaches were not preferred as they implied longer scanning times and only low-pixel images were normally measured. However, currently most modern AFM systems can obtain fast force-volumetric datasets of higher resolution grids because of better piezoelectric actuators and faster feedback and electronic control systems (see Section 2.5.3 for more details on 3D methods).

EFM is usually operated in Open Loop (OL) configuration (contrary to standard KPFM), which means that the signal is measured without a bias feedback loop control and also allows the simultaneous acquisition of more than one of the harmonic responses.[64] However, less common EFM variants like Scanning Polarization Force Microscopy (SPFM)[65] are operated in a closed-loop configuration by changing the tip-sample distance and using the magnitude of the second harmonic ($2\omega$) for the feedback control. SPFM opened up the possibility to image weakly adsorbed samples or liquid droplets.[66] Nevertheless, as the information on sample topography and polarizability is mixed, their images are usually challenging to analyze quantitatively.

### 2.2.3 Quantitative EFM

The convolution of both $V_{CPD}$ and capacitance terms in equations (2.4–2.7) complicates the quantification interpretation of an EFM measurement directly. Hence, in most applications EFM is considered a qualitative approach. That said, much progress has been made in extracting quantitative information on surface potential,[47,67,68] dielectric constant,[60,69–72] trapped charges,[73,74] and polarizability[75,76] from EFM measurements. Most applications of quantitative EFM have focused on mapping the variation in surface charge or dielectric constant, often by modeling the electric field at the biased tip and/or calibrating the precise tip shape. Another focus has been on mapping the variation in surface potential, which can be deduced from the functional force-bias dependence. However, it is important to consider, particularly for insulating materials, that the surface potential does not uniquely define the materials' electronic properties. Equally, the tip-sample forces in EFM are strongly coupled to variations in dielectric properties, as well as surface-bound and volume-trapped charges.[75,77] Quantifying such properties, however, requires a precise understanding of the EFM force or force gradient contrast. Such analysis is complicated due to an unknown capacitance term determined by the complex tip-sample geometry, as well as the dielectric properties of the tip-sample gap in which electric field propagates. Indeed, quantitative analysis of EFM data calls for detailed knowledge of the tip geometry as well as the exact nature of the tip-sample interaction.[78]

The use of EFM for quantification of the dielectric properties, also known as Scanning Dielectric Microscopy (SDM),[34] is based on inferring the dielectric constant of the material under study from the measurable electrostatic force acting on the probe, when the material of interest is put between the probe and a back electrode substrate. However, as previously exposed, the complex geometrical characteristics of the probe-sample systems makes it challenging to model the electric field accurately, resulting in few analytical models available[32,79] that are only applicable for thin and flat sample morphologies. Thus, normally one must resort to finite element numerical simulations that can calculate the electric field distribution, using a realistic geometry of the system.[80] In addition, the convolution of the $V_{CPD}$ with the capacitance gradient in equations (2.4–2.7) must be avoided. Hence, for dielectric quantification purposes, normally the $2\omega$-oscillation term is measured, as the dependence on the DC terms disappears (see equation 2.7) and the capacitance gradient can be directly linked to the modeled electrostatic force.

The modeling normally uses some tip-calibration procedure, where the specific shape and dimensions of the tip are determined (either by scanning electron microscopy imaging[81] or using a force-distance curve on top of a region of bare metal of the sample[34]). Afterward, the measured variations of the capacitive gradient during a force-distance curve (vertical scan) or an image (horizontal scan) can be fitted to the modeled tip electrostatic force (either using an analytical approximation or including the contribution of realistic probe-sample geometry by means of simulations) having the local sample dielectric constant as the only free fitting parameter. Usually only the real part of the dielectric permittivity is considered in the modeling, neglecting the effect of the dielectric losses (imaginary part of the permittivity), which for most samples have a negligible effect. Following this procedure, dielectric constant quantification has been mapped over a myriad of systems like microelectronic nanopatterned oxides,[80] nanoparticles,[39] 2D materials,[82] soft polymers,[83] nanoconfined water,[84] DNA,[38] proteins,[40,85] lipids,[86] and nanocomposites,[87] among others. Moreover, SDM has been successfully explored as a label-free alternative for subcellular biochemical composition mapping of both prokaryotic[42] and eukaryotic[44] cells, using the dielectric response of each biomolecule as a fingerprint for biochemical identification.

In addition, the penetration depth of the electric field below the sample surface during EFM operation allows the technique to be used for subsurface characterization. The position of buried nanostructures below the surface, which can appear to be flat or bumpy in the topography channel, can then be mapped in a non-destructive way via EFM.[50–53,88] If the dielectric properties of the different components forming the nanocomposite are known, the dielectric characterization procedure can be used in a similar manner, but using the position[53] or morphology[89] of the buried nanostructure as the fitting parameter. Therefore, we envision the application of EFM for non-destructive, tomographic characterization at the nanoscale[90] as a potential tool for studying other phenomena such as percolation or non-linear conduction mechanisms in energy-related materials. Furthermore, the use of ultrasmall cantilevers (with much faster mechanical response) has been proven to enlarge the frequency regime of the technique, allowing for dielectric spectroscopy up to the MHz range.[91]

The main bottleneck of SDM has usually been the non-automated, time-consuming process needed to accomplish the quantification via finite element numerical simulations. As a time-saving alternative, supervised machine learning approaches employing neural networks have been successfully applied to SDM[92] shortcutting the electric field modeling side and recently pushing forward toward in situ nanoscale dielectric imaging.[93]

Apart from dielectric characterization, EFM can also be used to map nanoscale surface charge distributions. In this case, also some relationship between the monitored electrostatic force and the surface charge must be modeled by taking into account the influence of the probe-sample geometry, either analytically,[94] phenomenologically (using some calibration specimen whose surface charge is well known[95]), or by means of simulations. However, contrary to the case of SDM (where the $2\omega$-oscillation of the tip is monitored that contains the dielectric information) for surface charge characterization, both the $\omega$ term and the $2\omega$ can be used depending on whether the surface charge changes are detected through changes in the capacitance or changes in the surface potential.

Following this approach, EFM has been used since the very beginning to study ferroelectric surfaces[96] and to map the charge density in small structures such as charged ferroelectric domain walls.[97,98] In addition, charge injection[55,99] and/or charge relaxation/diffusion[100] into the sample can be studied by using the tip to inject the charges into the sample by applying a voltage when the tip is in contact with the sample, and further by using the same tip to read electrostatic signals via EFM. Moreover, the doping profiles in semiconductors can be studied specifically,[101] and such studies have triggered the development of a specific modern modification of EFM named broadband electrostatic force microscopy (bb-EFM),[102] which enlarges the frequency range of the technique up to 10 GHz. This is achieved by modulating the carrier frequency signal with a low-frequency signal (in the kHz range), which is used for the detection. However, such modifications are not novel as they follow the same approach used previously for the application of EFM in liquid,[71] where higher frequencies are needed (more details in Section 2.6), and which has recently been used to decouple electrostatic artifacts from real electromechanical phenomena in Piezoresponse Force Microscopy (PFM) or electrochemical strain microscopy (ESM) experiments.[103]

## 2.3 KELVIN PROBE FORCE MICROSCOPY

### 2.3.1 Contact Potential Difference and the Kelvin Method

KPFM has become an invaluable technique for probing CPDs on the nanoscale. The name KPFM originates from the macroscopic current detection approach developed in 1898 by Lord Kelvin, commonly referred to as the Kelvin method or Kelvin Probe.[104] This tool can be used to measure an "electrochemical" or contact potential between a metal probe and a metal sample of different work functions ($\Phi_1$ and $\Phi_2$). The work function is defined as the amount of energy required to eject a bound electron in the Fermi level ($\varepsilon$) to the vacuum energy level ($\varepsilon_{vac}$). When an electrical connection is made between the dissimilar metals, as shown in Figure 2.3a, electrons flow from the metal with the lowest work function to the material with the highest work function. If the electrodes are in a parallel-plate capacitor configuration, equal and opposite surface charges will form on the capacitor plates (e.g., Figure 2.3a). The contact potential difference developed across the plates (i.e., CPD) is given by $V_{CPD} = \Phi_{tip} - \Phi_{sample} / -e$. The CPD can be measured by applying an external "backing" potential ($V_{DC}$) to the probe (or sample), until the point at which the current (or surface charges) between the probe and sample is nullified (see Figure 2.3a). In this equilibrated state, the applied backing potential is equal to the CPD or the work function difference ($\Delta\Phi$) between the metals, $V_{DC} = \pm V_{CPD}$. Note that the $\pm$ sign depends on whether $V_{DC}$ is applied to the sample (+) or the probe (−).[9]

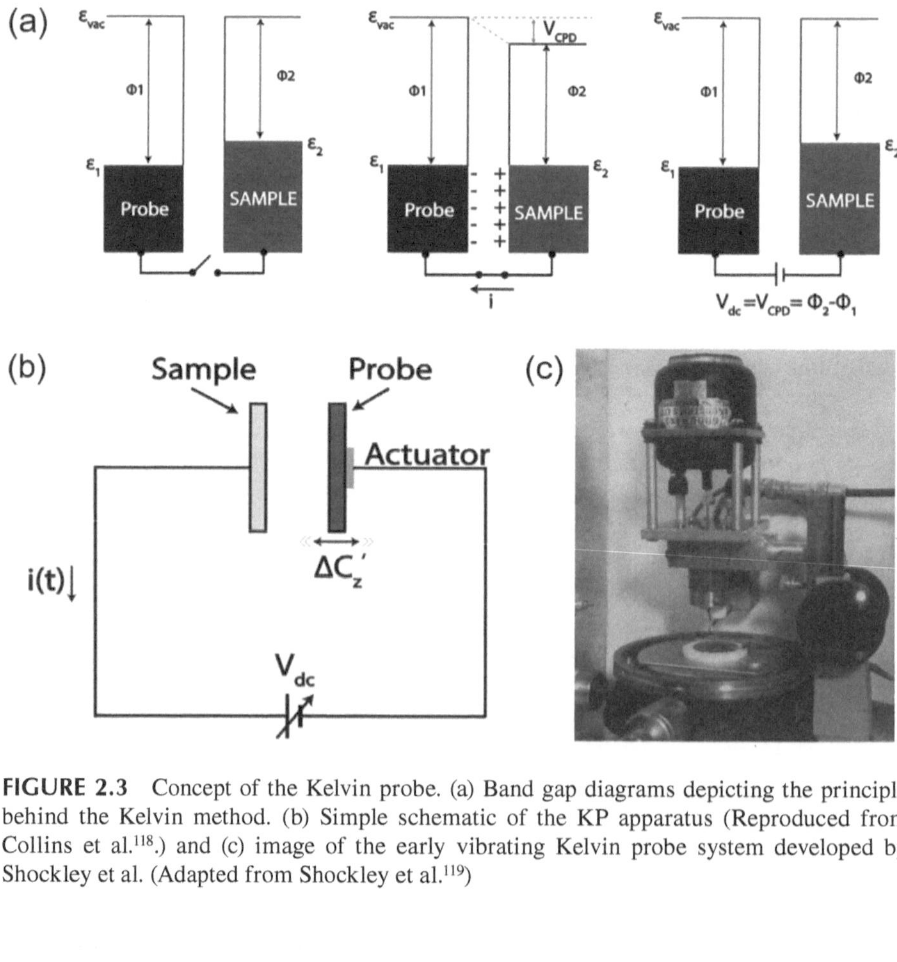


**FIGURE 2.3** Concept of the Kelvin probe. (a) Band gap diagrams depicting the principle behind the Kelvin method. (b) Simple schematic of the KP apparatus (Reproduced from Collins et al.[118].) and (c) image of the early vibrating Kelvin probe system developed by Shockley et al. (Adapted from Shockley et al.[119])

In 1932 Zisman et al. introduced an adapted method, which included modulation of the separation between the capacitor plates (see Figure 2.3b), producing a time varying capacitance and as a result current flows back and forth within the circuit.[105] The periodic vibration of the capacitor plates leads to an observable current that can be used as an input for bias feedback compensation. This current is reduced to zero by applying a DC-voltage to one of the plates. This voltage corresponds to the CPD of the two materials. An early example of a vibrating Kelvin probe apparatus is shown in Figure 2.3c. Since this time, decades of research have led to the continued improvement of the method.[106–111] Incorporation of XYZ piezo drivers as well as improvements in miniaturized probe design and fabrication have allowed the realization of the SKP,[27,110] a more versatile tool capable of spatially mapping micron scale variations in work function differences. The SKP is extremely sensitive to surface chemistry and condition and has been widely applied as a surface analysis technique for the investigation of corrosion,[112] metallurgy,[113] and semiconductors[114,115] having been used to explore phenomena involving adsorption kinetics[116] and surface photovoltage (SPV) spectroscopy,[117] among a myriad of applications.

### 2.3.2 Kelvin Probe Force Microscopy

A major drawback to current detection used in the macroscopic Kelvin probes is an ultimate limit on the size with which the probes can be decreased to before sensitivity deteriorates. So, while the SKP is a quantitative approach, it lacks sufficient spatial resolution as compared to EFM. At the same time, while EFM allows spatial resolution on much smaller scales, it is largely a qualitative method (with exceptions outlined in Section 2.2.3).

Not long after the realization of EFM, a few researchers had the idea to combine the best elements of EFM with the macroscopic Kelvin method.[2,4] In Figure 2.4 the AC voltage and DC bias dependencies on the electrostatic force interactions described in equations (2.5–2.7) are demonstrated experimentally. In these experiments, a conductive AFM probe was positioned 50 nm above a freshly cleaved highly oriented pyrolytic graphite (HOPG) surface as AC and DC voltages were applied between probe and sample. The static bending of the cantilever (equation 2.5) has a parabolic dependence on the DC bias and a square $V_{AC}$ voltage dependence (Figure 2.4a and b). The second term at the fundamental drive frequency has linear bias dependence (equation 2.6) and demonstrates a linear coupling with $V_{AC}$ (Figure 2.4c and d). From Figure 2.4e and f, the third term at $2\omega$ is shown to depend only on the capacitive coupling between tip and sample and is not influenced by the DC bias, having a square dependence on the $V_{AC}$ voltage.

Weaver et al.[2] were the first to explore the possibility of measuring potentials using EFM as an avenue to extend the scanning Kelvin method to the nanoscale. Noteworthy, the first harmonic response in Figure 2.4c shows a linear dependence on DC bias, dropping to zero when the tip bias $V_{dc} = -V_{CPD}$ (or $+V_{CPD}$ if the DC bias is applied to sample). In 1991, Nonnemacher et al.[4] coined the name KPFM to describe a similar approach used for determination of CPD. Contrasting KPFM to macroscopic Kelvin probe, KPFM measures electrostatic forces instead of currents as the controlling parameter of the bias feedback regulation. Compared to EFM, in KPFM

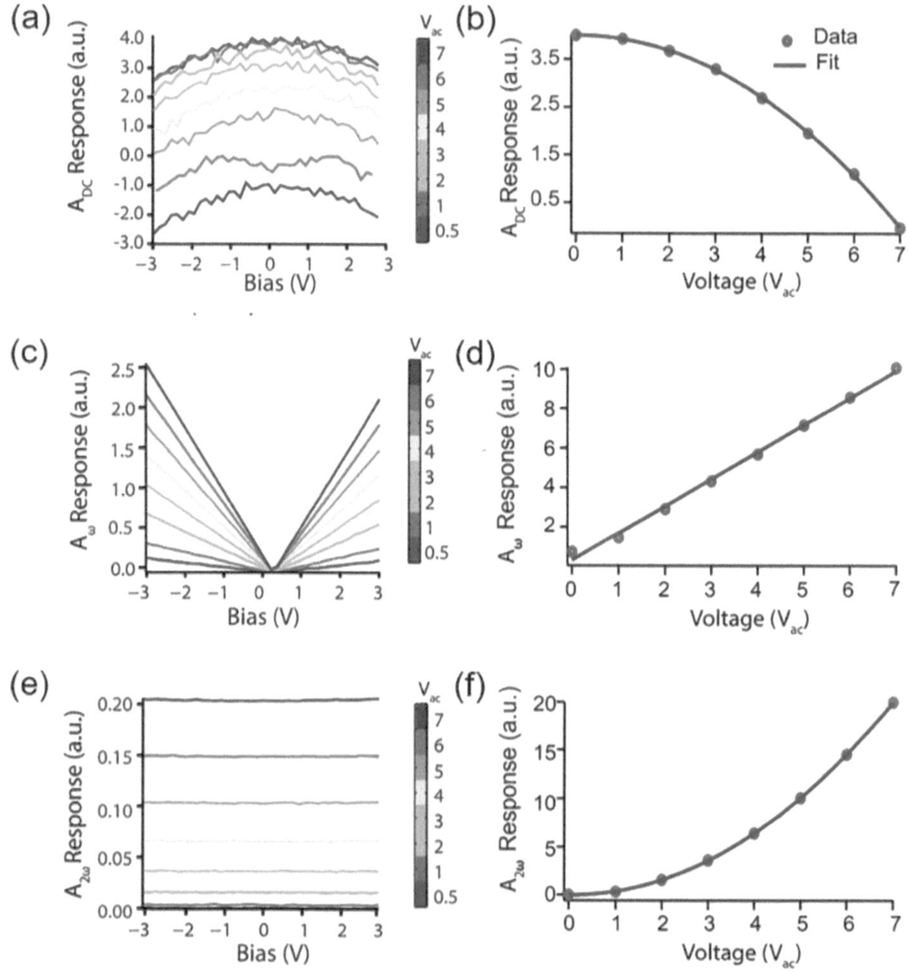


**FIGURE 2.4** (a,c,e) DC bias and (b,d,f) AC voltage dependence of the (a,b) static deflection, (c,d) first and (e,f) second harmonic amplitude response. Measured using a conductive AFM 50 nm above a grounded freshly cleaved highly oriented pyrolytic graphite (HOPG) surface. (Adapted from Collins et al.[64])

an additional DC voltage ($V_{DC}$) between tip and sample is applied and controlled by a feedback loop to minimize/nullify the electrostatic force at the frequency of the applied AC voltage (see equation 2.6). We note that today KPFM is often referred to as scanning surface potential microscopy (SSPM), SKP microscopy (SKPM), or Kelvin force microscopy (KFM).

### 2.3.3 Amplitude and Frequency Modulation

The first demonstration of KPFM was realized using AM detection,[4] in a similar fashion to that described schematically in Figure 2.5a. AM-KPFM is normally

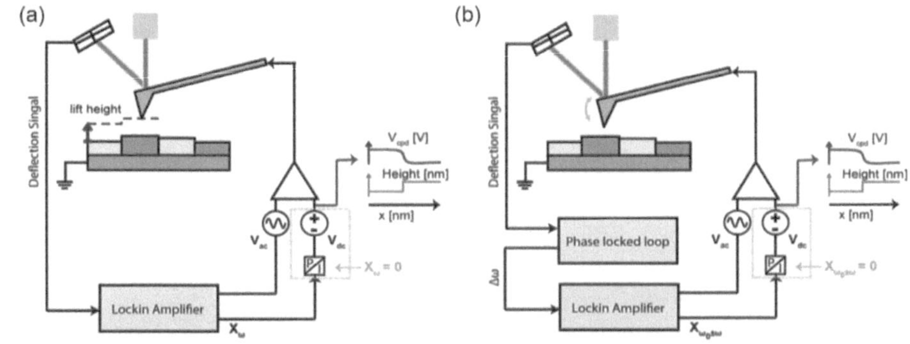


**FIGURE 2.5** Simple schematic describing an (a) AM- and (b) FM-KPFM experimental setup. (Reprinted from Collins et al.[118])

operated in lift mode (see Table 2.1) such that the topography is captured on the first line trace with the cantilever mechanically vibrating, and on the lift scan the probe is electrostatic excited by the tip bias ($V_{tip}$) without mechanical excitation. For detection purposes, the driving frequency of the AC voltage is tuned to be close the resonance frequency of the cantilever for maximum amplification, ensuring a high SNR. The relevant CPD information is at the first harmonic response of the modulation frequency (equation 2.6). An is typically used to detect the amplitude and phase of this force component, attenuating all other information regarding the tip-sample interactions outside of this frequency. As described previously, the amplitude at the fundamental drive frequency is minimized when $V_{DC}$ matches the magnitude of $V_{CPD}$. Note that if the sample is biased externally, the total KPFM nulling potential will be the sum of the $V_{CPD}$ and the external bias. While most samples can be grounded to exclude this additional contribution, externally biased devices or electrochemical devices with built-in voltages cannot be nullified (e.g., batteries and fuel cells) and will include contributions from the external bias with the intrinsic CPD.[120]

A bias feedback loop is used to adjust the constant component of the tip bias $V_{DC}$ until the amplitude from the LIA is nullified. Importantly, the in-phase quadrature component (X) rather than the amplitude is used as input to the feedback loop to account for the polarity of the CPD. While AM-KPFM is normally operated in lift mode, it is possible to operate in single-pass configuration if the mechanical and electrical excitation are separated in frequency space. Single-pass AM-KPFM operation offers the advantage of faster imaging, a more localized measurement, and electrostatic compensation for topography measurement[28,29]; however, careful consideration of the drive frequency is required to minimize possible CPD/topography crosstalk.[121]

As an alternative to AM-KPFM, in FM-KPFM (see Figure 2.5b), the tip is mechanically actuated at its fundamental resonance frequency and the signal from the photodetector is fed into a frequency detector, typically a PLL or a frequency demodulator. Simultaneously, a low-frequency electrical excitation is applied to the tip. The applied bias induces a modulation of the electrostatic force, producing a frequency shift ($\Delta\omega$) described by equations 2.3 and 2.4. The output of the frequency detector is used as the input to an LIA, and the magnitude of oscillation at

**TABLE 2.1**
**Table Summarizing the EFM Scanning Modes**

| **Scanning Mode** | **Single Pass** | **Lit Mode** | **Constant Plane** | **Constant $F_{el}$/ SPFM** | **EFVM/3D-EFM** |
|---|---|---|---|---|---|
| Schematics |  |  |  |  | 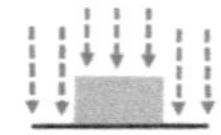 |
| 1st pass | Topo at $\omega_{mech}$ $F_{el}$ at $\omega_{el}/2\omega_{el}$ | Topo at $\omega_{mech}$ / Contact mode | Topo at $\omega_{mech}$ / Contact mode | Topo at $\omega_{mech}$ / Contact mode | Topo at setpoint $F_{el}$ at $\omega_{el}/2\omega_{el}$ |
| 2nd pass | OFF | $F_{el}$ at $\omega_{el}/2\omega_{el}$ | $F_{el}$ at $\omega_{el}/2\omega_{el}$ | Topo that fulfills $F_{el}$=Const. at $2\omega_{el}$ | OFF |
| EFM feedback | OFF | OFF | OFF | ON at $2\omega_{el}$ | OFF |
| Trajectory 2nd pass | OFF | Constant tip-sample distance | Constant tip-Substrate distance | Constant $F_{el}$ at $2\omega$ | OFF |
| Example Image |   | 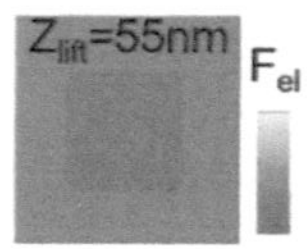  | 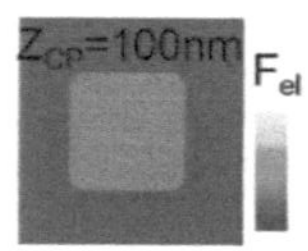  |   | 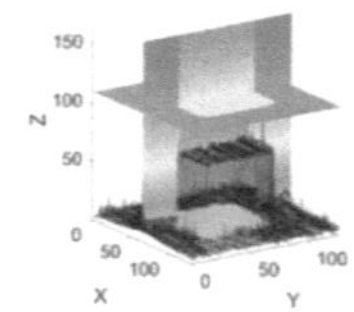  |
| Example Profile/ Curve | 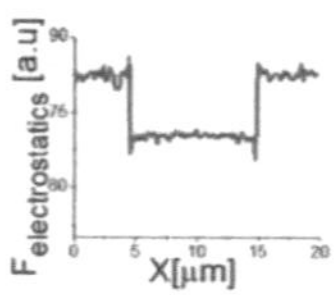  | 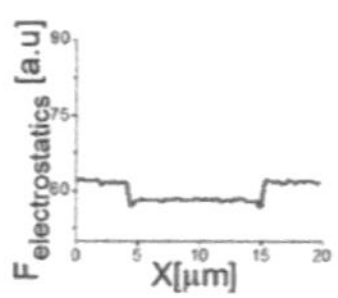  | 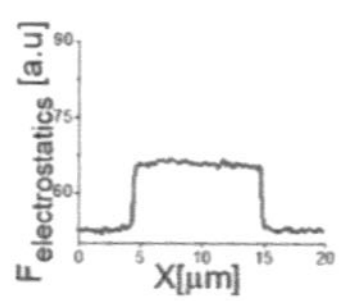  | 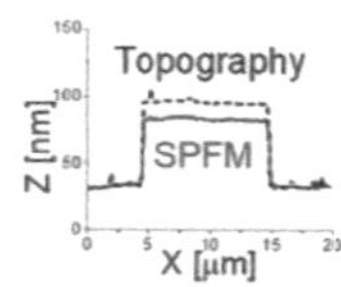  | 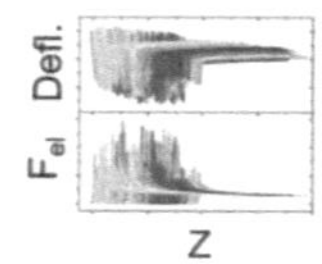  |
| Speed | Fast | Medium | Medium | Slow | Speed |

*Source:* Table is adapted from Checa.[63]

the frequency of the applied AC bias is detected. In this way, FM-KPFM is sensitive to electrostatic force gradient, as opposed to the electrostatic force. The in-phase (X) component serves as the input to the Kelvin bias feedback loop, which adjusts the DC bias to minimize the in-phase component to zero. The frequency of the AC excitation must be carefully chosen to be within the bandwidth of the PLL, while high enough to avoid interference with the topography feedback loop.

When operating KPFM, it is important to remember that the whole probe, tip apex, cone, and cantilever can contribute to the overall electrostatic interaction. In this way, the probe shape plays a significant role in KPFM imaging, and its influence will be dependent on experimental parameters and modes of operation. Several authors have investigated the influence of the long-range nature of the electrostatic interaction with the whole probe, as well as comparing the sensitivity and resolution between modes of operation. Colchero et al.[122] explored the electrostatic forces present between the tip/cantilever and the sample. They found that for most probe geometries, significant "background" or "stray capacitance" contributions from the cantilever/apex are present, even within a few nanometers from the sample surface. This stray capacitance can have the effect limiting the achievable lateral resolution and producing a CPD which is of weighted average of the area under the probe. Both experiment and simulations have been used to compare AM and FM-KPFM and have led to the consensus that in order to avoid a reduction in the spatial resolution due to stray capacitance, it is better to use FM-KPFM which is sensitive to the force gradient limiting the interaction to the tip apex, while avoiding contributions from the cantilever.[123,124] Zerweck et al.[123] demonstrated this by comparing AM and FM-KPFM measurements on a KCl island on a gold substrate. The experiment was further simulated using 3D finite element modeling showing good agreement (Figure 2.6a). FM-KPFM shows considerably sharper contrast and energy resolution, reaching the expected CPD difference (900 mV) between the materials within a short distance (~50 nm). On the other hand, AM-KPFM never recovers the full potential difference within the 400 nm transition. By contrast, AM-KPFM has the advantage of high potential sensitivity; therefore, the use of smaller AC voltages is possible compared to FM-KPFM.[124] This sensitivity vs. resolution trade-off spurred the development on heterodyne AM-KPFM to remove the effect of the stray capacitance on surface potential measurements while retaining the high sensitivity.[125]

Axt et al.[126] performed a thorough investigation of various excitation/detection schemes, shown in Figure 2.6b, specifically focused on the characterization of nanoscale electrical devices. Comparing AM-KPFM in lift mode with FM-KPFM performed in single-pass scanning, they concluded that lift mode AFM was wholly unsuitable for characterization of mesoscopic perovskite solar cell due to the spatial averaging effect from the cantilever contribution (see Figure 2.6c). They then went on to compare several different combinations of KPFM modes, including heterodyne KPFM,[125,127] on a reference structure consisting of an interdigitated electrode array mimicking a cross-sectional device, but having well-defined surface potentials. They concluded that FM methods provide more quantitative results and are less affected by the presence of stray electric fields compared to AM-KPFM methods. They noted that FM heterodyne KPFM[127] outperformed all others (slightly outperforming

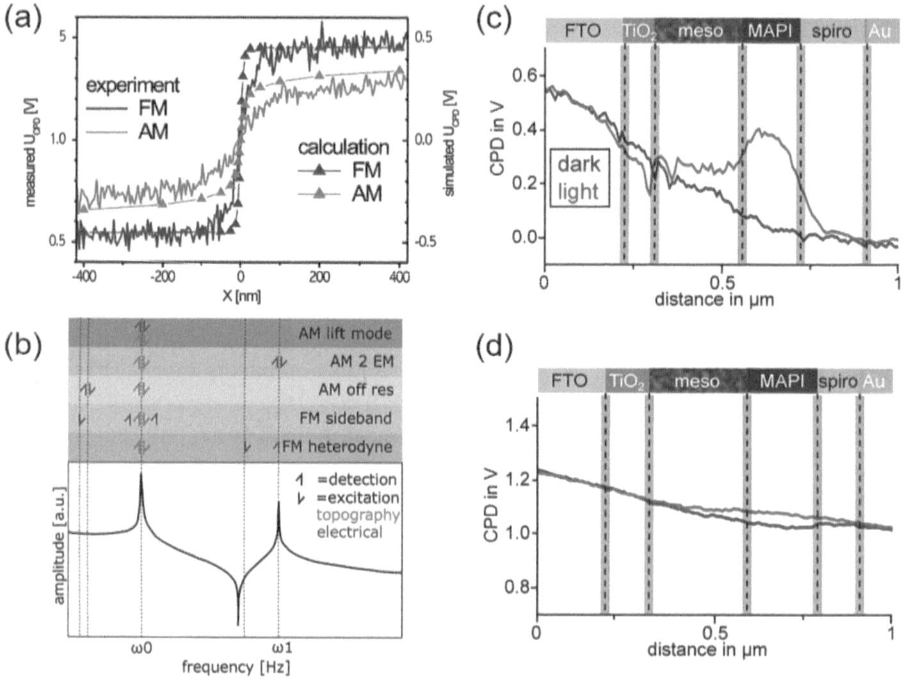


**FIGURE 2.6** (a) Experimental line profiles recorded in AM- (*gray*) and FM-mode (*black*) over a KCl island (left) on Au substrate (right). The simulation of the two modes is overlaid assuming a potential difference of 0.9V between the two regions. (Reproduced from Zerweck et al.[123]) (b) Comparison of excitation and detection frequencies for KPFM methods compared by Axt et al.[126] showing (lower) the transfer function of the cantilever as well as the (top) excitation (arrow upward) and detection (arrow downward) for the corresponding methods with the respective frequencies. Red is used for the topography signal and blue for the electrical excitation and detection. Modes investigated include AM-KPFM; AM lift mode; AM off resonance; AM second eigenmode; FM-KPFM; frequency modulation heterodyne. (c,d) CPD line profiles of two KPFM experiments on the same cross-section of a mesoscopic perovskite solar cell under short circuit conditions with and without illumination recorded by (c) frequency modulation -KPFM (FM sideband) and (d) amplitude modulation KPFM (AM lift mode) scans in lift mode with a tip-sample distance of 10 nm. (Adapted from Axt et al.[126].)

FM-KPFM with sideband detection), with the added benefit of having less bandwidth constraints compared to FM-KPFM using sideband detection.

### 2.3.4 Tip Calibration and Environmental Considerations

KPFM detects the CPD between the probe and sample; probe independent measurements of the sample require knowledge of the probe work function ($\Phi_{tip}$), and hence, additional calibration steps are required. This is usually achieved by calibrating the probe against a metal sample with a known work function. HOPG is a good option for such a calibration due to its chemical inertness in addition to the fact that it can be freshly cleaved revealing a pristine, contaminant-free surface. The work function of HOPG is relatively stable and is defined in ambient as $\Phi_{HOPG} = 4.475 \pm 0.005$ eV.[128] The $\Phi_{tip}$ can be calibrated by measuring the CPD difference between the probe and the freshly cleaved HOPG, where $\Phi_{tip} = eCPD_{HOPG} + \Phi_{HOPG}$. An important consideration in the calibration of work functions is the condition of the conductive probe itself. Often, the probe can pick up contamination, or the metal coating can degrade during scanning, introducing sudden offset jumps in the potential maps (can be 100 s of mV between adjacent scan lines). Such tip changes can complicate comparison between measurements, etc.

Importantly, since KPFM is a surface-sensitive technique, great care should be taken to the surface condition when interpretating the data. Indeed, the measured potential is likely a convolution of the bulk potential and the surface potential, modified by any surface reconstruction, surface defects, and surface charges. Under ambient conditions, additional contributions from the influence of inherent water layers, hydroxyl groups, contaminants, and adsorbed layers on the same work function must be considered. For quantitative studies of absolute work functions or surface potentials, operation in Ultra-High Vacuum (UHV) conditions is preferred.

### 2.3.5 Feedback Artifacts

Although quantitative measurements should be the goal of any KPFM measurement, in practice KPFM measurements of CPD have been known to be strongly influenced by the AC driving voltage,[129] feedback gains,[130] phase offset,[130] tip-sample distance,[131] and topography.[132,133] These numerous artifacts challenge the quantitative nature of KPFM, prompting significant efforts toward eliminating these issues.[132,134,135] Note that KPFM operation is only valid if this implicit assumption of a precise nulling of the first harmonic force is met. Under the condition of $F_{\omega} = 0$, the measured CPD can be measured directly (i.e., $V_{CPD} = -V_{DC}$, if nulling bias is applied to the tip), independent of $V_{AC}$ and the capacitance gradient. Practically, it has been shown that the measured CPD is often dependent on $1/V_{AC}$ and the capacitance gradient.[64,129,134]

It is also known that the nulling bias used to determine the CPD can be influenced by a variety of factors including electronic offsets and crosstalk (between $V_{AC}$ and the photodetector output and/or between $V_{AC}$ and the piezoactuator),[136,137] thermomechanical and electrical noise sources,[138] the choice of feedback gains,[130,134] $V_{AC}$ frequency, and LIA phase offset.[130] The net effect of these factors is that the input signal to the feedback loop contains contributions that are not associated with the electrostatic tip-sample forces. This point is demonstrated experimentally in Figure 2.7 using OL bias spectroscopy. Additional contributions at the input to the feedback loop cause a shift in the true CPD nulling point (see Figure 2.7b). Under these conditions, the feedback loop will attempt to minimize a mixed signal leading to an error in the measured CPD value.[139] As a result, the absolute (i.e., the real values for the system being measured) CPD measurements in closed-loop KPFM are subject to the feedback effect and can vary typically within an instrument-dependent ~ 1 V range.[136] Although some forms of parasitic signal components can be minimized,[132,137] it requires careful calibration of all electronic instrumentation and sufficiently shielded electronic cabling or active feedback compensation.[134,139] If present, the feedback effect can lead to dependences of the measured CPD on $V_{AC}$, $C'_z$,

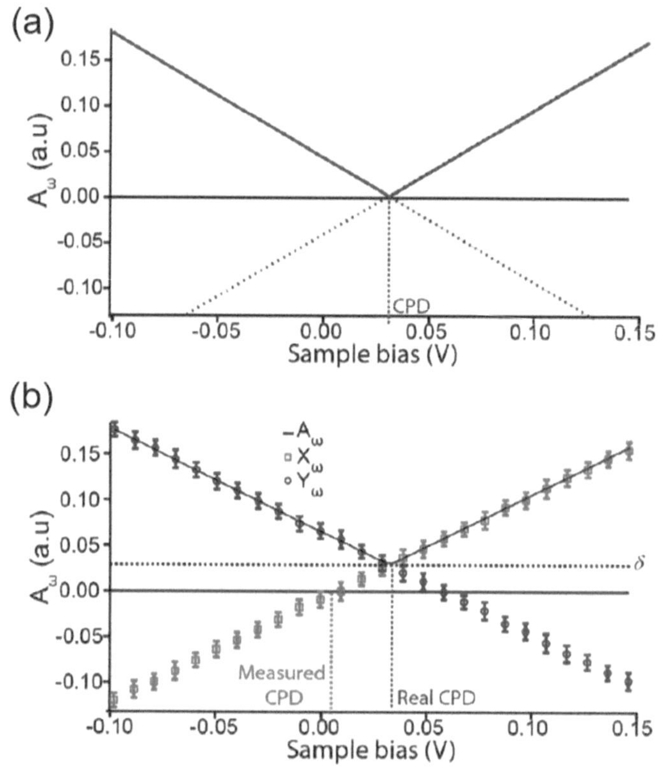


**FIGURE 2.7** Open loop bias spectroscopy and the feedback effect. An OLBS measurement showing the electrostatically excited cantilever oscillation amplitude ($A_\omega$) as a function of DC bias applied to the sample. (a) Schematic of the ideal behavior, i.e., zero oscillation when the DC bias corresponds to the CPD. (b) Experimentally measured behavior showing amplitude, in phase ($X_\omega$) and out of phase ($Y_\omega$). (Note the phase was adjusted by 90° between collection of $X_\omega$ and $Y_\omega$). (Adapted from Collins et al.[131])

and topography. Finally, although rarely discussed in applications involving KPFM, precise tuning of the LIA phase and feedback loop parameters is necessary to realize reliable and quantitative CPD measurements. These points are addressed in detail by Jacobs et al.[130] Briefly, they introduce a simple procedure to fine-tune the feedback by applying a square wave excitation bias to the surface of an electrode, effectively modulated the CPD by a known value, and tuning the LIA phase (in OL) and Kelvin feedback loop gains (in closed loop) until the square wave bias is correctly measured by the Kelvin controller.

## 2.4 EFM/KPFM APPLICATIONS FOR ENERGY RESEARCH

EFM and KPFM are gaining widespread attention in the area of energy research due to their unique ability to characterize the nanoscale electrochemical landscape of materials and devices. For example, EFM has been used to map nanoscale charge transfer[140–142] and optoelectronic performance[143] in solar cell devices, channel connectivity in proton-exchange[144,145] and anion-exchange membranes[146] for applications in fuel cells, or to examine the local electric field intensification in electrodeposited platinum nanodentrites used for methanol catalytic combustion.[147] Unfortunately, the nature of EFM makes the extraction of quantitative material properties challenging for regular AFM users. On the other hand, the ability of KPFM to directly quantify local potentials has established it as a state-of-the-art characterization tool for energy materials and devices, and having been widely adopted in the field of photovoltaics,[148–150] solid-state batteries,[151] fuel cells,[152–154] and electro/photo-catalysts[155] to name but a few.

In combination with illumination, EFM/KPFM methods have been extraordinarily successful for investigation of optoelectronic properties.[9,148,156] They have been widely applied to a whole host of solar cell materials and devices, ranging from polycrystalline silicon,[157,158] chalcopyrite,[159–162] and CdTe,[163] to organic thin-film solar cells.[148,164–166] Studies have varied from investigating the role of grain boundaries to illumination-assisted KPFM for mapping SPV or to potentiometric profiling on cross-sectioned solar cell devices.[156]

As an example, in Figure 2.8a–c, we highlight the work by Sadewasser et al.,[167] who utilized KPFM in UHV to map potential profiles across individual grain boundaries for a $CuGaSe_2$ thin film grown on Mo-covered glass by physical vapor deposition (PVD). The samples showed a typical granular topographic structure, and by comparing with KPFM potential captured under dark/light conditions, it was possible to deduce a reduced surface band bending upon illumination. In addition, as shown in Figure 2.8c, the work function dip with respect to the dark measurement is different for individual grain boundaries, suggesting differences in electronic behaviors exist.[167,168] In the context of solar cell operation, knowledge of the band bending and nanoscale variation is extremely important as it can have a real impact on the transfer and collection of the charge carriers. KPFM has also been extensively applied to map variations in SPV, i.e., the change in the work function with illumination ($\Phi_{illuminated} - \Phi_{dark}$). Glatzel et al.[164] performed KPFM under both dark and illuminated conditions necessary to extract locally resolved SPV for conjugated polymer/fullerene organic solar cells, as shown in Figure 2.8d. Recently, Garrett et al.[169] combined KPFM under illumination with high speed KPFM was realized using heterodyne KPFM (see Figure 2.8e). They demonstrated mapping of the SPV (or open circuit potential) on hybrid organic/inorganic perovskite films at an imaging rate of ~16 frames/second.

Regarding measurements on devices, Tanimoto and Vatel[35,170] demonstrated that KPFM can be used to map charge transport on laterally structured devices during device operation. In these experiments, the KPFM probe acts as a moving voltage electrode, providing a map of the internal potential distribution across active device from which corresponding resistances can be reconstructed (provided the device current is known).[171–174] Besides lateral devices, cross-sectional KPFM measurements are attractive due to the ability to map the internal field/potential distribution within practical devices. Such information is particularly valuable across the electrode contact interfaces and the p-n junctions.[175] Ballif et al.[140] were among the first to demonstrate a cross-sectional study of a solar cell. As shown in Figure 2.8f, they used EFM to investigate a polycrystalline $CdTe/CdS/SnO_2$/glass solar cell that was cleaved in such a way that the device was operational with nominal photovoltaic efficiencies, making it possible to differentiate surface (work function, band bending, etc.) from bulk device properties. EFM revealed a potential drop close to the CdTe/CdS interface compatible with an *n*-CdS/*p*-CdTe heterojunction model. Similar cross-sectional studies using KPFM have been widely adopted for a broad plethora of solar cell materials.[167] As an example, Figure 2.8g shows a cross-sectional KPFM study performed by Bergmann et al.[176] on high efficiency mesoscopic methylammonium lead tri-iodide solar cells. They demonstrated unbalanced charge transport in the device such that on illumination (under short-circuit conditions), holes accumulated in front of the hole-transport layer because of unbalanced charge transport in the device. Consequently, after turning off the light illumination, charges remained trapped inside the active device layers. The avoidance of the charge trapping and unbalanced charge transport could be an avenue for improved performance in such devices and help our understanding of anomalous hysteresis in such devices.[176]

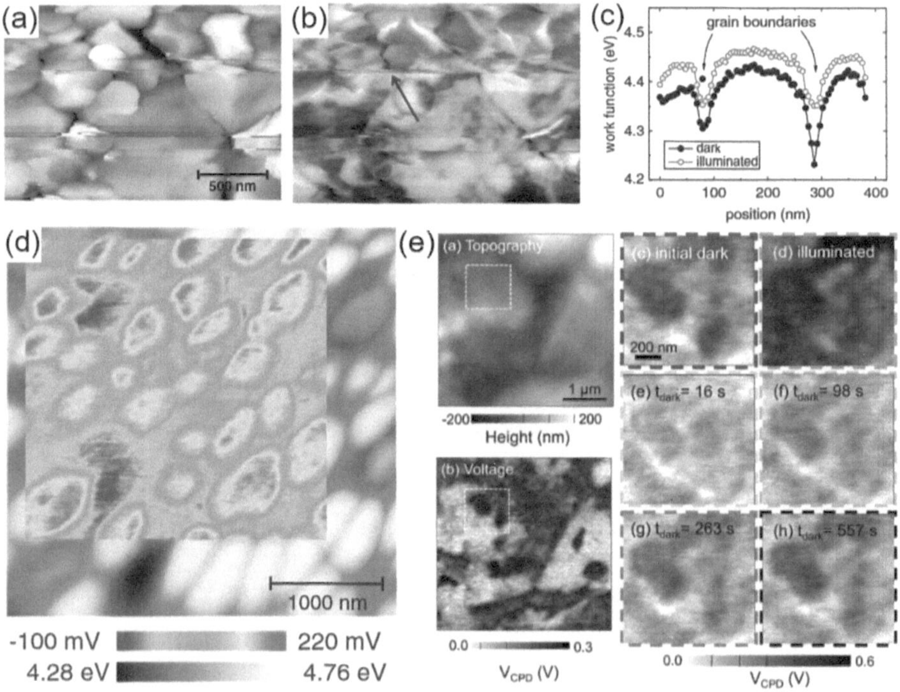


**FIGURE 2.8** Example applications of KPFM for solar cell research. (a–c) UHV-KPFM used to study the influence of grain boundaries in chalcopyrite ($CuGaSe_2$) thin films showing (a) topography (z scale = 360 nm), (b) work function under dark conditions (color scale 0.423–0.45 eV), and (c) cross-sectional profile comparing work function across two grain boundaries (location shown by arrow in (b)) measured under dark and sub-bandgap illumination. (Adapted from Sadewasser et al.[167].) (d) Locally resolved surface photovoltage (SPV) of toluene-cast MDMO-PPV:PCBM-blended organic thin film. The image in the background represents the work function image under illumination. (Adapted from Glatzel et al.[164]) (e) Time-resolved surface photovoltage imaging of a hybrid organic–inorganic perovskite film based on methylammonium lead ($MAPbI_3$) captured using fast heterodyne KPFM at an imaging speed of 16 s/scan. (Adapted from Sadewasser et al.[167])

*(Continued)*

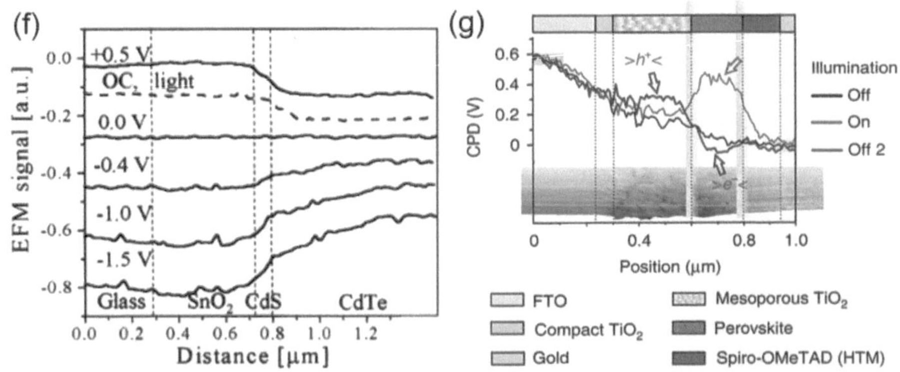


**FIGURE 2.8 (*CONTINUED*)** (f) EFM line scans taken at different sample biases along a Glass/SnO2/CdS/CdTe cross section solar cell showing a drop of the EFM signal at the CdTe/CdS interface. (Adapted from Ballif et al.[140]) (g) KPFM potential profiles captured from a highly efficient mesoscopic lead methylammonium tri-iodide solar cell for short-circuit conditions before illumination (black line), during illumination (red line), and immediately after illumination (blue line). The increase in potential after turning on the illumination corresponds to an accumulation of holes inside the capping layer (red arrow). After switching off the light, trapped holes become visible inside the mesoporous structure with an increase in potential ($>h^+<$) and trapped electrons become visible in the perovskite capping layer ($>e^-<$). (Adapted from Bergmann et al.[176].)

As KPFM can provide spatially resolved SPV, it has also been applied on photocatalysts to visualize their built-in electric field in the surface space charge region, which determines the charge separation and transport processes of the photogenerated carriers.[177–181] Zhu et al.[177] monitored the SPV on different facets of a single crystal $BiVO_4$ photocatalyst via KPFM and Spatial Resolved Surface Photovoltaic Spectrometry (SRSPS) and found that the charge transfer is highly anisotropic. The SPV signal on the {011} facet is 70 times stronger than that on the {010} facet, which is related to the different built-in electric fields in the space charge region (SCR) of different facets. The same research group further applied this approach to investigate the role of co-catalyst and dual co-catalysts in photocatalysis[178] by imaging the SPV across bare, $MnO_x$–, and $MnO_x$/Pt-loaded $BiVO_4$ single crystals. KPFM clearly visualized the CPD differences, and hence, the work function differences between the $BiVO_4$ {011} and {011} facets changed after selective deposition of a $MnO_x$ co-catalyst on the {011} facets from 30 to –70 mV (Figure 2.9a and b). The potential difference between {011} and {011} facets can be tuned by varying the size of the $MnO_x$ co-catalysts. The role of dual co-catalysts was investigated by selective deposition of $MnO_x$ on {011} facets and Pt on {010} facets. Adding Pt on the {010} facets not only increased the SPV signal by 40% compared to that of the single $MnO_x$ co-catalyst, but also made the photogenerated electrons and holes separate distinctively toward the {010} and {011} facets, respectively (Figure 2.9c). This work demonstrated that careful selection of the type of co-catalyst and the location for co-catalyst deposition can strongly increase the interfacial charge transfer and even alter the direction of built-in electric fields to enhance charge separation (Figure 2.9d).

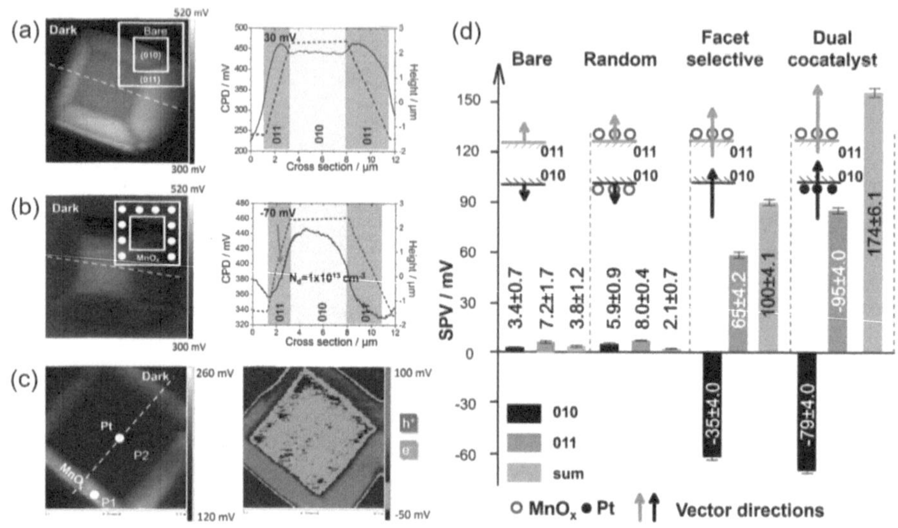


**FIGURE 2.9** Effect of co-catalysts in photocatalysis. (a) and (b) Dark state KPFM image (left) and height and potential profile across the dash lines (right) of a single $BiVO_4$ photocatalyst particle (a) before and (b) after selective deposition of $MnO_x$ nanoparticles on {011} facets; (c) Dark state KPFM image (left) and spatial distribution of the SPV signals (right) of a single $BiVO_4$ photocatalyst particle with MnOx and Pt dual co-catalysts selectively deposited at the {011} and {010} facets, respectively; (d) Comparison of SPV signals measured at {010} and {011} facets of bare and selective co-catalyst-coated $BiVO_4$. Orange and black arrows indicate the built-in electric fields toward external surface of {011} and {010} facets, respectively. (Adapted from Zhu et al.[178])

In battery research, KPFM has been useful in studying cathode/anode electrode materials,[182–186] interfaces,[187] and dynamic processes in situ/operando for solid-state batteries.[188] KPFM has been used to explore differences between aged and non-aged $LiFePO_4$ electrodes[182] and to study $LiCoO_2$ electrodes with different charge/discharge cycles.[183] It should be noted that for alkali and alkaline earth metal batteries, KPFM under highly controlled environmental conditions either within a sealed environmental cell or glove box is required.[189]

Conventional electrochemical characterization methods are bulk techniques and cannot provide local potential information. To overcome this, Masuda et al.[187] performed KPFM measurements on a solid-state Li-ion battery (SS-LIB) prepared in cross-section by Ar ion milling. This, and a later study,[188] focused on a composite cathode composed of $LiCoPO_4$ (LCP), $Li_{1+x}Al_xTi_{2-x}(PO_4)_3$ (LATP) solid electrolyte, and a Pd conductive additive, investigated by operando KPFM under nitrogen atmosphere (see Figure 2.10a). KPFM was used to visualize the internal potential distributions within solid-state batteries under different states of charge.[187] Energy dispersive X-ray spectrometry (EDS) was performed on the same area to provide local compositional information in addition to KPFM potential maps. Note that when the battery is charged, the measured potential is a convolution of the electrostatic potential and change in the work function of the material due to delithiation. The authors noted

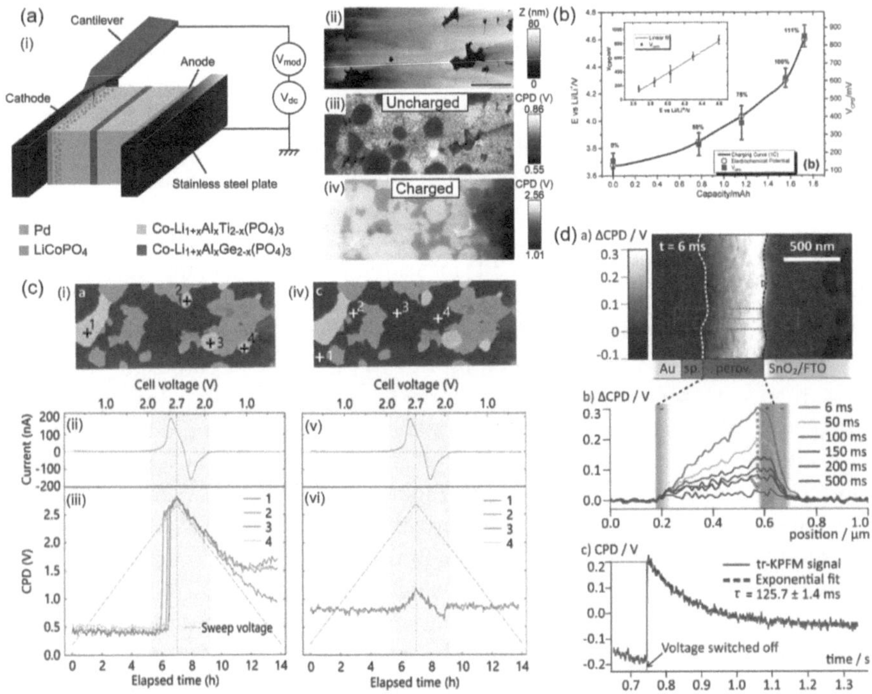


**FIGURE 2.10** (a) (i) Schematic illustration of the structure of the cross-sectional KPFM setup used by Masuda et al.[187] to visualize the internal nanoscale potentials in a SS-LIB (measurements were performed in a $N_2$ flow glove box). (ii) Topography image and KPFM CPD image collected (iii) before charging and (iv) after charge. (Adapted from Masuda et al.[187].) (b) Measured KPFM potential vs. charging capacity for NCA cathode material harvested from non-aged commercial pouch cells. Inset describes the linear correlation between the surface potential measured by KPFM (i.e., CPD) and the electrochemical potential, E. (Adapted from Schmutz and Frankel.[14]) (c) Real-time KPFM measurement of a cathode composite region (as shown in (a-i)) measured during CV operation. (ii, v) Current from CV data and (iii, vi) CPD data is plotted as a function of elapsed time during device cycling (cell voltage is shown in top x-axis). Temporal profiles are extracted from (ii, iii) Pd and (v, vi) LATP particles in specific locations indicated in CPD maps shown in (i) and (iv). (Adapted from Masuda et al.[188]) (d) (Top) ΔCPD map measured 6 ms after switching off the external voltage supplied to the perovskite solar cell device. Interfaces of the perovskite layer are marked with dotted lines. (Middle) Spatially averaged cross-section graphs obtained from the region indicated by the red box at different time intervals after switching off the external voltage. Regions with non-linear ΔCPD profiles correspond to non-zero local charge density, indicated by red/blue shaded boxes according to polarity of the charges. TR-KPFM trace recorded at a position close to the perovskite/SnO2 interface, showing a voltage overshoot following the switching of the FTO voltage from −0.5 V to 0 V. The overshoot decay was fitted with a single exponential function. (Adapted from Weber et al.[191])

a large potential drop (0.8–1 V) at the composite cathode/solid electrolyte interface (right side of Figure 2.10a-iv). Such insight is important since the potential drops across the various interfaces comprising the battery device will dictate the battery performance via the internal resistance of the cell. It is noteworthy that KPFM is one of the few techniques that are able to directly probe such potential drops on cross-sectional devices.[120]

While KPFM proves to be a valuable characterization tool for solid-state batteries, its application to conventional liquid batteries is complicated by the finite conductivity and charge screening in solution, and the fundamental lack of appropriate liquid KPFM techniques. The status and current state of the art pertaining to liquid EFM/KPFM measurements are dealt with in Section 2.6. As an intermediate, correlations have been discovered between the CPD measured by KPFM and the electrochemical potential in electrolytes. Schmutz et al.[14] demonstrated a linear dependence between CPD and open circuit potentials for a variety of metals. In a different study, Stone et al.[190] correlated the CPD measured by KPFM on amine or carboxyl groups after exposure to water with their expected charge for a wide range of pH values. Applied to batteries, Zhu et al.[186] explored the relationship between the measured KPFM potential and electrochemical potential of a $LiNi_{0.80}Co_{0.15}Al_{0.05}O_2$ (NCA) cathode material at various states of charge (see Figure 2.10b). The NCA cathode materials were harvested from a non-aged commercial pouch cells with 5 Ah capacity. The harvested 18 mm electrodes were tested in three-electrode electrochemical cells with the NCA electrode as working electrode, which were then charged to predefined state of charge. The cycled electrodes were rinsed three times with dimethyl carbonate and dried under vacuum before being characterized by KPFM. Interestingly, they found a direct correlation between the local CPD measured by KPFM in a gaseous environment and the electrochemical potential measured in electrolyte. While such correlations offer the possibility to indirectly investigate the local behavior of battery material and link them to in situ/operando performance, directly probing the potential in situ during battery operation would be preferred if such an approach was possible at the solid-liquid interface. As disassembling and rinsing the electrodes might remove the thin solid electrolyte interphase (SEI) layer, a passivation layer is formed between electrode and electrolyte, which is crucial to battery cyclability, hence the measured electrode surface is different from the surface in a real device.

Complementary to the above-presented studies has been the development of dynamic or time-resolved methods by which further information on electronic[35,45–47] and ionic[192–195] charge transport can be extracted. The development of real-time or even time-resolved (TR) EFM and KPFM has the potential to boost energy applications by further extending them to dynamic/kinetic phenomena. For battery applications, much of the dynamic processes are slow (minutes-hours), and as such, conventional KPFM can be used to monitor changes in the potential over time. As an example (Figure 2.10c), KPFM measurements have recently been used to capture dynamic information on the change of internal potential distribution of a composite cathode (same as that described in Figure 2.10a) in real time during Cyclic Voltammetry (CV) of the device.[188] The results suggested that the formation and collapse of electronic conductive pathways within the composite cathode were strongly correlated with current flow as CPD did not change monotonically with the applied potential but changed only when there was current passing through the electrode. This suggests that the electronic conductive network is only formed when Li-ions were moving in and out of the active materials (LCP). The potential changes in

LATP regions during CV does not depend on location within the electrode, whereas the potential changes in Pd particles vary with location. The origin of different location dependencies between different materials is still unknown. Figure 2.10d, TR-KPFM in conjunction with voltage- or light pulsing was used to uncover underlying mechanism of the current–voltage hysteresis in a hybrid lead-halide perovskite solar cell.[191] Using this approach, it was possible to observe the formation and trapping of a localized interfacial charge at the anode interface that screened most of the electric field in the cell, thereby explaining the reported phenomena of hysteresis in perovskite solar cells.[191] Investigating faster processes (<<ms) requires the adoption of novel TR-EFM/KPFM approaches that are discussed in Section 2.5.5 and have been reviewed elsewhere.[16,17]

## 2.5 ADVANCED MODES OF EFM/KPFM OPERATION

### 2.5.1 Open-Loop Modes of KPFM Operation

Classical KPFM implementations rely on a closed bias feedback loop for compensating the electrostatic force between probe and samples and hence determination of the CPD. OL techniques refer to methods that do not require a bias feedback loop. OL-KPFM techniques can largely be split into OL DC bias spectroscopy approaches (see Figure 2.7) or pure dynamic AC approaches that we consider in this section. OL-KPFM have been investigated for their simpler implementation, reduced sensitivity to electronic offsets, and electronic crosstalk instrumentation,[64,137,139] as well as operation in liquid environments.[196–200] Takeuchi et al.[201] first demonstrated imaging by the purely dynamic AC-modulated OL-KPFM technique. Their method was operated in UHV and demonstrated mapping the CPD of semiconductors without requiring application of DC bias feedback.[201] Since then, this approach has become known as Dual Harmonic (DH)-KPFM as it relies on the detection of the harmonic responses of a voltage-modulated probe.

As shown in Figure 2.11a, in DH-KPFM, the cantilever is excited with an AC voltage at a single drive frequency, $\omega$, without the need for a DC bias (i.e., $V_{DC} = 0$). The electrostatic force acting on the tip produces harmonic responses of the cantilever, at $\omega$ and $2\omega$, described by equations (2.8) and (2.9):

$$A_{\omega} = G(\omega)F_{\omega} = G(\omega)\left|C_z'(V_{CPD})\right|V_{AC}, \tag{2.8}$$

$$A_{2\omega} = G(2\omega)F_{2\omega} = G(2\omega)\left|C_z'\right|\frac{V_{ac}^2}{4}, \tag{2.9}$$

where $G(\omega)$ and $G(2\omega)$ are the gains due to the cantilever transfer function at $\omega$ and $2\omega$, respectively. Key to the realization of DH-KPFM is that the capacitance gradient term can be eliminated by utilizing the ratio of the harmonics. Provided that the $V_{AC}$ and cantilever transfer function gain, $X_{gain}$ (where $X_{gain} = G(\omega)/G(2\omega)$), are known or can be determined, the $V_{CPD}$ can be calculated as:

$$V_{CPD} = \frac{A_{\omega}\cos(\phi_{\omega})}{A_{2\omega}}\frac{V_{ac}}{4X_{gain}}, \tag{2.10}$$

The polarity of $V_{CPD}$ is given by the phase ($\phi_\omega$) of the cantilever response at $\omega$. In general, DH-KPFM garnered little attention until several years after its first realization when it was implemented under both ambient[202] and liquid[196–200] environments and was used to realize atomic resolution imaging of Si(111)7×7 surface potential imaging in UHV.[203]

Collins et al.[202] first extended this approach to ambient demonstrated on a ferroelectric bismuth ferrite thin film noting that while both DH- and conventional KPFM measured similar relative surface potentials between positively and negatively poled domains, an offset in the absolute potential measured by closed-loop and DH-KPFM was recorded, which was prescribed to feedback artifacts in the case of CL-KPFM.[131,202] Polak et al.[137] demonstrated that after careful calibration of parasitic signals for both cases, CPD measured by KPFM and DH-KPFM can be made independent of both frequency and amplitude of the excitation signal. Kilpatrick et al.[139] demonstrated a "setpoint correction" for CL-KPFM to account for non-idealities of the bias feedback loop, which allowed quantitative agreement between CL-KPFM and DH-KPFM within 1%. In a different, albeit related approach, Borgani et al.[204,205] configured multifrequency intermodulation AFM for measuring surface potentials. Intermodulation EFM is operated during single-pass imaging and consists of driving the cantilever with two pure drive tones close to the resonance frequency of the cantilever, one drive used to mechanically actuate the cantilever and another to modulate the electrostatic force. Since the responses of interest all lie close to the resonance frequency, this approach leads to increased SNRs over off-resonance DH-KPFM and is expected to allow for higher spatial resolution as the tip is closer to the sample when operated in single pass.

### 2.5.2 Multifrequency and Multidimensional KPFM

Over the last decade, there has been a tremendous shift toward multifrequency methods for more precise functional mapping.[206] In contrast to single frequency, multifrequency methods involve the excitation and/or detection of the cantilever deflection at two or more frequencies. Concomitantly, there has been a steady rise toward the capture of multidimensional datasets, which include additional information in either space, voltage, and/or time dimensions (e.g., Force Volume (FV) imaging AFM,[207–209] conductive AFM IV spectroscopy,[210] switching,[211] and relaxation,[212] spectroscopy and PFM).

Multifrequency and multidimensional methods can be used to separate different force contributions or to explore their dependencies (e.g., voltage, distance, time). Band Excitation (BE) KPFM has been implemented in several different forms.[131,213,214] Contrary to single frequency techniques, in BE, dynamic excitation is performed using a chirp waveform at many frequencies simultaneously (see Figure 2.11b). Typically, the excited band has a central frequency close to the cantilever resonance frequency, $\omega_0$, and a bandwidth several 10s or 100s of kHz in width. Excitation and detection of signals within this band of frequencies yields the transfer function of the cantilever, i.e., the segment of the response–frequency curve near the mechanical resonance. This curve is then collected at each spatial point (or each point of parameter space for complex spectroscopic imaging methods), at the rate comparable

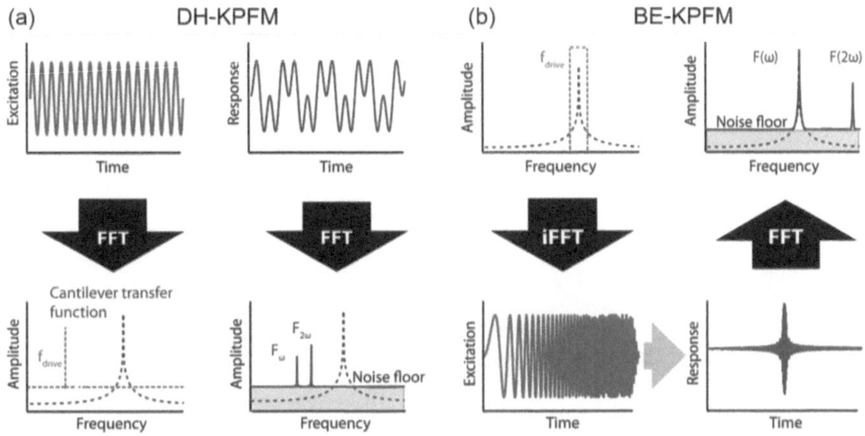


**FIGURE 2.11** Schematic describing (a) single frequency and (b) BE excitation schemes and responses. In single frequency methods (a) the cantilever is driven with a simple sinusoidal function at a fixed frequency and has a dynamic response at the fundamental drive and a frequency-doubled response. For BE (b), an excitation signal with a predetermined amplitude density in a frequency band around the cantilever resonance frequency is selected. This excitation is inverse Fourier transformed into the time domain and used to drive the cantilever. The response of the cantilever to this signal is Fourier transformed to reconstruct the cantilever transfer function. (Adapted from Collins et al.[131])

to classical lock-in detection. It usually requires additional external equipment to achieve the fast data reading and digitization as well as supplementary programming code for the signal generation/collection (such as Matlab or Python) and the AFM control (such as Labview). In an offline analysis, the response curve is transformed into classical physical observables by fitting the contact resonance peak to a simple harmonic oscillator (SHO) model.

The BE method has been applied to KPFM in a variety of different forms. Guo et al.[215] demonstrated BE-KPFM in an OL DC bias spectroscopy measurement. A schematic describing the method for a single location is shown in Figure 2.12. Practically, BE-KPFM consists of a 4D dataset consisting of the amplitude (and phase) response of the cantilever vs. frequency, bias, and XY position. Figure 2.12b and c shows the recorded cantilever response amplitude vs. frequency for a series of DC bias values. The values for $A_0$ cos($\phi$) and $\omega_0$ obtained from SHO fitting are shown in Figure 2.12d as a function of bias.[213] $A_0$ cos($\phi$) is a linear function of bias while $\omega_0$ exhibits a parabolic bias dependence as expected. In this way, BE-KPFM is sensitive to the electrostatic force and the force gradient simultaneously, combining the advantages inherent in both AM- and FM-KPFM.[213] Note that the relative noise in $\omega_0(V)$ is larger in the vicinity of the nulling potential as a result of smaller response amplitudes. Later, this poor sensitivity around the nulling bias was overcome by adopting photothermal excitation of the cantilever leading to higher sensitivity and improved spatial resolution compared to force detection.[214] Finally, half harmonic BE (HHBE)–KPFM was implemented,[216] which did not require application of a DC

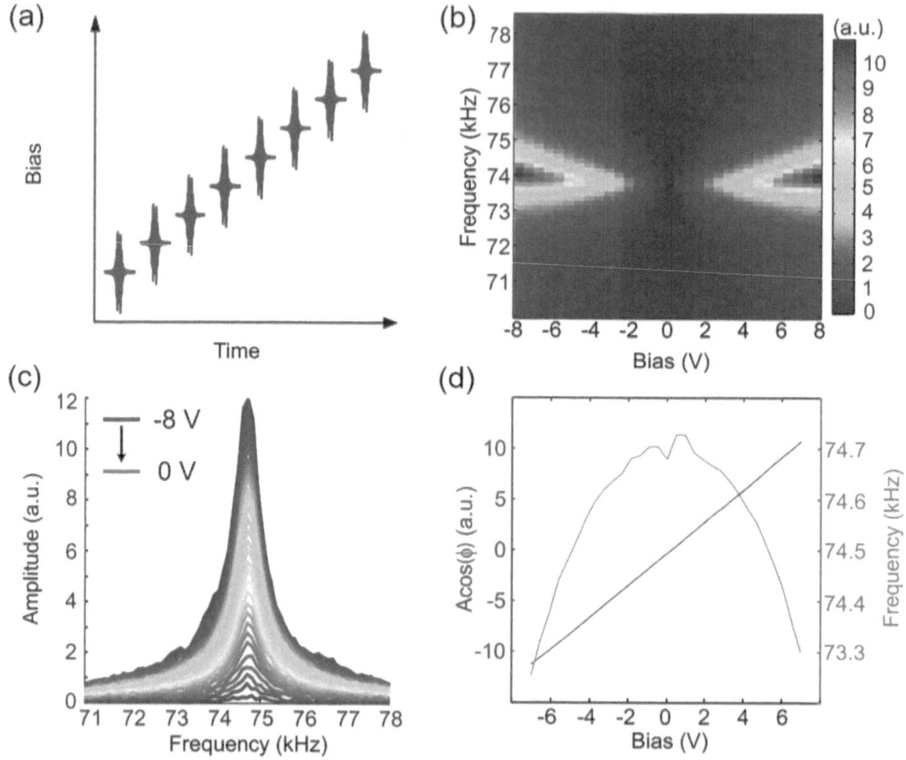


**FIGURE 2.12** BE-KPFM operating principle. (a) Train of BE waveforms superimposed on a linear DC voltage ramp. (b) Cantilever response amplitude vs. bias and frequency. (c) Cantilever response amplitude vs. frequency for negative bias values. (d) Bias dependence of cantilever mixed amplitude response (left y axis) and resonance frequency (right y axis). (Adapted from Collins et al.[213].)

bias as a promising approach for applications involving voltage-sensitive materials or operation in liquids.

### 2.5.3 Three-Dimensional EFM/KPFM

A major consideration when studying electrostatic forces is the long-range nature of the interaction. This poses a complication in interpretation of EFM/KPFM measurements and ultimately limits the maximum achievable resolution of the technique. In addition, while characterization of structured materials including nanowires, nanoparticles, or simply rough surfaces is of great interest for many applications, operation of traditional KPFM is limited to largely topography-free or smooth surfaces. These factors have motivated research into the development of 3D mapping techniques to capture not only the spatial variation in the X-Y plane, but also as a function of tip-sample distance. Bayerl et al.[217] applied KPFM in a 3D distance spectroscopy approach developed for artifact-free mapping of the potential distribution on surfaces with large topography variations. Rather than raster scanning in the plane of the surface as with conventional scanning probe techniques, a fine grid of points is specified on the surface and the tip is gently approached toward the surface at each point. They applied their 3D KPFM approach to characterize the piezoelectric potential generated by laterally strained ZnO microwires (MWs) by extracting the potential asymmetry from opposite sides of the MW.[217] More recently, FVBE-KPFM[213] was developed as an OL multidimensional bias spectroscopy alternative. This is an extension of BE-KPFM[131] where a 5D data set comprising electrostatic response (amplitude, phase) vs. frequency, bias, and position (X, Y, Z) is acquired. An example of a bias vs. distance spectroscopy measurement showing both amplitude and frequency is given in Figure 2.13a and b, respectively, for a single location on a reference gold metal structure. The multidimensional data was compressed by fitting the bias dependence of the electrostatic force (linear) and electrostatic force gradient (parabolic), respectively. The fitting parameters set can be used to determine the 3D CPD and capacitance gradient from the electrostatic force and force gradient, respectively. Figure 2.13c and d depicts maps of the CPD for metal/insulator test

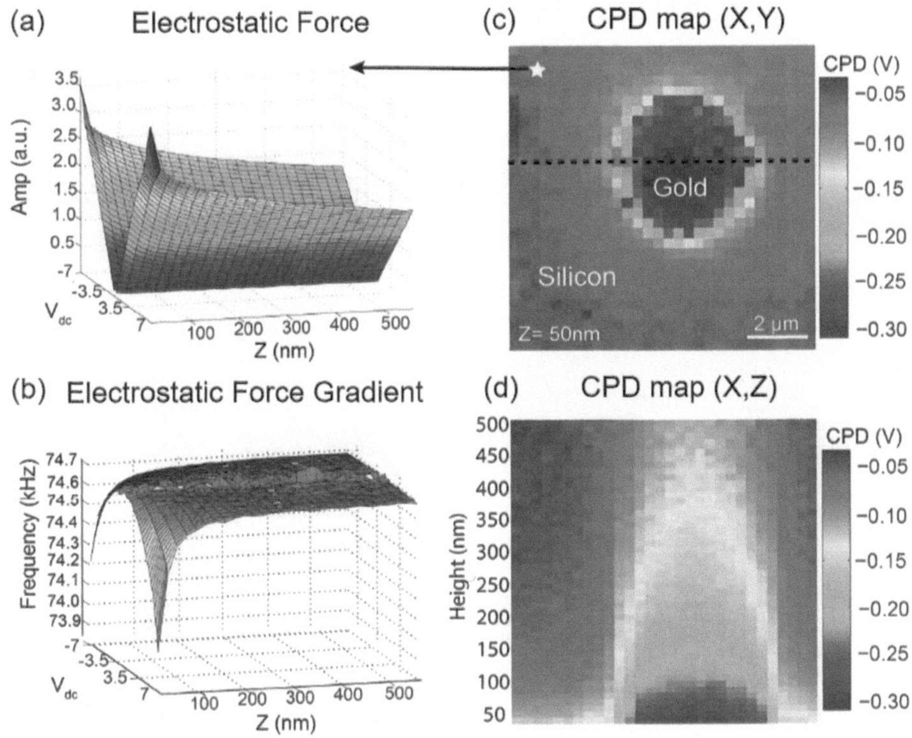


**FIGURE 2.13** 3D CPD mapping using force volume BE-KPFM. Single point (a) electrostatic force and (b) electrostatic force gradient measurement as a function of tip sample distance (Z). Spatial mapping of the CPD in the (c) X-Y plane (Z = 50 nm) and the (d) X-Z plane from (X plane indicated by a dashed black line in image (c)) determined from fitting to the electrostatic force of a 45×45×50 (X,Y,Z) grid recorded on a gold/silicon test structure. (Adapted from Collins et al.[213])

sample, showing a single XY plane as well as single XZ plane from a 45×45×50 grid, providing a wealth of information on the overall tip-sample interactions present in KPFM measurements. In this regard, FVBE-KPFM can offer a useful insight between KPFM experiment and modeling and potentially be used to completely deconvolute the effect of the probe geometry.

In parallel to such 3D KPFM modes, force volume EFM has also been developed (see Figure 2.14). Scanning Dielectric FV Microscopy (SDFVM) has been used to obtain label-free maps of dielectric constant in topographically complex and heterogeneous samples like prokaryotic[42] and eukaryotic[44] cells. Such approaches (which are fully in line with the modern data-driven trends on SPM) enable the reconstruction of all imaging distances and modes (i.e., lift mode, constant plane, scanning polarization microscopy) by further post-processing of a richer multidimensional dataset sometimes referred to as data cube (Figure 2.14b). Not only that, but it has also been proven that the dielectric constant maps can be used to obtain nanoscale biomolecular composition maps as the main biomolecular components of

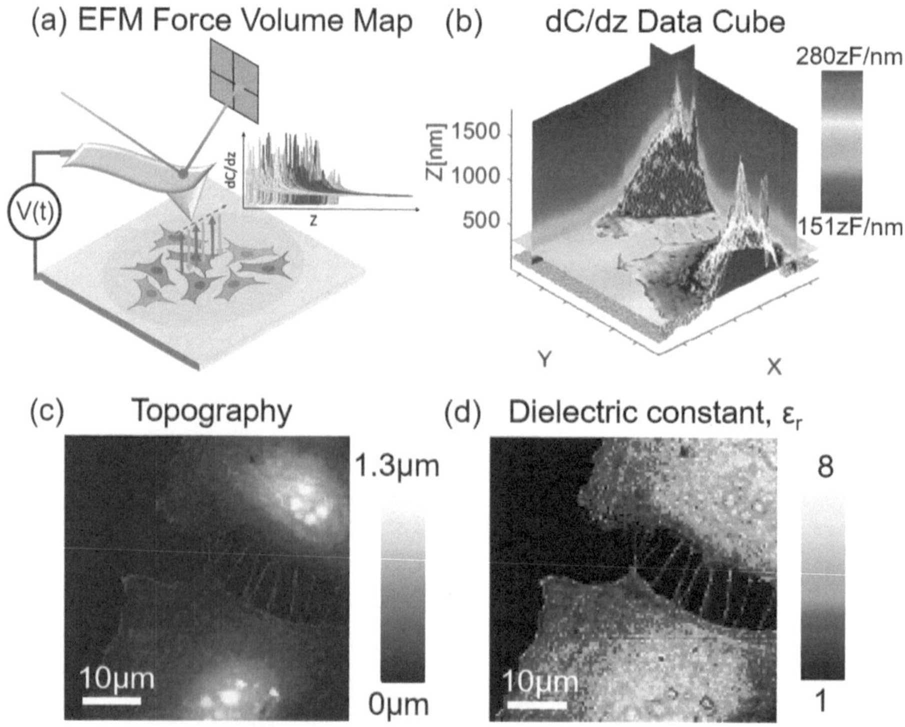


**FIGURE 2.14** (a) Experimental sketch of the application of SDFVM to dielectric constant mapping of eukaryotic cells. (Inset) Set of 2$\omega$-EFM force-distance curves acquired at each pixel of the sample. (b) Reconstruction of the dC/dz Data Cube. (c) Topography of a dry and fixed HeLa cell. (d) Subsequent dielectric constant map obtained via fitting of each pixel's 2$\omega$-EFM force-distance curve giving the local dielectric constant that can be used for label-free nanoscale biomolecular compositional mapping. (Adapted from Checa et al.[44])

cells display a different dielectric response (in lipids $\varepsilon \approx 2$, in proteins $\varepsilon \approx 4$, and in DNA $\varepsilon \approx 8$). In addition, SDFVM increases the accuracy and automatization of dielectric constant measurement (Figure 2.14d) as it allows the fitting of a full electrostatic force-distance curve in each pixel, necessary for dielectric constant mapping. Moreover, such methods are especially interesting for soft and weekly absorbed samples (like polymers or biological samples) as they avoid applying lateral forces to them, increasing the measurement stability, reducing sample damage, and opening the door to perform simultaneously mechanical and electrical characterization (as the contact part of the force-distance curve can be also used for mechanical mapping).

### 2.5.4 Contact and Pulsed Force Techniques

As discussed in the previous sections, electrostatic forces between tip and sample are typically measured at a certain tip-sample distance in a two-pass technique or upon mechanical and electrical excitation at different frequencies in a single-pass measurement. However, measuring electrostatic interactions while the tip is statically or dynamically in contact with the sample at a set interaction force can provide information on the charge distribution and (junction) CPD. These techniques do not require mechanical excitation at a cantilever resonance and provide a higher spatial resolution than other techniques that can be limited by long-range electrostatic and adhesive forces or high AC voltages.

#### 2.5.4.1 Contact Mode Electrostatic Force Microscopy

Electrostatic forces between the tip and the sample surface can also be present when the tip is in contact with the sample surface. These electrostatic interactions can be utilized to probe electrical properties, surface potentials, and charge-transfer dynamics.[218,219] The main advantage of contact mode compared to non-contact techniques is the achievable higher lateral resolution, which increases with decreasing tip-sample distance. However, in tapping mode techniques, the minimum achievable distance is severely limited by adhesive forces.

This limitation was overcome by Hong et al. in 1999 by establishing dynamic contact mode EFM, which provided an improved spatial resolution and decoupling of topography and charge properties.[218] In this technique, the tip is scanned across the sample at a constant force setpoint while AC and DC voltages are applied. The sample topography is derived from the repulsive atomic force and the DC part of the electrostatic force, whereas detection of AC components at the excitation frequency provides information on the distribution of surface charges and surface potentials. While dynamic contact mode EFM can be used to study materials with insulating native thin oxide layers such as semiconductors, it is less suitable for metals and conducting samples that exhibit current flow.[219] In 2020, Minj et al.[219] reported an improved contact-resonance EFM technique that achieves high sensitivity and resolution on a wider range of materials comprising semiconductors, metals, and dielectrics. In this method, a small tip-sample distance corresponding to an insulating gap is achieved either through organic contamination of the sample or encapsulating the tip with a polymer particle. To enhance the SNR and thus the probing sensitivity of electrostatic forces, the electric excitation is performed at the mechanical contact resonance.

### 2.5.4.2 Contact Kelvin Probe Force Microscopy

The above described contact EFM techniques can be extended to contact Kelvin probe force microscopy (cKPFM) where in addition to probing electrostatic forces with an AC voltage at contact resonance, DC read and write pulses are applied.[220] In this spectroscopic method, the tip is in contact with the sample and a sequence of triangular-square DC write pulses is used to change the surface potential of the sample, as commonly performed in piezoresponse switching spectroscopy.[211] Between those write pulses, DC read voltages are applied and varied sequentially with each cycle to change electrostatic forces during the read-out (see Figure 2.15a). In Figure 2.15b, the measured cKPFM signal (y-axis) derived from the cantilever oscillation at contact resonance on amorphous $HfO_2$ is plotted as a function of the DC read voltage (x-axis) and changes dependent on the DC write pulse (color scale). The y-intercept of this graph corresponds to the signal without a DC read voltage and, therefore, is equivalent to the signal measured in PFM or contact EFM. The x-intercept, however, provides information on the junction contact potential difference (jCPD) between the probed material and the tip. For metal-metal contacts, jCPD and CPD are the same. However, other types of junctions and the existence of passivating surface layers, chemical interactions and charge-induced processes underpinned by ionic motion, charge injection, and electrochemistry leads to differences in jCPD and CPD. Figure 2.15c shows that the jCPD on amorphous $HfO_2$ extracted from panel (b) changes hysteretically as a function of the applied DC write voltage, indicating surface charge dynamics of physical, chemical, or electrochemical origin. The values of jCPD can be strongly time dependent based on life-time and mobility of surface and injected charges, as well as charge trapping and de-trapping that can be affected even by scanning the tip across the sample. As in contact EFM techniques, cKPFM allows for a higher spatial resolution than non-contact electrostatic methods and provides information on the jCPD. Moreover, cKPFM is often used to distinguish between genuine ferroelectric switching from signals of non-ferroelectric origin such as electrostatic interactions that are a common artifact in AFM-based piezoelectric electromechanical characterization.[220]

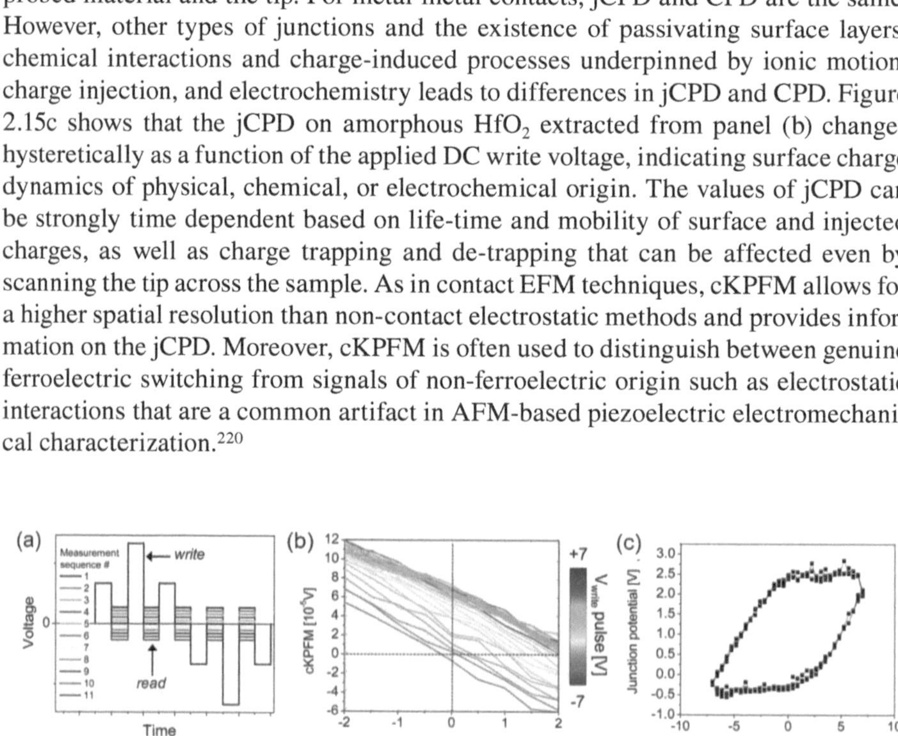


**FIGURE 2.15** (a) Sequence of read and write DC voltage pulses in cKPFM. (b) Electrostatic cKPFM signal measured on amorphous $HfO_2$. (c) Junction potential extracted from x-intercept in panel (b) for each DC write voltage. (Adapted from Balke et al.[220])

### 2.5.4.3 Pulsed Force KPFM

Unlike contact EFM and cKPFM techniques, where charge dynamics and electrostatic forces are probed while the tip is permanently in contact with the sample surface, pulsed force Kelvin probe force microscopy (PF-KPFM)[221] is conducted in pulsed force or peak force mode. In this method, the tip approaches the sample surface at each pixel until a certain maximum interaction force, as indicated by the cantilever deflection, is reached. This force is typically set to low values and the tip is subsequently retracted.[222,223] Approaching and retracing the cantilever is performed away from any mechanical cantilever resonances. PF-KPFM, developed by Jakob et al.,[221] utilizes Coulomb forces generated from intrinsic Fermi level differences instead of applying external voltages. These electrostatic forces are proportional to $V_{CPD}$ (see equation 2.2) and arise if the tip and sample are spatially separated and electrically connected. The electrical connection is mediated by a field-effect transistor situated between the tip and sample (Figure 2.16a). In the conductive state, spatially separated charges form between the tip and sample, which results in an electrostatic force that is monitored through cantilever oscillations. When the electrical connection between tip and sample is turned off as mediated by the transistor, the existing charges redistribute between the tip-sample capacitor and the residual capacitance of the field-effect transistor. Therefore, periodically switching the conductivity states between tip and sample ON and OFF causes a periodical electrostatic force and leads to oscillations in the deflection signal without the need for externally applied voltages (Figure 2.16b). Fast-Fourier transformation of the deflection signal yields characteristic peaks in the frequency domain and provides a measure for the Coulomb force and CPD. Once the tip and sample come into contact during each pulsed force/ peak force cycle, the system is restored as charges at the transistor are recombined and cantilever oscillations are damped. With this single-pass technique, a spatial resolution of <10 nm has been achieved in ambient conditions and can therefore be used to map boundaries between different materials or domains.

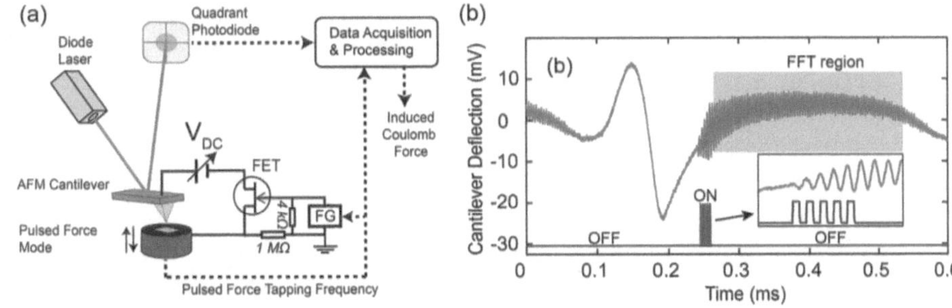


**FIGURE 2.16** (a) Experimental setup of PF-KPFM consisting of an AFM cantilever operated in pulsed force mode and an electronic circuit based on a field effect transistor that allows for conductivity between tip and sample to be switched ON or OFF. (b) Cantilever deflection trace (red curve) as a function of time indicating mechanical oscillations emanating from Coulomb forces between tip and sample. Blue curve shows pulsed ON and OFF states of the electrical connection. (Adapted from Jakob et al.[221])

### 2.5.5 Time-Resolved EFM/KPFM Methods

EFM and KPFM have been used to aid understanding of nanoscale electronic properties and the structure-property relationship in numerous energy-related materials and devices. However, the classical implementation of these methods utilizes the combination of homo/heterodyne detection, as well as feedback regulation in the case of KPFM, ultimately limiting the measurement time on the millisecond to second timescales. Therefore, EFM/KPFM can be used to measure the system dynamics if the time-dependent phenomena are sufficiently slow (>~1 ms), whereas faster dynamics will be masked by the measurement scheme. That said, both KPFM and EFM have in the past been combined with illumination pulses to monitor slow dynamic changes in photovoltage charging on the millisecond to minute timescales. Sadewasser et al.[224] demonstrated such an approach on a $CuGaSe_2$ semiconductor at a single location as a light source was switched ON and OFF slowly over the course of several minutes. Similar dynamic KPFM measurements after illumination have been realized on a variety of solar cell materials.[225,226] Besides optical excitation, EFM/KPFM combined with bias pulsing across lateral electrodes has become a promising approach to monitor charge transport in a variety of materials/devices including organic field-effect transistors (OFETs),[227–229] conjugated polyelectrolytes,[230] and hybrid organic-inorganic perovskites.[230–232] Ng et al. demonstrated the possibility of using EFM in a time-resolved fashion to quantify the kinetics of trap formation in OFETs.[227]

While the above methods involve a simple application of EFM/KPFM, others have sought to improve the time resolution and informational content, which can be limited by the need for bias feedback. Strelcov et al.[193] developed an OL, time-resolved KPFM method that allowed mapping the charge dynamics as a function of voltage, time, and space. This method has been successfully applied on memristive[192] and ferroelectric[194] samples revealing the details of polarization switching as well as being used for probing electrochemical reactions on nanostructured ceria.[233] Recently, Collins et al. improved on this method by adopting a DH detection scheme enabling the capture of multidimensional bias and time-resolved information.[234] The working principle behind hyperspectral KPFM is shown in Figure 2.17a. Figure 2.17b and c demonstrates its applicability to investigate transport of mobile surface charges across a three-unit-cell $LaAlO_3/SrTiO_3$ oxide heterostructure[234] as a function of the applied device bias (Figure 2.17b) or redistribution of charge as a function of time after switching OFF the device bias (Figure 2.17c).

In most of these applications, the accessible time-dependent processes are on the order of a fraction of a millisecond or greater. While such information is useful for investigating relatively slow dynamics, for many applications of interest knowledge of the underlying fast dynamics/kinetics of the electronic or ionic processes is of critical importance. This deficit has driven the development of fast time-resolved modes of operation that extend the time resolution to even faster timescales.

Initially, intensity-modulated KPFM was developed to capture faster kinetics than the Kelvin bias feedback could track in existing light-assisted KPFM measurements.[158] This method is described in detail elsewhere.[16] The approach can be used to capture a temporally averaged photovoltage spectrum as a function of the modulation frequency of the illumination.[158] Operated in a grid spectroscopy

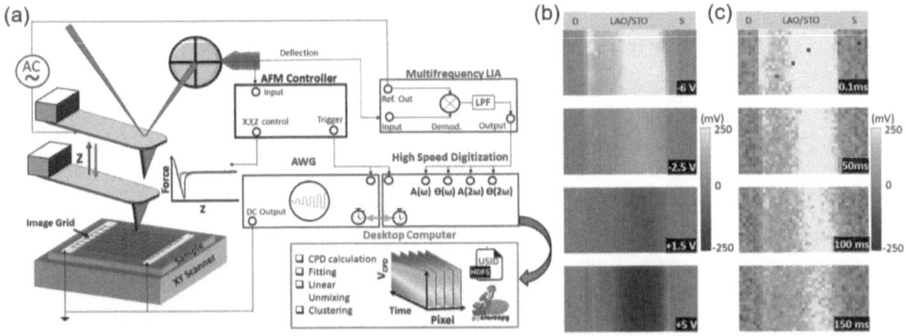


**FIGURE 2.17** Schematic describing the working operation of hyperspectral Kelvin probe force microscopy (HS-KPFM). HS-KPFM uses a grid-based imaging approach to precisely position the tip above the sample surface. Only AC voltages are applied to the probe, while DC biasing of the device is synchronized using trigger pulses and an arbitrary wave generator (AWG). High-speed digitization and storage of the demodulator channels allows CPD calculation (offline) using dual harmonic detection scheme described previously. (b,c) HS-KPFM of an LAO/STO device with an electrode separation of 50 µm (d-drain, S-Source). (b) Time-averaged $\Delta V_{CPD}$ data after switching OFF biases to the device (depolarization). (c) $\Delta V_{CPD}$ for series of time steps directly after (depolarization) by −6V, showing clear redistribution of charge within the channel gap. (Adapted from et al.[234])

mode, intensity-modulated KPFM has been used to visualize the carrier lifetime in polycrystalline Si solar cells[158] as well as organic donor/acceptor blends for solar applications.[235,236]

A major breakthrough in the capture of fast temporal dynamics for photoresponsive materials was the development of time-resolved EFM (trEFM) by the group of Ginger.[237] The basic operation of trEFM is shown in Figure 2.18a and b. In this seminal work, upon illumination of a nanostructured donor/acceptor interfaces of 50:50 blend films of poly-(9,9-dioctylfluorene-co-benzothiadiazole) (F8BT) and poly-(9,9-dioctylfluorene-co-bis-N, N-(4-butylphenyl)-bis-N, N-phenyl-1,4-phenylenediamine) (PFB), charge carriers in the photovoltaic material are generated by the photoelectric effect. The accumulation of photo-generated charge carriers results in changes of the capacitance and electrostatic force gradient, in turn causing a resonance frequency shift according to equation (2.3). Fitting the gradual change in the resonance frequency shift to a single exponent allows extraction of the charging rate, which is attributed to the charge dissociation in the tip-sample capacitor forming a potential well created by the tip bias. It was shown that the charging rate was linearly dependent on light intensity (Figure 2.18c). From the maps of charging rate (Figure 2.18e), it was discovered that the interface between the donor/acceptor materials showed that the charging rates were reduced by 30%–50%. Further, the charging rates measured by trEFM correlated well with the macroscopic external quantum efficiencies of the films.

In its initial implementation, trEFM was reported to have a spatial/time resolution of ~100 nm/~100 µs.[237] The maximum achievable temporal resolution was limited by the signal/noise stability limit of frequency shift feedback loop. In recent works

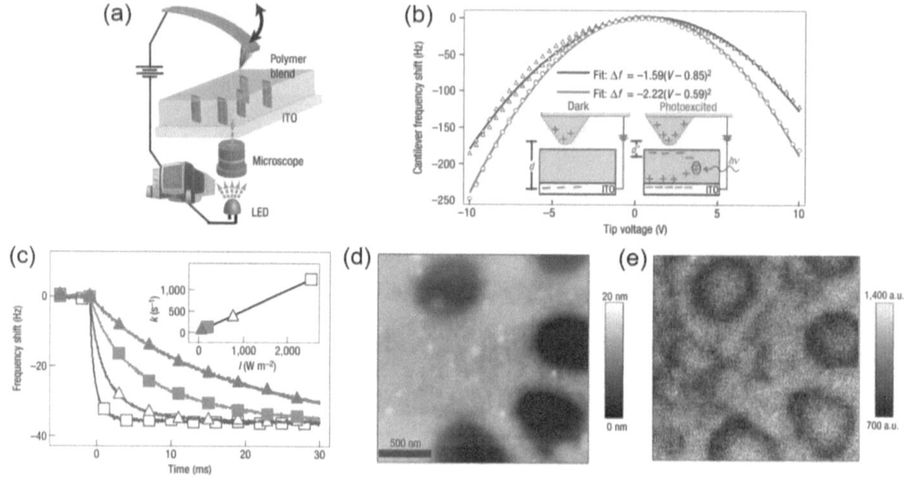


**FIGURE 2.18** (a) Schematic diagram of trEFM measurement setup involving bottom-up illumination of the sample by a pulsed light-emitting diode (LED). Charge accumulation results in a change in the frequency shift ($\Delta f$) experienced by the bias cantilever. (b) Bias spectroscopy curves in dark (triangles) and under illumination (circles) with 405 nm LED on a 50:50 F8BT/PFB blend film. Shifts in parabola maxima indicated charge separation under illumination (inset). (c) Normalized charge build-up during illumination of the F8BT/PFB film with different light intensities leading to a linear dependence of the charging rate on light intensity (inset). (d) AFM height image of a 50:50 film of F8BT/PFB and (e) photo-induced charging rate map generated by plotting the inverse exponential time constant at each point. Dark rings indicate regions of slower charging. (Reproduced from Coffey and Ginger.[237])

via the same group,[238,239] the authors overcame the necessity for frequency feedback by adopting a big data capture approach in which the full cantilever oscillation is digitized at up to 50 MHz. In fast-free trEFM,[219] the cantilever oscillation is digitized while a light pulse is triggered to initiate the local dynamics of interest. A triggering circuit is employed to phase lock the trigger events to the cantilever motion so that the trigger always occurs at the same point in the cantilever oscillation cycle, thereby improving the efficacy of signal averaging and thus the time resolution. In a post-processing step, the cantilever motion is analyzed using numerical demodulation of the digitized cantilever signal. The instantaneous phase (and instantaneous frequency) is then extracted from the demodulated signal via a Hilbert transform of the cantilever position vs. time data. Further improvements on this method were reported by controlling the cantilever phase at the perturbation time as well as incorporation of photothermal excitation, making it possible to differentiate rise times down to 10 ns.[239] More recently, Dwyer and co-workers implemented a related and improved technique called phase kick EFM.[240]

Collins et al. developed a different big data capture approach for time-resolved KPFM called general acquisition mode (G-Mode) KPFM. In a similar way to fast-free trEFM, G-Mode KPFM works by capturing, storing, and compressing the AFM photodetector signal at the sampling rate limit (~4–10 MHz). Combining full data acquisition with adaptive filtering and machine learning-based denoising methods (e.g., principal component analysis (PCA)) has been shown to be an effective method for recovering a densely sampled record of the dynamic cantilever trajectory. Once the real-time cantilever dynamics have been filtered, the CPD can be recovered in an offline analysis using digital LIAs[64] or by direct force conversion.[30] Recently, both G-Mode KPFM and fast-free trEFM were used to study light-induced dynamics in thin films comprising Ruddlesden–Popper phases of the layered 2D perovskite, $(C_4H_9NH_3)_2PbI_4$. Both methods provided complementary information, which was used to unveil a slower evolution at grain boundaries due to a combination of ion migration occurring between $PbI_4$ planes, as well as electronic carriers traversing grain boundary traps, thereby changing the time-dependent band unbending at grain boundaries.[241]

Recently, a bias-based pump–probe approach (pp-KPFM) was developed.[242,243] In pp-KPFM, short probing pulses are modulated by a slower sinusoidal envelope and synchronized to electric pump pulses applied to the device under investigation. A second feedback loop is used to minimize the average electrostatic force, providing a measure of the time-averaged potential distribution as in classical KPFM. This scheme has been used to measure the charge-carrier dynamics with a time resolution of 2 μs in pentacene-based OFETs.[242,243] However, a time resolution of ~1 ps has been demonstrated using optical pp-KPFM in UHV applied to measure the charge-carrier lifetime in GaAs grown at low temperatures.[244] A more detailed review of these fast time-resolved methods and others can be found elsewhere.[16,17]

## 2.6 APPLICATIONS AT THE SOLID-LIQUID INTERFACE

As we have highlighted throughout this chapter, the electrostatic actuation of a conductive AFM probe is the base of operation of KPFM and EFM, but such an excitation strategy in liquid environment triggers plenty of complex physicochemical phenomena (electrochemical surface stress, ionic migration, chemical reactions, etc.), which have classically restricted their applicability to dry environments. However, often different energy-related processes in batteries, ionic liquids, or fuel cells will occur in liquid environments. Hence, pushing forward toward KPFM/EFM applicability to liquid environments is of crucial importance, especially to map electrochemical processes at the solid-liquid interface, complementing the already well-established in situ AFM measurements of *operando* liquid devices[18–20] and current-sensing techniques such as SECM.[21,22] In this section, we will give an overview of the last technical advances regarding KPFM and EFM applications in wet contexts and their main applications.

### 2.6.1 Measuring Electrostatic Forces with SPM at the Solid-Liquid Interface

The first attempts of to correlate surface charge with measured forces at the solid-liquid interface consisted of measuring the force acting on a chemically charged insulating probe (usually with a colloidal tip attached) that could be used to distinguish the nature of the different phenomena (electrostatic forces, van der Waals, osmotic pressure, etc.) occurring at the solid-liquid interface.[245,246] Such studies were carried out at different salt concentrations and pH conditions, finding them to be in agreement with Derjaguin, Landau, Verwey, and Overbeek (DLVO) theory.[247] Following this approach, the charge density and Debye length of different systems could be measured,[248] as well as the surface charge in DNA[249] or bacteriorhodopsin membranes.[250]

The possibility of using voltage-modulated approaches opens the door to the more interesting case of study of dynamic and frequency-dependent electrostatic phenomena, or electrified interfaces. Nevertheless, such voltage-modulated force-detection based SPM electrical modes (like KPFM and EFM) suffer from ionic charge screening when operated in polar electrolyte solutions. In a nutshell, when the electric field is applied to the probe, the ions in the electrolyte migrate toward the electrode's surface, following the field and forming the so-called electric double layer (EDL). Such charge accumulation at the solid-liquid interface can induce electrochemical surface stress[251] or trigger chemical reactions, especially when low-frequency (<~1 kHz) and high voltages (>~1 V) are applied and activate a complicated force-frequency response.[252,253] As a result, the probe-sample potential drop that KPFM and EFM use to map the electrostatic force occurs (in liquid environment) only in a small region at the close vicinity of the electrode's surface along the characteristic Debye length, $\lambda_D = \sqrt{\frac{k_B T \varepsilon_0 \varepsilon_r}{2 n_0 e^2}}$, which depends mainly on the electrolyte molarity, $n_0$. Therefore, dynamics are governed by the fast characteristic times of the EDL charging ($\tau_{EDL} \approx$ ns-µs) so that any electrostatic force occurring between two bodies in an electrolyte solution at distances further than $\lambda_D$ and within times slower than $\tau_{EDL}$ will be screened.[257,258]

Throughout the last decade, different strategies have been explored to shortcut the charge screening effect necessary to apply voltage-modulated electrostatic force-based SPM measurements (KPFM and EFM) at the solid-liquid interface, as listed below.

1. KPFM in non-polar solvents,[254,258,266] which entirely avoids complications associated with charge screening from mobile ions and the finite conductivity of polar solvents.
2. KPFM/EFM of encapsulated liquids,[255,84] which takes advantage of the electric field penetration inside the sample, setting up the experiment with the tip in air/UHV (avoiding the problems related to the tip-electrolyte interface) but performing the measurements over thin micro/nano-channels that are filled with the liquid of interest and are either covered/capped, being able to map either the dielectric constant of encapsulated liquids or the double layer interfacial potentials at the solid-liquid interface.
3. Electrochemical force microscopy (EcFM),[256,257] which monitors the transient behavior during the formation of the EDL, and resolves the different dynamic characteristic processes at the solid-liquid interface (EDL charging, ionic diffusion, and bulk relaxation).

4. Dual harmonic (DH-KPFM),[196] which takes advantage of positioning the tip at a distance shorter than $\lambda_D$ with respect to the sample. In this situation, the overlapping of the EDLs gives rise to a measurable force, which allows for the calculation of local contact potentials using a similar setup to the one in air. However, such studies are normally implemented in extremely low molar concentrations (e.g., ultrapure Milli-Q water), where the characteristic length scale of the system is still quite large ($\lambda_D \approx 500$nm), which makes it experimentally possible.
5. In-liquid EFM[62,71,258–262] and dual-frequency mode open-loop electric potential microscopy (DF-OL-EPM),[197,198] which explore the application of high-frequency voltage waves (higher than the characteristic relaxation frequency of the electrolyte, ($f_{RC} = \frac{\sigma}{2\pi\varepsilon_0\varepsilon_r}$), to avoid EDL formation. This extends the applicability of such techniques to larger molarities (~10 mM), as for such cases the relaxation frequency is found to be in the MHz range. Under these conditions, the migration of the ionic species does not take place; therefore, the EDL dynamics are avoided and the derivative of the capacitance (dC/dz) and/or the local potential of the system can be measured.

### 2.6.2 Applications of EFM in Liquid

Among the previously introduced approaches for measuring electrostatic forces in electrolyte solutions, much progress has been made in the direction of using high excitation frequencies (>1 MHz), especially interesting for organic, biological, and electrochemical applications. Under this configuration, the conductive contributions of the ions are neglectable (or at least not dominant) and the system behaves like a pure dielectric (or a dielectric with losses). Such an approach allows to have a significant voltage drop between the probe and the sample (different to what happens in the lower frequency range), but at the same time represents a challenge for the detection scheme as the MHz range is well beyond the mechanical resonance of standard commercially available cantilevers. Therefore, the experimental setup needs to be adapted to detect the high-frequency signal. This can be done by applying a high-frequency signal, $\omega_{el}$, whose amplitude is modulated by a low frequency, $\omega_{\text{mod}}$, in the kHz range:

$$V(t) = \frac{V_{\text{AC}}}{2}\left(1+\cos\left(\omega_{\text{mod}}t\right)\right)\cos\left(\omega_{el}t\right) \tag{2.11}$$

Such heterodyne scheme allows for the use of $\omega_{\text{mod}}$ for force detection but does not introduce any low-frequency component in the electric field between the probe and the sample, so it does not "activate back" the ionic migration.[71,260,261] As previously explained, this methodology has also been applied in air conditions, whose advantage, in this case, is to enlarge the frequency bandwidth of EFM up to the GHz range.[102]

Regarding the quantification of EFM in a liquid environment, modeling needs to be performed which takes into account the additional voltage drops the probe-electrolyte and sample-electrolyte interfaces induced by the presence of Stern layers or other adsorbates. Those can be represented by additional interfacial capacitances that are a priori unknown and thus must be left free as an additional fitting parameter during the quantification or calibrated somehow.[261,262] In addition, depending on the frequency and molarity used, conductivity effects can also play a role and should also be considered, as it is possible to have a mixed resistive/capacitive response.[62,71,260,261]

In terms of applications, such an approach has allowed mapping the capacitance of ultrathin (<1 nm) organic SAMs at the metal/electrolyte interface reducing drastically the spatial resolution of other (current-based) techniques, reaching sub-100nm spatial resolution.[261] Furthermore, modeling of the technique allows quantification of the charge accumulation at the SAM interface as a function of the number of molecular carbons forming the SAM (see Figure 2.19), which we envision will have tremendous application potential for mapping electrode interfaces in batteries and supercapacitors.

Not only that, but several other applications have been recently inspected in biological samples, such as the study of biological membranes,[259] where it was found that the inclusion of cholesterol in the membrane-induced dielectric changes[262] the obtention of internal information of encapsulated liposomes[263] such as the number of lamella they contain or the dielectric mapping eukaryotic cells in low molarity media.[93]

Finally, and interestingly for energy research fields, the same approach has been successfully applied to mapping the conductivity and interfacial capacitance of an electrolyte-gated organic field-effect transistor (EGOFET) under operation,[62] whose

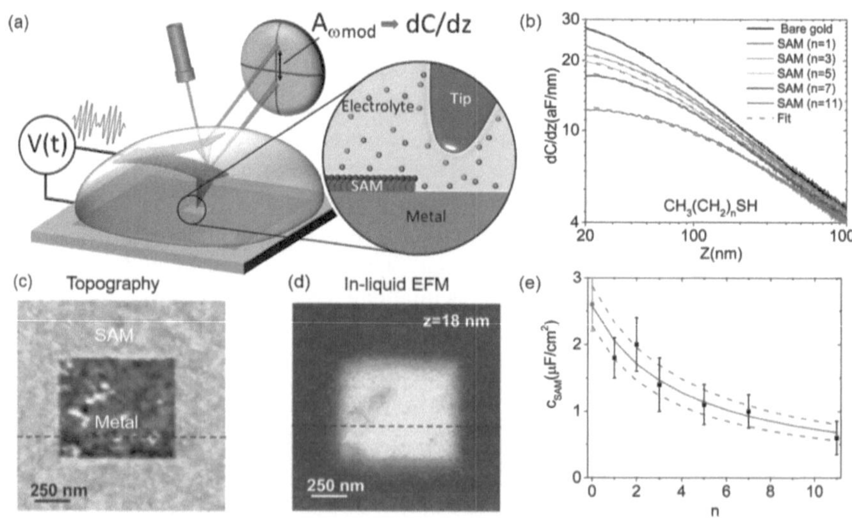


**FIGURE 2.19** (a) Schematics of experimental setup of in-liquid EFM. (b) Capacitance gradient force-distance curves for SAMs with different number of carbons, highlighting the technique resolution to molecular changes at the solid-liquid interface. (c) Topography and (d) in-liquid EFM images of the ultrathin (<1nm) SAM with sub-100nm spatial resolution. (e) Quantified interfacial capacitance of the monolayer as a function of the number of carbons in the SAM. (Adapted from Millan-Solsona et al.[261])

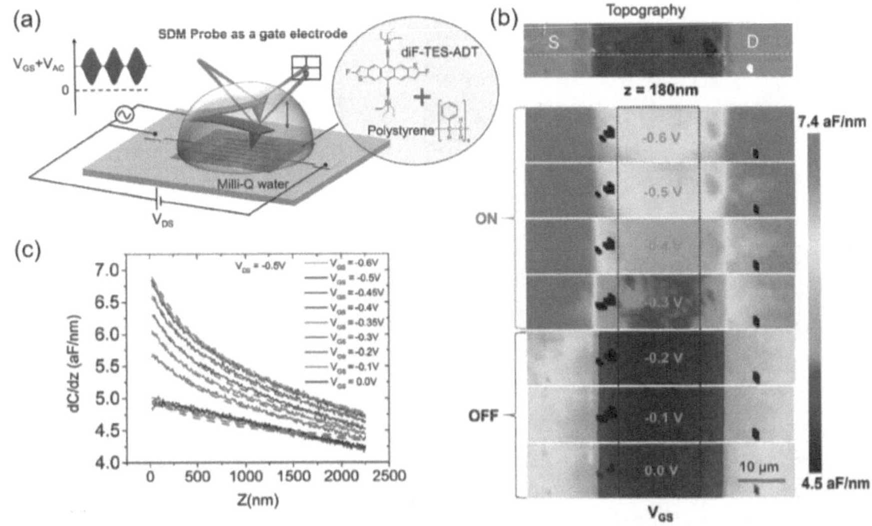


**FIGURE 2.20** (a) Schematics of experimental setup of in-liquid EFM for conductivity and interfacial capacitance mapping of the operando EGOFET channel. (b) AFM topography of the EGOFET channel (top) and in-liquid EFM images of EGOFET channel for different gating voltages (bottom). (c) Capacitance gradient force-distance curves acquired on the channel of the transistor for different gating voltages. (Adapted from Kyndiah et al.[62])

macroscopic ON-OFF mechanism depends on the nanoscale EDL formation at the channel-electrolyte interface. In this case, the EFM tip is used simultaneously as a gating electrode and as a probe (see Figure 2.20), which can correlate the global transistor response with the channel's nanometric conductivity and interfacial capacitance, which is intrinsically linked to the heterogeneous organic channel morphology.

### 2.6.3 Applications of KPFM in Liquid

KPFM has been widely used to map nanoscale surface potentials of materials in ambient and UHV environments. However, to study and ultimately understand charge-related processes, e.g., to further improve energy storage devices such as electrochemical batteries, supercapacitors, or electrocatalysts, nanoscale surface potential measurements in liquid environments are required. The recent efforts involving extending KPFM to liquids have been reviewed in depth elsewhere.[118,264] The first major breakthrough came when Domanski et al.[254] demonstrated the possibility of operating classical closed-loop KPFM in electrically insulating nonpolar solvents, in this case, decane. They studied changes in the work function of gold substrates upon hexadecanethiol chemisorption (Figure 2.21a). The authors achieved a quantitative understanding of the adsorption process, finding good agreement with UV photoelectron spectroscopy measurements in UHV. In a subsequent work, Umeda et al. reported surface potential measurements of a p-n patterned silicon sample imaged in a fluorocarbon liquid.[265] These measurements represented the first examples of KPFM performed in liquids and built on prior work carried out over the previous

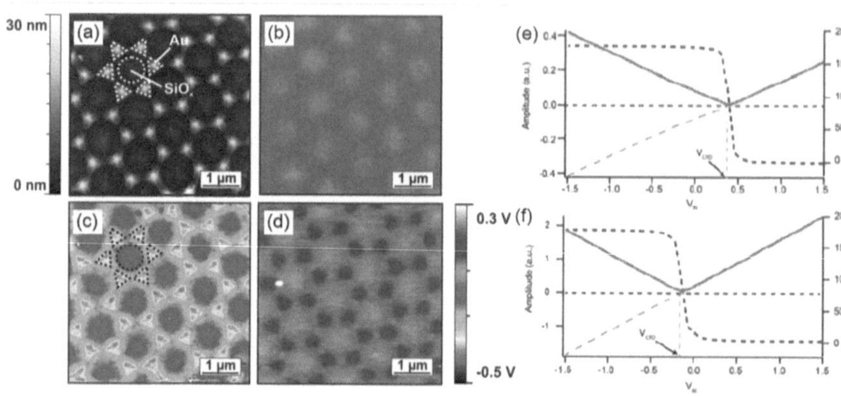

FIGURE 2.21 (a) Topography image of an SiOx/Au structure prepared by nanosphere lithography. KPFM images of the CPD measured in (b) air and (c,d) decane. (c) Before and (d) after exposure to hexadecanethiol for 90 minutes. (Reproduced from Domanski et al.[254]) KPFS showing the oscillation amplitude (red line), phase (dashed blue line), and in-phase response (dashed grey line) in (e) ambient and (f) decane. KPFS was performed using a specified tip-sample separation of 200 nm, and voltage amplitude of 1 V with a frequency of 20 kHz applied to the tip. (Adapted from Collins et al.[264].)

four decades when Fort et al. investigated the feasibility of operating macroscopic Kelvin probe in a variety of polar and non-polar solvents.[266] In that study, the authors noted that Kelvin probe was not possible in partially conducting or non-polar liquids.

Kobayashi et al.[196] were the first to demonstrate quantitative surface potential mapping in polar liquid using OL-KPFM, described here as DH-KPFM (see Section 2.5.1), which they referred to as open-loop electric potential (OLEP) microscopy. In its first liquid implementation, OLEP was operated in conjunction with FM-AFM[196] (later with AM-AFM[200]) as topographic feedback while the two electrical response signals were detected using LIAs. The authors noted that by operating at a relatively high drive frequency (30 kHz), it was possible to avoid spurious forces, such as bias-induced surface stress or electrochemical reactions. They demonstrated that the ability to map local variations in the potential distribution at a solid-liquid interface could be imaged in a weak electrolyte solution (<3 mM). Potential contrast between a dodecylamine thin film and the HOPG substrate in 1 mM NaCl solution[196] or between latex beads with different charges and sizes were measured.[200] These results indicate that, at low molarities, OLEP/DH-KPFM can be used for quantitative, or at least semi-quantitative potential measurements.

Collins et al.[199] successfully monitored the change in work function of a graphene on copper sample going from ambient to liquid environment. As shown in Figure 2.22, using DH-KPFM it was possible to demonstrate that the work function of the copper substrate changed upon immersion in deionized water while the graphene surface remained unchanged, which could indicate that graphene was acting as an effective corrosion inhibitor for the copper substrate.[199]

A limiting factor for DH-KPFM was the requirement to apply the electrical excitation below the natural resonance frequency of the cantilever to account for the cantilever transfer function gain (see Section 2.5.1). Kobayashi et al. developed a

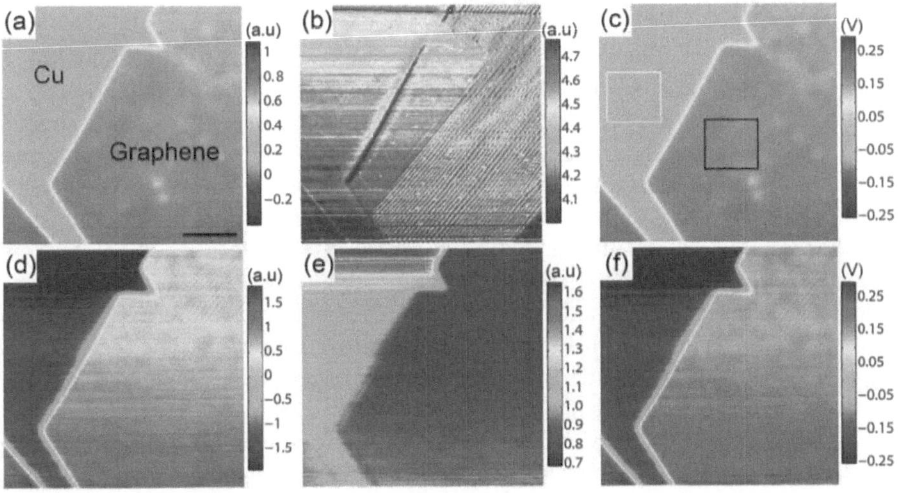


**FIGURE 2.22** DH-KPFM images of single-layer graphene on copper foil in (a–c) air and (d–f) deionized water using a lift height of 50 nm (scale bar = 5 μm). (a, d) First and (b,e) second harmonic images and (c, f ) CPD. (Reproduced from Collins et al.[199])

similar method that they called DF-OLEP-EFM.[267] In this method, a heterodyne detection scheme is adopted, allowing electrical excitation at higher frequencies and where the mixing products are used to determine the CPD. This method was used to map potential variations of nanoparticles on a graphite surface in 1 mM and 10 mM NaCl solutions. This method was also used to investigate the nanoscale corrosion behavior of fine Cu wires and duplex stainless steel in situ.[198] Temporal variations in consecutive potential images exhibited nanoscale dynamics, thus allowing real-time identification of local corrosion sites even when the surface topography remained largely unchanged (see Figure 2.23). Such approaches will likely be useful for investigating reactions under surface oxide layers or highly corrosion-resistant materials.

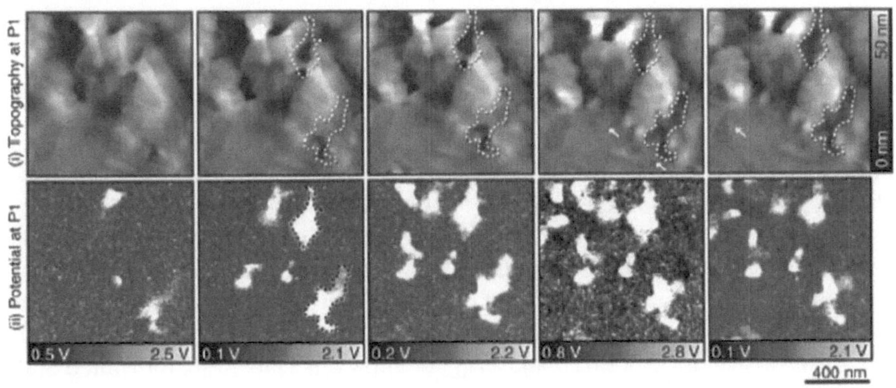


**FIGURE 2.23** Temporal variation in surface potential of corrosion sites on duplex steel using DF-OLEP-KPFM. (Reproduced from Honbo et al.[198])

## 2.7 CONCLUSIONS AND FUTURE PERSPECTIVE

EFM and KPFM methods have contributed significantly to the understanding of nanoscale electronic properties in numerous fields of research and hold promise for understanding structure-property relationships in energy generation and storage in optoelectronic materials and device systems. The key for the continued improvement and future optimization of energy material/device performance is exploring local charge dynamics under relevant conditions. For example, the performance of batteries is based on, and ultimately limited by, the rate and localization of ion flows through the device on different length scales ranging from atoms to grains to interfaces. Therefore, further advancement and application of time-resolved KPFM methods will require acquiring information across multiple length and time scales for the improvement of current and development of future battery technologies among other energy applications.

With increased information flowing from time-resolved or multidimensional EFM/KPFM methods, machine learning and artificial intelligence approaches should be leveraged to quickly and easily discern regions of interest, denoise experimental signals, and perform fast and accurate predictions of material properties.[30,268,269] Such modern and data-driven analysis tools will also be useful for linking functional properties captured from multimodal or correlated approaches, which could remain hidden using classical analysis tools. Indeed, energy applications will greatly benefit from multimodal frameworks that can be used to help correlate electronic information from KPFM to other structural or chemical information obtained by, e.g., time-of-flight secondary ion microscopy (TOF-SIMS)[270] or electron backscatter diffraction.[271]

Finally, liquid KPFM and EFM methods provide opportunities to explore energy materials under more relevant conditions. Further progress will depend on the development of specialized probes (e.g., insulated probes[272,273] and co-axial probes[274]) and drive schemes that will be necessary to extend the types of electrolytes within which measurements can be performed. Equally, multidimensional time-resolved methods offer the possibility to go beyond work function mapping, allowing the exploration of local EDL dynamics including charge screening, ion diffusion, and ultimately, reactivity among a myriad of phenomena critical to energy research.[256,257]

## ACKNOWLEDGMENTS

We would like to thank Karren L. More for the careful proofreading of the manuscript. This work was conducted at the Center for Nanophase Materials Sciences, which is a DOE Office of Science User Facility. This manuscript has been authored by UT-Battelle, LLC under Contract No. DE-AC05-00OR22725 with the U.S. Department of Energy. The United States Government retains and the publisher, by accepting the article for publication, acknowledges that the United States Government retains a non-exclusive, paid-up, irrevocable, world-wide license to publish or reproduce the published form of this manuscript, or allow others to do so, for United States Government purposes. The Department of Energy will provide public access to these results of federally sponsored research in accordance with the DOE Public Access Plan (http://energy.gov/downloads/doe-public-access-plan).

## REFERENCES


1. Binnig, G.; Quate, C. F.; Gerber, C., Atomic force microscope. *Physical Review Letters* 1986, *56* (9), 930.
2. Weaver, J.; Abraham, D. W., High resolution atomic force microscopy potentiometry. *Journal of Vacuum Science & Technology B: Microelectronics and Nanometer Structures Processing, Measurement, and Phenomena* 1991, *9* (3), 1559–1561.
3. Martin, Y.; Abraham, D. W.; Wickramasinghe, H. K., High-resolution capacitance measurement and potentiometry by force microscopy. *Applied Physics Letters* 1988, *52* (13), 1103–1105.
4. Nonnenmacher, M.; o'Boyle, M.; Wickramasinghe, H. K., Kelvin probe force microscopy. *Applied Physics Letters* 1991, *58* (25), 2921–2923.
5. Rahe, P.; Söngen, H., Imaging static charge distributions: A comprehensive KPFM theory. In Sadewasser, S., Glatzel, T. (eds.) *Kelvin Probe Force Microscopy*, Springer, Cham: 2018; pp. 147–170.
6. Söngen, H.; Rahe, P.; Bechstein, R.; Kühnle, A., Interpretation of KPFM data with the weight function for charges. In Sadewasser, S., Glatzel, T. (eds.) *Kelvin Probe Force Microscopy*, Springer, Cham: 2018; pp. 171–200.
7. Rosenwaks, Y.; Saraf, S.; Tal, O.; Schwarzman, A.; Glatzel, T.; Lux-Steiner, M. C., Kelvin probe force microscopy of semiconductors. In Kalinin, S., Gruverman, A. (eds.) *Scanning Probe Microscopy*, Springer, New York: 2007; pp. 663–689.
8. Sommerhalter, C.; Matthes, T. W.; Glatzel, T.; Jäger-Waldau, A.; Lux-Steiner, M. C., High-sensitivity quantitative Kelvin probe microscopy by noncontact ultra-high-vacuum atomic force microscopy. *Applied Physics Letters* 1999, *75* (2), 286–288.
9. Melitz, W.; Shen, J.; Kummel, A. C.; Lee, S., Kelvin probe force microscopy and its application. *Surface Science Reports* 2011, *66* (1), 1–27.
10. Rodriguez, B.; Kalinin, S. V., KPFM and PFM of biological systems. In Sadewasser, S., Glatzel, T. (eds.) *Kelvin Probe Force Microscopy*, Springer, Cham: 2012; pp. 243–287.
11. Sinensky, A. K.; Belcher, A. M., Label-free and high-resolution protein/DNA nanoarray analysis using Kelvin probe force microscopy. *Nature Nanotechnology* 2007, *2* (10), 653.
12. Shi, Y.; Collins, L.; Feng, R.; Zhang, C.; Balke, N.; Liaw, P. K.; Yang, B., Homogenization of AlxCoCrFeNi high-entropy alloys with improved corrosion resistance. *Corrosion Science* 2018, *133*, 120–131.
13. Rohwerder, M.; Turcu, F., High-resolution Kelvin probe microscopy in corrosion science: Scanning Kelvin probe force microscopy (SKPFM) versus classical scanning Kelvin probe (SKP). *Electrochimica Acta* 2007, *53* (2), 290–299.
14. Schmutz, P.; Frankel, G., Characterization of AA2024-T3 by scanning Kelvin probe force microscopy. *Journal of the Electrochemical Society* 1998, *145* (7), 2285.
15. Shvebelman, M. M.; Agronin, A. G.; Urenski, R. P.; Rosenwaks, Y.; Rosenman, G. I., Kelvin probe force microscopy of periodic ferroelectric domain structure in KTiOPO4 crystals. *Nano Letters* 2002, *2* (5), 455–458.
16. Sadewasser, S.; Nicoara, N., Time-resolved electrostatic and kelvin probe force microscopy. In Sadewasser, S., Glatzel, T. (eds.) *Kelvin Probe Force Microscopy*, Springer, Cham: 2018; pp. 119–143.
17. Mascaro, A.; Miyahara, Y.; Enright, T.; Dagdeviren, O. E.; Grütter, P., Review of time-resolved non-contact electrostatic force microscopy techniques with applications to ionic transport measurements. *Beilstein Journal of Nanotechnology* 2019, *10* (1), 617–633.
18. Breitung, B.; Baumann, P.; Sommer, H.; Janek, J.; Brezesinski, T., In situ and operando atomic force microscopy of high-capacity nano-silicon based electrodes for lithium-ion batteries. *Nanoscale* 2016, *8*, 14048–14056.

19. Virwani, K.; Ansari, Y.; Nguyen, K.; Moreno-Ortiz, F. J. A.; Kim, J.; Giammona, M. J.; Kim, H. C.; La, Y. H., In situ AFM visualization of Li-$O_2$ battery discharge products during redox cycling in an atmospherically controlled sample cell. *Beilstein Journal of Nanotechnology* 2019, *10*, 930–940.
20. Wen, R.; Hong, M.; Byon, H. R., In situ AFM imaging of Li-$O_2$ electrochemical reaction on highly oriented pyrolytic graphite with ether-based electrolyte. *Journal of the American Chemical Society* 2013, *135*, 10870–10876.
21. Mahankali, K.; Thangavel, N. K.; Reddy Arava, L. M., In situ electrochemical mapping of lithium-sulfur battery interfaces using AFM-SECM. *Nano Letters* 2019, *19* (8), 5229–5236.
22. Ventosa, E.; Schuhmann, W., Scanning electrochemical microscopy of Li-ion batteries. *Physical Chemistry Chemical Physics* 2015, *17* (43), 28441–50.
23. Stern, J.; Terris, B.; Mamin, H.; Rugar, D., Deposition and imaging of localized charge on insulator surfaces using a force microscope. *Applied Physics Letters* 1988, *53* (26), 2717–2719.
24. Terris, B.; Stern, J.; Rugar, D.; Mamin, H., Contact electrification using force microscopy. *Physical Review Letters* 1989, *63* (24), 2669.
25. Binnig, G.; Rohrer, H.; Gerber, C.; Weibel, E., Surface studies by scanning tunneling microscopy. *Physical Review Letters* 1982, *49* (1), 57.
26. Parker Jr., J. H.; Warren, R. W., Kelvin device to scan large areas for variations in contact potential. *Review of Scientific Instruments* 1962, *33* (9), 948–950.
27. Mäckel, R.; Baumgärtner, H.; Ren, J., The scanning Kelvin microscope. *Review of Scientific Instruments* 1993, *64* (3), 694–699.
28. Sadewasser, S.; Lux-Steiner, M. C., Correct height measurement in noncontact atomic force microscopy. *Physical Review Letters* 2003, *91* (26), 266101.
29. Ziegler, D.; Rychen, J.; Naujoks, N.; Stemmer, A., Compensating electrostatic forces by single-scan Kelvin probe force microscopy. *Nanotechnology* 2007, *18* (22), 225505.
30. Collins, L.; Ahmadi, M.; Wu, T.; Hu, B.; Kalinin, S. V.; Jesse, S., Breaking the time barrier in kelvin probe force microscopy: Fast free force reconstruction using the G-mode platform. *ACS Nano* 2017, *11*, 8717–8729.
31. Collins, L.; Belianinov, A.; Somnath, S.; Balke, N.; Kalinin, S. V.; Jesse, S., Full data acquisition in Kelvin Probe Force Microscopy: Mapping dynamic electric phenomena in real space. *Scientific Reports* 2016, *6*, 1–11.
32. Gomila, G.; Toset, J.; Fumagalli, L., Nanoscale capacitance microscopy of thin dielectric films. *Journal of Applied Physics* 2008, *104*, 024315.
33. Girard, P., Electrostatic force microscopy: Principles and some applications to semiconductors. *Nanotechnology* 2001, *12*, 485–490.
34. Fumagalli, L.; Gomila, G., Probing dielectric constant at the nanoscale with scanning probe microscopy. In Li, J. V., Ferrari, G. (eds.) *Capacitance Spectroscopy of Semiconductors*, Pan Stanford Publishing, Singapore: 2018.
35. Tanimoto, M.; Vatel, O., Kelvin probe force microscopy for characterization of semiconductor devices and processes. *Journal of Vacuum Science and Technology B* 1996, *14* (2), 1547–1551.
36. Fujihira, M., Structural study of Langmuir–Blodgett films by scanning surface potential microscopy. *Journal of Vacuum Science & Technology B: Microelectronics and Nanometer Structures* 1994, *12* (3), 1604–1604.
37. Fujihira, M., Kelvin probe force microscopy of molecular surfaces. *Annual Reviews in Materials Science* 1999, *29* (1), 353–380.
38. Cuervo, A.; Dans, P. D.; Carrascosa, J. L.; Orozco, M.; Gomila, G.; Fumagalli, L., Direct measurement of the dielectric polarization properties of DNA. *Proceedings of the National Academy of Sciences of the United States of America* 2014, *111* (35), E3624–30.

39. Fumagalli, L.; Esteban-Ferrer, D.; Cuervo, A.; Carrascosa, J. L.; Gomila, G., Label-free identification of single dielectric nanoparticles and viruses with ultraweak polarization forces. *Nature Materials* 2012, *11*, 808–816.
40. Lozano, H.; Fabregas, R.; Blanco-cabra, N.; Millán-solsona, R.; Torrents, E.; Fumagalli, L.; Gomila, G., Dielectric constant of flagellin proteins measured by scanning dielectric microscopy. *Nanoscale* 2018, 19188–19194.
41. Mikamo-Satoh, E.; Yamada, F.; Takagi, A.; Matsumoto, T.; Kawai, T., Electrostatic force microscopy: Imaging DNA and protein polarizations one by one. *Nanotechnology* 2009, *20* (14), 145102.
42. Checa, M.; Millan-solsona, R.; Blanco, N.; Torrents, E.; Fabregas, R.; Gomila, G., Mapping the dielectric constant of a single bacterial cell at the nanoscale with scanning dielectric force volume microscopy. *Nanoscale* 2019, *11*, 20809–20819.
43. Esteban-Ferrer, D.; Edwards, M. A.; Fumagalli, L.; Juarez, A.; Gomila, G., Electric polarization properties of single bacteria measured with electrostatic force microscopy. *ACS Nano* 2014, *8*, 9843–9849.
44. Checa, M.; Millan-Solsona, R.; Mares, A. G.; Pujals, S.; Gomila, G., Fast label-free nanoscale composition mapping of eukaryotic cells via scanning dielectric force volume microscopy and machine learning. *Small Methods* 2021, *5*, 2100279.
45. Huey, B. D.; Bonnell, D. A., Nanoscale variation in electric potential at oxide bicrystal and polycrystal interfaces. *Solid State Ionics* 2000, *131* (1), 51–60.
46. Kalinin, S. V.; Bonnell, D. A., Surface potential at surface-interface junctions in $SrTiO_3$ bicrystals. *Physical Review B* 2000, *62* (15), 10419.
47. Kalinin, S.; Bonnell, D., Local potential and polarization screening on ferroelectric surfaces. *Physical Review B* 2001, *63* (12), 125411–125411.
48. Woods, C. R.; Ares, P.; Nevison-Andrews, H.; Holwill, M. J.; Fabregas, R.; Guinea, F.; Geim, A. K.; Novoselov, K. S.; Walet, N. R.; Fumagalli, L., Charge-polarized interfacial superlattices in marginally twisted hexagonal boron nitride. *Nature Communications* 2021, *12*, 1–7.
49. Ares, P.; Cea, T.; Holwill, M.; Wang, Y. B.; Roldán, R.; Guinea, F.; Andreeva, D. V.; Fumagalli, L.; Novoselov, K. S.; Woods, C. R., Piezoelectricity in monolayer hexagonal boron nitride. *Advanced Materials* 2020, *32*, 1–6.
50. Cadena, M. J.; Misiego, R.; Smith, K. C.; Avila, A.; Pipes, B.; Reifenberger, R.; Raman, A., Sub-surface imaging of carbon nanotube–polymer composites using dynamic AFM methods. *Nanotechnology* 2013, *24*, 135706.
51. Cadena, M. J.; Reifenberger, R. G.; Raman, A., High resolution subsurface imaging using resonance-enhanced detection in 2nd-harmonic KPFM. *Nanotechnology* 2018, *29*, 405702.
52. Reifenberger, R.; Raman, A.; Avila, A.; Casta, O. A.; Engineering, E.; Engineering, M.; Lafayette, W.; States, U., Depth-sensitive subsurface imaging of polymer nanocomposites using second harmonic Kelvin probe force. *ACS Nano* 2015, *9*, 2938–2947.
53. Balakrishnan, H.; Millán, R.; Checa, M.; Fabregas, R.; Fumagalli, L.; Gomila, G., Depth mapping of metallic nanowire polymer nanocomposites by scanning dielectric microscopy. *Nanoscale* 2021, *13*, 10116–10126.
54. Costi, R.; Cohen, G.; Salant, A.; Rabani, E.; Banin, U., Electrostatic force microscopy study of single Au-CdSe hybrid nanodumbbells: Evidence for light-induced charge separation. *Nano Letters* 2009, *9*, 2031–2039.
55. Yalcin, S. E.; Galande, C.; Kappera, R.; Yamaguchi, H.; Martinez, U.; Velizhanin, K. A.; Doorn, S. K.; Dattelbaum, A. M.; Chhowalla, M.; Ajayan, P. M.; Gupta, G.; Mohite, A. D., Direct imaging of charge transport in progressively reduced graphene oxide using electrostatic force microscopy. *ACS Nano* 2015, *9*, 2981–2988.

56. Melin, T.; Deresmes, D.; Stiévenard, D., Charge injection in individual silicon nanoparticles deposited on a conductive substrate. *Applied Physics Letters* 2002, *81* (26), 5054–5056.
57. Barbet, S.; Mélin, T.; Diesinger, H.; Deresmes, D.; Stiévenard, D., Charge-injection mechanisms in semiconductor nanoparticles analyzed from force microscopy experiments. *Physical Review B* 2006, *73* (4), 045318.
58. McMorrow, J. J.; Cress, C. D.; Affouda, C. A., Charge injection in high-κ gate dielectrics of single-walled carbon nanotube thin-film transistors. *ACS Nano* 2012, *6* (6), 5040–5050.
59. Nakamura, M.; Yamada, H., Electrostatic force microscopy. In Morita, S. (ed.) *Roadmap of Scanning Probe Microscopy*, Springer, Heidelberg: 2007; pp. 43–51.
60. Fumagalli, L.; Ferrari, G.; Sampietro, M.; Gomila, G., Quantitative nanoscale dielectric microscopy of single-layer supported biomembranes. *NanoLetters* 2009, *9*, 1604.
61. Van Der Hofstadt, M.; Fabregas, R.; Biagi, M. C.; Fumagalli, L.; Gomila, G., Nanoscale dielectric microscopy of non-planar samples by lift-mode electrostatic force microscopy. *Nanotechnology* 2016, *27*, 405706.
62. Kyndiah, A.; Checa, M.; Leonardi, F.; Millan-Solsona, R.; Di Muzio, M.; Tanwar, S.; Fumagalli, L.; Mas-Torrent, M.; Gomila, G., Nanoscale mapping of the conductivity and interfacial capacitance of an electrolyte-gated organic field-effect transistor under operation. *Advanced Functional Materials* 2021, *31*, 1–8.
63. Checa, M., Una nova tècnica de microscòpia de forces atòmiques per a l'estudi de les propietats nanoelèctriques en cèl·lules. 2020.
64. Collins, L.; Belianinov, A.; Somnath, S.; Rodriguez, B. J.; Balke, N.; Kalinin, S. V.; Jesse, S., Multifrequency spectrum analysis using fully digital G Mode-Kelvin probe force microscopy. *Nanotechnology* 2016, *27*, 105706.
65. Hu, J.; Xiao, X. D.; Salmeron, M., Scanning polarization force microscopy: A technique for imaging liquids and weakly adsorbed layers. *Applied Physics Letters* 1995, *67* (4), 476–478.
66. Hu, J.; Xiao, X. D.; Ogletree, D. F.; Salmeron, M., Imaging the condensation and evaporation of molecularly thin films of water with nanometer resolution. *Science* 1995, *268*, 267–269.
67. Lei, C.; Das, A.; Elliott, M.; Macdonald, J. E., Quantitative electrostatic force microscopy-phase measurements. *Nanotechnology* 2004, *15* (5), 627.
68. Bridger, P.; Bandić, Z.; Piquette, E.; McGill, T., Measurement of induced surface charges, contact potentials, and surface states in GaN by electric force microscopy. *Applied Physics Letters* 1999, *74* (23), 3522–3524.
69. Sacha, G.; Gómez-Navarro, C.; Sáenz, J.; Gómez-Herrero, J., Quantitative theory for the imaging of conducting objects in electrostatic force microscopy. *Applied Physics Letters* 2006, *89* (17), 173122.
70. Gramse, G.; Casuso, I.; Toset, J.; Fumagalli, L.; Gomila, G., Quantitative dielectric constant measurement of thin films by DC electrostatic force microscopy. *Nanotechnology* 2009, *20* (39), 395702.
71. Gramse, G.; Edwards, M. A.; Fumagalli, L.; Gomila, G., Dynamic electrostatic force microscopy in liquid media. *Applied Physics Letters* 2012, *101*, 213108.
72. Schwartz, G. A.; Riedel, C.; Arinero, R.; Tordjeman, P.; Alegría, A.; Colmenero, J., Broadband nanodielectric spectroscopy by means of amplitude modulation electrostatic force microscopy (AM-EFM). *Ultramicroscopy* 2011, *111* (8), 1366–1369.
73. Buh, G.; Chung, H.; Kuk, Y., Real-time evolution of trapped charge in a $SiO_2$ layer: An electrostatic force microscopy study. *Applied Physics Letters* 2001, *79* (13), 2010–2012.
74. Jespersen, T. S.; Nygård, J., Charge trapping in carbon nanotube loops demonstrated by electrostatic force microscopy. *Nano Letters* 2005, *5* (9), 1838–1841.

75. Cherniavskaya, O.; Chen, L.; Weng, V.; Yuditsky, L.; Brus, L. E., Quantitative noncontact electrostatic force imaging of nanocrystal polarizability. *The Journal of Physical Chemistry B* 2003, *107* (7), 1525–1531.
76. Ben-Porat, C. H.; Cherniavskaya, O.; Brus, L.; Cho, K.-S.; Murray, C. B., Electric fields on oxidized silicon surfaces: Static polarization of PbSe nanocrystals. *The Journal of Physical Chemistry A* 2004, *108* (39), 7814–7819.
77. Kalinin, S. V., Nanoscale electric phenomena at oxide surfaces and interfaces by scanning probe microscopy. *arXiv preprint cond-mat/0209599* 2002.
78. Belaidi, S.; Girard, P.; Leveque, G., Electrostatic forces acting on the tip in atomic force microscopy: Modelization and comparison with analytic expressions. *Journal of Applied Physics* 1997, *81* (3), 1023–1030.
79. Hudlet, S.; Saint Jean, M.; Guthmann, C.; Berger, J., Evaluation of the capacitive force between an atomic force microscopy tip and a metallic surface. *European Physical Journal B* 1998, *2*, 5–10.
80. Gomila, G.; Gramse, G.; Fumagalli, L., Finite-size effects and analytical modeling of electrostatic force microscopy applied to dielectric films. *Nanotechnology* 2014, *25*, 255702.
81. Fumagalli, L.; Edwards, M. A.; Gomila, G., Quantitative electrostatic force microscopy with sharp silicon tips. *Nanotechnology* 2014, *25*, 495701.
82. Jeon, D.; Kang, Y.; Kim, T., Observing the layer-number-dependent local dielectric response of WSe2 by electrostatic force microscopy. *Journal of Physical Chemistry Letters* 2020, *11*, 6684–6690.
83. Riedel, C.; Arinero, R.; Tordjeman, P.; Leveque, G.; Schwartz, G. A.; Alegria, A.; Colmenero, J., Nanodielectric mapping of a model polystyrene-poly(vinyl acetate) blend by electrostatic force microscopy. *Physical Review E* 2010, *81*, 1–4.
84. Fumagalli, L.; Esfandiar, A.; Fabregas, R.; Hu, S.; Ares, P.; Janardanan, A.; Yang, Q.; Radha, B.; Taniguchi, T.; Watanabe, K.; Gomila, G.; Novoselov, K. S.; Geim, A. K., Anomalously low dielectric constant of confined water. *Science* 2018, *360* (6395), 1339–1342.
85. Gramse, G.; Schönhals, A.; Kienberger, F., Nanoscale dipole dynamics of protein membranes studied by broadband dielectric microscopy. *Nanoscale* 2019, *11*, 4303–4309.
86. Dols-Perez, A.; Gramse, G.; Calò, A.; Gomila, G.; Fumagalli, L., Nanoscale electric polarizability of ultrathin biolayers on insulating substrates by electrostatic force microscopy. *Nanoscale* 2015, *7*, 18327–18336.
87. El Khoury, D.; Arinero, R.; Laurentie, J. C.; Bechelany, M.; Ramonda, M.; Castellon, J., Electrostatic force microscopy for the accurate characterization of interphases in nanocomposites. *Beilstein Journal of Nanotechnology* 2018, *9*, 2999–3012.
88. Riedel, C.; Schwartz, G. A.; Arinero, R.; Colmenero, J.; Sa, J. J., On the use of electrostatic force microscopy as a quantitative subsurface characterization technique: A numerical study. *Applied Physics Letters* 2011, *99*, 227–229.
89. Lozano, H.; Millán-Solsona, R.; Fabregas, R.; Gomila, G., Sizing single nanoscale objects from polarization forces. *Scientific Reports* 2019, *9* (1), 1–12.
90. Fabregas, R.; Gomila, G., Dielectric nanotomography based on electrostatic force microscopy: A numerical analysis. *Journal of Applied Physics* 2020, *127*, 024301.
91. Cadena, M. J.; Sung, S. H.; Boudouris, B. W.; Reifenberger, R.; Raman, A., Nanoscale mapping of dielectric properties of nanomaterials from kilohertz to megahertz using ultrasmall cantilevers. *ACS Nano* 2016, *10* (4), 4062–4071.
92. Sacha, G. M.; Varona, P.; Sacha, G. M.; Cardellach, M., An inverse problem solution for undetermined electrostatic force. *Nanotechnology* 2009, *20*, 5–10.
93. Checa, M.; Millan-Solsona, R.; Glinkowska Mares, A.; Pujals, S.; Gomila, G., Dielectric imaging of fixed HeLa cells by in-liquid scanning dielectric force volume microscopy. *Nanomaterials* 2021, *11* (6), 1402.

94. Mélin, T.; Diesinger, H.; Deresmes, D.; Stiévenard, D., Electric force microscopy of individually charged nanoparticles on conductors: An analytical model for quantitative charge imaging. *Physical Review B - Condensed Matter and Materials Physics* 2004, *69*, 1–8.
95. Qi, G.; Yang, Y.; Yan, H.; Guan, L.; Li, Y.; Qiu, X.; Wang, C., Quantifying surface charge density by using an electric force microscope with a referential structure. *The Journal of Physical Chemistry C* 2009, *113*, 204–207.
96. Saurenbach, F.; Terris, B. D., Imaging of ferroelectric domain walls by force microscopy. *Applied Physics Letters* 1990, *56*, 1703–1705.
97. Schaab, J.; Cano, A.; Lilienblum, M.; Yan, Z.; Bourret, E.; Ramesh, R.; Fiebig, M.; Meier, D., Optimization of electronic domain-wall properties by aliovalent cation substitution. *Advanced Electronic Materials* 2016, *2*, 1500195.
98. Schoenherr, P.; Shapovalov, K.; Schaab, J.; Yan, Z.; Bourret, E. D.; Hentschel, M.; Stengel, M.; Fiebig, M.; Cano, A.; Meier, D., Observation of uncompensated bound charges at improper ferroelectric domain walls. *Nano Letters* 2019, *19*, 1659–1664.
99. Han, B.; Chang, J.; Song, W.; Sun, Z.; Yin, C.; Lv, P.; Wang, X., Study on micro interfacial charge motion of polyethylene nanocomposite based on electrostatic force microscope. *Polymers (Basel)* 2019, *11* (12), 2035.
100. Ng, C. Y.; Chen, T. P.; Lau, H. W.; Liu, Y.; Tse, M. S.; Tan, O. K.; Lim, V. S. W., Visualizing charge transport in silicon nanocrystals embedded in $SiO_2$ films with electrostatic force microscopy. *Applied Physics Letters* 2004, *85*, 2941–2943.
101. Gupta, S.; Williams, O. A.; Bohannan, E., Electrostatic force microscopy studies of boron-doped diamond films. *Journal of Materials Research* 2007, *22*, 3014–3028.
102. Gramse, G.; Kölker, A.; Škereň, T.; Stock, T. J. Z.; Aeppli, G.; Kienberger, F.; Fuhrer, A.; Curson, N. J., Nanoscale imaging of mobile carriers and trapped charges in delta doped silicon p–n junctions. *Nature Electronics* 2020, *3*, 531–538.
103. Badur, S.; Renz, D.; Göddenhenrich, T.; Ebeling, D.; Roling, B.; Schirmeisen, A., Voltage- and frequency-based separation of nanoscale electromechanical and electrostatic forces in contact resonance force microscopy: Implications for the analysis of battery materials. *ACS Applied Nano Materials* 2020, *3*, 7397–7405.
104. Thomson, W., Contact electricity of metals. *Philosophical Magazine* 1898, *46*, 82–120.
105. Zisman, W., A new method of measuring contact potential differences in metals. *Review of Scientific Instruments* 1932, *3* (7), 367–370.
106. Baikie, I.; Van Der Werf, K.; Oerbekke, H.; Broeze, J.; van Silfhout, A., Automatic Kelvin probe compatible with ultrahigh vacuum. *Review of Scientific Instruments* 1989, *60* (5), 930–934.
107. Baikie, I.; Smith, P.; Porterfield, D.; Estrup, P., Multitip scanning bio-Kelvin probe. *Review of Scientific Instruments* 1999, *70* (3), 1842–1850.
108. Baikie, I.; Estrup, P., Low cost PC based scanning Kelvin probe. *Review of Scientific Instruments* 1998, *69* (11), 3902–3907.
109. Mignolet, J., Studies in contact potentials. II. Vibrating cells for the vibrating condenser method. *Discussions of the Faraday Society* 1950, *8*, 326–331.
110. Baumgärtner, H.; Liess, H., Micro Kelvin probe for local work-function measurements. *Review of Scientific Instruments* 1988, *59* (5), 802–805.
111. Saito, S.; Soumura, T.; Maeda, T., Improvements of the piezoelectric driven Kelvin probe. *Journal of Vacuum Science & Technology A: Vacuum, Surfaces, and Films* 1984, *2* (3), 1389–1391.
112. Jönsson, M.; Thierry, D.; LeBozec, N., The influence of microstructure on the corrosion behaviour of AZ91D studied by scanning Kelvin probe force microscopy and scanning Kelvin probe. *Corrosion Science* 2006, *48* (5), 1193–1208.
113. Uhlig, H. H., Volta potentials of the copper-nickel alloys and several metals in air. *Journal of Applied Physics* 1951, *22* (12), 1399–1403.

114. Kim, J.; Lägel, B.; Moons, E.; Johansson, N.; Baikie, I.; Salaneck, W. R.; Friend, R.; Cacialli, F., Kelvin probe and ultraviolet photoemission measurements of indium tin oxide work function: A comparison. *Synthetic Metals* 2000, *111*, 311–314.
115. Lägel, B.; Baikie, I. D.; Petermann, U., A novel detection system for defects and chemical contamination in semiconductors based upon the Scanning Kelvin Probe. *MRS Online Proceedings Library (OPL)* 1998, *510*, 619–626.
116. Fort Jr., T.; Wells, R. L., Adsorption of water on clean aluminum by measurement of work function changes. *Surface Science* 1972, *32* (3), 543–553.
117. Castaldini, A.; Cavalcoli, D.; Cavallini, A.; Rossi, M., Scanning Kelvin probe and surface photovoltage analysis of multicrystalline silicon. *Materials Science and Engineering: B* 2002, *91*, 234–238.
118. Collins, L.; Kilpatrick, J. I.; Kalinin, S. V.; Rodriguez, B. J., Towards nanoscale electrical measurements in liquid by advanced KPFM techniques: A review. *Reports on Progress in Physics* 2018, *81* (8), 086101.
119. Shockley, W.; Hooper, W.; Queisser, H.; Schroen, W., Mobile electric charges on insulating oxides with application to oxide covered silicon pn junctions. *Surface Science* 1964, *2*, 277–287.
120. Kalinin, S. V.; Dyck, O.; Balke, N.; Neumayer, S.; Tsai, W.-Y.; Vasudevan, R.; Lingerfelt, D.; Ahmadi, M.; Ziatdinov, M.; McDowell, M. T., Toward electrochemical studies on the nanometer and atomic scales: Progress, challenges, and opportunities. *ACS Nano* 2019, *13* (9), 9735–9780.
121. Li, G.; Mao, B.; Lan, F.; Liu, L., Practical aspects of single-pass scan Kelvin probe force microscopy. *Review of Scientific Instruments* 2012, *83* (11), 113701.
122. Colchero, J.; Gil, A.; Baró, A., Resolution enhancement and improved data interpretation in electrostatic force microscopy. *Physical Review B* 2001, *64* (24), 245403.
123. Zerweck, U.; Loppacher, C.; Otto, T.; Grafström, S.; Eng, L. M., Accuracy and resolution limits of Kelvin probe force microscopy. *Physical Review B* 2005, *71* (12), 125424.
124. Glatzel, T.; Sadewasser, S.; Lux-Steiner, M. C., Amplitude or frequency modulation-detection in Kelvin probe force microscopy. *Applied Surface Science* 2003, *210* (1), 84–89.
125. Sugawara, Y.; Kou, L.; Ma, Z.; Kamijo, T.; Naitoh, Y.; Jun Li, Y., High potential sensitivity in heterodyne amplitude-modulation Kelvin probe force microscopy. *Applied Physics Letters* 2012, *100* (22), 223104.
126. Axt, A.; Hermes, I. M.; Bergmann, V. W.; Tausendpfund, N.; Weber, S. A., Know your full potential: Quantitative Kelvin probe force microscopy on nanoscale electrical devices. *Beilstein Journal of Nanotechnology* 2018, *9* (1), 1809–1819.
127. Garrett, J. L.; Munday, J. N., Fast, high-resolution surface potential measurements in air with heterodyne Kelvin probe force microscopy. *Nanotechnology* 2016, *27* (24), 245705.
128. Hansen, W. N.; Hansen, G. J., Standard reference surfaces for work function measurements in air. *Surface Science* 2001, *481* (1–3), 172–184.
129. Wu, Y.; Shannon, M. A., ac driving amplitude dependent systematic error in scanning Kelvin probe microscope measurements: Detection and correction. *Review of Scientific Instruments* 2006, *77*, 043711.
130. Jacobs, H.; Knapp, H.; Stemmer, A., Practical aspects of Kelvin probe force microscopy. *Review of Scientific Instruments* 1999, *70*, 1756.
131. Collins, L.; Kilpatrick, J.; Weber, S. A.; Tselev, A.; Vlassiouk, I.; Ivanov, I.; Jesse, S.; Kalinin, S.; Rodriguez, B., Open loop Kelvin probe force microscopy with single and multi-frequency excitation. *Nanotechnology* 2013, *24* (47), 475702.
132. Mélin, T.; Barbet, S.; Diesinger, H.; Théron, D.; Deresmes, D., Note: Quantitative (artifact-free) surface potential measurements using Kelvin force microscopy. *Review of Scientific Instruments* 2011, *82* (3), 036101-1–036101-3.

133. Okamoto, K.; Sugawara, Y.; Morita, S., The elimination of the 'artifact' in the electrostatic force measurement using a novel noncontact atomic force microscope/electrostatic force microscope. *Applied Surface Science* 2002, *188* (3–4), 381–385.
134. Diesinger, H.; Deresmes, D.; Mélin, T., Capacitive crosstalk in AM-mode KPFM. In Sadewasser, S., Glatzel, T. (eds.) *Kelvin Probe Force Microscopy*, Springer, Cham: 2012; pp. 25–44.
135. Diesinger, H.; Deresmes, D.; Nys, J.-P.; Melin, T., Kelvin force microscopy at the second cantilever resonance: An out-of-vacuum crosstalk compensation setup. *Ultramicroscopy* 2008, *108* (8), 773–781.
136. Barbet, S.; Popoff, M.; Diesinger, H.; Deresmes, D.; Théron, D.; Mélin, T., Cross-talk artefacts in Kelvin probe force microscopy imaging: A comprehensive study. *Journal of Applied Physics* 2014, *115* (14), 144313.
137. Polak, L.; de Man, S.; Wijngaarden, R. J., Note: Switching crosstalk on and off in Kelvin probe force microscopy. *Review of Scientific Instruments* 2014, *85* (4), 046111.
138. Butt, H.-J.; Jaschke, M., Calculation of thermal noise in atomic force microscopy. *Nanotechnology* 1995, *6* (1), 1.
139. Kilpatrick, J. I.; Collins, L.; Weber, S. A.; Rodriguez, B. J., Quantitative comparison of closed-loop and dual harmonic Kelvin probe force microscopy techniques. *Review of Scientific Instruments* 2018, *89* (12), 123708.
140. Ballif, C.; Moutinho, H. R.; Al-Jassim, M. M., Cross-sectional electrostatic force microscopy of thin-film solar cells. *Journal of Applied Physics* 2001, *89*, 1418–1424.
141. Chen, W.; Zhou, Y.; Wang, L.; Wu, Y.; Tu, B.; Yu, B.; Liu, F.; Tam, H. W.; Wang, G.; Djurišić, A. B.; Huang, L.; He, Z., Molecule-doped nickel oxide: Verified charge transfer and planar inverted mixed cation perovskite solar cell. *Advanced Materials* 2018, *30*, 1–9.
142. Jiang, C. S.; Moutinho, H. R.; Geisz, J. F.; Friedman, D. J.; Al-Jassim, M. M., Direct measurement of electrical potentials in GaInP2 solar cells. *Applied Physics Letters* 2002, *81*, 2569–2571.
143. Yan, H.; Li, D.; Li, C.; Lu, K.; Zhang, Y.; Wei, Z.; Yang, Y.; Wang, C., Bridging mesoscopic blend structure and property to macroscopic device performance via in situ optoelectronic characterization. *Journal of Materials Chemistry* 2012, *22*, 4349–4355.
144. Barnes, A. M.; Buratto, S. K., Imaging channel connectivity in nafion using electrostatic force microscopy. *Journal of Physical Chemistry B* 2018, *122*, 1289–1295.
145. Son, B.; Park, J.; Kwon, O., Analysis of ionic domains on a proton exchange membrane using a numerical approximation model based on electrostatic force microscopy. *Polymers* 2021, *13* (8), 1258.
146. Barnes, A. M.; Du, Y.; Liu, B.; Zhang, W.; Seifert, S.; Coughlin, E. B.; Buratto, S. K., Effect of surface alignment on connectivity in phosphonium-containing diblock copolymer anion-exchange membranes. *Journal of Physical Chemistry C* 2019, *123*, 30819–30826.
147. Liu, J.; Wang, X.; Lin, Z.; Cao, Y.; Zheng, Z.; Zeng, Z.; Hu, Z., Shape-controllable pulse electrodeposition of ultrafine platinum nanodendrites for methanol catalytic combustion and the investigation of their local electric field intensification by electrostatic force microscope and finite element method. *Electrochimica Acta* 2014, *136*, 66–74.
148. Berger, R.; Domanski, A. L.; Weber, S. A., Electrical characterization of organic solar cell materials based on scanning force microscopy. *European Polymer Journal* 2013, *49* (8), 1907–1915.
149. Howard, J. M.; Lahoti, R.; Leite, M. S., Imaging metal halide perovskites material and properties at the nanoscale. *Advanced Energy Materials* 2020, *10* (26), 1903161.
150. Grévin, B., Kelvin probe force microscopy characterization of organic and hybrid perovskite solar cells. In Sadewasser, S., Glatzel, T. (eds.) *Kelvin Probe Force Microscopy*, Springer, Cham: 2018; pp. 331–365.

151. Wang, S.; Liu, Q.; Zhao, C.; Lv, F.; Qin, X.; Du, H.; Kang, F.; Li, B., Advances in understanding materials for rechargeable lithium batteries by atomic force microscopy. *Energy & Environmental Materials* 2018, *1* (1), 28–40.
152. Zhu, J.; Lee, J.-W.; Lee, H.; Xie, L.; Pan, X.; De Souza, R. A.; Eom, C.-B.; Nonnenmann, S. S., Probing vacancy behavior across complex oxide heterointerfaces. *Science Advances* 2019, *5* (2), eaau8467.
153. Nonnenmann, S. S.; Bonnell, D. A., Miniature environmental chamber enabling in situ scanning probe microscopy within reactive environments. *Review of Scientific Instruments* 2013, *84* (7), 073707.
154. Zhu, J.; Wang, J.; Mebane, D. S.; Nonnenmann, S. S., In situ surface potential evolution along Au/Gd: $CeO_2$ electrode interfaces. *APL Materials* 2017, *5* (4), 042503.
155. Yu, W.; Fu, H. J.; Mueller, T.; Brunschwig, B. S.; Lewis, N. S., Atomic force microscopy: Emerging illuminated and operando techniques for solar fuel research. *The Journal of Chemical Physics* 2020, *153* (2), 020902.
156. Sadewasser, S., Optoelectronic studies of solar cells. In Sadewasser, S., Glatzel, T. (eds.) *Kelvin Probe Force Microscopy*, Springer, Cham: 2012, pp. 151–174.
157. Takihara, M.; Igarashi, T.; Ujihara, T.; Takahashi, T., Photovoltage mapping on polycrystalline silicon solar cells by Kelvin probe force microscopy with piezoresistive cantilever. *Japanese Journal of Applied Physics* 2007, *46* (8S), 5548.
158. Takihara, M.; Takahashi, T.; Ujihara, T., Minority carrier lifetime in polycrystalline silicon solar cells studied by photoassisted Kelvin probe force microscopy. *Applied Physics Letters* 2008, *93* (2), 021902.
159. Leendertz, C.; Streicher, F.; Lux-Steiner, M. C.; Sadewasser, S., Evaluation of Kelvin probe force microscopy for imaging grain boundaries in chalcopyrite thin films. *Applied Physics Letters* 2006, *89* (11), 113120.
160. Jiang, C.-S.; Noufi, R.; Ramanathan, K.; AbuShama, J.; Moutinho, H.; Al-Jassim, M., Does the local built-in potential on grain boundaries of Cu (In, Ga) $Se_2$ thin films benefit photovoltaic performance of the device? *Applied Physics Letters* 2004, *85* (13), 2625–2627.
161. Jiang, C.-S.; Noufi, R.; Ramanathan, K.; Moutinho, H.; Al-Jassim, M., Electrical modification in Cu (In, Ga) $Se_2$ thin films by chemical bath deposition process of CdS films. *Journal of Applied Physics* 2005, *97* (5), 053701.
162. Sadewasser, S., Surface potential of chalcopyrite films measured by KPFM. *Physica Status Solidi (A)* 2006, *203* (11), 2571–2580.
163. Visoly-Fisher, I.; Cohen, S. R.; Cahen, D., Direct evidence for grain-boundary depletion in polycrystalline CdTe from nanoscale-resolved measurements. *Applied Physics Letters* 2003, *82* (4), 556–558.
164. Glatzel, T.; Hoppe, H.; Sariciftci, N. S.; Lux-Steiner, M. C.; Komiyama, M., Kelvin probe force microscopy study of conjugated polymer/fullerene organic solar cells. *Japanese Journal of Applied Physics* 2005, *44* (7S), 5370.
165. Hoppe, H.; Glatzel, T.; Niggemann, M.; Hinsch, A.; Lux-Steiner, M. C.; Sariciftci, N., Kelvin probe force microscopy study on conjugated polymer/fullerene bulk heterojunction organic solar cells. *Nano Letters* 2005, *5* (2), 269–274.
166. Palermo, V.; Ridolfi, G.; Talarico, A. M.; Favaretto, L.; Barbarella, G.; Camaioni, N.; Samorì, P., A Kelvin probe force microscopy study of the photogeneration of surface charges in all-thiophene photovoltaic blends. *Advanced Functional Materials* 2007, *17* (3), 472–478.
167. Sadewasser, S.; Glatzel, T.; Schuler, S.; Nishiwaki, S.; Kaigawa, R.; Lux-Steiner, M. C., Kelvin probe force microscopy for the nano scale characterization of chalcopyrite solar cell materials and devices. *Thin Solid Films* 2003, *431*, 257–261.
168. Marrón, D. F.; Sadewasser, S.; Meeder, A.; Glatzel, T.; Lux-Steiner, M. C., Electrical activity at grain boundaries of Cu (In, Ga) $Se_2$ thin films. *Physical Review B* 2005, *71* (3), 033306.

169. Garrett, J. L.; Tennyson, E. M.; Hu, M.; Huang, J.; Munday, J. N.; Leite, M. S., Real-time nanoscale open-circuit voltage dynamics of perovskite solar cells. *Nano Letters* 2017, *17* (4), 2554–2560.
170. Vatel, O.; Tanimoto, M., Kelvin probe force microscopy for potential distribution measurement of semiconductor-devices. *Journal of Applied Physics* 1995, *77* (6), 2358–2362.
171. Huey, B. D.; Bonnell, D. A., Spatially localized dynamic properties of individual interfaces in semiconducting oxides. *Applied Physics Letters* 2000, *76* (8), 1012–1014.
172. Bonnell, D. A.; Huey, B.; Carroll, D., In-situ measurement of electric-fields at individual grain-boundaries in $TIO_2$. *Solid State Ionics* 1995, *75*, 35–42.
173. Freitag, M.; Johnson, A. T.; Kalinin, S. V.; Bonnell, D. A., Role of single defects in electronic transport through carbon nanotube field-effect transistors. *Physical Review Letters* 2002, *89* (21), 216801.
174. Kalinin, S. V.; Shin, J.; Jesse, S.; Geohegan, D.; Baddorf, A. P.; Lilach, Y.; Moskovits, M.; Kolmakov, A., Electronic transport imaging in a multiwire $SnO_2$ chemical field-effect transistor device. *Journal of Applied Physics* 2005, *98* (4), 044503.
175. Shikler, R., Electronic surface properties of semiconductor surfaces and interfaces. In Sadewasser, S., Glatzel, T. (eds.) *Kelvin Probe Force Microscopy*, Springer, Cham: 2012; pp. 101–115.
176. Bergmann, V. W.; Weber, S. A.; Ramos, F. J.; Nazeeruddin, M. K.; Grätzel, M.; Li, D.; Domanski, A. L.; Lieberwirth, I.; Ahmad, S.; Berger, R., Real-space observation of unbalanced charge distribution inside a perovskite-sensitized solar cell. *Nature communications* 2014, *5* (1), 1–9.
177. Zhu, J.; Fan, F.; Chen, R.; An, H.; Feng, Z.; Li, C., Direct imaging of highly anisotropic photogenerated charge separations on different facets of a single $BiVO_4$ photocatalyst. *Angewandte Chemie* 2015, *127* (31), 9239–9242.
178. Zhu, J.; Pang, S.; Dittrich, T.; Gao, Y.; Nie, W.; Cui, J.; Chen, R.; An, H.; Fan, F.; Li, C., Visualizing the nano cocatalyst aligned electric fields on single photocatalyst particles. *Nano Letters* 2017, *17* (11), 6735–6741.
179. Chen, R.; Zhu, J.; An, H.; Fan, F.; Li, C., Unravelling charge separation via surface built-in electric fields within single particulate photocatalysts. *Faraday Discussions* 2017, *198*, 473–479.
180. Chen, R.; Pang, S.; An, H.; Zhu, J.; Ye, S.; Gao, Y.; Fan, F.; Li, C., Charge separation via asymmetric illumination in photocatalytic $Cu_2O$ particles. *Nature Energy* 2018, *3* (8), 655–663.
181. Gao, Y.; Nie, W.; Zhu, Q.; Wang, X.; Wang, S.; Fan, F.; Li, C., The polarization effect in surface-plasmon-induced photocatalysis on $Au/TiO_2$ nanoparticles. *Angewandte Chemie* 2020, *132* (41), 18375–18380.
182. Nagpure, S. C.; Bhushan, B.; Babu, S., Surface potential measurement of aged Li-ion batteries using Kelvin probe microscopy. *Journal of Power Sources* 2011, *196* (3), 1508–1512.
183. Wu, J.; Yang, S.; Cai, W.; Bi, Z.; Shang, G.; Yao, J., Multi-characterization of $LiCoO_2$ cathode films using advanced AFM-based techniques with high resolution. *Scientific Reports* 2017, *7* (1), 1–9.
184. Zhu, J.; Zeng, K.; Lu, L., In-situ nanoscale mapping of surface potential in all-solid-state thin film Li-ion battery using Kelvin probe force microscopy. *Journal of Applied Physics* 2012, *111* (6), 063723.
185. Luchkin, S. Y.; Amanieu, H.-Y.; Rosato, D.; Kholkin, A. L., Li distribution in graphite anodes: A kelvin probe force microscopy approach. *Journal of Power Sources* 2014, *268*, 887–894.

186. Zhu, X.; Revilla, R. I.; Hubin, A., Direct correlation between local surface potential measured by Kelvin probe force microscope and electrochemical potential of LiNi0. 80Co0. 15Al0. 05O2 cathode at different state of charge. *The Journal of Physical Chemistry C* 2018, *122* (50), 28556–28563.
187. Masuda, H.; Ishida, N.; Ogata, Y.; Ito, D.; Fujita, D., Internal potential mapping of charged solid-state-lithium ion batteries using in situ Kelvin probe force microscopy. *Nanoscale* 2017, *9* (2), 893–898.
188. Masuda, H.; Matsushita, K.; Ito, D.; Fujita, D.; Ishida, N., Dynamically visualizing battery reactions by operando Kelvin probe force microscopy. *Communications Chemistry* 2019, *2* (1), 1–6.
189. Kempaiah, R.; Vasudevamurthy, G.; Subramanian, A., Scanning probe microscopy based characterization of battery materials, interfaces, and processes. *Nano Energy* 2019, *65*, 103925.
190. Stone, A. D.; Mesquida, P., Kelvin-probe force microscopy of the pH-dependent charge of functional groups. *Applied Physics Letters* 2016, *108* (23), 233702.
191. Weber, S. A.; Hermes, I. M.; Turren-Cruz, S.-H.; Gort, C.; Bergmann, V. W.; Gilson, L.; Hagfeldt, A.; Graetzel, M.; Tress, W.; Berger, R., How the formation of interfacial charge causes hysteresis in perovskite solar cells. *Energy & Environmental Science* 2018, *11* (9), 2404–2413.
192. Strelcov, E.; Kim, Y.; Jesse, S.; Cao, Y.; Ivanov, I. N.; Kravchenko, I. I.; Wang, C.-H.; Teng, Y.-C.; Chen, L.-Q.; Chu, Y. H., Probing local ionic dynamics in functional oxides at the nanoscale. *Nano Letters* 2013, *13* (8), 3455–3462.
193. Strelcov, E.; Jesse, S.; Huang, Y.-L.; Teng, Y.-C.; Kravchenko, I. I.; Chu, Y.-H.; Kalinin, S. V., Space-and time-resolved mapping of ionic dynamic and electroresistive phenomena in lateral devices. *ACS Nano* 2013, *7* (8), 6806–6815.
194. Strelcov, E.; Ievlev, A. V.; Jesse, S.; Kravchenko, I. I.; Shur, V. Y.; Kalinin, S. V., Direct probing of charge injection and polarization-controlled ionic mobility on ferroelectric $LiNbO_3$ surfaces. *Advanced Materials* 2014, *26* (6), 958–963.
195. Freitag, M.; Radosavljevic, M.; Zhou, Y.; Johnson, A.; Smith, W. F., Controlled creation of a carbon nanotube diode by a scanned gate. *Applied Physics Letters* 2001, *79* (20), 3326–3328.
196. Kobayashi, N.; Asakawa, H.; Fukuma, T., Nanoscale potential measurements in liquid by frequency modulation atomic force microscopy. *Review of Scientific Instruments* 2010, *81*.
197. Hirata, K.; Kitagawa, T.; Miyazawa, K.; Okamoto, T.; Fukunaga, A.; Takatoh, C.; Fukuma, T., Visualizing charges accumulated in an electric double layer by three-dimensional open-loop electric potential microscopy. *Nanoscale* 2018, *10*, 14736–14746.
198. Honbo, K.; Ogata, S.; Kitagawa, T.; Okamoto, T.; Kobayashi, N.; Sugimoto, I.; Shima, S.; Fukunaga, A.; Takatoh, C.; Fukuma, T., Visualizing nanoscale distribution of corrosion cells by open-loop electric potential microscopy. *ACS Nano* 2016, *10*, 2575–2583.
199. Collins, L.; Kilpatrick, J. I.; Vlassiouk, I. V.; Tselev, A.; Weber, S. A.; Jesse, S.; Kalinin, S. V.; Rodriguez, B. J., Dual harmonic Kelvin probe force microscopy at the graphene–liquid interface. *Applied Physics Letters* 2014, *104* (13), 133103.
200. Kobayashi, N.; Asakawa, H.; Fukuma, T., Quantitative potential measurements of nanoparticles with different surface charges in liquid by open-loop electric potential microscopy. *Journal of Applied Physics* 2011, *110* (4), 044315.
201. Takeuchi, O.; Ohrai, Y.; Yoshida, S.; Shigekawa, H., Kelvin probe force microscopy without bias-voltage feedback. *Japanese Journal of Applied Physics* 2007, *46* (8S), 5626.

202. Collins, L.; Kilpatrick, J.; Bhaskaran, M.; Sriram, S.; Weber, S. A.; Jarvis, S.; Rodriguez, B. J., Dual harmonic Kelvin probe force microscopy for surface potential measurements of ferroelectrics. In *2012 International Symposium Applications of Ferroelectrics held jointly with 2012 European Conference on the Applications of Polar Dielectrics and 2012 International Symposium Piezoresponse Force Microscopy and Nanoscale Phenomena in Polar Materials (ISAF/ECAPD/PFM)*, IEEE: 2012; pp. 1–4.
203. Kou, L.; Ma, Z.; Li, Y. J.; Naitoh, Y.; Komiyama, M.; Sugawara, Y., Surface potential imaging with atomic resolution by frequency-modulation Kelvin probe force microscopy without bias voltage feedback. *Nanotechnology* 2015, *26* (19), 195701.
204. Borgani, R.; Forchheimer, D.; Bergqvist, J.; Thorén, P.-A.; Inganäs, O.; Haviland, D. B., Intermodulation electrostatic force microscopy for imaging surface photo-voltage. *Applied Physics Letters* 2014, *105* (14), 143113.
205. Borgani, R.; Pallon, L. K.; Hedenqvist, M. S.; Gedde, U. W.; Haviland, D. B., Local charge injection and extraction on surface-modified $Al_2O_3$ nanoparticles in LDPE. *Nano Letters* 2016, *16* (9), 5934–5937.
206. Garcia, R.; Herruzo, E. T., The emergence of multifrequency force microscopy. *Nature Nanotechnology* 2012, *7* (4), 217.
207. Söngen, H.; Nalbach, M.; Adam, H.; Kühnle, A., Three-dimensional atomic force microscopy mapping at the solid-liquid interface with fast and flexible data acquisition. *Review of Scientific Instruments* 2016, *87* (6), 063704.
208. Parpura, V.; Haydon, P. G.; Henderson, E., Three-dimensional imaging of living neurons and glia with the atomic force microscope. *Journal of Cell Science* 1993, *104* (2), 427–432.
209. Fukuma, T.; Ueda, Y.; Yoshioka, S.; Asakawa, H., Atomic-scale distribution of water molecules at the mica-water interface visualized by three-dimensional scanning force microscopy. *Physical Review Letters* 2010, *104* (1), 016101.
210. Frammelsberger, W.; Benstetter, G.; Kiely, J.; Stamp, R., Thickness determination of thin and ultra-thin $SiO_2$ films by C-AFM IV-spectroscopy. *Applied Surface Science* 2006, *252* (6), 2375–2388.
211. Jesse, S.; Baddorf, A. P.; Kalinin, S. V., Switching spectroscopy piezoresponse force microscopy of ferroelectric materials. *Applied Physics Letters* 2006, *88* (6), 062908.
212. Vasudevan, R. K.; Zhang, S.; Baris Okatan, M.; Jesse, S.; Kalinin, S. V.; Bassiri-Gharb, N., Multidimensional dynamic piezoresponse measurements: Unraveling local relaxation behavior in relaxor-ferroelectrics via big data. *Journal of Applied Physics* 2015, *118* (7), 072003.
213. Collins, L.; Okatan, M. B.; Li, Q.; Kravenchenko, I. I.; Lavrik, N. V.; Kalinin, S. V.; Rodriguez, B. J.; Jesse, S., Quantitative 3D-KPFM imaging with simultaneous electrostatic force and force gradient detection. *Nanotechnology* 2015, *26* (17), 175707.
214. Collins, L.; Jesse, S.; Balke, N.; Rodriguez, B. J.; Kalinin, S.; Li, Q., Band excitation Kelvin probe force microscopy utilizing photothermal excitation. *Applied Physics Letters* 2015, *106* (10), 104102.
215. Guo, S.; Kalinin, S. V.; Jesse, S., Open-loop band excitation Kelvin probe force microscopy. *Nanotechnology* 2012, *23* (12), 125704–125704.
216. Guo, S.; Kalinin, S. V.; Jesse, S., Half-harmonic Kelvin probe force microscopy with transfer function correction. *Applied Physics Letters* 2012, *100* (6), 063118.
217. Bayerl, D. J.; Wang, X., Three-dimensional Kelvin probe microscopy for characterizing in-plane piezoelectric potential of laterally deflected ZnO micro-/nanowires. *Advanced Functional Materials* 2012, *22* (3), 652–660.
218. Hong, J.; Park, S.-I.; Khim, Z., Measurement of hardness, surface potential, and charge distribution with dynamic contact mode electrostatic force microscope. *Review of Scientific Instruments* 1999, *70* (3), 1735–1739.

219. Minj, A.; Serron, J.; Celano, U.; Paredis, K., Surface contamination: A natural way toward high-resolution electric force microscopy in contact-resonant mode. *The Journal of Physical Chemistry C* 2020, *124* (46), 25331–25340.
220. Balke, N.; Maksymovych, P.; Jesse, S.; Kravchenko, I. I.; Li, Q.; Kalinin, S. V., Exploring local electrostatic effects with scanning probe microscopy: Implications for piezoresponse force microscopy and triboelectricity. *ACS Nano* 2014, *8* (10), 10229–10236.
221. Jakob, D. S.; Wang, H.; Xu, X. G., Pulsed force Kelvin probe force microscopy. *ACS Nano* 2020, *14* (4), 4839–4848.
222. Rosa-Zeiser, A.; Weilandt, E.; Hild, S.; Marti, O., The simultaneous measurement of elastic, electrostatic and adhesive properties by scanning force microscopy: Pulsed-force mode operation. *Measurement Science and Technology* 1997, *8* (11), 1333.
223. Pittenger, B.; Erina, N.; Su, C., Quantitative mechanical property mapping at the nanoscale with PeakForce QNM. *Application Note Veeco Instruments Inc* 2010, *1*, 1–11.
224. Sadewasser, S.; Glatzel, T.; Rusu, M.; Jäger-Waldau, A.; Lux-Steiner, M. C., Surface photo voltage measurements for the characterization of the $CuGaSe_2$/ZnSe interface using Kelvin probe force microscopy. In *Proceedings of 17th EU Photovoltaics Solar Energy Conference*, 2001; p. 1155.
225. Henning, A.; Günzburger, G.; Jöhr, R.; Rosenwaks, Y.; Bozic-Weber, B.; Housecroft, C. E.; Constable, E. C.; Meyer, E.; Glatzel, T., Kelvin probe force microscopy of nanocrystalline $TiO_2$ photoelectrodes. *Beilstein Journal of Nanotechnology* 2013, *4* (1), 418–428.
226. Beu, M.; Klinkmüller, K.; Schlettwein, D., Use of Kelvin probe force microscopy to achieve a locally and time-resolved analysis of the photovoltage generated in dye-sensitized ZnO electrodes. *Physica Status Solidi (A)* 2014, *211* (9), 1960–1965.
227. Ng, T. N.; Marohn, J. A.; Chabinyc, M. L., Comparing the kinetics of bias stress in organic field-effect transistors with different dielectric interfaces. *Journal of Applied Physics* 2006, *100* (8), 084505.
228. Bürgi, L.; Richards, T.; Chiesa, M.; Friend, R. H.; Sirringhaus, H., A microscopic view of charge transport in polymer transistors. *Synthetic Metals* 2004, *146* (3), 297–309.
229. Melzer, C.; Siol, C.; von Seggern, H., Transit phenomena in organic field-effect transistors through Kelvin-probe force microscopy. *Advanced Materials* 2013, *25* (31), 4315–4319.
230. Collins, S. D.; Mikhnenko, O. V.; Nguyen, T. L.; Rengert, Z. D.; Bazan, G. C.; Woo, H. Y.; Nguyen, T. Q., Observing ion motion in conjugated polyelectrolytes with Kelvin probe force microscopy. *Advanced Electronic Materials* 2017, *3* (3), 1700005.
231. Birkhold, S. T.; Precht, J. T.; Giridharagopal, R.; Eperon, G. E.; Schmidt-Mende, L.; Ginger, D. S., Direct observation and quantitative analysis of mobile frenkel defects in metal halide perovskites using scanning Kelvin probe microscopy. *The Journal of Physical Chemistry C* 2018, *122* (24), 12633–12639.
232. Ahmadi, M.; Collins, L.; Higgins, K.; Kim, D.; Lukosi, E.; Kalinin, S. V., Spatially resolved carrier dynamics at MAPbBr3 single crystal–electrode interface. *ACS Applied Materials & Interfaces* 2019, *11* (44), 41551–41560.
233. Ding, J.; Strelcov, E.; Kalinin, S. V.; Bassiri-Gharb, N., Spatially resolved probing of electrochemical reactions via energy discovery platforms. *Nano Letters* 2015, *15* (6), 3669–3676.
234. Collins, L.; Vasudevan, R. K.; Sehirlioglu, A., Visualizing charge transport and nanoscale electrochemistry by hyperspectral Kelvin probe force microscopy. *ACS Applied Materials & Interfaces* 2020, *12* (29), 33361–33369.
235. Shao, G.; Glaz, M. S.; Ma, F.; Ju, H.; Ginger, D. S., Intensity-modulated scanning Kelvin probe microscopy for probing recombination in organic photovoltaics. *ACS Nano* 2014, *8* (10), 10799–10807.

236. Fernández Garrillo, P. A.; Borowik, Ł.; Caffy, F.; Demadrille, R.; Grévin, B., Photo-carrier multi-dynamical imaging at the nanometer scale in organic and inorganic solar cells. *ACS Applied Materials & Interfaces* 2016, *8* (45), 31460–31468.
237. Coffey, D. C.; Ginger, D. S., Time-resolved electrostatic force microscopy of polymer solar cells. *Nature Materials* 2006, *5*, 735–740.
238. Giridharagopal, R.; Rayermann, G. E.; Shao, G.; Moore, D. T.; Reid, O. G.; Tillack, A. F.; Masiello, D. J.; Ginger, D. S., Submicrosecond time resolution atomic force microscopy for probing nanoscale dynamics. *Nano Letters* 2012, *12* (2), 893–898.
239. Karatay, D. U.; Harrison, J. S.; Glaz, M. S.; Giridharagopal, R.; Ginger, D. S., Fast time-resolved electrostatic force microscopy: Achieving sub-cycle time resolution. *Review of Scientific Instruments* 2016, *87* (5), 053702.
240. Dwyer, R. P.; Nathan, S. R.; Marohn, J. A., Microsecond photocapacitance transients observed using a charged microcantilever as a gated mechanical integrator. *Science Advances* 2017, *3* (6), e1602951.
241. Giridharagopal, R.; Precht, J. T.; Jariwala, S.; Collins, L.; Jesse, S.; Kalinin, S. V.; Ginger, D. S., Time-resolved electrical scanning probe microscopy of layered perovskites reveals spatial variations in photoinduced ionic and electronic carrier motion. *ACS Nano* 2019, *13* (3), 2812–2821.
242. Murawski, J.; Graupner, T.; Milde, P.; Raupach, R.; Zerweck-Trogisch, U.; Eng, L., Pump-probe Kelvin-probe force microscopy: Principle of operation and resolution limits. *Journal of Applied Physics* 2015, *118* (15), 154302.
243. Murawski, J.; Mönch, T.; Milde, P.; Hein, M.; Nicht, S.; Zerweck-Trogisch, U.; Eng, L., Tracking speed bumps in organic field-effect transistors via pump-probe Kelvin-probe force microscopy. *Journal of Applied Physics* 2015, *118* (24), 244502.
244. Schumacher, Z.; Spielhofer, A.; Miyahara, Y.; Grutter, P., The limit of time resolution in frequency modulation atomic force microscopy by a pump-probe approach. *Applied Physics Letters* 2017, *110* (5), 053111.
245. Butt, H.-J., Analyzing electric double layers with the atomic force microscope. In *Encyclopedia of Electrochemistry*, 2007. DOI:10.1002/9783527610426.bard010205.
246. Hillier, A. C.; Kim, S.; Bard, A. J., Measurement of double-layer forces at the electrode/electrolyte interface using the atomic force microscope: Potential and anion dependent interactions. *The Journal of Physical Chemistry A* 1996, *100*, 18808–18817.
247. Ducker, W. A.; Senden, T. J.; Pashley, R. M., Direct measurement of colloidal forces using an atomic force microscope. *Nature* 1991, *353*, 239–241.
248. Raiteri, R.; Grattarola, M.; Butt, H.-J., Measuring electrostatic double-layer forces at high surface potentials with the Atomic Force Microscope. *The Journal of Physical Chemistry A* 1996, *100*, 16700–16705.
249. Sotres, J.; Baro, A. M., AFM imaging and analysis of electrostatic double layer forces on single DNA molecules. *Biophysical Journal* 2010, *98* (9), 1995–2004.
250. Butt, H.-J., Measuring local surface charge densities in electrolyte solutions with a scanning force microscope. *Biophysical Journal* 1992, *63*, 578–582.
251. Raiteri, R.; Butt, H.-J., Measuring electrochemically induced surface stress with an atomic force microscope. *The Journal of Physical Chemistry A* 1995, *99*, 15728–15732.
252. Umeda, K. I.; Kobayashi, K.; Matsushige, K.; Yamada, H., Direct actuation of cantilever in aqueous solutions by electrostatic force using high-frequency electric fields. *Applied Physics Letters* 2012, *101*, 1–5.
253. Umeda, K. I.; Kobayashi, K.; Oyabu, N.; Hirata, Y.; Matsushige, K.; Yamada, H., Analysis of capacitive force acting on a cantilever tip at solid/liquid interfaces. *Journal of Applied Physics* 2013, *113*, 154311.
254. Domanski, A. L.; Sengupta, E.; Bley, K.; Untch, M. B.; Weber, S. A.; Landfester, K.; Weiss, C. K.; Butt, H.-J.; Berger, R., Kelvin probe force microscopy in nonpolar liquids. *Langmuir* 2012, *28* (39), 13892–13899.

255. Strelcov, E.; Arble, C.; Guo, H.; Hoskins, B. D.; Yulaev, A.; Vlassiouk, I. V.; Zhitenev, N. B.; Tselev, A.; Kolmakov, A., Nanoscale mapping of the double layer potential at the graphene–electrolyte interface. *Nano Letters* 2020, *20* (2), 1336–1344.
256. Collins, L.; Jesse, S.; Kilpatrick, J. I.; Tselev, A.; Okatan, M. B.; Kalinin, S. V.; Rodriguez, B. J., Kelvin probe force microscopy in liquid using electrochemical force microscopy. *Beilstein Journal of Nanotechnology* 2015, *6*, 201–214.
257. Collins, L.; Jesse, S.; Kilpatrick, J. I.; Tselev, A.; Varenyk, O.; Okatan, M. B.; Weber, S. A. L.; Kumar, A.; Balke, N.; Kalinin, S. V.; Rodriguez, B. J., Probing charge screening dynamics and electrochemical processes at the solid-liquid interface with electrochemical force microscopy. *Nature Communications* 2014, *5*, 1–8.
258. Checa, M.; Millan-Solsona, R.; Gomila, G., Frequency-dependent force between ac voltage biased plates in electrolyte solutions. *Physical Review E* 2019, *100*, 022604.
259. Gramse, G.; Dols-Perez, A.; Edwards, M. A.; Fumagalli, L.; Gomila, G., Nanoscale measurement of the dielectric constant of supported lipid bilayers in aqueous solutions with electrostatic force microscopy. *Biophysical Journal* 2013, *104*, 1257–1262.
260. Gramse, G.; Edwards, M. A.; Fumagalli, L.; Gomila, G., Theory of amplitude modulated electrostatic force microscopy for dielectric measurements in liquids at MHz frequencies. *Nanotechnology* 2013, *24*, 415709.
261. Millan-Solsona, R.; Checa, M.; Fumagalli, L.; Gomila, G., Mapping the capacitance of self-assembled monolayers at metal/electrolyte interfaces at the nanoscale by in-liquid scanning dielectric microscopy. *Nanoscale* 2020, *12*, 20658–20668.
262. Di Muzio, M.; Millan-Solsona, R.; Borrell, J. H.; Fumagalli, L.; Gomila, G., Cholesterol effect on the specific capacitance of submicrometric DOPC bilayer patches measured by in-liquid scanning dielectric microscopy. *Langmuir* 2020, *36*, 12963–12972.
263. Di Muzio, M.; Millan-Solsona, R.; Dols-Perez, A.; Borrell, J. H.; Fumagalli, L.; Gomila, G., Dielectric properties and lamellarity of single liposomes measured by in-liquid scanning dielectric microscopy. *Journal of Nanobiotechnology* 2021, *19* (1), 1–14.
264. Collins, L.; Weber, S. A.; Rodriguez, B. J., Applications of KPFM-based approaches for surface potential and electrochemical measurements in liquid. In Sadewasser, S., Glatzel, T. (eds.) *Kelvin Probe Force Microscopy*, Springer, Cham: 2018; pp. 391–433.
265. Umeda, K.-I.; Kobayashi, K.; Oyabu, N.; Hirata, Y.; Matsushige, K.; Yamada, H., Practical aspects of Kelvin-probe force microscopy at solid/liquid interfaces in various liquid media. *Journal of Applied Physics* 2014, *116* (13), 134307.
266. Fort Jr., T.; Wells, R. L., Measurement of contact potential difference between metals in liquid environments. *Surface Science* 1968, *12* (1), 46–52.
267. Kobayashi, N.; Asakawa, H.; Fukuma, T., Dual frequency open-loop electric potential microscopy for local potential measurements in electrolyte solution with high ionic strength. *Review of Scientific Instruments* 2012, *83* (3), 033709.
268. Belianinov, A.; Vasudevan, R.; Strelcov, E.; Steed, C.; Yang, S. M.; Tselev, A.; Jesse, S.; Biegalski, M.; Shipman, G.; Symons, C., Big data and deep data in scanning and electron microscopies: Deriving functionality from multidimensional data sets. *Advanced Structural and Chemical Imaging* 2015, *1* (1), 1–25.
269. Collins, L.; Belianinov, A.; Proksch, R.; Zuo, T.; Zhang, Y.; Liaw, P. K.; Kalinin, S. V.; Jesse, S., G-mode magnetic force microscopy: Separating magnetic and electrostatic interactions using big data analytics. *Applied Physics Letters* 2016, *108* (19), 193103.
270. Liu, Y.; Borodinov, N.; Collins, L.; Ahmadi, M.; Kalinin, S. V.; Ovchinnikova, O. S.; Ievlev, A. V., Role of decomposition product ions in hysteretic behavior of metal halide perovskite. *ACS Nano* 2021, *15*, 9017–9026.
271. Leonhard, T.; Schulz, A. D.; Röhm, H.; Wagner, S.; Altermann, F. J.; Rheinheimer, W.; Hoffmann, M. J.; Colsmann, A., Probing the microstructure of methylammonium lead iodide perovskite solar cells. *Energy Technology* 2019, *7* (3), 1800989.

272. Noh, J. H.; Nikiforov, M.; Kalinin, S. V.; Vertegel, A. A.; Rack, P. D., Nanofabrication of insulated scanning probes for electromechanical imaging in liquid solutions. *Nanotechnology* 2010, *21* (36), 365302.
273. Akiyama, T.; Gullo, M. R.; de Rooij, N. F.; Tonin, A.; Hidber, H.-R.; Frederix, P. L.; Engel, A.; Staufer, U., Development of insulated conductive probes with platinum silicide tips for atomic force microscopy in cell biology. *Japanese Journal of Applied Physics* 2004, *43* (6S), 3865.
274. Brown, K. A.; Berezovsky, J.; Westervelt, R. M., Coaxial atomic force microscope probes for imaging with dielectrophoresis. *Applied Physics Letters* 2011, *98* (18), 183103.

# 3 Piezoresponse Force Microscopy and Electrochemical Strain Microscopy

*Qibin Zeng and Kaiyang Zeng*
National University of Singapore

**CONTENTS**



DOI: 10.1201/9781003174042-3

Coupling between the electric field and mechanical strain is a nearly universal phenomenon in a broad variety of materials, such as the piezoelectrics, ferroelectrics, biomaterials such as collagens, energy storage materials, and many others.[1,2] Understanding the electric field-strain (E-S) coupling phenomena can be of great importance for the fundamental science and application studies of these materials. However, limited by the spatial resolution, traditional macroscopic E-S coupling characterization methods, such as the laser interferometry-based E-S measurement,[3,4] can only provide the information from a massive sample volume, and thus to perform a more in-depth analysis for these E-S coupling properties is highly challenging. The invention of the Piezoresponse Force Microscopy (PFM), as well as its derivative, the Electrochemical Strain Microscopy (ESM), marked a dramatic change in this research field by providing an unprecedently high-resolution measurement to explore the nanoscale E-S coupling phenomena.[1,5–10] In fact, PFM and ESM are technically identical, and both of them are developed at the basis of the contact-mode Atomic Force Microscopy (AFM). Based on the powerful AFM platform, PFM/ESM does manifest unique advantages of high spatial resolution, easy implementation, high sample compatibility, and multiple functions, which make them quickly become a mainstream and yet almost indispensable characterization techniques in the studies of E-S coupling of the materials, especially the piezo/ferro-electrics and energy storage materials.[1,2,5–25]

In this chapter, we will focus our discussion on a number of technical aspects of the PFM/ESM, as the specific applications of PFM/ESM in energy research field will be systematically introduced in the other chapters of this book. Firstly, the basic technical principles of the PFM/ESM which includes the Vertical-PFM and ESM, Lateral-PFM, Vector-PFM as well as the PFM/ESM spectroscopy are introduced and then followed by a summary of the key functions of the PFM/ESM, including the important E-S coupling detection, surface domain characterization and manipulation, as well as the voltage spectroscopy measurement. Next, the challenges and issues faced by current PFM/ESM techniques will be systemically discussed, where a number of affecting factors, such as contact-mode scanning, electrostatic force effect, and multi signal source as well as the challenges in spatial resolution and quantification, are discussed. Finally, the most important developments in PFM/ESM technique during the past decades, such as the contact resonance, resonance tracking, metrological, dynamic contact, and multi-frequency heterodyne PFM/ESM, will be briefly reviewed.

## 3.1 PRINCIPLE OF PFM AND ESM

The basic setup of conventional PFM was firstly proposed by Güthner and Dransfeld in 1992.[5] They used an AFM tip to locally pole domains on a ferroelectric polymer film and subsequently image the generated domain patterns by detecting the out-of-plane piezoelectric strain, which initiated the PFM-based studies of nanoscale ferroelectricity.[5] Soon afterwards, Eng et al. demonstrated to use the similar PFM configuration but to measure the in-plane domain structures by monitoring the in-plane piezoelectric strain.[26] To differentiate these two types of PFM measurements, the former was named as Vertical-PFM while the latter was referred as

Lateral-PFM.[27–29] Based on the Vertical- and Lateral-PFM, approaches to reconstruct the 3D and 2D polarization distributions at the sample surface had been developed sequentially,[29–31] which was then collectively named as Vector-PFM.[29] At the same time, the PFM-based spectroscopy measurement, namely the Piezoresponse Force Spectroscopy (PFS), had also been developed as a very effective way to study the local hysteresis behaviors of the ferroelectrics.[32,33] Thereafter, the Vertical-, Lateral-, and Vector-PFM as well as the PFS constitute the major detection form of the PFM which has been extensively utilized up to now.

Motivated by the successful development and applications of the PFM in classical ferroelectric or piezoelectric studies, PFM has been further extended to explore the general E-S coupling phenomena in a broad variety of functional materials.[34–40] A branch of particular importance is the application of PFM (mainly the Vertical-PFM) in characterizing the electro-chemo-mechanical coupling of mobile ion-containing materials, such as the electrode and electrolyte for Lithium-ion battery and the materials for solid-oxide fuel cells.[8,9,19,21,22] For those materials, the periodically biased conductive tip concentrates the electric field in a small volume of the sample, resulting in redistribution of mobile ions through diffusion or electromigration mechanisms. Then the strong coupling between ionic concentration and the lattice parameter causes the molar volume and strain (i.e., Vegard strain) change and finally results in a periodic surface displacement.[9,10] This special electrochemical Vegard strain has successfully been explored to investigate the ionic diffusion properties, leading to the development of ESM.[8–10,17–25] In fact, ESM and PFM are technically identical, and the only difference between them is the signal source, i.e., PFM measures the electric field-induced piezoelectric strain while the ESM measures the electric field-induced electrochemical Vegard strain. Therefore, in this chapter, the PFM and ESM share the same technical discussions by default if not otherwise specified.

### 3.1.1 Vertical-PFM and ESM

Vertical-PFM and ESM have the same instrument configuration, and both of them measure the tip voltage-induced out-of-plane surface strain via detecting the flexural deflection of the AFM cantilever.[23,41] Note that the Vertical-PFM can sometimes be employed to probe the in-plane piezoelectric strain by monitoring the flexural bulking of the cantilever,[6,41,42] but here we focus on discussing its most common application, i.e., detecting the out-of-plane piezoelectric deformation. Figure 3.1a and b schematically shows the basic setup of the Vertical-PFM and ESM, respectively. In the PFM/ESM experiment, a voltage with the form of equation (3.1) is applied between the conductive probe and the substrate to stimulate the sample:

$$V = V_{\mathrm{DC}} + V_{\mathrm{AC}} \sin\left(2\pi ft\right) \tag{3.1}$$

Here, $V_{\mathrm{DC}}$ is an optional DC bias, $V_{\mathrm{AC}}$ and $f$ are amplitude and frequency of the AC bias, and $t$ is time. Under this tip voltage, an out-of-plane piezoelectric strain with the displacement of $Z_p$ will be generated underneath the tip in Vertical-PFM due to the inverse piezoelectric effect.[6,27] Since the piezoelectric displacement is proportional to the applied voltage,[28] the surface displacement $Z_p$ in Vertical-PFM is usually expressed as[28,43]:

$$Z_p = d_{\text{eff}} V = Z_{p0} + A_p \sin\left(2\pi f t + \phi_p\right) \tag{3.2}$$

where $d_{\text{eff}}$ is an artificially introduced effective piezoelectric coefficient, $Z_{p0} = d_{\text{eff}} V_{\text{DC}}$ is the DC displacement, $A_p = |d_{\text{eff}}| V_{\text{AC}}$ is the amplitude of the AC displacement, and $\phi_p$ is the phase difference between the AC displacement and AC bias. As the electric field under the nano-tip is highly nonuniform, $d_{\text{eff}}$ is usually different with the real piezoelectric coefficient $d_{33}$, but theoretically $d_{\text{eff}}$ is dominated by the $d_{33}$.[28] The phase difference $\phi_p$ is determined by the sign of the $d_{\text{eff}}$ or $d_{33}$. For ferroelectric materials, applying positive tip voltage results in the expansion of the $c^-$ domains, i.e., the polarization vector orients out-of-plane with pointing downward (Figure 3.1a), and the AC displacement is in phase with the tip voltage, thus $\phi_p = 0°$, and for $c^+$ ferroelectric domains, $\phi_p = 180°$.[44]

Similar with the Vertical-PFM, an out-of-plane electrochemical Vegard strain with surface displacement of $Z_V$ will be stimulated in ESM due to the ionic motion-induced molar volume changes (Figure 3.1b).[23,45] However, in contrast to the linear piezoelectric coupling, the relationship between the Vegard strain and applied voltage is much more complicated.[8–10] Generally, from the phenomenology point of view, the surface displacement $Z_V$ in ESM can be described by the form of:

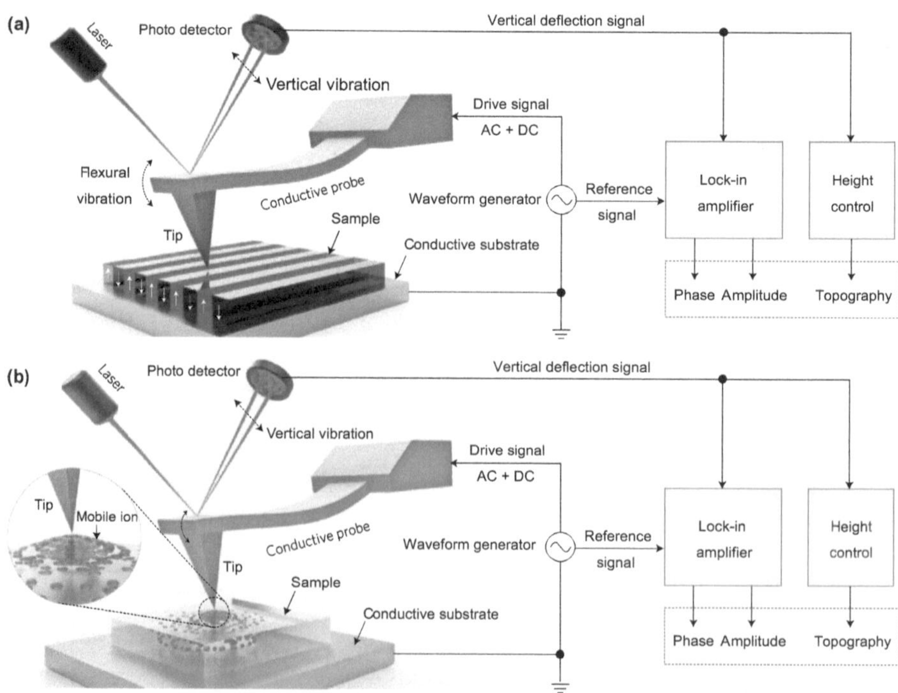


**FIGURE 3.1** Schematic diagram of the (a) vertical-PFM and (b) ESM setup.

$$Z_V \approx Z_{V0} + A_V \sin\left(2\pi ft + \phi_V\right) \tag{3.3}$$

where $Z_{V0}$ is a static DC displacement, $A_V$ and $\phi_V$ are amplitude and phase of the first harmonic AC displacement. Here, $A_V$ and $\phi_V$ are related to both the material's properties and the driving frequency. For the most frequently used ESM (with frequency of ~$10^5$ Hz), the amplitude $A_V$ can be approximately calculated by[8,10]:

$$A_V \approx \sqrt{\frac{2D}{\pi f}} \frac{(1+\nu)\beta}{\eta} V_{ac} \tag{3.4}$$

where $\nu$, $\beta$, and $D$ are the Poisson's ratio, effective Vegard coefficient (determined by the relationship between lattice size and ionic concentration), and the diffusion coefficient of the ion, respectively, and $\eta$ represents the linear relation between chemical potential and applied electric field. However, for the phase $\phi_V$, its analytic expression and physical significance have seldom been systematically discussed in previous studies due to complexity of the electro-chemo-mechanical coupling.[8–10,18,19,24,46,47]

Although both the DC and AC displacements in equations (3.2) and (3.3) can reflect the sample's physical properties, only the AC displacement is measured in Vertical-PFM and ESM because the magnitude of $Z_p$ and $Z_V$ are typically very small (within picometer range) and using AC measurement scheme we can obtain a much better sensitivity and signal-to-noise ratio (SNR).[28] To detect the AC displacement, the Vertical-PFM and ESM are operated in repulsive contact mode; thus the out-of-plane surface displacement underneath the tip can stimulate a flexural vibration on the cantilever via the tip-sample interaction,[1,44] indicating that the AC displacement can finally be measured by monitoring the vibration of the cantilever. Typically, the deformation of the AFM cantilever is detected by a four-quadrant photo detector (FQPD) using a laser beam lever, where the flexural vibration signal is contained in the vertical deflection output of the FQPD (Figure 3.1). By using a lock-in amplifier to demodulate the first harmonic component of the vertical deflection signal, the amplitude ($A_{C\text{-}F}$) and phase ($\phi_{C\text{-}F}$) of the flexural vibration of the cantilever can be obtained. In a typical sample-tip-cantilever system such as those used in the Vertical-PFM and ESM, the vertical cantilever displacement, induced by the out-of-plane surface displacement, can be calculated by:

$$D_c(f) = H(f) \cdot D_s(f) \tag{3.5}$$

where $D_c(f)$ and $D_s(f)$ are the displacements of the cantilever and surface in frequency domain, respectively, and $H(f)$ is the transfer function of the sample-tip-cantilever system. Equation (3.5) implies that for a constant $H(f)$, the cantilever displacement can directly reflect the variation of the surface displacement. Therefore, by simply monitoring the $A_{C\text{-}F}$ and $\phi_{C\text{-}F}$, the nanoscale AC surface displacement in the Vertical-PFM and ESM can be measured. Meanwhile, as the DC component of the vertical deflection signal is used as the height feedback signal in contact-mode operation, the topography is usually imaged simultaneously with the out-of-plane AC displacement in the Vertical-PFM and ESM (Figure 3.1).

### 3.1.2 Lateral-PFM and Vector-PFM

At the beginning, the PFM studies were limited to the form of Vertical-PFM, i.e., measuring the out-of-plane piezoelectric strain based on the flexural vibration of the cantilever, thereby only the vertical deflection signal of the FQPD was used in the earlier studies.[5,16,48,49] A few years later, Eng et al. demonstrated to measure the in-plane piezoelectric strain by monitoring the torsional vibration of cantilever using the lateral deflection signal of FQPD, which was then named as Lateral-PFM.[26] The Lateral-PFM almost has a same instrument configuration with that of the Vertical-PFM, and the major differences between them are the cantilever's vibration form and signal acquisition channel. Figure 3.2 shows the basic setup of Lateral-FPM. Similarly, in Lateral-PFM, a voltage with the form of equation (3.1) is applied in between the conductive tip and substrate to stimulate the sample. For samples that generate non-zero in-plane piezoelectric strains underneath the tip, such as the ferroelectrics with in-plane distributed *a*-domains (see Figure 3.2), the corresponding in-plane DC and AC surface displacements will cause in-plane DC and AC friction forces to the tip apex thus the cantilever is driven to vibrate.[28] There are two possible vibration forms that can be excited by the in-plane friction force, i.e., the buckling and torsional vibrations which are excited by the forces with direction along and normal to the cantilever axis, respectively.[6,41] As mentioned above, the buckling vibration can be detected by Vertical-PFM, while in Lateral-PFM, only the torsional vibration is measured. To measure the torsional motion of the cantilever, the lateral deflection output of the QFPD is monitored, and at the same time, the vertical deflection signal is acquired for simultaneous topography imaging (Figure 3.2). By using a lock-in amplifier, the amplitude ($A_{C\text{-}T}$) and phase ($\phi_{C\text{-}T}$) of the first harmonic component of the lateral deflection signal can be obtained. As the torsional vibration of the cantilever and in-plane AC surface displacement have a similar relationship as shown in equation (3.5), for a constant sample-tip-cantilever system, $A_{C\text{-}T}$ and $\phi_{C\text{-}T}$ can unambiguously reveal the distribution of in-plane AC displacement and thereby the piezoelectric or ferroelectric properties, such as the shear piezoelectric activity distribution and domain structure of the *a*-domains.[6,28,41] Note that the ESM can also have the same lateral detection form in theory, but nearly all of the ESM

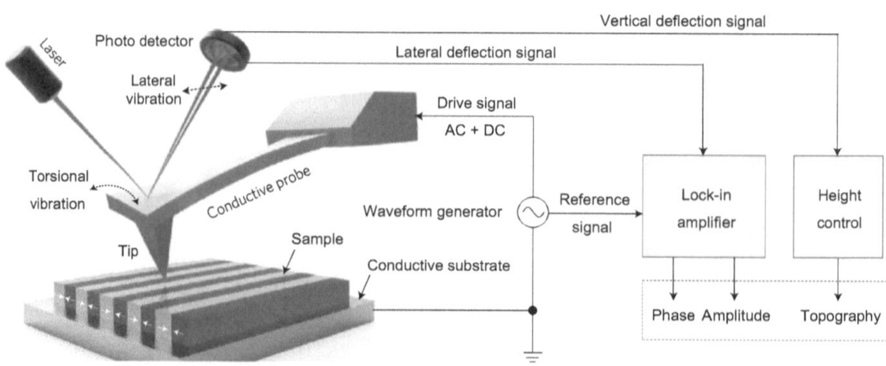


**FIGURE 3.2** Schematic diagram of the lateral-PFM setup.

measurements are based on the vertical detection and thus the concept of lateral ESM is rarely mentioned.

According to the technical principle of the Vertical- and Lateral-PFM, it is easy to understand that both of them can only provide the piezoelectric information along specific polarization directions, which may satisfy the characterization requirements of samples with matched polarization vectors, such as the single crystals and thin films with the polarization vector along the out-of-plane or in-plane directions. However, for samples with randomly oriented polarization vectors, such as the polycrystal ferroelectric ceramics, neither Vertical- nor Lateral-PFM can independently give an accurate measurement for the domain structures and polarization orientations.[29] By systematically analyzing the components of tip electric field-induced piezoelectric strain, Eng et al. developed an approach to reconstruct the three-dimensional (3D) polarization distribution of the sample surface by using a combination of the Vertical- and Lateral-PFM measurements,[30] leading to the development of Vector-PFM.[29] Vector-PFM is completely based on the basic Vertical- and Lateral-PFM; thus no additional instrument modifications are required, and in fact, the key procedurals of Vector-PFM measurement are mainly about the calibration and mathematical operation of the obtained Vertical- and Lateral-PFM data.[29–31] With the calibrated Vertical- and Lateral-PFM data, Vector-PFM allows reconstructing the two-dimensional (2D) or 3D polarization vector maps for arbitrarily oriented domains using two (one Vertical-PFM image + one Lateral-PFM image, or two Lateral-PFM images) or three (one Vertical-PFM image + two Lateral-PFM images) data sets, respectively.[29] A detailed introduction about implementing the Vector-PFM measurement can be found in the literature,[29] in which the calibration and mathematical processing of the Vertical- and Lateral-PFM data sets have been systematically discussed.

### 3.1.3 PFM and ESM Spectroscopy

In addition to the high-resolution mapping capability, the unique instrument configuration of the PFM/ESM makes it perfectly support the DC bias-based spectroscopy measurement. Referring to the macroscopic P-E loop measurement,[50] the PFM spectroscopy (i.e., the PFS) was firstly proposed as an analogue to measure the local ferroelectric hysteresis loop at a single point,[33] and then this method has been further developed into a 2D mapping form, i.e., the switching spectroscopy PFM (SS-PFM),[51] where the hysteresis loop is acquired in a 2D spot-array with the point-by-point measurements. After the inception of ESM, exactly the same spectroscopy method, including both the single-point and 2D mapping forms, has been also used in the ESM studies.[8,19,52] Since the applied tip voltage in PFM/ESM has the form of equation (3.1), the spectroscopy measurement can be easily implemented by applying different $V_{DC}$ to modulate the local status of the sample, such as the polarization, and at the same time, monitoring the amplitude and phase of the AC surface displacement as a function of $V_{DC}$. When performing the PFM/ESM spectroscopy measurement, the tip is fixed at a point and an AC drive superimposed on a predefined DC probing wave is applied to the tip, while the amplitude and phase of the AC displacement are acquired in synchronization with the DC wave. According to the different measurement demands, the DC probing wave can be specifically designed.[19,46,53–56] In general, there are two strategies to set the DC probing wave, including the continuous and pulsed DC modes,[15,53,54] which are schematically shown in Figure 3.3. The continuous DC mode (Figure 3.3a) is very similar with the conventional macroscopic P-E loop measurement, in which the AC surface displacement is measured under a continuously changed DC bias. Since the DC bias is continuous, the possible relaxation effects, such as the polarization back-switching,[53,54] can be avoided during the spectroscopy measurement. However, as the surface displacement is always measured under the DC bias in the continuous DC mode, the large electrostatic force will significantly affect the surface displacement signal and then cause remarkable distortion and artifacts to the final spectrum (which will be discussed later).[15,53,57] Given this severe issue of the continuous DC spectroscopy mode, a pulse DC mode is proposed.[15,53,54] In the pulse DC mode (Figure 3.3b), the desired DC bias is alternatively applied for a determined period of time (on state) and subsequently released to zero for a specific period (off state). During the measurement, the AC surface displacement signal can be acquired at both on and off states, which generates the on-field and off-field spectra, respectively.[58] However, the on-field and off-field spectra are different. For the on-field spectrum, the electrostatic force effect can be significant as it is measured when the DC bias is on, whereas for the off-field spectrum, the electrostatic force is substantially minimized since the DC bias is off during the off state measurement. Therefore, the on-field and off-field spectra are usually analyzed comparatively to avoid the electrostatic force-induced incorrect interpretation of the spectroscopy results.[43,53,59] Although the pulsed DC method can obtain the spectrum with minimized electrostatic force effects, the alternative on and off states in fact make the DC poling process discontinuous, and thus the abovementioned relaxation behaviors can be pronounced in this pulsed DC mode.[53,54] However, sometimes the relaxation behaviors are important information in material studies, such as investigating the polarization of relaxor ferroelectrics; to this end, PFM/ESM spectroscopy can further be performed with a time-resolved scheme to

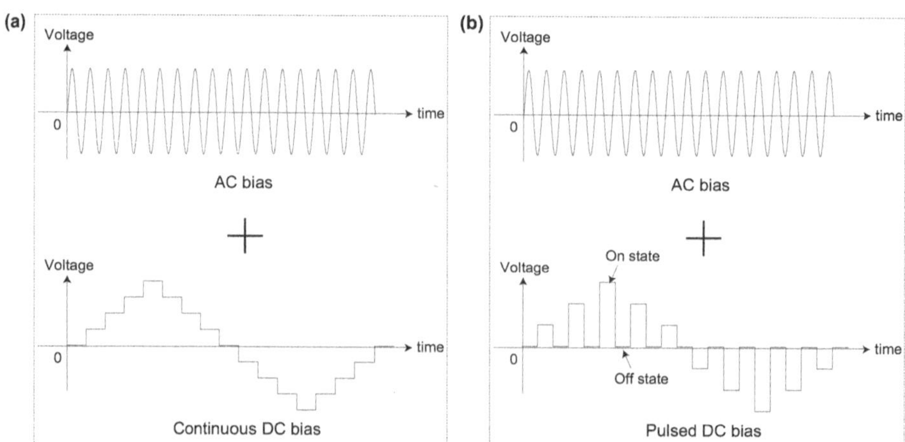


**FIGURE 3.3** Schematic illustration of the excitation signal used in PFM/ESM spectroscopy. (a) The continuous DC mode and (b) the pulsed DC mode PFM/ESM spectroscopy.

trace the relaxation processes.[19,20,60–62] In the time-resolved PFM/ESM spectroscopy, the AC displacement signal after each DC bias is continuously recorded for a period of time (typically $10^1$ to $10^3$ ms), implying that a time-resolved spectrum is acquired at each DC bias, and thus the voltage poling processes can be analyzed with both voltage and time resolutions.

## 3.2 FUNCTIONS OF PFM AND ESM

Although the PFM/ESM configuration is designed for ferroelectric studies in the beginning, the subsequent developments and applications have clearly demonstrated that this technique can be broadly used to study multiple physical properties with E-S coupling behavior, such as the piezoelectricity, ferroelectricity, solid state electrochemistry, electrostriction, and flexoelectricity.[2,12–14,23,28,45,63–65] Despite the extensive applications in diverse research topics, PFM/ESM in general provides three fundamental functions in these measurements, including the E-S coupling detection, surface domain characterization and manipulation, as well as the voltage spectroscopy measurement. With these basic functions of the PFM/ESM, a wealth of physical properties behind the E-S coupling can be investigated, such as the domain wall growth kinetics, ferroelectric polarization dynamics, electrochemical activity, and even nanomechanics.[66–68] In this section, we will discuss the fundamental functions of the PFM/ESM, and the specific studies of material properties by using PFM/ESM will be introduced in other chapters of this book.

### 3.2.1 Electric Field-Strain Coupling Detection

The E-S coupling detection is the most fundamental function of the PFM/ESM, and all of the PFM/ESM-based physical property studies are actually implemented via detecting the tip electric field-induced surface strain. Figure 3.4 shows the basic schemes of using the PFM/ESM for E-S coupling measurements. Firstly, shown in Figure 3.4a–c are single-point PFM measurements of piezoelectric, electrostriction, and converse flexoelectric strains, respectively.[65,69,70] The single-point detection is the simplest operation form in the PFM/ESM measurements, and by measuring the amplitude of the AC surface displacement (typically under low frequency) at a fixed point, it is frequently used to quantify the E-S coupling parameters, such as the piezoelectric coefficient $d_{33}$.[69–71] 2D mapping the distribution of the electric field-induced strain is another common application of the PFM/ESM, in which the inhomogeneity of the E-S coupling properties can be clearly revealed. Figure 3.4d and e shows the typical 2D amplitude mapping of the piezoelectric and electrochemical strains using PFM and ESM, respectively. With the 2D PFM/ESM mapping, the piezoelectricity of CdS nanoplate (Figure 3.4d) and the nonuniform lithium-ion diffusion and intercalation behaviors (Figure 3.4e) can be clearly observed.[8,70]

### 3.2.2 Surface Domain Characterization and Manipulation

Analyzing the surface domain structures with high spatial resolution is the most important function of the PFM/ESM. By mapping the 2D distributions of the surface

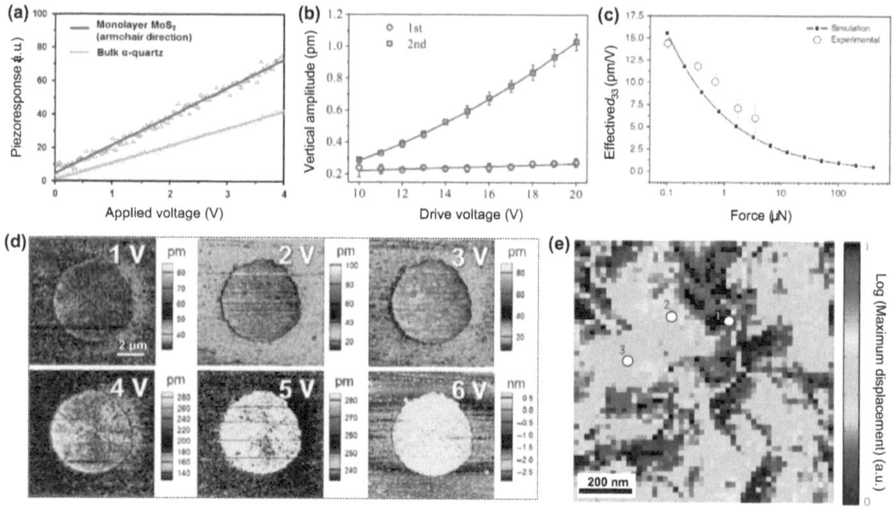


**FIGURE 3.4** Electric field-strain coupling detection by PFM/ESM. (a) Single-point detection of the piezoelectric strain of $MoS_2$ and quartz, (b) the electrostriction strain of silicon, and (c) the converse flexoelectric strain of $SrTiO_3$ by PFM. (d) 2D mapping of the piezoelectric strain of CdS and (e) electrochemical strain of $LiCoO_2$ using PFM and ESM, respectively. ((a) Reprinted with permission from Kim et al.[69] Copyright 2016 Elsevier. (b) Reprinted with permission from Yu et al.[65] Copyright 2018 American Institute of Physics. (c) Adapted from Abdollahi et al.[64] under the terms of the CC-BY 4.0 license. (d) Adapted from Wang et al.[70] under the terms of the CC-BY 4.0 license. (e) Adapted with permission from Ref. Balke et al.[8] Copyright 2010 Springer Nature.)

vibration amplitude and phase, PFM/ESM can unambiguously define the domain structures with the lateral resolution up to sub-10 nm.[1,13] In addition, the localized electric field-based probing scheme makes PFM/ESM has unprecedent feasibility to manipulate the surface domains at nanoscale. Since the ferroelectric domain is the most common form of the domain formed in materials with E-S coupling, the vast majority of the domain studies are focused on ferroelectric materials with utilizing PFM.[1,6,7,14,41,63,72] However, interesting domain structures studied by PFM/ESM have also been reported from several non-ferroelectrics or debatable ferroelectrics.[21,40,73–75] Figure 3.5a–c shows the basic domain mapping of ferroelectrics by using PFM. From the PFM amplitude (Figure 3.5a) and phase (Figure 3.5b) images, the domain structure of periodically poled lithium niobate (PPLN), including the domain shape, domain wall, and polarization orientation, can be clearly observed. Figure 3.5c shows a typical 3D domain characterization by Vector-PFM, in which the final 3D domain image is reconstructed from three PFM data sets, including one Vertical-FPM and two Lateral-PFM images (usually marked as *x*Lateral- and *y*Lateral-PFM).[29] In addition to domain mapping, domain manipulation is also extensively performed by PFM/ESM with single-point and 2D writing methods. Figure 3.5d and e shows the PFM amplitude and phase images of a domain array fabricated artificially via the single-point writing method.[76] By applying DC writing pulse to a fixed point, a point domain can be created and its radius can be controlled by adjusting the pulse

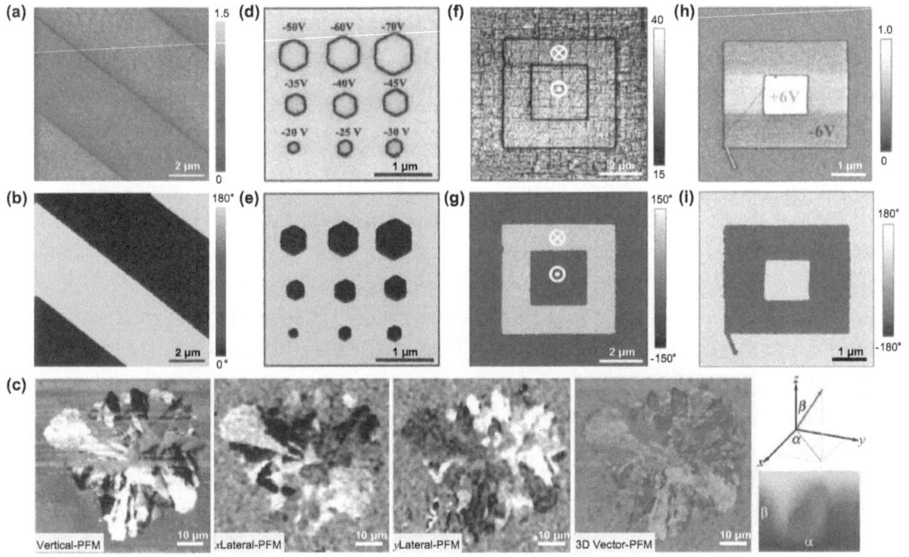


**FIGURE 3.5** Surface domain characterization and manipulation by PFM/ESM. (a) PFM amplitude and (b) phase images of the PPLN sample. (c) 3D vector-PFM mapping of $LaBGeO_5$ crystallite. (Adapted with permission from Kalinin et al.[29] Copyright 2006 Microscopy Society of America.) (d) PFM amplitude and (e) phase images of the ferroelectric domains fabricated on $LiNbO_3$ with single-point writing method. (Adapted with permission from Rodriguez et al.[76] Copyright 2005 American Institute of Physics.) (f) PFM amplitude and (g) phase images of the ferroelectric domains fabricated on PZT with 2D writing method. (Adapted from Tan et al.[77] under the terms of the CC-BY 4.0 license.) (h) PFM/ESM amplitude and (i) phase images of non-ferroelectric domains fabricated on $LaAlO_3$ with 2D writing method. (Adapted with permission from Sharma et al.[74] Copyright 2015 Wiley-VCH.)

amplitude and width. This single-point domain writing scheme can potentially be used for ferroelectric-based high-density data storage as it can generate very small domain patterns.[12] In contrast, the 2D writing method is usually used to create much larger domains, in which the domain is fabricated by performing 2D scanning on sample surface while a DC writing bias is applied to tip. As the scanning area and DC writing bias at each position can be manually controlled, the 2D writing method can imitate the conventional lithography to produce surface domains with desired size and shape.[2,13] Figure 3.5f–i demonstrates the domain patterns fabricated on ferroelectric $Pb(Zr, Ti)O_3$ (PZT) (Figure 3.5f and g) and non-ferroelectric $LaAlO_3$ thin films (Figure 3.5h and i) by PFM/ESM with the 2D writing method.[74,77] Note that for some non-ferroelectrics or debatable ferroelectrics, such as the $LaAlO_3$ and $CH_3NH_3PbI_3$ perovskite, both the PFM/ESM signal origination and the underlying physics of the domain formation are still under discussion or debate,[40,67,68,73–75,78] and thus it is difficult to determine whether it is PFM or ESM used in these studies, which is exactly a significant challenge faced by PFM/ESM when they are used to characterize the properties of unknown materials (will be discussed later).

### 3.2.3 Voltage Spectroscopy Measurement

Exploring the variation of material properties, such as ferroelectricity and solid-state ionic transportation, under external electric field is of great importance for many electric field-based device applications.[13,23] With the capabilities of local electric field poling and in-situ E-S coupling detection, PFM/ESM uniquely provides the voltage spectroscopy function to access these research topics.[7,13,14,23,45,56] For PFM spectroscopy (i.e., the PFS), the most important application is to measure the local ferroelectric hysteresis loop, which can help understanding the physical mechanisms of ferroelectric polarization switching from nanoscale.[13,27,51,54,55,72,79] Figure 3.6a and b shows a typical PFS measurement on ferroelectric PZT using the pulsed DC method, and both the on-field and off-field PFM amplitude (Figure 3.6a) and phase (Figure 3.6b) exhibit the standard hysteretic variation with the applied voltage. On the other hand, the ESM spectroscopy is mainly used to investigate the solid-state ionic transportation properties.[23,45] Shown in Figure 3.6c is an ESM spectroscopy measurement on $LiCoO_2$, in which an electrochemical hysteresis of the material can be observed when the applied bias changes.[8] In addition to the

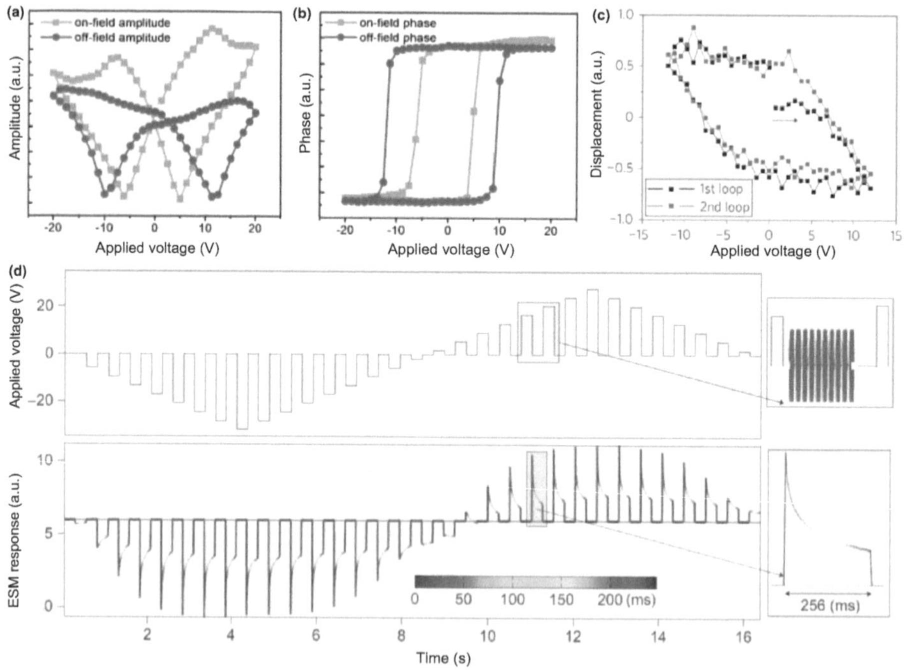


**FIGURE 3.6** Voltage spectroscopy measurement by PFM/ESM. (a) PFM amplitude and (b) phase hysteresis loops measured on PZT. (Reprinted with permission from Anbusathaiah et al.[184] Copyright 2010 Elsevier.) (c) ESM response as a function of DC bias measured on $LiCoO_2$. (Reprinted with permission from Balke et al.[8] Copyright 2010 Springer Nature.) (d) Time-resolved ESM spectroscopy measurement on YSZ. (Adapted with permission from Kumar et al.[19] Copyright 2011 Springer Nature.)

basic spectroscopy mode, the PFM/ESM spectroscopy can be easily performed by a time-resolved manner. Figure 3.6d demonstrates a typical time-resolved ESM spectroscopy measurement on yttrium-stabilized zirconia (YSZ), in which the oxygen vacancy relaxation processes after each DC bias are measured in an individual time-resolved spectrum, and thus finally the whole spectrum provides both voltage and time resolutions for analyzing the local transportation properties of the oxygen vacancies in YSZ.[19]

## 3.3 CHALLENGES IN PFM AND ESM

Since its inception in 1992, PFM/ESM has evolved from a relatively obscure and at times controversial scanning probe technique to a broadly accepted tool for nanoscale E-S coupling detection, domain characterization and manipulation, as well as voltage spectroscopy measurement.[1,2] Although PFM and ESM have attained great success in the studies of electromechanical and electro-chemo-mechanical coupling properties, with the increasing width and depth of the research for variety of materials, the limitations of the PFM/ESM, such as the abovementioned electrostatic force and signal source issues, are gradually emerging as bottlenecks which have been significantly impeding its analysis and further applications.[1,2,43,80,81] In this section, the most important challenges faced by PFM/ESM will be discussed, including the contact-mode operation, electrostatic force effect, multi signal source issue, spatial resolution limitation and quantification.

### 3.3.1 Contact-Mode Operation

Since the measurement of the tiny surface displacement in PFM/ESM relies on the local tip-sample interaction, PFM/ESM has to be operated at the static contact mode to maintain a strong enough yet stable tip-sample interaction.[44] However, this classical contact-mode operation does become one of the most fundamental limitations of the PFM/ESM, as the contact-mode operation can induce pronounced modifications to both the tip and sample surface, such as the tip abrasion, contamination, and damage,[1,82–87] as well as the damage and contamination in the sample surface,[1,85–88] and all of these will affect the PFM/ESM signal, thereby resulting in significant reproducibility problems and artifacts in the measurements. Figure 3.7 schematically shows the common situations of tip and sample modifications caused by contact-mode operation, including the tip abrasion (Figure 3.7a), tip contamination (Figure 3.7b), tip damage (Figure 3.7c), sample damage (Figure 3.7d), and sample contamination (Figure 3.7e).

Figure 3.8 shows several experimental evidences obtained by Scanning Electron Microscopy (SEM) (Figure 3.8a–c)[83,87,89] and Time-of-Flight Secondary Ion Mass Spectrometry (ToF-SIMS) (Figure 3.8d),[88] which clearly demonstrate the occurrence of tip contamination (Figure 3.8a),[87] abrasion (Figure 3.8(b))[89] and damage (Figure 3.8c),[83] as well as the contamination and chemical modification of the sample surface (Figure 3.8d)[88] after contact-mode scanning. Since the sample and tip modifications are almost inevitable during the contact-mode PFM/ESM scanning, by far researchers can only have very limited solutions to reduce such effects, such

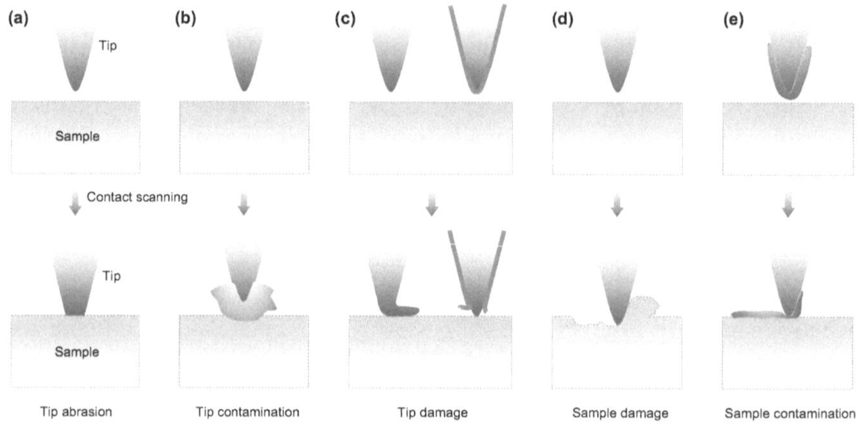


**FIGURE 3.7** Schematic illustrations of the possible tip and sample modifications caused by contact-mode operation. (a) The abrasion and (b) contamination of tip. (c) The tip damage, including the tip geometry change (left) and conductive coating damage (right). (d) The sample surface damage caused by the scratch of tip. (e) The sample surface contamination induced by using contaminated tip.

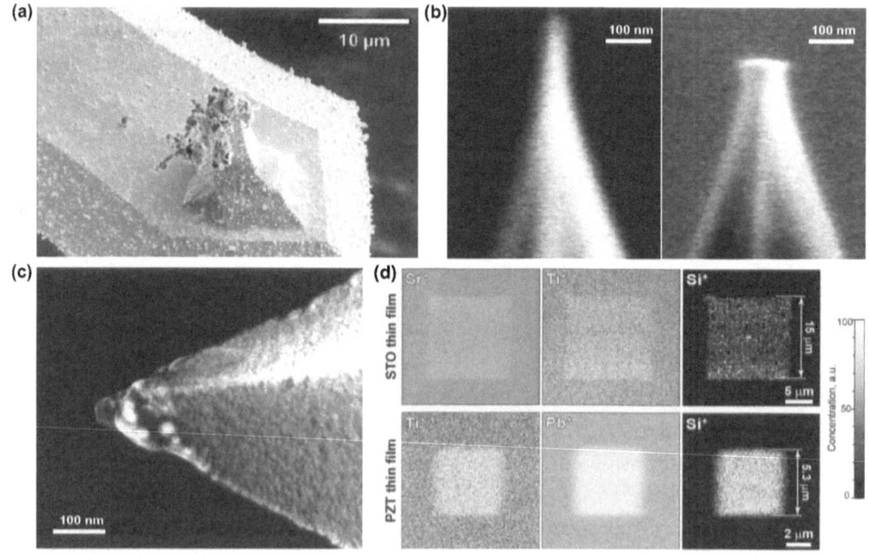


**FIGURE 3.8** Experimental evidences of tip and sample modifications after contact-mode scanning. (a) SEM images that show the tip contamination, (b) tip abrasion, and (c) damage of the conductive coating. ((a) Reprinted with permission from Ref. Saadi et al.[87] Copyright 2020 IOP Publishing. (b) Adapted with permission from Kopycinska-Müller et al.[89] Copyright 2006 Elsevier. (c) Adapted with permission from Lantz et al.[83] Copyright 1998 American Institute of Physics.) (d) ToF-SIMS maps of the spatial distributions of $Sr^+$, $Ti^+$, and $Si^+$ on the surface of $SrTiO_3$ (STO) and $Ti^+$, $Pb^+$, and $Si^+$ on the surface of PZT after contact-mode scanning using silicon tip with Pt-Ir coating. (Adapted with permission from Ievlev et al.[88] Copyright 2018 American Chemical Society.)

as decreasing the setpoint of the force feedback, using softer cantilever, and mapping with the point-by-point manner.[85,90,91]

In addition to the modifications, the tip-sample contact force usually introduces a significant stress field in the sample, which can potentially lead to a remarkable flexoelectric polarization[92–94] and affect the ferroelectric polarization switching behavior.[95,96] Meanwhile, experimental studies have shown that the contact-mode scanning is prone to induce charge injection[97,98] and triboelectrification[99] on the sample surface, which will change the local surface potential and finally cause artifacts in the PFM/ESM measurements via the electrostatic force effect. Furthermore, as there is a contact area formed at the tip-sample junction which contains a large number of interactional atoms, the contact-mode operation fundamentally limits the spatial resolution of the PFM/ESM to the best of sub-10 nm,[1,6,7] and a higher spatial resolution, such as the molecular or atomic resolution, can hardly be achieved. Another important issue closely related to the contact-mode operation is the resonance tracking. As the surface strain induced by tip electric field is typically with the order of picometers, resonance amplification is usually necessitated in PFM/ESM to enhance the SNR and reduce the AC drive needed.[100–103] However, a significant concern with the resonance enhancement is that the resonance is highly sensitive to the tip-sample contact status, and a slight change of the contact status may cause an apparent influence to the PFM/ESM signal.[43,100,104] Given the very high possibility of the occurrence of tip and sample modifications in contact-mode operation, once resonance enhancement is utilized in the PFM/ESM measurements, the resonance tracking issue is usually inevitable even when the sample surface is flat.

### 3.3.2 Electrostatic Force Effect

Electrostatic force is one of the most intractable issues which continuously affects the PFM/ESM signal for a long time since its invention. Because the existence of electrostatic force can give rise to significant artifacts or misinterpretations in PFM/ESM studies,[7,15,43,53,59,80,101,105–109] a large amount of research work has been implemented to separate, quantify, or eliminate the contribution of the electrostatic force from PFM/ESM results.[1,2,15,27,41,44,53,57,59,80,97,101,106,110–119] As the PFM/ESM system is composed of a conducting tip in contact with the dielectric surface on a conductive substrate, which can be considered as a capacitor structure (Figure 3.9a). As a consequence, when an external voltage with the form of equation (3.1) is applied between the AFM probe and the conductive substrate, an additional electrostatic force, $F_{\text{EF}}$, will be generated along the whole probe. Due to the structure of the AFM probe, $F_{\text{EF}}$ is usually considered as the summation of two parts, i.e., the local ($F_{\text{L-EF}}$) and distribute ($F_{\text{D-EF}}$) electrostatic force (Figure 3.9a), which are defined by the following equation[1,44]:

$$\begin{aligned} F_{\text{EF}} &= F_{\text{L-EF}} + F_{\text{D-EF}} \\ &= \frac{1}{2}C'_L\left[V_{\text{dc}} - V_{\text{L-cpd}} + V_{\text{ac}}\sin(2\pi ft)\right]^2 + \frac{1}{2}C'_D\left[V_{\text{dc}} - V_{\text{D-cpd}} + V_{\text{ac}}\sin(2\pi ft)\right]^2 \end{aligned} \quad (3.6)$$

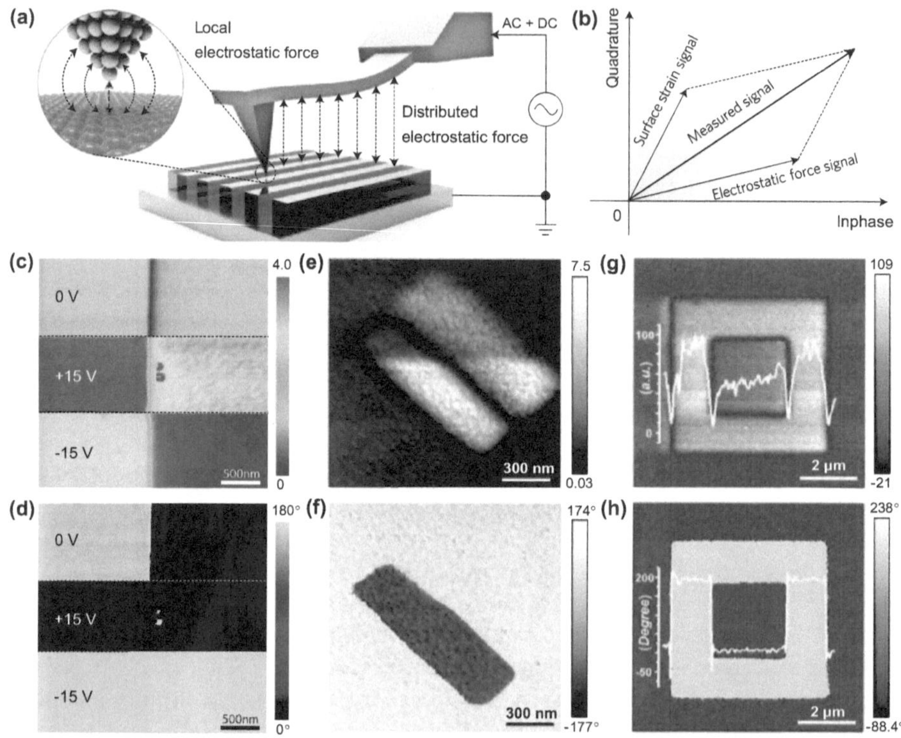


**FIGURE 3.9** Electrostatic force and its influence on PFM/ESM domain characterization. (a) Schematic of the local and distributed electrostatic forces in PFM/ESM. (b) Vector diagram that shows the relationship between surface strain signal, electrostatic force signal, and measured signal in PFM/ESM. (c) PFM amplitude and (d) phase images of PPLN measured with 0, and ±15 V DC bias applied. (Adapted from Zeng et al.[120] under the terms of the CC-BY 4.0 license.) (e) PFM amplitude and (f) phase images measured on non-ferroelectric amorphous $HfO_2$ after DC poling. (Adapted with permission from Balke et al.[59] Copyright 2015 American Chemical Society.) (g) PFM amplitude and (h) phase images measured on ferroelectric $BaTiO_3$ ultrathin film after DC poling. (Adapted with permission from Yao et al.[98] Copyright 2020 Elsevier.)

where $C'_L$ and $V_{\text{L-cpd}}$ are local capacitance gradient and contact potential difference between the tip apex and sample, respectively. $C'_D$ and $V_{\text{D-cpd}}$ are average capacitance gradient and contact potential difference between the cantilever and sample, respectively. By expanding the equation (3.6), the first harmonic component of the electrostatic force, $F_{\text{EF-1}\omega}$, can be obtained, which is given by

$$F_{\text{EF-1}\omega} = \left[C'_L(V_{\text{DC}} - V_{\text{L-cpd}}) + C'_D(V_{\text{DC}} - V_{\text{D-cpd}})\right]V_{\text{AC}}\sin(2\pi ft) \tag{3.7}$$

Obviously, $F_{\text{EF-1}\omega}$ can drive the cantilever to vibrate at the same frequency of the surface strain. Therefore, in PFM/ESM, the cantilever is actually excited by the surface strain and electrostatic force simultaneously, and the corresponding signal composition can be understood by the vector diagram shown in Figure 3.9b. It can be clearly seen that the final measured signal is always the vector sum of the real surface strain signal and the electrostatic force signal, and therefore the influence of the electrostatic force is very difficult to be eliminated in PFM/ESM where the standard first harmonic detection approach is adopted.[1] Figure 3.9c and d demonstrates the typical influences of the electrostatic force in PFM domain characterization of the PPLN sample,[120] and it is clear that the PFM amplitude (Figure 3.9c) and phase (Figure 3.9d) contrasts can be totally dominated by the electrostatic force. Since the electrostatic force can significantly affect the domain image, the charge injection effect may cause pronounced artifacts in PFM/ESM domain manipulation experiment, as the introduction of surface charge will largely change the $V_{\text{L-cpd}}$ and thereby the electrostatic force. Shown in Figure 3.9e–h are pseudo ferroelectric domains obtained by performing 2D domain writing on non-ferroelectric $HfO_2$ and ferroelectric $BaTiO_3$ thin films, respectively.[59,98] Although the PFM amplitude (Figure 3.9e and g) and phase (Figure 3.9f and h) images all will manifest the ferroelectric domain features, further studies have confirmed that these domain contrasts actually indicate the charge injection-induced electrostatic force instead of the ferroelectric polarization.[59,98] In addition to the domain imaging, the PFM/ESM spectroscopy is also significantly affected by the electrostatic force. According to equation (3.7), $F_{\text{EF-1}\omega}$ has a linear relationship with the applied $V_{\text{DC}}$, implying that the electrostatic force can get a substantial enhancement during the PFM/ESM spectroscopy measurement, especially when continuous DC mode is used.

Figure 3.10a and b shows the PFM amplitude and phase as a function of DC bias measured on PPLN using continuous DC mode.[120] The V-shaped amplitude curve (Figure 3.10a) and near 180° phase change (Figure 3.10b) indicate that the first harmonic electrostatic force is affecting the PFM signal significantly.[111,112,120] Figure 3.10c and d displays the PFM hysteresis loops of the PZT obtained by using continuous and pulsed DC spectroscopy modes, respectively.[53,57] The continuous and on-field hysteresis loops in Figure 3.10c and d clearly show that even on a material with strong piezoelectricity, the shape of the loop can still be largely distorted by the electrostatic force. Although the pulsed DC spectroscopy mode can obtain the spectrum with minimized electrostatic force effects at the off state, the charge injection that occurs at the on state can still induce pronounced electrostatic force at the off state. Shown in Figure 3.10e and f are off-field PFM hysteresis loops measured on non-ferroelectric $HfO_2$ and ferroelectric $BaTiO_3$ thin films, respectively.[59,98] Despite the typical PFM hysteretic features of ferroelectrics that can be observed (Figure 3.10e and f), these amplitude and phase loops actually reflect the charge injection-induced variation of the electrostatic force.[59,98]

After the continuous effort in addressing electrostatic force issue, it has been proved that the electrostatic force contribution can be relatively minimized by using probes with high spring constant or imaging the materials with strong E-S coupling.[1,41,53,63,80,113] In addition, introducing special shielded probes to screen the electric field of the cantilever and applying a DC offset to compensate the contact potential difference are also proposed to minimize the electrostatic force.[2,63,111] However, high spring constant probe will decrease the detection sensitivity of the surface strain and

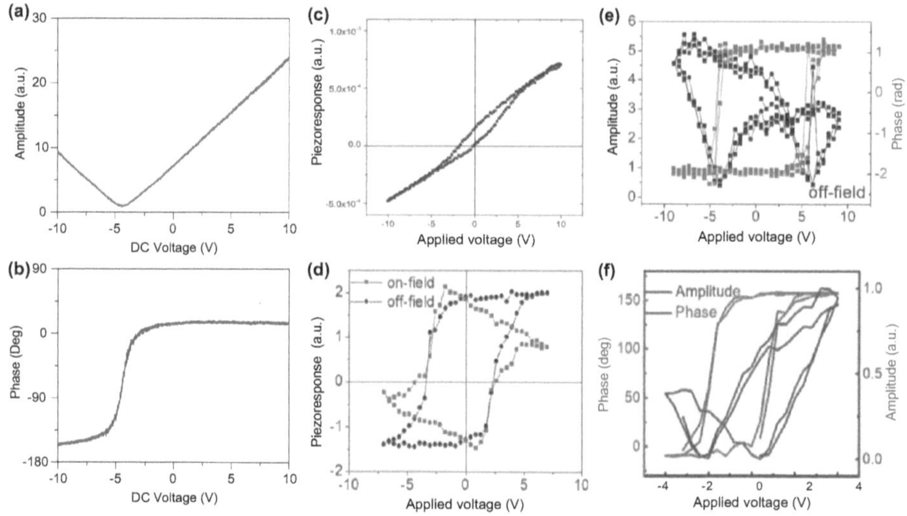


**FIGURE 3.10** Influences of the electrostatic force in PFM spectroscopy measurement. (a) PFM amplitude and (b) phase as a function of DC bias measured on PPLN sample. (Reprinted with permission from Zeng et al.[120] under the terms of the CC-BY 4.0 license.) (c) PFM hysteresis loop measured on PZT using continuous DC spectroscopy method. (Reprinted with permission from Hong et al.[53] Copyright 2001 American Institute of Physics.) (d) PFM hysteresis loop measured on PZT using pulsed DC spectroscopy method. (Reprinted with permission from Balke et al.[57] Copyright 2015 American Institute of Physics.) (e) Off-field PFM amplitude and phase hysteresis loops of non-ferroelectric $HfO_2$. (Reprinted with permission from Balke et al.[59] Copyright 2015 American Chemical Society.) (f) PFM amplitude and phase hysteresis loops measured on $BaTiO_3$ ultrathin film using soft cantilever. (Reprinted with permission from Yao et al.[98] Copyright 2020 Elsevier.)

tend to introduce a larger tip-sample force which increases the possibility of surface and tip damages.[2,41,121] The shielded probe and DC compensation approaches can be effective, but the microfabrication of this special probe is highly challengeable and with higher cost,[2,122] and the nonuniform distribution of surface potential and the tip bias-induced charge injection or even sample status modification (especially in ESM) also make the DC compensation method difficult to be implemented.[106,116,123,124] On the other hand, Kalinin and colleagues have recently developed the contact Kelvin Probe Force Microscopy (cKPFM) to quantify the contribution of electrostatic force in PFM,[59,97,117] but this method still cannot eliminate the electrostatic force when scanning PFM image. High-frequency PFM working at the high-order eigenmodes of the cantilever has also been put forward to minimize the electrostatic force contribution.[1,112,118,119] But the associated decrease of detection sensitivity, laser spot size effect, and large bandwidth requirement for both photodetector and lock-in amplifier make the application of high-frequency PFM quite limited.[100,112,118,125] Hence, introducing more versatile approaches toward eliminating the electrostatic force issue is of fundamental importance for future development of PFM and nanoscale ferroelectric research.

### 3.3.3 Multi-Signal Sources

Another concern of great importance for PFM/ESM is the signal source of the surface strain. For materials with explicitly known properties, such as the conventional piezo/ferro-electrics and Lithium-ion battery materials, the origin of the surface strain in PFM/ESM can be unambiguously determined, and the only concern of the measured signal is usually about the electrostatic force. However, it has been proved that the PFM/ESM setup can directly respond to a broad variety of tip electric field-induced surface deformations, including the piezoelectric strain, electrochemical Vegard strain, electrostrictive strain, Joule thermal strain, Maxwell stress-induced strain, and converse flexoelectric strain.[43,64,80,105,126] In addition, it is also recognized extensively that a series of tip bias-induced phenomena, such as the charge injection, chemical dipole, field effect, irreversible reactions, and direct flexoelectric effect, can cause apparent PFM/ESM responses through the indirect manners, e.g., the charge injection, chemical dipole, and field effect can contribute to the PFM/ESM signal via the electrostriction and electrostatic force effect.[43,65,80,98,106,127,128] Figure 3.11 schematically demonstrates these common effects that can directly or indirectly induce the PFM/ESM responses.[43] Due to these widely existed tip bias-induced effects, it is obvious that when using PFM/ESM setup to characterize the materials with unknown or debatable E-S coupling properties, large ambiguities or artifacts may be introduced to the determination of the signal sources as well as the interpretation of the measurement results.

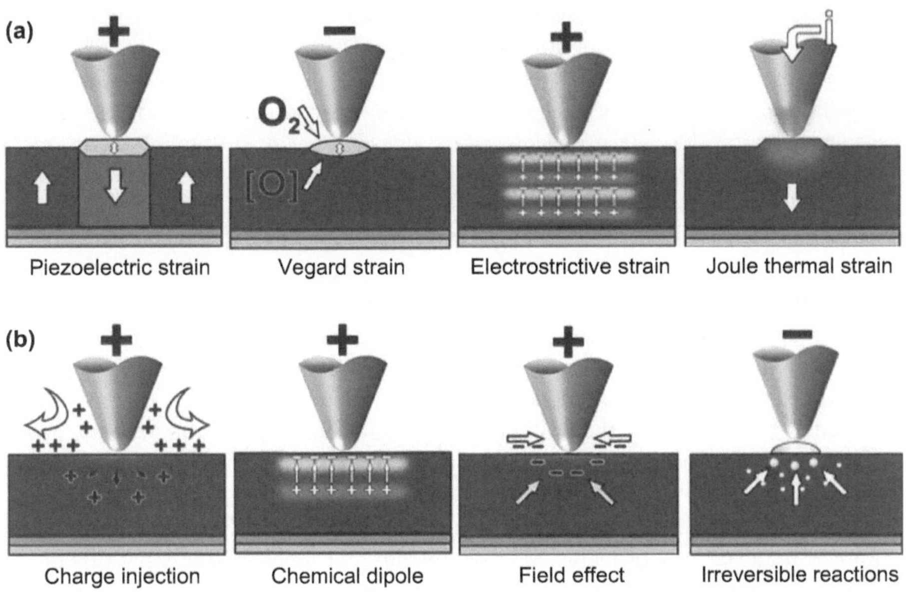


**FIGURE 3.11** Tip bias-induced effects that can contribute to PFM/ESM signal. (a) Tip bias-induced surface strains that can directly contribute to PFM/ESM signal. (b) Tip bias-induced phenomena that can indirectly contribute to PFM/ESM signal. (Adapted with permission from Vasudevan et al. [43] Copyright 2017 American Institute of Physics.)

The confusion between PFM and ESM is one of the most representative ambiguities caused by the signal source issues of the PFM/ESM. As the only difference between PFM and ESM is the signal source, if the local properties of the sample are unknown beforehand, it is difficult to distinguish the real method used.[7,20,74,78,101,129–131] Seol et al. have used the frequency-dependent PFM to detect the frequency dispersion phenomenon of lithium-ion conducting glass ceramic and standard PZT samples, their results showed that the intrinsic piezoelectric signal were not heavily dependent on the probing frequency while ESM signal had shown heavy frequency dependence.[43,132] However, this approach may be subject to polycrystalline ferroelectrics and especially capacitor-based devices, where frequency dispersion can emerge as a result of domain wall dynamics within the probing volumes.[133,134] Chen et al. proposed to use the spectroscopy of PFM/ESM to judge the sources of the signal based on the nonpolar and bipolar characteristics of the Vegard and piezoelectric strains, respectively. As DC field can only manipulate the magnitude of ESM signal, not its phase, while both the magnitude and phase of the PFM signal can be altered by the DC field for ferroelectrics.[131] This approach, however, can be remarkably affected by electrostatic force,[53,59,106] and it will be invalid when the sample has only piezoelectricity. In addition, clear 180° phase switching in non-ferroelectric silicon wafers and silica glasses has been observed via DC spectroscopy of the PFM/ESM,[129,130] indicating the ambiguity of this spectroscopy-based discrimination method. Therefore, fundamental progresses are still necessitated to effectively differentiate the PFM and ESM signals.

In addition to the ambiguity between the piezoelectric and Vegard strains, the universally existed electrostriction effect is another common form of the E-S coupling that can cause huge ambiguities to the PFM/ESM measurement.[65,131,135] It is well known that the electrostrictive effect is present in all materials, regardless of the symmetry,[80,136,137] and the interaction between the charged dipoles in the dielectric materials and the external electric field can also result in the enhancement of electrostriction.[43,65,131] Since the electrostriction shows a quadratic relationship with respect to the applied electric field,[3,136] its influence on PFM is similar to that of the electrostatic force. Kalinin and colleagues have theoretically calculated the contribution of electrostriction to the PFM/ESM responses. They pointed out that this contribution could not be readily separated from the PFM/ESM signal, and furthermore, it was strongly dependent on the properties of the measured materials.[110] Li and colleagues have also proposed to use the second harmonic PFM response to distinguish the piezoelectric strain from the electrostrictive strain.[65,131] However, this method is expected to be affected by the first and second harmonic electrostatic forces, which was not addressed in these studies, thereby its validity may need further investigations to verify.

According to the discussion above, it is expected that the PFM/ESM signal of an unknown material can probably stem from multiple mechanisms simultaneously, the obtained results thereby the conclusions drawn will be highly prone to encounter significant controversies. A striking illustration of this scenario is the ongoing investigation and discussion of the ferroic nature of methylammonium lead iodide ($CH_3NH_3PbI_3$ or $MAPbI_3$) perovskite.[38,40,67,68,75,138–145] To determine the ferroic nature of the $MAPbI_3$ perovskite, PFM/ESM setup has been extensively utilized to perform domain mapping and hysteresis loop measurements. However, completely contradictory conclusions, i.e., $MAPbI_3$ is ferroelectric[40,75,139,142,143,145] or non-ferroelectric,[38,67,68,138,144,146] has been reached. Shown in Figure 3.12 are typical PFM results measured from $MAPbI_3$ perovskite samples, in which the ferroelectric-like domain structures have been obtained by using both Vertical- and Lateral-PFM (Figure 3.12a and b),[75] while non-domain features have been revealed by using PFM with laser Doppler vibrometer (LDV-PFM),[67] and at the same time, hysteretic behaviors have been demonstrated by PFS (Figure 3.12).[38] Since there are many contradictory experimental evidences reported, by far the real ferroic nature of the $MAPbI_3$ is still on hot debate. In fact, the complexity of this topic is, to a large extent, caused by a simple yet fundamental question, i.e., the PFM signal source of the $MAPbI_3$. Since $MAPbI_3$ has demonstrated the behaviors of semiconductor,[75] piezoelectricity,[147] electrostriction,[4,148] and ionic migration[149] simultaneously, there are at least five possible mechanisms that can lead to PFM responses on $MAPbI_3$ sample, including the piezoelectric strain, electrostrictive strain, Vegard strain, Joule thermal strain (especially for thin films), and the electrostatic force effect, and in reality, perhaps a mixture of more than one of these signal sources can contribute to the observation from PFM. Obviously, such a complex situation does make the PFM signal sources of the $MAPbI_3$ cannot be unambiguously determined, thereby leading to large controversies in defining the ferroic nature of the $MAPbI_3$ based on PFM measurements.

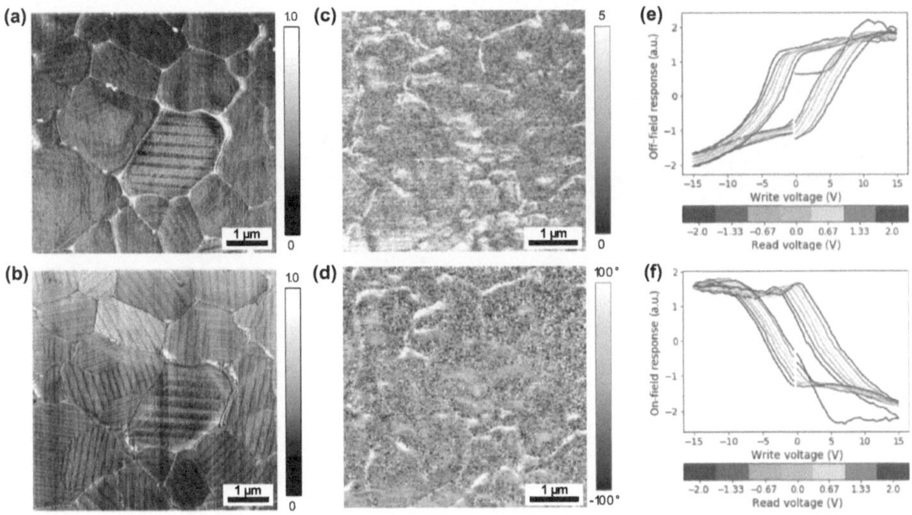


**FIGURE 3.12** Debate about the ferroic nature of $MAPbI_3$. (a) Amplitude images of $MAPbI_3$ measured by Vertical-PFM and (b) Lateral-PFM. (Adapted with permission from Röhm et al.[75] Copyright 2019 Wiley-VCH.) (c) Amplitude and (d) phase images of $MAPbI_3$ obtained by using LDV-PFM. (Adapted with permission from Liu et al.[67] Copyright 2018 Springer Nature.) (e) Off-field and (f) on-field PFM hysteresis loops of $MAPbI_3$. (Reprinted from Garten et al.[38] under the terms of the CC-BY 4.0 license.)

### 3.3.4 Spatial Resolution

It is well-known that, for any microscopy technique, one of the crucial parameters is the spatial resolution, as it intuitively reflects the minimum features that a microscopy can differentiate. Although many other scanning probe techniques, such as the dynamic AFM and Scanning Tunneling Microscopy, can easily achieve an ultra-high resolution imaging,[150,151] the technical principle and detection scheme of the PFM and ESM have limited their spatial resolution to the best of sub-10 nm.[1,2,10,13,19,23,24,152,153] It has been widely believed that the resolution in PFM/ESM measurement is determined by multiple factors, such as the tip radius, tip-sample interaction force, electrostatic force, capillary force, and sample's dielectric and elastic properties, thereby to rigorously define the spatial resolution of the PFM/ESM is not easy.[1,2,6,153,154] Kalinin and colleagues have proposed a resolution theory to help establishing a consistent definition of the spatial resolution in PFM.[13,155] For simplification, the spatial resolution (or lateral resolution) of PFM in real space is determined from the measured width of the domain wall between the antiparallel *c* domains,[6,155] while the resolution of ESM is defined as the half width of the excited spatial region at half maximum.[10] Figure 3.13a shows the determination of the PFM spatial resolution by using the full width at half maximum (FWHM) of the domain wall in $LiNbO_3$ sample, which indicates a resolution of ~25 nm by using the tip with radius of ~20 nm.[6,154] Figure 3.13b displays the measured domain wall width (or spatial resolution) as a function of the tip radius, which shows that the spatial resolution of PFM is almost linearly related to the tip radius.[6,154] To minimize the influences of the long-range electrostatic force and capillary interaction, Rodriguez

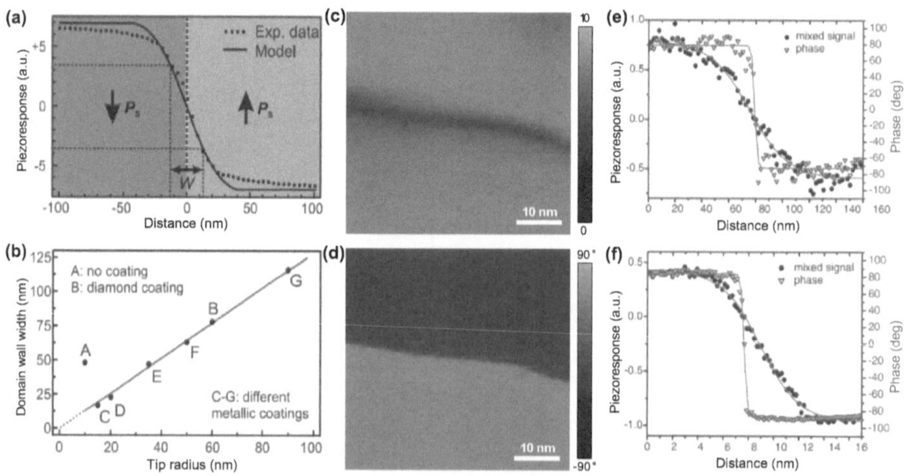


**FIGURE 3.13** Spatial resolution of PFM. (a) Spatial resolution defined by the FWHM of the domain wall between antiparallel *c* domains. (b) Measured domain wall width as a function of the nominal tip radius. (Adapted with permission from Soergel et al.[6] Copyright 2011 IOP Publishing.); (c) High-resolution PFM amplitude and (d) phase images of PZT acquired in distilled water; (e) PFM mixed signal and phase profile across a typical domain wall in ambient and (f) liquid environments. (Adapted with permission from Rodriguez et al.[153] Copyright 2006 American Physical Society.)

and colleagues have performed the PFM measurement on etched PZT sample in liquid environment, where an impressive resolution of ~3 nm has been obtained,[153] and Figure 3.13c and d shows these high-resolution PFM amplitude and phase images acquired in their studies. Figure 3.13e and f demonstrates the remarkable differences between the resolution capabilities of ambient and liquid PFM, which clearly show the significant improvement of spatial resolution in liquid environment.[153] However, even in this liquid environment, the resolution of PFM still fails to break the sub-10 nm bottleneck to reach an ultra-high level. According to the prediction of Kalinin et al., there is no fundamental limitation on achieving sub-nanometer and potentially atomic resolution in PFM, and the primary difficulty in achieving this goal is still about the minimization of the electrostatic response to the PFM signal and the precise control of tip-surface contact area required to achieve molecular and atomic resolution.[1,2]

### 3.3.5 Quantification

Quantifying the tip electric field-induced surface displacement, E-S coupling coefficient, domain wall width, 2D or 3D polarization orientation as well as the hysteretic information is an essential branch of the PFM/ESM measurements.[29,62,69–71,156,157] However, a detailed analysis of the sample responses and cantilever motion has revealed that the quantification of the PFM/ESM measurements is subject to multiple limitations, including the nonuniform distribution of the tip-generated electro-elastic field, contributions from the electrostatic force effect and system inherent background, lack of the tip geometry information, and complex dynamics of the sample-tip-cantilever system.[6,7,44,81,101,158–160] Furthermore, the calibration procedures also significantly affect the quantification of the PFM/ESM observations.[1,6,29] Therefore, it is generally believed that PFM/ESM can only, to a certain extent, provide quantitative or semi-quantitative information,[6] even in the simplest quantification measurements such as measuring the effective piezoelectric coefficient ($d_{eff}$) as shown in equation (3.2).[69–71] Figure 3.14a and b shows the measurement of $d_{eff}$ on standard X-cut quartz sample under different DC bias. It is obvious that the PFM amplitude and measured $d_{eff}$ vary significantly with the DC offset, indicating that the surface strain signal is largely contributed by the electrostatic force effect.[158] In addition to the parasitic effects, the instrument configuration and the operation of the PFM/ESM, such as the optical beam deflection (OBD) system and the selection of the cantilever, drive frequency, and laser spot position, can also induce remarkable influences to the surface strain detection.[81,101,121,156] The OBD method is the most widely used displacement detection scheme in PFM/ESM and other scanning probe techniques due to its advantages in configuration, operation, and performance.[81] However, the deflection signal of OBD actually represents the deflection angle of the cantilever rather than the cantilever displacement.[160] Therefore, the OBD-based PFM/ESM intrinsically has difficulty in detection of the cantilever's real displacement. To improve the quantification capability of PFM/ESM, the direct displacement-based detection method, i.e., the laser Doppler vibrometer (LDV) (also called interferometric displacement sensor (IDS)[101,161]), has been introduced to PFM/ESM.[81,101] Figure 3.14c–e show the comparison between the traditional OBD and LDV methods in quantifying the $d_{eff}$.[81,101] It is evident that the $d_{eff}$ measured by using OBD-based PFM shows a high

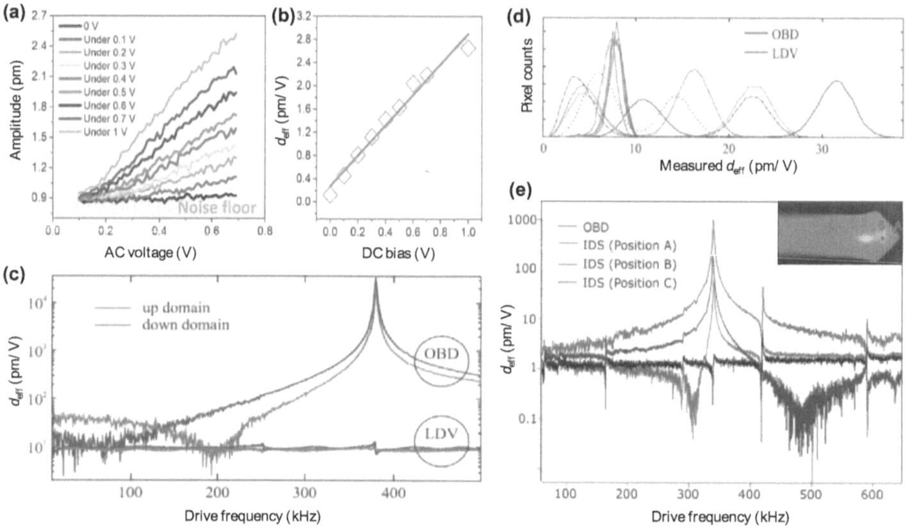


**FIGURE 3.14** Quantification of piezoelectric coefficient by PFM. (a) PFM amplitude as a function of AC voltage measured with different DC bias. (b) The $d_{eff}$ obtained by linearly fitting the results of (a). (Adapted with permission from Seol et al.[158] Copyright 2019 Elsevier.) (c) Measured $d_{eff}$ as a function of drive frequency by using OBD and LDV. (d) Histograms of the $d_{eff}$ for five different cantilevers measured with OBD and LDV. (Adapted with permission from Labuda et al.[81] Copyright 2015 American Institute of Physics.) (e) Measured $d_{eff}$ of X-cut quartz as a function of drive frequency by using OBD and IDS (i.e., LDV), where the laser spot of IDS measurement is placed at the position A, B, and C of the cantilever (top right). (Adapted with permission from Collins et al.[101] Copyright 2019 American Chemical Society.)

dependence in drive frequency (Figure 3.14c and e), and at the same time, vary from cantilever to cantilever (Figure 3.14d), which indicates significant concerns in the reproducibility and accuracy of the quantification. In contrast, by placing the laser spot right above the tip, the results provided by LDV-based PFM show an almost constant trend when changing the drive frequency (Figure 3.14c and e) and cantilever (Figure 3.14d). However, Figure 3.14e also demonstrates that, once the laser spot of the LDV deviates from the position of the tip, the results of LDV-based PFM will be similarly affected by the drive frequency. Although the LDV approaches allows directly measuring the tip displacement, there still exists a noticeable difference between the measured and theoretical values, implying that even this advanced PFM/ESM mode does not ensure an accurate quantification of the E-S coupling coefficients.[7,81,101] Therefore, substantial developments are still needed to further improve the quantitative measurement of the PFM/ESM.

## 3.4 ADVANCES IN PFM AND ESM

During the past ~30 years, PFM has evolved from a technique with controversies to a gradually accepted and then broadly utilized technique for nanoscale characterization of E-S coupling phenomena of the various functional materials. At the beginning,

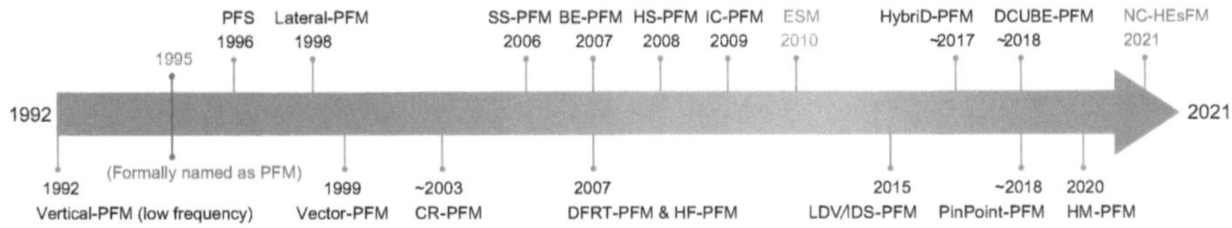


**FIGURE 3.15** History and milestones of the development of PFM and ESM.

PFM (generally called AFM or scanning force microscopy during that time[5,16,162]) was challenged by the interpretation of the observed signal due to the significant contribution of electrostatic force effect,[43,102] and thus it had even been referred to as dynamic contact Electrostatic Force Microscopy.[163,164] In 1995, Gruverman coined the terminology "piezoresponse" to describe the PFM signal and named this technique,[43,49] and henceforth, the name "PFM" has been gradually accepted and used by the research community up to now. Although with certain controversies in the beginning, PFM has still attracted a broad variety of attentions for both fundamental science and technology studies, leading to multiple significant developments in PFM/ESM techniques during these decades.[8,26,29,33,51,81,85,91,102,118,120,165–172] Figure 3.15 briefly summarizes the most important technical progresses achieved in PFM/ESM since 1992. In general, these technical progresses can be classified into two categories, one is about the developments in expanding the capability of PFM/ESM, including the Lateral-PFM,[26] Vector-PFM,[29] PFS,[33] and ESM;[8] and another one is about the advanced PFM/ESM modes which provide optimized characterization performance, including the contact resonance PFM (CR-PFM),[102] switching spectroscopy PFM (SS-PFM),[51] band-excitation PFM (BE-PFM),[165] dual-frequency resonance tracking PFM (DFRT-PFM),[166] high-frequency PFM (HF-PFM),[118] high speed PFM (HS-PFM),[167] intermittent contact PFM (IC-PFM),[168] LDV/IDS-PFM,[81] HybriD-PFM,[85] data cube PFM (DCUBE-PFM),[172] PinPoint-PFM,[169] Heterodyne Megasonic PFM (HM-PFM),[120,170] and Non-Contact Heterodyne Electrostrain Force Microscopy (NC-HEsFM).[171,173] In this section, several key achievements in the advances of PFM/ESM, including the contact resonance PFM/ESM, resonance tracking PFM/ESM, metrological PFM/ESM, dynamic contact PFM/ESM, HM-PFM, and NC-HEsFM, will be briefly reviewed and each technique's advantages and disadvantages will also be discussed.

### 3.4.1 Contact Resonance PFM/ESM

In the very early studies, the working frequency of PFM was selected within low-frequency band (1~10 kHz) to avoid the influence of contact resonance.[5,16,48,71] At low frequencies, the frequency dispersion of the cantilever transfer function is relatively small thus allowing better quantitative measurement of the piezoelectric properties.[43] However, non-resonance PFM usually suffers from the problem of low SNR, thereby large AC drive voltage is generally required to obtain enough SNR, but large AC drive is highly prone to change the sample status or even cause electrical damage of the samples. Therefore, to reduce the AC drive voltage while keeping a good SNR, CR-PFM has been developed.[7,102,103,174,175] By virtue of the resonance amplification,

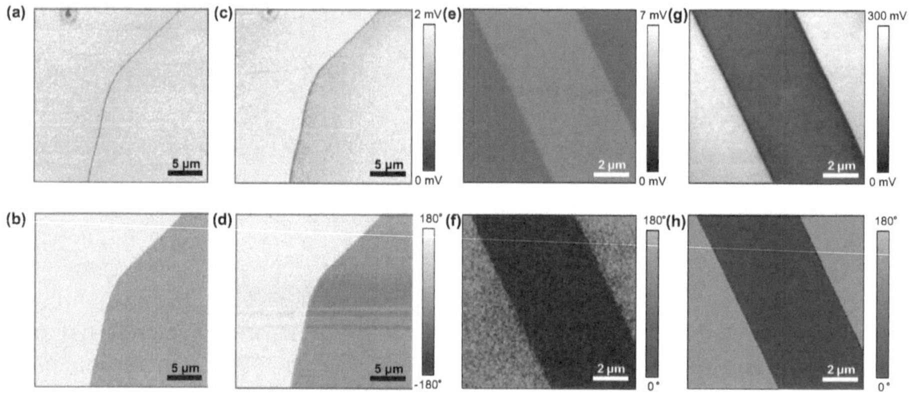


**FIGURE 3.16** Comparisons between PFM and CR-PFM. (a, b) PFM amplitude and phase; (c, d) CR-PFM amplitude and phase images of Mn-doped $PbTiO_3$ measured with applying 4.0 and 0.1 $V_{pp}$ AC drive, respectively. (Adapted with permission from Okino et al.[103] Copyright 2003 The Japan Society of Applied Physics.) (e, f) PFM amplitude and phase and (g, h) CR-PFM amplitude and phase images of PPLN measured with using drive frequencies of 40 and 330 kHz, respectively. (Adapted with permission from Proksch[121] Copyright 2015 American Institute of Physics.)

CR-PFM has a significantly enhanced SNR and thus the AC drive voltage required is far smaller than that in non-resonance PFM.[102,103] This exclusive advantage has made contact resonance widely used in various E-S coupling measurements, especially for the materials with weak E-S coupling or thin film samples.[34,70,92,176] Typically, due to the very small Vegard strain and high electrical damage tendency of ionic materials, ESM is usually operated in contact resonance mode to obtain better performance.[8,19,24] Figure 3.16 shows the comparison between PFM and CR-PFM. When the contact resonance is not used (Figure 3.16a and b), a large AC drive (4.0 $V_{pp}$) is needed to obtain PFM images with good SNR on Mn-doped $PbTiO_3$ sample, whereas by using CR-PFM, applying 0.1 $V_{pp}$ AC drive is enough to generate PFM images with the similar SNR (Figure 3.16c and d).[103] Meanwhile, when the same AC drive is used, CR-PFM can provide PFM images with much better SNR than that of the PFM without using resonance amplification (Figure 3.16e–h).[121] Although much better SRN can be achieved through the resonance amplification, the signal of CR-PFM/ESM becomes extremely related to the dynamics of the sample-tip-cantilever system, because any minor changes of the tip-sample contact condition may result in pronounced resonance frequency shift and the variation of the quality ($Q$) factor.[43,100,160]

### 3.4.2 Resonance Tracking PFM/ESM

As discussed in the above section, resonance amplification brings superior SNR for PFM/ESM measurement, but it simultaneously poses a new challenge for PFM/ESM, that is, to keep a constant resonance status during the contact-mode scanning. However, the contact resonance is extremely sensitive to the tip-sample contact status. Any slight variation of the contact status, such as the contact area and contact force,

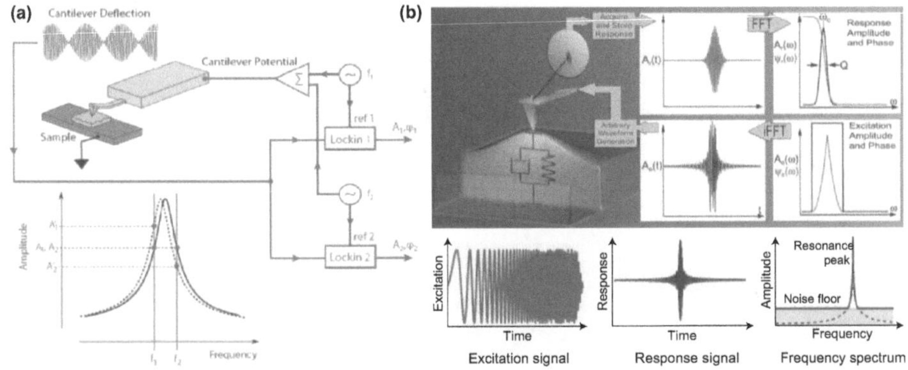


**FIGURE 3.17** Principles of DFRT-PFM/ESM and BE-PFM/ESM. (a) Schematic diagrams of the DFRT-PFM/ESM setup. (Adapted with permission from Rodriguez et al.[166] Copyright 2007 IOP Publishing.) (b) Schematic illustration of the signal generation and processing in BE-PFM/ESM. (Adapted with permission from Jesse et al.[165] and Collins et al.[185] Copyright 2007 and 2013 IOP Publishing.)

may cause the contact resonance shift pronouncedly, resulting in apparent crosstalk to the PFM/ESM signal.[100] Therefore, many efforts have been made by researchers to overcome this drawback, and the proposed yet now widely used strategy is to track the resonance in real time by using resonance tracking techniques. For example, dual-frequency resonance-tracking (DFRT) and band-excitation (BE) techniques have been developed to achieve real-time resonance tracking,[165,166] leading to the emergence of DFRT-PFM/ESM and BE-PFM/ESM.[8,19,23,165,166,177] Figure 3.17a and b shows the technical principles of DFRT-PFM/ESM and BE-PFM/ESM, respectively.[165,166] Both DFRT and BE are based on the ideal simple-harmonic-oscillator (SHO) model to extract the resonance parameters, including the resonance amplitude, frequency, and phase, and the difference is that DFRT uses two while BE uses multiple frequency points to fit the nonlinear SHO model.[90,104,165,166] In DFRT mode, the amplitudes of two frequency points are used as the feedback of the resonance tracking loop (Figure 3.17a), and during scanning, the resonance tracking loop dynamically adjusts the two drive frequencies to keep the contact resonance tracked.[166]

Figure 3.18a–c shows a typical data set obtained by using DFRT-PFM, including resonance amplitude, phase, and frequency shift images.[166] In contrast to the closed-loop operation of DFRT, BE method uses an open-loop scheme to track the resonance. In BE mode, the 2D mapping is implemented by a point-by-point manner, and at each point, the sample is excited by a band signal (Figure 3.17b) which contains multiple frequency components (i.e., a frequency band), and then the sample's response signal is continuously acquired for a specific period of time. By applying Fourier transformation to the acquired response signal, the frequency spectrum of each pixel is obtained, then SHO model is used to fit the frequency spectrum, and the resonance amplitude, phase, frequency, and $Q$ factor can be extracted. Shown in Figure 3.18d–g are typical images obtained by using BE-PFM, which includes the images of resonance amplitude, phase, frequency, and $Q$ factor.[90] Generally speaking, BE method can provide a much more accurate resonance tracking than that

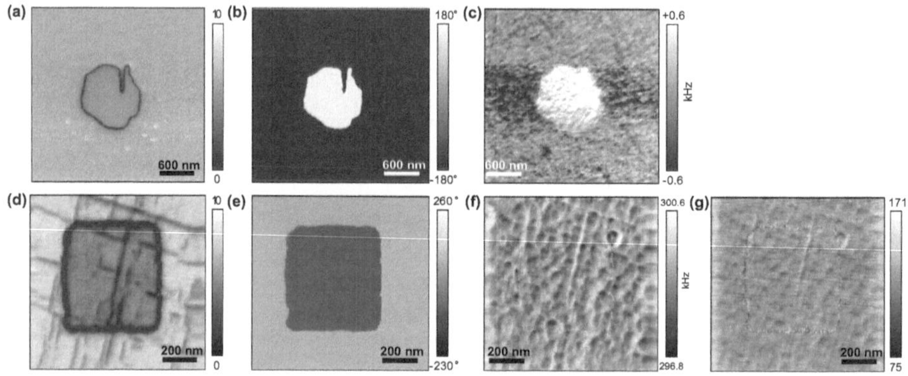


**FIGURE 3.18** Ferroelectric domain images obtained by DFRT-PFM and BE-PFM. (a) DFRT-PFM amplitude, (b) phase, and (c) resonance frequency shift images of PPLN. (Adapted with permission from Rodriguez et al.[166] Copyright 2007 IOP Publishing.) (d) BE-PFM amplitude, (e) phase offset, (f) resonance frequency, and (g) $Q$ factor images measured on PZT. (Adapted with permission from Jesse et al.[90] Copyright 2010 IOP Publishing.)

of the DFRT method as it uses multiple frequency points to fit the nonlinear SHO model, but the cost is that BE mode usually takes much longer imaging time than that of the DFRT mode.[104] Although DFRT and BE techniques have already been extensively utilized in PFM/ESM studies, a fundamental limitation of DFRT and BE techniques is that both of these methods use the ideal SHO model to extract the resonance parameters. For resonance curves that show standard shape of the SHO model, DFRT and BE methods can give a good resonance tracking accuracy. However, if the contact resonance shows large deviation from the standard SHO model, which indeed frequently occurs in many practical experiments, using DFRT and BE techniques (especially the DFRT) may lead to large errors in the resonance tracking results.

### 3.4.3 Metrological PFM/ESM

To obtain quantitative information of the local E-S coupling is sometimes essential in PFM/ESM studies. Typically, conventional PFM/ESM uses the OBD method to detect the cantilever displacement and thereby the tip electric field-induced surface displacement. The OBD system usually has a large laser spot, and it measures the cantilever displacement via detecting the deflection angle of the cantilever. Since in PFM/ESM measurement, the vertical displacement and deflection angle of the cantilever vary with the longitudinal position of the cantilever, using conventional OBD method to precisely quantify the tiny surface displacement (typically within picometer range) is usually difficult and may introduce large errors. To obtain a much more precise detection of the surface displacement underneath the tip, Labuda and colleagues have introduced the LDV (or IDS) system to PFM/ESM.[81] Based on the interferometric detection, LDV can directly measure the displacement of the cantilever with a much higher precision. Meanwhile, by virtue of the large numerical aperture, the laser spot of the LDV can be focused down to ~2 μm, which allows high-resolution measuring the local displacement of the cantilever. Figure 3.19a shows the comparison between the setups of the OBD and LDV, in which the significantly reduced LDV laser spot size can be clearly observed.[81] With such a small spot size, LDV can precisely locate the laser spot to the position directly above the tip, which allows to directly measure the surface displacement underneath the tip without the coupling from the cantilever dynamics.[81] Figure 3.19b shows the results of LDV-PFM measurement on PPLN standard sample.[81] It is clear that the obtained amplitude images changes with the position of the probing laser spot, and ideal domain image can only be acquired when the spot is located directly above the sharp tip, implying that the cantilever dynamics can significantly affect the imaging and quantification of the E-S coupling.[81] Figure 3.19c demonstrates the application of LDV-PFM in detecting the true tip electric field-induced surface strain on $MAPbI_3$ sample.[101] By setting the laser spot at either side of the tip, or directly above the tip, dramatically different amplitude images can be obtained (Figure 3.19c), which clearly indicates that, in PFM measurements, the frequently observed domain features of the $MAPbI_3$ samples[40,75] can be totally induced by the stray cantilever dynamics instead of the true piezoelectric strain.[101] With the superior capability in measuring the true surface displacement, LDV-PFM does show a substantial improvement in the quantification

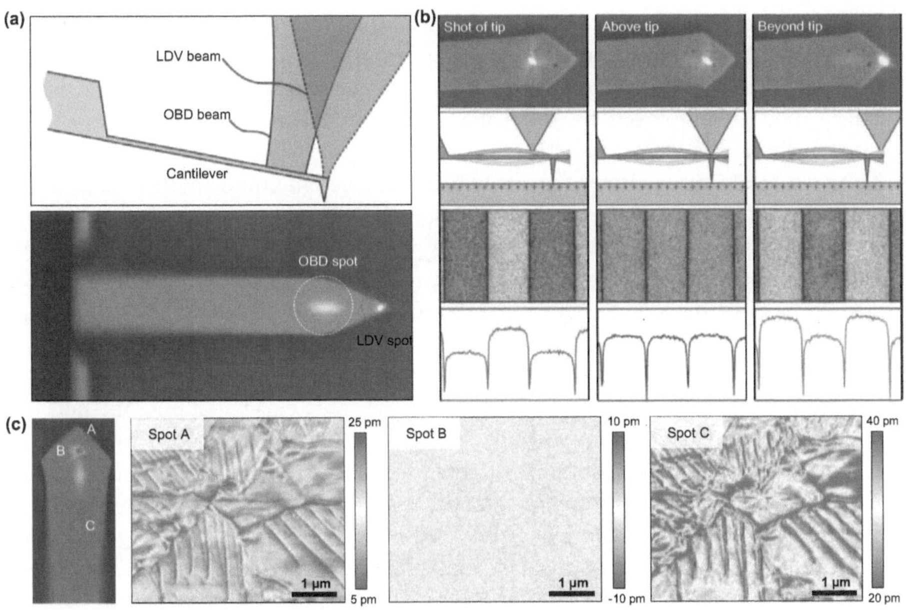


**FIGURE 3.19** LDV system and the results of LDV-PFM. (a) Comparison between LDV and conventional OBD. (b) LDV-PFM amplitude images of PPLN obtained by setting the laser spot at different positions. (Adapted with permission from Labuda et al.[81] Copyright 2015 American Institute of Physics.) (c) LDV-PFM amplitude images of $MAPbI_3$ obtained by setting the laser spot at different positions. (Adapted with permission from Collins et al.[101] Copyright 2019 American Chemical Society.)

of $d_{eff}$ compared with that of the conventional PFM method (Figure 3.14c and d).[81] However, even using the high-precision interferometer, LDV-PFM still consistently gives piezoelectric coefficients of X-cut quartz and PPLN two to three times lower than the reported bulk values, implying that further studies are needed to elucidate the quantification of E-S coupling coefficients by PFM/ESM technique.[81,101]

### 3.4.4 Dynamic Contact PFM/ESM

Typically, the tip electric field-induced strain has a surface displacement at picometer level, and to directly probe this tiny displacement within such a small tip-sample interaction region is almost impractical. Conventional PFM/ESM utilizes an indirect manner to achieve this detection, i.e., the dynamic surface displacement is transferred to the cantilever motion via local tip-sample interaction. Then the cantilever motion can be easily measured by monitoring its deflection using an optical lever system. As both the tip electric field strength and tip-sample interaction decay with increasing tip-sample separation dramatically, to provide a strong enough yet stable tip electric field and tip-sample interaction, PFM/ESM is always operated at the static contact mode to obtain the best performance.[6,7,23] However, as discussed in the previous sections, this contact-mode operation does become a fundamental issue of the PFM/ESM technique, causing a series of challenges to the PFM/ESM measurement.[1,85] To address the contact-mode issue, Rodriguez and colleagues have firstly proposed to perform the PFM measurement with an intermittent contact mode, i.e., the IC-PFM, by simultaneously driving the first and second flexural modes of the cantilever, where the first flexural mode is used for height control and the second flexural mode is employed for PFM measurement.[168] However, due to the significant contribution of electrostatic force as well as the decreased detection sensitivity, effective IC-PFM measurement has only been achieved in liquid environment, which has largely limited the application of this method.[168] Furthermore, with the development of force spectroscopy-based AFM imaging mode, such as the PeakForce mode (Bruker, USA),[178] Fast Force Mapping mode (Oxford Instruments, USA),[179] PinPoint mode (Park Systems, Korea),[180] and HybriD mode (NT-MDT, Russia),[85] another type of dynamic contact PFM/ESM mode based on the force curve mapping scheme has been developed, such as the HybriD-PFM (NT-MDT),[85] DCUBE-PFM (Bruker),[172] and PinPoint-PFM (Park Systems).[169] By using the force curve mapping method, the tip is intermittently in contact with the surface via force spectroscopy method, and the PFM/ESM signal acquisition is triggered when the tip-sample force has reached the predefined set-point (in repulsive force region). Figure 3.20 shows the working principle of HybriD-PFM and DCUBE-PFM.[85,91,172] In HybriD-PFM (Figure 3.20a), the tip is continuously doing the fast force spectroscopy, and PFM measurement (including both Vertical- and Lateral-PFM) is performed within the time window (region "3", Figure 3.20b) when the tip is in contact with the sample surface during the force spectroscopy.[85] In contrast to the continuous force spectroscopy used in the HybriD-PFM, DCUBE-PFM uses an intermittent force spectroscopy scheme,[91,172] in which the force spectroscopy is performed with a point-by-point manner, and during the force spectroscopy at each pixel, the tip will be kept fixed at the predefined force set-point for a specific period to conduct the PFM measurement (Figure 3.20b).[91]

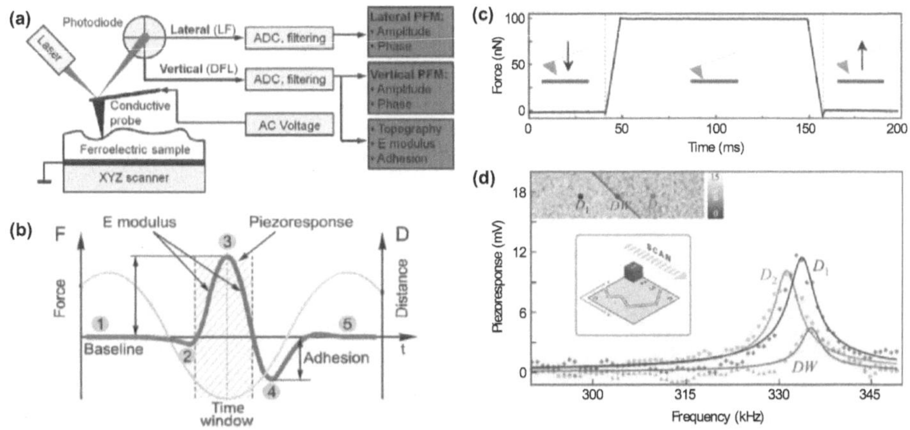


**FIGURE 3.20** Principles of HybriD-PFM and DCUBE-PFM. (a) Schematic illustration of the functional principle of HybriD-PFM. (b) The force control strategy used in HybriD-PFM. (Reprinted with permission from Kalinin et al.[85] Copyright 2018 Elsevier.) (c) Typical force-time curve and (d) frequency spectrum in DCUBE-PFM measurement, and the inset (middle) of (d) schematically shows the point-by-point data acquisition scheme used in DCUBE-PFM. (Adapted with permission from Cui et al.[91], Copyright 2019 IOP Publishing; and De Wolf et al.[172] Copyright 2018 Cambridge University Press.)

To obtain better SNR, DCUBE-PFM uses the resonance amplification,[91,169] where a frequency spectrum around the resonance frequency is measured using frequency sweeping method and then SHO model is used to fit the frequency spectrum to extract the resonance parameters (Figure 3.20c).[91]

Figure 3.21 displays the typical PFM images obtained by using dynamic contact PFM.[168,169,172] From Figure 3.21a and b, it is evident that IC-PFM cannot provide effective piezoresponse characterization under ambient conditions. Although domain contrast can be observed from the IC-PFM measurement in liquid environment (Figure 3.21c and d), the amplitude and phase images do manifest a bad SNR and significant contribution of other signals (mainly the electrostatic force effect).[168] Figure 3.21e–h shows topography and piezoresponse amplitude images obtained by using conventional PFM and PinPoint-PFM.[169] In conventional contact-mode PFM, artifacts in both topography and amplitude images can be seen throughout the entire scanning area (Figure 3.21e and f), indicating that the tip is repeatedly scratching on the sample surface. By contrast, the phenanthrene polymer sample is well-distinguished both in the topography and amplitude images given by the dynamic contact PinPoint-PFM (Figure 3.21g and h), implying a substantial improvement in the contact-mode issue of conventional PFM. Figure 3.21i and j shows the typical phase and amplitude data cubes obtained by DCUBE-PFM,[172] in which the phase-frequency and amplitude-frequency spectra are recorded at each pixel and thus two 3D data cubes are formed. By using SHO model to fit the recorded frequency spectra, the 2D distributions of the resonance amplitude, phase, frequency, and $Q$ factor can be attained accordingly,[91,172] which is very similar with the BE-PFM.[165] Although conducting PFM/ESM measurement with the force spectroscopy mode

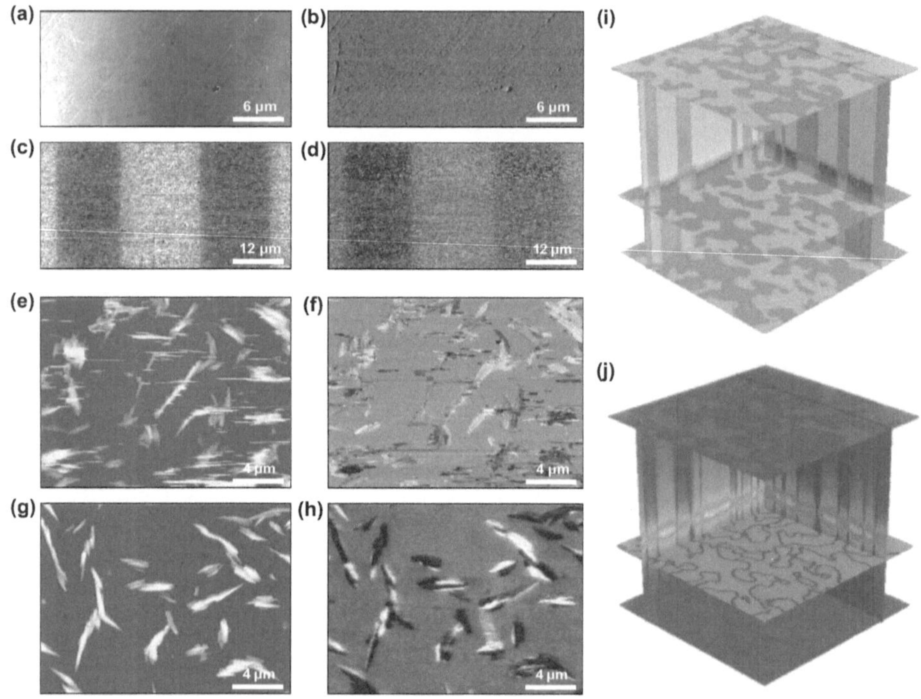


**FIGURE 3.21** Ferroelectric characterization using dynamic contact PFM. (a, c) IC-PFM amplitude and (b, d) phase images of PLLN, (a, b) is in ambient and (c, d) is in liquid environments. (Adapted with permission from Rodriguez et al.[168], Copyright 2009 IOP Publishing.) (e, g) Topography and (f, h) piezoresponse amplitude images of phenanthrene film, (e, f) is using conventional PFM and (g, h) is using PinPoint-PFM. (Adapted with permission from Shi et al.[169], Copyright 2018 Park Systems.) (i) DCUBE-PFM phase and (j) amplitude data cubes of $LiTaO_3$. (Reprinted with permission from De Wolf et al.[172] Copyright 2018 Cambridge University Press.)

can significantly improve the contact-mode issues, the associated new problems, such as the complicated scanning control, unstable tip-sample contact status (due to the continuous approach and retract operation), and longer imaging time, will also limit the application of this method, and in fact, the PFM/ESM measurement at each point is still performed under the tip-sample contact status.

### 3.4.5 Heterodyne Megasonic Piezoresponse Force Microscopy

According to the discussion in the previous sections, the extensive application of PFM has unambiguously revealed numerous challenges and concerns about this technique, which is now greatly challenging its validity in many piezo/ferro-electric studies.[7,14,40,43,67,68,80,138] Among these issues, the signal source-related issues, such as the electrostatic force effect,[7,43,53,59,80,101,106,158,181] Vegard strain,[7,20,74,101,129–131] electrostrictive strain,[65,131,135] Joule thermal strain, and many others,[7,21,43,65,80,92,93,98,110,126,127,131,181,182] are pressingly pending to be addressed because the sources of the signal are of fundamental importance for reaching correct interpretation of the PFM results.[7,14,43,80] In order to address the signal source issue of the conventional PFM, especially the electrostatic force effect, high-frequency excitation, as an effective solution, has been introduced to PFM. Seal and colleagues firstly demonstrated the HF-PFM in 2007, in which the high eigenmodes of the cantilever with eigenfrequencies of 1–10 MHz were employed to detect the piezoelectric strain.[118] MacDonald and colleagues have also proven that HF-PFM working at high eigenmodes of the cantilever can effectively minimize the electrostatic force effect.[112] Although the electrostatic force issue can get a good solution in HF-PFM, the associated issues, such as decreasing the detection sensitivity, laser spot size effect, large bandwidth requirement for both photodetector and lock-in amplifier, can in turn make the application of HF-PFM very limited.[112,118] However, using high-frequency excitation to detect piezoelectric strain does provide a meaningful instruction for the development of advanced PFM.

By performing an overall assessment about the high-frequency excitation in PFM, Zeng and colleagues recently introduce several substantial improvements in PFM which can be achieved simultaneously, including minimizing the electrostatic force-induced cantilever vibration, attenuating electrochemical Vegard strain and electrostriction effects as well as reducing the influence of dynamic electrochemical processes during the measurement.[120,170] In order to break the technical limitations and realizing PFM measurements at much higher frequency, heterodyne detection scheme has been introduced to the PFM, and an advanced PFM technique which focuses on detecting piezoelectric strain with high to very high frequency has been developed, i.e., the HM-PFM.[120,170] HM-PFM utilizes both electrical and mechanical drives with MHz frequency to probe the surface piezoelectric vibration at nanoscale via heterodyne detection method. Figure 3.22 shows the experimental setup of the HM-PFM.[120] In the HM-PFM, the tip is stimulated by the holder transducer via holder drive $V_{\text{holder}}$ at frequency $f_t$, while the sample vibration is excited by the sample drive $V_s$ at frequency $f_s$. The difference-frequency piezoresponse (DFP), with frequency $f_{\text{diff}} = f_s - f_t$, generated from the heterodyne process is detected by the photodetector and then demodulated by the lock-in amplifier. Figure 3.23 shows the typical ferroelectric characterization on typical ferroelectrics, including PPLN, $Pb(Zn_{1/3}Nb_{2/3})$

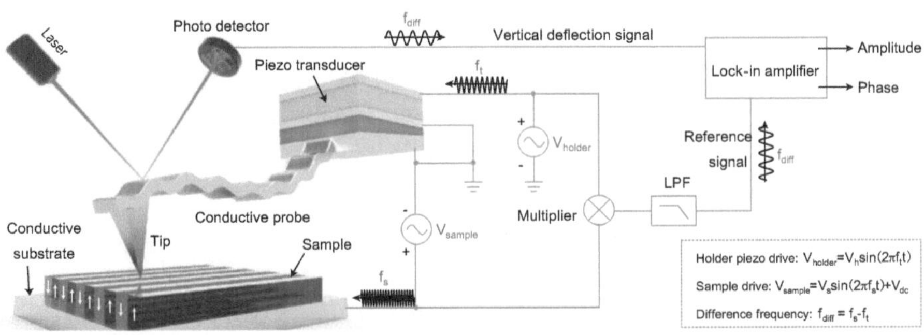


**FIGURE 3.22** Schematic diagram of the HM-PFM setup. (Adapted from Zeng et al.[120] under the terms of the CC-BY 4.0 license.)

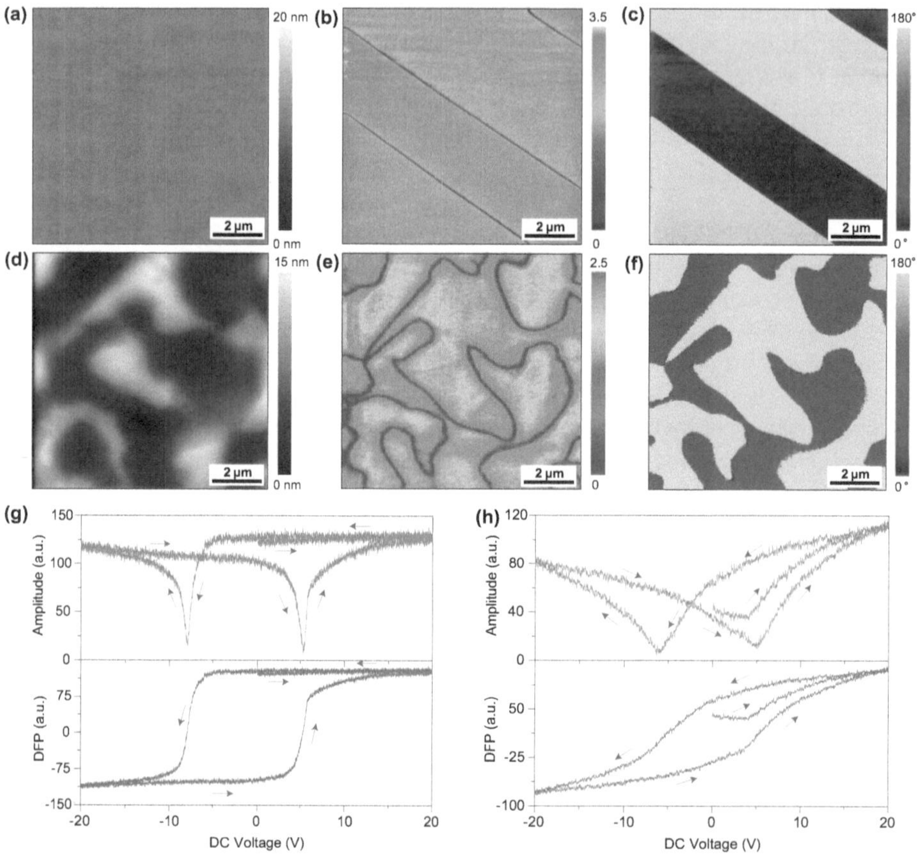


**FIGURE 3.23** Ferroelectric characterization with HM-PFM. (a) Topography, (b) HM-PFM amplitude, and (c) phase images of PPLN. (d) Topography, (e) HM-PFM amplitude, and (f) phase image of the PZN-9%PT. (g) Hysteresis loops of PZT film and (h) PZN-9%PT measured by HM-PFM switching spectroscopy using the continuous DC method. Measurement conditions: (b, c) $f_t = 8.356$ MHz and $f_s = 8.63959$ MHz; (e, f) $f_t = 8.395$ MHz and $f_s = 8.6813$ MHz; (g) $f_t = 13.516$ MHz and $f_s = 13.82158$ MHz; (h) $f_t = 15.451$ MHz and $f_s = 15.75504$ MHz. (Adapted from Zeng et al.[120] under the terms of the CC-BY 4.0 license.)

$O_3$_9%$PbTiO_3$ (PZN-9%PT), and PZT, by using HM-PFM.[120] From the amplitude and phase images of the DFP signal, the domain walls between the two adjacent domains can be clearly observed in the amplitude images (Figure 3.23b and e), and the domains with upward and downward polarization are distinctly revealed in the phase images (Figure 3.23c and f). Furthermore, Figure 3.23b,e,c, and f indicates a uniform amplitude distribution and a nearly 180° phase difference between the domains with opposite polarization, which agrees well with the characteristics of the proposed "ideal" PFM measurements of ferroelectric materials.[121] Figure 3.23g and h displays the attained HM-PFM hysteresis loops of the PZT and PZN-9%PT by using the continuous DC spectroscopy method, in which both the amplitude loops (blue curves) and DFP loops (red curves, calculated by amplitude × cos(phase)) show

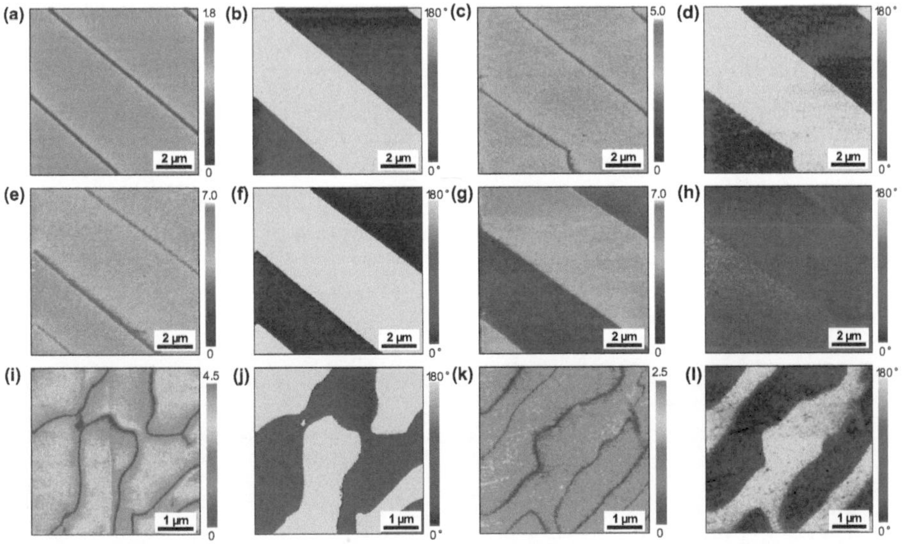


**FIGURE 3.24** Ferroelectric domain characterization with high-frequency excitation. (a, c, e, g) HM-PFM amplitude and (b, d, f, h) the respective phase images of PPLN sample. (i, k) HM-PFM amplitude and (j, l) the respective phase images of PZN-9%PT sample. Measurement conditions: (a, b) $f_t$ = 27.9 MHz and $f_s$ = 28.1882 MHz; (c, d) $f_t$ = 40.195 MHz and $f_s$ = 40.4823 MHz; (e, f) $f_t$ = 62.255 MHz and $f_s$ = 62.61 MHz; (g, h) $f_t$ = 108.95 MHz and $f_s$ = 109.137 MHz; (i, j) $f_t$ = 27.871 MHz and $f_s$ = 28.13966 MHz; (k, l) $f_t$ = 42.9 MHz and $f_s$ = 42.975 MHz. (Adapted from Zeng et al.[120] under the terms of the CC-BY 4.0 license.)

the expected characteristics of ferroelectricity.[120] As HM-PFM uses the heterodyne method to detect the piezoelectric strain, the excitation frequency for sample is no longer limited by the bandwidth of conventional optical lever system.

Figure 3.24 displays the HM-PFM amplitude and phase images acquired with drive frequencies ranging from ~30 MHz to ~110 MHz.[120] Surprisingly, it is found that an effective ferroelectric domain characterization can still be achieved even when the drive frequency is increased to $f_s$ = 109.137 MHz (Figure 3.24g and h). This ~110 MHz drive frequency is about 100–1,000 times higher than the frequency used in conventional PFM and more than ten times higher than the highest excitation frequency (8.4 MHz[118]) reported in the previous HF-PFM experiments.

One of the most important advantages of the HM-PFM is to minimize the electrostatic force contribution. Figure 3.25 shows the comparison between the electrostatic force contributions in the conventional PFM and HM-PFM measurements.[120] It is evident that when ±15 V DC bias is applied, in the conventional PFM, both amplitude and phase contrasts have been changed remarkably (Figure 3.25a and b), and at the same time, the PFM amplitude and phase clearly show the V-shaped and near 180° phase changes with the DC bias (Figure 3.25g and h), respectively, which all clearly indicate the significant contribution from electrostatic force.[111–113] In contrast, the HM-PFM amplitude and phase images obtained at two different excitation frequencies show almost no change when ±15 V DC bias is applied (Figure 3.25c–f), and even using various excitation frequencies, nearly all of the HM-PFM amplitude and phase signals keep constant with the applied DC bias (Figure 3.25i and j), implying that the contribution from the electrostatic force in the HM-PFM measurement has been significantly minimized.[120] Further theoretical analysis shows that, by introducing heterodyne detection to PFM, the conventional piezoresponse signal generation mechanism has been changed and thus the direct coupling between the piezoelectric and electrostatic force signals has been broken, which greatly supports to minimize the electrostatic force contribution by using high-frequency excitation.[120]

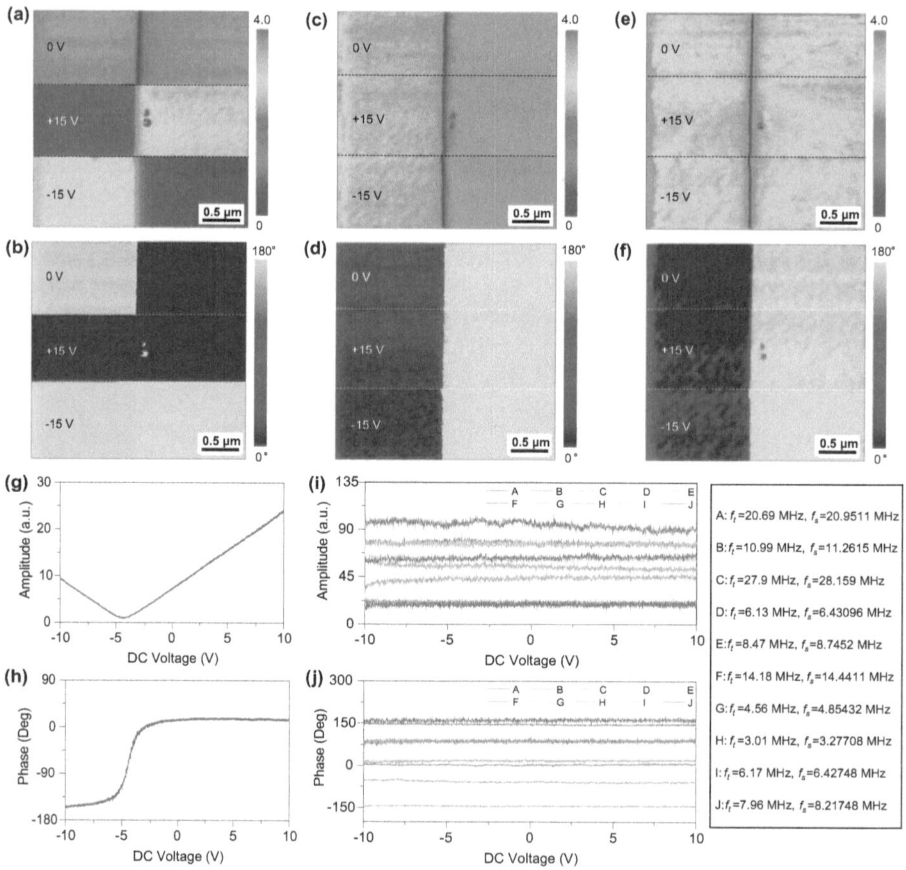


**FIGURE 3.25** Electrostatic force contribution in the measurement of PFM and HM-PFM on PPLN. (a) Conventional PFM amplitude and (b) phase images. (c, e) HM-PFM amplitude and (d, f) the respective phase images. (g) PFM amplitude and (h) phase as a function of DC voltage. (i) HM-PFM amplitude and (j) phase as a function of DC voltage under various excitation frequencies. Measurement conditions: (c, d) $f_t$ = 8.14 MHz and $f_s$ = 8.422 MHz; (e, f) $f_t$ = 7.176 MHz and $f_s$ = 7.45915 MHz. (Adapted from Zeng et al.[120] under the terms of the CC-BY 4.0 license.)

Based on the advantages of high-frequency excitation, HM-PFM provides a brand-new measurement, called the difference-frequency piezoresponse frequency spectrum (DFPFS), which can be used as a powerful evidence when identifying the piezoelectricity of unknown materials.[120,170] Since the high-frequency and heterodyne detection scheme can minimize both the electrostatic force effect and Vegard strain, materials with no noticeable electromechanical couplings, such as the piezoelectricity, will be difficult to induce observable DFP response peaks in the DFPFS measurements. Figure 3.26 shows the DFPFS measured on seven different materials with tip frequency ranging from 2 to 14 MHz.[120] Obviously, a large number of resonance-like peaks can be seen in the entire spectrum of the DFPFS for PPLN and PZN-9%PT samples due to the strong piezoelectricity, and even for the X-cut quartz with weak piezoelectricity, several peaks can still be clearly observed (Figure 3.26). In contrast, the DFPFS obtained from the materials without electromechanical couplings, including the glass, $SiO_2$, $LiMn_2O_4$, and $LiCoO_2$, shows almost no noticeable peaks in the spectrum, indicating that the DFPFS can be used as the evidence of identifying true electromechanical coupling.

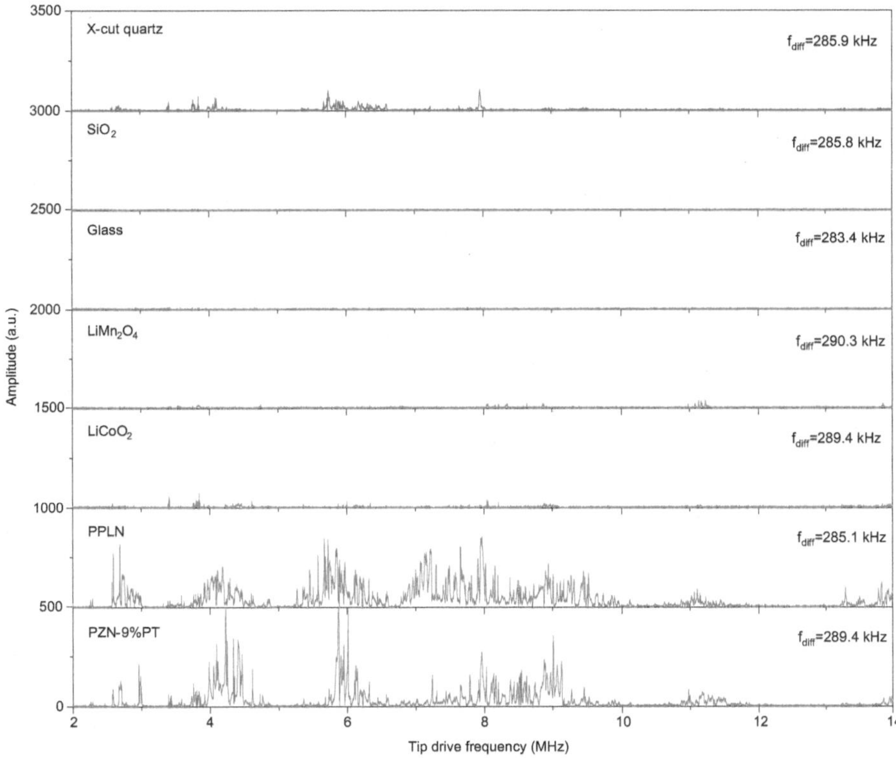


**FIGURE 3.26** HM-PFM DFPFS measurements on different materials. Measurement conditions: sample drive frequency $f_s = f_t + f_{\text{diff}}$ and $f_{\text{diff}}$ is set as the first-order contact resonance frequency of the cantilever on each sample, respectively. Note all the spectrums are offset for clarity. (Adapted from Zeng et al.[120] under the terms of the CC-BY 4.0 license.)

### 3.4.6 Non-Contact Heterodyne Electrostrain Force Microscopy

From Figure 3.15, it can be seen that, during the past decades, the major developments of PFM are actually promoted by the issues and limitations of the conventional PFM, such as the DFRT/BE-PFM is developed for addressing the resonance tracking issue, and the HM-PFM is developed for solving the signal source problems. However, due to the intrinsic limitations of PFM techniques, by far each of these advanced PFM modes can only provide certain limited solutions to optimize the characterization performance of the PFM. For instance, the DCUBE-PFM and PinPoint-PFM can solve a part of contact-mode operation and resonance tracking issues, but the electrostatic force effect in these modes can still be significant; HM-PFM can solve the issues of electrostatic force effect and Vegard strain, but it still faced with the contact-mode operation problems. Therefore, to achieve a comprehensive improvement in the nanoscale E-S coupling characterization, a more ideal PFM or PFM-like method which can simultaneously address multiple issues is highly necessitated.

By comprehensively analyzing the instrument configuration of PFM, it is found that several fundamental issues of the PFM, such as the contact-mode operation, resonance tracking, electrostatic force effect, and spatial resolution limitation, are related to the cantilever-based atomic force sensor and homodyne detection scheme used in the conventional PFM.[171] Whereas in most of the advanced PFM modes, such as the CR-PFM, DFRT-PFM, BE-PFM, HF-PFM, IC-PFM, HybriD-PFM, DCUBE-PFM, and PinPoint-PFM, both the cantilever force sensor and homodyne detection have also been used without any modifications.[26,85,91,102,118,165,166,168,169,172] Therefore, each of these advanced PFM modes can only provide limited solutions to improve the measurement. To break the limitations of the classic PFM technique architecture, Zeng and colleagues have firstly introduced a new system with using the quartz tuning fork (QTF) force sensor, heterodyne detection, and multi-frequency operation schemes to measure the nanoscale E-S coupling, and this new system is an excellent candidate for substituting the role of the conventional PFM, which is named as Non-Contact Heterodyne Electrostrain Force Microscopy (NC-HEsFM).[171,173] Similar to that of the PFM, NC-HEsFM also measures the tip electric field-induced surface strain, but via using the brand-new QTF force sensor and heterodyne detection scheme, a true noncontact yet nondestructive surface strain measurement with significantly minimized electrostatic force effect have been successfully achieved in the NC-HEsFM. Since the terminology "piezoresponse", coined by Gruverman et al. in 1995,[43,49] may indeed cause some misunderstandings when discussing the E-S coupling phenomena in non-piezoelectric materials, a more general term, "electrostrain", is therefore used here to describe the surface strain (or displacement) induced by an external electric field.

The complete setup of the NC-HEsFM is schematically shown in Figure 3.27, in which a sharp conductive AFM probe is glued at the end of the QTF prong.[171] During the measurement, the QTF is mechanically excited to vibrate at its second and a higher $n$th ($n \geq 3$) flexural mode (anti-symmetric vibration) simultaneously by the holder drive at frequency $f_2$ and $f_h$, respectively, while the electrostrain of the sample is stimulated by the sample drive at frequency $f_s$. Due to the nonlinear tip-sample interaction, a difference-frequency electrostrain (DFE) signal, with the frequency of

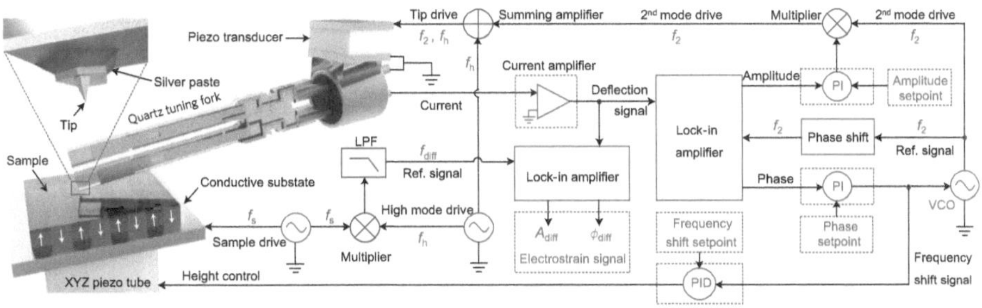


**FIGURE 3.27** Schematic diagram of the NC-HEsFM setup. (Adapted with permission from Zeng et al.[171] Copyright 2021 Wiley-VCH.)

$f_{\mathrm{diff}} = f_s - f_h$, will be generated. To maximize the SNR of this DFE signal, $f_{\mathrm{diff}}$ is set to near the resonance frequency of the first flexural mode of the QTF. A phase locked loop is used to track the resonance frequency shift ($\Delta f_2$) of the second flexural mode, and this $\Delta f_2$ is used as the height feedback signal for constant frequency shift scanning, indicating that the tip-sample force gradient in NC-HEsFM is kept constant so that the resonance can be tracked automatically during the scanning. Figure 3.28 shows the results of ferroelectric characterization for several classic ferroelectric materials obtained by using NC-HEsFM. From the amplitude and phase images of the DFE signal, the domain walls between the two adjacent domains can be clearly observed in the amplitude images (Figure 3.28b and e), and the domains with upward and downward polarization are distinctly revealed in the phase images (Figure 3.28c and f). In addition, the amplitude and phase images show an almost uniform distribution (Figure 3.28b and e) and a nearly 180° phase difference between the domains with opposite polarization (Figure 3.28c and f), respectively, which agrees well with the characteristics of the proposed "ideal" ferroelectric characterizations.[121,183] Similar to that of the conventional PFM, NC-HEsFM also allows manipulating the ferroelectric domain by applying DC electric field. Figure 3.28g–i shows the structure of an artificial domain on PZT film written by NC-HEsFM, which indicates that NC-HEsFM can perform similar ferroelectric lithography but with a noncontact manner. Figure 3.28j–l displays a high-resolution NC-HEsFM mapping of the domain wall at ambient condition, where the position of zero amplitude and 180° phase jump are defined within a ~2 nm range (Figure 3.28k and l), and such a high-resolution domain wall images have hardly been reported by using the conventional PFM in the same ambient environment. In addition, NC-HEsFM also allows the measurement of ferroelectric hysteresis loop (switching spectroscopy), and a set of hysteresis loops measured on PZT film is shown in Figure 3.28 m, where the amplitude, phase, and DFE (calculated by amplitude × cos(phase)) loops all show the hysteretic characteristics of ferroelectricity.[43,50,59]

With using the QTF-based noncontact scanning and heterodyne detection schemes, the contact-mode and electrostatic force issues that have plagued PFM for many years now can all be eliminated in the newly designed NC-HEsFM system.[171,173] Figure 3.29 shows the comparison of the surface modification situations between the conventional PFM and NC-HEsFM measurements. The topography

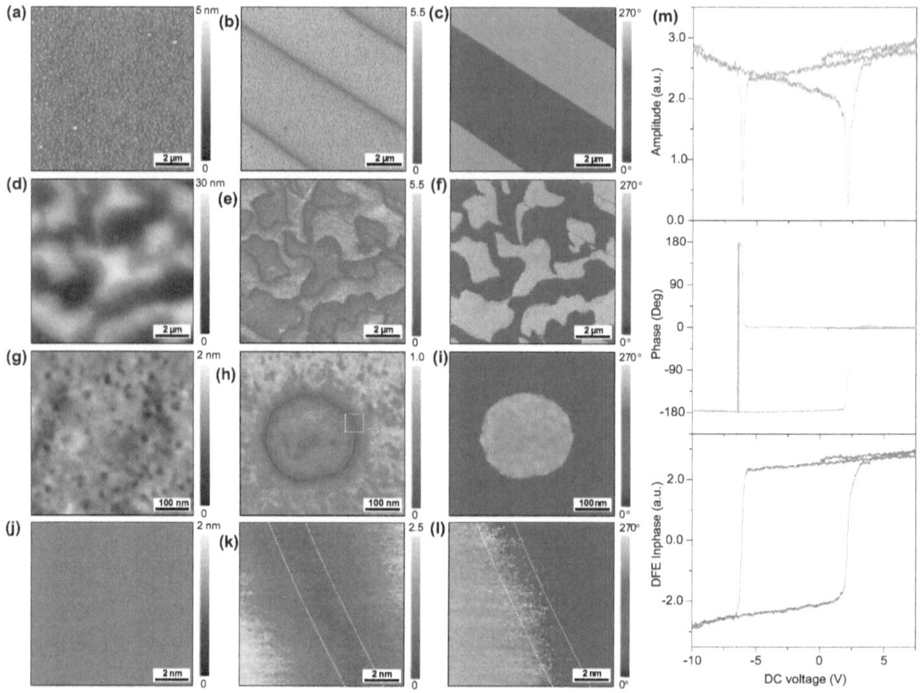


**FIGURE 3.28** Ferroelectric characterization by NC-HEsFM. (a) Topography, (b) DFE amplitude and (c) phase images of the PPLN sample. (d) Topography, (e) DFE amplitude, and (f) phase images of the PZN-9%PT sample. (g) Topography, (h) DFE amplitude, and (i) phase images of an artificial domain created on PZT film. (j) Topography, (k) DFE amplitude, and (l) phase image obtained around the PZT domain wall (indicated by the white box in (H)). (m) Amplitude (top), phase (middle), and in-phase (bottom) hysteresis loops of the PZT film. Adapted with permission from Zeng et al.[171] Copyright 2021 Wiley-VCH.)

images (Figure 3.29a–h) obtained by tapping-mode AFM and contact-mode PFM clearly demonstrate the highly destructive scanning of the conventional PFM, in which the fine features, even including the large protrusions, on the PPLN surface have gradually been scratched and removed during the contact-mode PFM scanning. In contrast, situations are dramatically different in the NC-HEsFM measurements. Figure 3.29i–p shows the topography images obtained by NC-HEsFM. Clearly, no observable zoomed-in scanning traces and surface modifications can be found during the continuous noncontact scanning, which unambiguously indicates that the NC-HEsFM has achieved a real noncontact and nondestructive measurement.[171]

To examine the influence of electrostatic force, DC spectroscopy has been performed in conventional PFM and NC-HEsFM, and the results are plotted in Figure 3.30. Figure 3.30a and b shows the PFM amplitude and phase as a function of DC bias, where the typical electrostatic force-induced V-shaped amplitude curve and ~180° phase change can be observed clearly, implying that the electrostatic force has significantly affected the conventional PFM signal. In contrast, completely different variation trends of the DFE amplitude and phase can be observed in the measurement

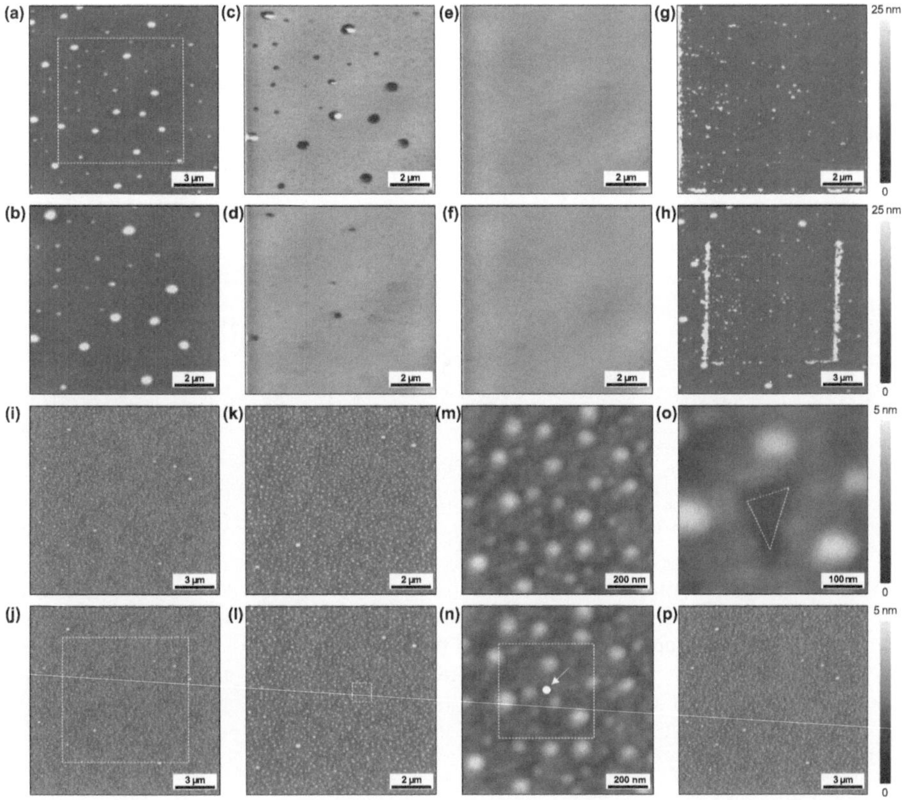


**FIGURE 3.29** Comparison of the surface modification situations between the conventional PFM and NC-HEsFM measurements. (a) Initial and (b) zoomed-in topography images measured by tapping-mode AFM. (c–f) The zoomed-in topography images obtained from the first to the fourth PFM scanning, respectively. (g, h) The topography images measured after the PFM scanning using tapping-mode AFM. (i) Initial topography images measured by the first and (j) the second NC-HEsFM scanning. (k) The topography of the first and (l) the second zoomed-in NC-HEsFM scanning. (m) The topography of the third and (n) the forth zoomed-in NC-HEsFM scans. (o) Topography image of in-situ NC-HEsFM scanning after the weak indentation. (p) Final topography image measured by NC-HEsFM after zoomed-in scanning. Adapted with permission from Zeng et al.[171] Copyright 2021 Wiley-VCH.)

of NC-HEsFM (Figure 3.30c and d). It is evident that, even though a large DC bias scanning range (±30 V) is used, all of the DFE amplitude and phase signals measured by six different flexural modes can keep a constant magnitude with changing DC bias, which unambiguously indicates that the electrostatic force effects, including both the local and distributed part, have almost been completely eliminated in the NC-HEsFM. With this superior capability of minimizing electrostatic force effect, the domain wall position can thereby be determined accurately from the DFE amplitude and phase images (Figure 3.28k and l), and at the same time, standard ferroelectric hysteresis loop (Figure 3.28m) can be obtained even though continuous DC method is used in the switching spectroscopy measurement.[171] In brief, the two most recently

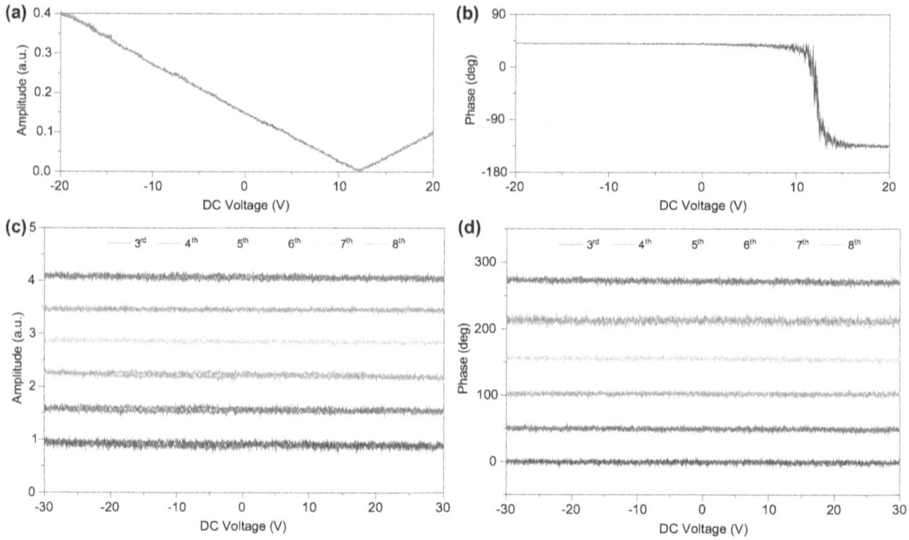


**FIGURE 3.30** Electrostatic force contribution in conventional PFM and NC-HEsFM. (a) The PFM amplitude; and (b) phase as a function of DC voltage measured by PFM on PPLN sample. (c) The DFE amplitude; and (d) DFE phase as a function of DC voltage measured by NC-HEsFM on PPLN sample using the third to the eight flexural mode. Note that all the spectrums in (c) and (d) are offset for clarity. (Adapted with permission from Zeng et al.[171] Copyright 2021 Wiley-VCH.)

developed PFM techniques, i.e., the HM-PFM and NC-HEsFM, have opened the new possibilities to characterize the E-S coupling behavior in many unknown materials.

## 3.5 SUMMARY

In this chapter, the technical principles, functions, challenges as well as the advances of PFM and ESM have been discussed. As an advanced scanning probe technique, PFM/ESM unprecedentedly provides researchers a high-resolution and versatile way to study the E-S coupling as well as the related properties at nanoscale. However, at the same time, extensive studies have also revealed that these powerful characterization capabilities of the PFM/ESM are actually challenged by a series of inherent issues and limitations of the PFM/ESM per se. Referring to the opinion from Gruverman et al., we are now standing at an inflection point in the PFM/ESM development and application.[7] On the one hand, owing to the unique advantages, PFM/ESM is an excellent characterization technique in the studies of E-S coupling phenomena for various material systems. On the other hand, an increasing number of experiments and studies have indeed demonstrated the issues and limitations of the current PFM/ESM techniques, and now it has been widely believed that a careful analysis of the PFM/ESM results is highly necessitated to distinguish between science and artifacts.[7,43,80] Although faced with numerous challenges, it is highly reasonable to believe that the PFM/ESM can keep playing the important role in future nanoscale E-S coupling studies, because the rapid technical developments have been continuously improving the characterization performance of the PFM/ESM, making the PFM/ESM measurements more and more unambiguous, stable, reproducible, and accurate.

## REFERENCES


1. Kalinin, S. V., A. Rar, and S. Jesse. 2006. A decade of piezoresponse force microscopy: Progress, challenges, and opportunities. *IEEE Trans. Ultrason. Ferroelectr. Freq. Control.* 53 (12):2226–2252.
2. Kalinin, S. V., B. J. Rodriguez, S. Jesse, et al. 2007. Nanoscale electromechanics of ferroelectric and biological systems: A new dimension in scanning probe microscopy. *Ann. Rev. Mater. Res.* 37 (1):189–238.
3. Newnham, R. E., V. Sundar, R. Yimnirun, J. Su, and Q. M. Zhang. 1997. Electrostriction: Nonlinear electromechanical coupling in solid dielectrics. *J. Phys. Chem. B* 101 (48):10141–10150.
4. Li, W., Z. Man, and J. Zeng, et al. 2021. Poling effect on the electrostrictive and piezoelectric response in $CH_3NH_3PbI_3$ single crystals. *Appl. Phys. Lett.* 118 (15):151905.
5. Güthner, P., and K. Dransfeld. 1992. Local poling of ferroelectric polymers by scanning force microscopy. *Appl. Phys. Lett.* 61 (9):1137–1139.
6. Soergel, E. 2011. Piezoresponse force microscopy (PFM). *J. Phys. D: Appl. Phys.* 44 (46):464003.
7. Gruverman, A., M. Alexe, and D. Meier. 2019. Piezoresponse force microscopy and nanoferroic phenomena. *Nat. Commun.* 10 (1):1661.
8. Balke, N., S. Jesse, A. N. Morozovska, et al. 2010. Nanoscale mapping of ion diffusion in a lithium-ion battery cathode. *Nat. Nanotech.* 5 (10):749–754.
9. Morozovska, A. N., E. A. Eliseev, and S. V. Kalinin. 2010. Electromechanical probing of ionic currents in energy storage materials. *Appl. Phys. Lett.* 96 (22):222906.
10. Morozovska, A. N., E. A. Eliseev, N. Balke, and S. V. Kalinin. 2010. Local probing of ionic diffusion by electrochemical strain microscopy: Spatial resolution and signal formation mechanisms. *J. Appl. Phys.* 108 (5):053712.
11. Gruverman, A., O. Auciello, and H. Tokumoto. 1998. Imaging and control of domain structures in ferroelectric thin films via scanning force microscopy. *Annu. Rev. Mater. Sci.* 28 (1):101–123.
12. Bonnell, D. A., S. V. Kalinin, A. L. Kholkin, and A. Gruverman. 2009. Piezoresponse force microscopy: A window into electromechanical behavior at the nanoscale. *MRS Bull.* 34 (9):648–657.
13. Kalinin, S. V., A. N. Morozovska, L. Q. Chen, and B. J. Rodriguez. 2010. Local polarization dynamics in ferroelectric materials. *Rep. Prog. Phys.* 73 (5):056502.
14. Kwon, O., D. Seol, H. Qiao, and Y. Kim. 2020. Recent progress in the nanoscale evaluation of piezoelectric and ferroelectric properties via scanning probe microscopy. *Adv. Sci.* 7 (18):1901391.
15. Hong, S. 2021. Single frequency vertical piezoresponse force microscopy. *J. Appl. Phys.* 129 (5):051101.
16. Franke, K., J. Besold, W. Haessler, and C. Seegebarth. 1994. Modification and detection of domains on ferroelectric PZT films by scanning force microscopy. *Surf. Sci.* 302 (1–2):L283–L288.
17. Zhu, J., L. Lu, and K. Zeng. 2013. Nanoscale mapping of lithium-ion diffusion in a cathode within an all-solid-state lithium-ion battery by advanced scanning probe microscopy techniques. *Acs Nano* 7 (2):1666–1675.

18. Balke, N., S. Jesse, Y. Kim, et al. 2010. Real space mapping of Li-ion transport in amorphous Si anodes with nanometer resolution. *Nano Lett.* 10 (9):3420–3425.
19. Kumar, A., F. Ciucci, A. N. Morozovska, S. V. Kalinin, and S. Jesse. 2011. Measuring oxygen reduction/evolution reactions on the nanoscale. *Nat. Chem.* 3 (9):707.
20. Kim, Y., A. N. Morozovska, A. Kumar, et al. 2012. Ionically-mediated electromechanical hysteresis in transition metal oxides. *ACS Nano* 6 (8):7026–7033.
21. Kumar, A., T. M Arruda, Y. Kim, et al. 2012. Probing surface and bulk electrochemical processes on the $LaAlO_3$-$SrTiO_3$ interface. *ACS Nano* 6 (5):3841–3852.
22. Balke, N., S. Kalnaus, N. J. Dudney, C. Daniel, S. Jesse, and S. V. Kalinin. 2012. Local detection of activation energy for ionic transport in lithium cobalt oxide. *Nano Lett.* 12 (7):3399–3403.
23. Kalinin, S. V., O. Dyck, N. Balke, et al. 2019. Toward electrochemical studies on the nanometer and atomic scales: Progress, challenges, and opportunities. *ACS Nano* 13 (9):9735–9780.
24. Kumar, A., D. Leonard, S. Jesse, et al. 2013. Spatially resolved mapping of oxygen reduction/evolution reaction on solid-oxide fuel cell cathodes with sub-10 nm resolution. *ACS Nano* 7 (5):3808–3814.
25. Strelcov, E., Y. Kim, S. Jesse, et al. 2013. Probing local ionic dynamics in functional oxides at the nanoscale. *Nano Lett.* 13 (8):3455–3462.
26. Eng, L. M., H. J. Güntherodt, G. Rosenman, et al. 1998. Nondestructive imaging and characterization of ferroelectric domains in periodically poled crystals. *J. Appl. Phys.* 83 (11):5973–5977.
27. Gruverman, A., and A. Kholkin. 2006. Nanoscale ferroelectrics: Processing, characterization and future trends. *Rep. Prog. Phys.* 69 (8):2443–2474.
28. Gruverman, A., and S. V. Kalinin. 2006. Piezoresponse force microscopy and recent advances in nanoscale studies of ferroelectrics. *J. Mater. Sci.* 41 (1):107–116.
29. Kalinin, S. V., B. J. Rodriguez, S. Jesse, et al. 2006. Vector piezoresponse force microscopy. *Microsc. Microanal.* 12 (3):206–220.
30. Eng, L. M., H.-J. Güntherodt, G. A. Schneider, U. Köpke, and J. Muñoz Saldaña. 1999. Nanoscale reconstruction of surface crystallography from three-dimensional polarization distribution in ferroelectric barium–titanate ceramics. *Appl. Phys. Lett.* 74 (2):233–235.
31. Rodriguez, B. J., A. Gruverman, A. I. Kingon, R. J. Nemanich, and J. S. Cross. 2004. Three-dimensional high-resolution reconstruction of polarization in ferroelectric capacitors by piezoresponse force microscopy. *J. Appl. Phys.* 95 (4):1958–1962.
32. Alexe, M., A. Gruverman, C. Harnagea, et al. 1999. Switching properties of self-assembled ferroelectric memory cells. *Appl. Phys. Lett.* 75 (8):1158–1160.
33. Gruverman, A., O. Auciello, and H. Tokumoto. 1996. Nanoscale investigation of fatigue effects in $Pb(Zr, Ti)O_3$ films. *Appl. Phys. Lett.* 69 (21):3191–3193.
34. Liu, Y., Y. Zhang, M.-J. Chow, Q. N. Chen, and J. Li. 2012. Biological ferroelectricity uncovered in aortic walls by piezoresponse force microscopy. *Phys. Rev. Lett.* 108 (7):078103.
35. Li, T., and K. Zeng. 2013. Nanoscale piezoelectric and ferroelectric behaviors of seashell by piezoresponse force microscopy. *J. Appl. Phys.* 113 (18):187202.
36. Martin, D., J. Müller, T. Schenk, et al. 2014. Ferroelectricity in Si-doped $HfO_2$ revealed: A binary lead-free ferroelectric. *Adv. Mater.* 26 (48):8198–8202.
37. Yang, S. M., S. J. Moon, T. H. Kim, and Y. S. Kim. 2014. Observation of ferroelectricity induced by defect dipoles in the strain-free epitaxial $CaTiO_3$ thin film. *Curr. Appl. Phys.* 14 (5):757–760.
38. Garten, L. M., D. T. Moore, S. U. Nanayakkara, et al. 2019. The existence and impact of persistent ferroelectric domains in $MAPbI_3$. *Sci. Adv.* 5 (1):eaas9311.

39. Yuan, S. Z., X. J. Meng, J. L. Sun, et al. 2011. Ferroelectricity of ultrathin ferroelectric Langmuir–Blodgett polymer films on conductive $LaNiO_3$ electrodes. *Mater. Lett.* 65 (12):1989–1991.
40. Röhm, H., T. Leonhard, M. J. Hoffmann, and A. Colsmann. 2017. Ferroelectric domains in methylammonium lead iodide perovskite thin-films. *Energy Environ. Sci.* 10 (4):950–955.
41. Denning, D., J. Guyonnet, and B. J. Rodriguez. 2016. Applications of piezoresponse force microscopy in materials research: From inorganic ferroelectrics to biopiezoelectrics and beyond. *Int. Mater. Rev.* 61 (1):46–70.
42. Nath, R., S. Hong, J. A. Klug, et al. 2010. Effects of cantilever buckling on vector piezoresponse force microscopy imaging of ferroelectric domains in $BiFeO_3$ nanostructures. *Appl. Phys. Lett.* 96 (16):163101.
43. Vasudevan, R. K., N. Balke, P. Maksymovych, S. Jesse, and S. V. Kalinin. 2017. Ferroelectric or non-ferroelectric: Why so many materials exhibit "ferroelectricity" on the nanoscale. *Appl. Phys. Rev.* 4 (2):021302.
44. Kalinin, S. V. and D. A. Bonnell. 2002. Imaging mechanism of piezoresponse force microscopy of ferroelectric surfaces. *Phys. Rev. B* 65 (12):125408.
45. Zhao, W., W. Song, L.-Z. Cheong, et al. 2019. Beyond imaging: Applications of atomic force microscopy for the study of Lithium-ion batteries. *Ultramicroscopy* 204:34–48.
46. Yang, N., A. Belianinov, E. Strelcov, et al. 2014. Effect of doping on surface reactivity and conduction mechanism in samarium-doped ceria thin films. *ACS Nano* 8 (12):12494–12501.
47. Giridharagopal, R., L. Q. Flagg, J. S. Harrison, et al. 2017. Electrochemical strain microscopy probes morphology-induced variations in ion uptake and performance in organic electrochemical transistors. *Nat. Mater.* 16 (7):737–742.
48. Gruverman, A., O. Auciello, J. Hatano, and H. Tokumoto. 1996. Scanning force microscopy as a tool for nanoscale study of ferroelectric domains. *Ferroelectrics* 184 (1):11–20.
49. Gruverman, A., O. Auciello, and H. Tokumoto. 1996. Scanning force microscopy for the study of domain structure in ferroelectric thin films. *J. Vac. Sci. Technol. B* 14 (2):602–605.
50. Chen, P. J., and S. T. Montgomery. 1980. A macroscopic theory for the existence of the hysteresis and butterfly loops in ferroelectricity. *Ferroelectrics* 23 (1):199–207.
51. Jesse, S., A. P. Baddorf, and S. V. Kalinin. 2006. Switching spectroscopy piezoresponse force microscopy of ferroelectric materials. *Appl. Phys. Lett.* 88 (6):062908.
52. Balke, N., S. Jesse, Y. Kim, et al. 2010. Decoupling electrochemical reaction and diffusion processes in ionically-conductive solids on the nanometer scale. *ACS Nano* 4 (12):7349–7357.
53. Hong, S., J. Woo, H. Shin, et al. 2001. Principle of ferroelectric domain imaging using atomic force microscope. *J. Appl. Phys.* 89 (2):1377–1386.
54. Guo, H. Y., J. B. Xu, I. H. Wilson, et al. 2002. Study of domain stability on $(Pb_{0.76}Ca_{0.24})TiO_3$ thin films using piezoresponse microscopy. *Appl. Phys. Lett.* 81 (4):715–717.
55. Vasudevan, R. K., S. Jesse, Y. Kim, A. Kumar, and S. V. Kalinin. 2012. Spectroscopic imaging in piezoresponse force microscopy: New opportunities for studying polarization dynamics in ferroelectrics and multiferroics. *MRS Commun.* 2 (3):61–73.
56. Kim, Y., A. Kumar, O. Ovchinnikov, et al. 2012. First-order reversal curve probing of spatially resolved polarization switching dynamics in ferroelectric nanocapacitors. *ACS Nano* 6 (1):491–500.
57. Balke, N., S. Jesse, Q. Li, et al. 2015. Current and surface charge modified hysteresis loops in ferroelectric thin films. *J. Appl. Phys.* 118 (7):072013.
58. Alexe, M., and A. Gruverman. 2004. *Nanoscale Characterisation of Ferroelectric Materials: Scanning Probe Microscopy Approach*. Springer: Berlin Heidelberg.

59. Balke, N., P. Maksymovych, S. Jesse, et al. 2015. Differentiating ferroelectric and non-ferroelectric electromechanical effects with scanning probe microscopy. *ACS Nano* 9 (6):6484–6492.
60. Kalinin, S. V., B. J. Rodriguez, S. Jesse, A. N. Morozovska, A. A. Bokov, and Z. G. Ye. 2009. Spatial distribution of relaxation behavior on the surface of a ferroelectric relaxor in the ergodic phase. *Appl. Phys. Lett.* 95 (14):142902.
61. Kalinin, S. V., B. J. Rodriguez, J. D. Budai, et al. 2010. Direct evidence of mesoscopic dynamic heterogeneities at the surfaces of ergodic ferroelectric relaxors. *Phys. Rev. B* 81 (6):064107.
62. Kumar, A., O. S. Ovchinnikov, H. Funakubo, S. Jesse, and S. V. Kalinin. 2011. Real-space mapping of dynamic phenomena during hysteresis loop measurements: Dynamic switching spectroscopy piezoresponse force microscopy. *Appl. Phys. Lett.* 98 (20):202903.
63. Kalinin, S. V., S. Jesse, B. J. Rodriguez, et al. 2007. Recent advances in electromechanical imaging on the nanometer scale: Polarization dynamics in ferroelectrics, biopolymers, and liquid imaging. *Jpn. J. Appl. Phys.* 46 (9A):5674–5685.
64. Abdollahi, A., N. Domingo, I. Arias, and G. Catalan. 2019. Converse flexoelectricity yields large piezoresponse force microscopy signals in non-piezoelectric materials. *Nat. Commun.* 10 (1):1266.
65. Yu, J., E. N. Esfahani, Q. Zhu, et al. 2018. Quadratic electromechanical strain in silicon investigated by scanning probe microscopy. *J. Appl. Phys.* 123 (15):155104.
66. Zhu, Q., K. Pan, S. Xie, Y. Liu, and J. Li. 2019. Nanomechanics of multiferroic composite nanofibers via local excitation piezoresponse force microscopy. *J. Mech. Phys. Solids* 126:76–86.
67. Liu, Y., L. Collins, R. Proksch, et al. 2018. Chemical nature of ferroelastic twin domains in $CH_3NH_3PbI_3$ perovskite. *Nat. Mater.* 17 (11):1013.
68. Strelcov, E., Q. Dong, T. Li, et al. 2017. $CH_3NH_3PbI_3$ perovskites: Ferroelasticity revealed. *Sci. Adv.* 3 (4):e1602165.
69. Kim, S. K., R. Bhatia, T.-H. Kim, et al. 2016. Directional dependent piezoelectric effect in CVD grown monolayer $MoS_2$ for flexible piezoelectric nanogenerators. *Nano Energy* 22:483–489.
70. Wang, X., X. He, H. Zhu, et al. 2016. Subatomic deformation driven by vertical piezoelectricity from CdS ultrathin films. *Sci. Adv.* 2 (7):e1600209.
71. Christman, J. A., R. R. Woolcott Jr., A. I. Kingon, and R. J. Nemanich. 1998. Piezoelectric measurements with atomic force microscopy. *Appl. Phys. Lett.* 73 (26):3851–3853.
72. Morozovska, A. N., S. V. Svechnikov, E. A. Eliseev, S. Jesse, B. J. Rodriguez, and S. V. Kalinin. 2007. Piezoresponse force spectroscopy of ferroelectric-semiconductor materials. *J. Appl. Phys.* 102 (11):114108.
73. Bark, C. W., P. Sharma, Y. Wang, et al. 2012. Switchable induced polarization in $LaAlO_3/SrTiO_3$ heterostructures. *Nano Lett.* 12 (4):1765–1771.
74. Sharma, P., S. Ryu, Z. Viskadourakis, et al. 2015. Electromechanics of ferroelectric-like behavior of $LaAlO_3$ thin films. *Adv. Funct. Mater.* 25 (41):6538–6544.
75. Röhm, H., T. Leonhard, A. D. Schulz, S. Wagner, M. J. Hoffmann, and A. Colsmann. 2019. Ferroelectric properties of perovskite thin films and their implications for solar energy conversion. *Adv. Mater.* 31 (26):1806661.
76. Rodriguez, B. J., R. J. Nemanich, A. Kingon, et al. 2005. Domain growth kinetics in lithium niobate single crystals studied by piezoresponse force microscopy. *Appl. Phys. Lett.* 86 (1):012906.
77. Tan, Z., L. Hong, Z. Fan, et al. 2019. Thinning ferroelectric films for high-efficiency photovoltaics based on the Schottky barrier effect. *NPG Asia Mater.* 11 (1):20.
78. Borowiak, A. S., N. Baboux, D. Albertini, et al. 2014. Electromechanical response of amorphous $LaAlO_3$ thin film probed by scanning probe microscopies. *Appl. Phys. Lett.* 105 (1):012906.

79. Bdikin, I. K., A. L. Kholkin, A. N. Morozovska, S. V. Svechnikov, S.-H. Kim, and S. V. Kalinin. 2008. Domain dynamics in piezoresponse force spectroscopy: Quantitative deconvolution and hysteresis loop fine structure. *Appl. Phys. Lett.* 92 (18):182909.
80. Seol, D., B. Kim, and Y. Kim. 2017. Non-piezoelectric effects in piezoresponse force microscopy. *Curr. Appl. Phys.* 17 (5):661–674.
81. Labuda, A., and R. Proksch. 2015. Quantitative measurements of electromechanical response with a combined optical beam and interferometric atomic force microscope. *Appl. Phys. Lett.* 106 (25):253103.
82. Hurley, D. C., M. Kopycinska-Müller, A. B. Kos, and R. H. Geiss. 2005. Nanoscale elastic-property measurements and mapping using atomic force acoustic microscopy methods. *Meas. Sci. Technol.* 16 (11):2167–2172.
83. Lantz, M. A., S. J. O'Shea, and M. E. Welland. 1998. Characterization of tips for conducting atomic force microscopy in ultrahigh vacuum. *Rev. Sci. Instrum.* 69 (4):1757–1764.
84. Bhushan, B., and K. J. Kwak. 2008. The role of lubricants, scanning velocity and operating environment in adhesion, friction and wear of Pt–Ir coated probes for atomic force microscope probe-based ferroelectric recording technology. *J. Phys. Condens. Matter.* 20 (32):325240.
85. Kalinin, A., V. Atepalikhin, O. Pakhomov, A. L. Kholkin, and A. Tselev. 2018. An atomic force microscopy mode for nondestructive electromechanical studies and its application to diphenylalanine peptide nanotubes. *Ultramicroscopy* 185:49–54.
86. Calahorra, Y., M. Smith, A. Datta, H. Benisty, and S. Kar-Narayan. 2017. Mapping piezoelectric response in nanomaterials using a dedicated non-destructive scanning probe technique. *Nanoscale* 9 (48):19290–19297.
87. Saadi, M. A. S. R., B. Uluutku, C. H. Parvini, and S. D. Solares. 2020. Soft sample deformation, damage and induced electromechanical property changes in contact- and tapping-mode atomic force microscopy. *Surf. Topogr. Metrol. Prop.* 8 (4):045004.
88. Ievlev, A. V., C. Brown, M. J. Burch, et al. 2018. Chemical phenomena of atomic force microscopy scanning. *Anal. Chem.* 90 (5):3475–3481.
89. Kopycinska-Müller, M., R. H. Geiss, and D. C. Hurley. 2006. Contact mechanics and tip shape in AFM-based nanomechanical measurements. *Ultramicroscopy* 106 (6):466–474.
90. Jesse, S., S. Guo, A. Kumar, B. J. Rodriguez, R. Proksch, and S. V. Kalinin. 2010. Resolution theory, and static and frequency-dependent cross-talk in piezoresponse force microscopy. *Nanotechnology* 21 (40):405703.
91. Cui, A., P. De Wolf, Y. Ye, et al. 2019. Probing electromechanical behaviors by datacube piezoresponse force microscopy in ambient and aqueous environments. *Nanotechnology* 30 (23):235701.
92. Lu, H., C.-W. Bark, D. Esque de los Ojos, et al. 2012. Mechanical writing of ferroelectric polarization. *Science* 336 (6077):59–61.
93. Kalinin, S. V., S. Jesse, W. Liu, and A. A. Balandin. 2006. Evidence for possible flexoelectricity in tobacco mosaic viruses used as nanotemplates. *Appl. Phys. Lett.* 88 (15):153902.
94. Wang, L., S. Liu, X. Feng, et al. 2020. Flexoelectronics of centrosymmetric semiconductors. *Nat. Nanotech.* 15 (8):661–667.
95. Kholkin, A. L., V. V. Shvartsman, A. Yu Emelyanov, R. Poyato, M. L. Calzada, and L. Pardo. 2003. Stress-induced suppression of piezoelectric properties in $PbTiO_3$: La thin films via scanning force microscopy. *Appl. Phys. Lett.* 82 (13):2127–2129.
96. Abplanalp, M., J. Fousek, and P. Günter. 2001. Higher order ferroic switching induced by scanning force microscopy. *Phys. Rev. Lett.* 86 (25):5799–5802.
97. Balke, N., P. Maksymovych, S. Jesse, I. I. Kravchenko, Q. Li, and S. V. Kalinin. 2014. Exploring local electrostatic effects with scanning probe microscopy: Implications for piezoresponse force microscopy and triboelectricity. *ACS Nano* 8 (10):10229–10236.

98. Yao, J., M. Ye, Y. Sun, et al. 2020. Atomic-scale insight into the reversibility of polar order in ultrathin epitaxial Nb: $SrTiO_3/BaTiO_3$ heterostructure and its implication to resistive switching. *Acta Mater.* 188:23–29.
99. Zhou, Y. S., Y. Liu, G. Zhu, et al. 2013. In situ quantitative study of nanoscale triboelectrification and patterning. *Nano Lett.* 13 (6):2771–2776.
100. Jesse, S., B. Mirman, and S. V. Kalinin. 2006. Resonance enhancement in piezoresponse force microscopy: Mapping electromechanical activity, contact stiffness, and Q factor. *Appl. Phys. Lett.* 89 (2):022906.
101. Collins, L., Y. Liu, O. S. Ovchinnikova, and R. Proksch. 2019. Quantitative electromechanical atomic force microscopy. *ACS Nano* 13 (7):8055–8066.
102. Harnagea, C., M. Alexe, D. Hesse, and A. Pignolet. 2003. Contact resonances in voltage-modulated force microscopy. *Appl. Phys. Lett.* 83 (2):338–340.
103. Okino, H., J. Sakamoto, and T. Yamamoto. 2003. Contact-resonance piezoresponse force microscope and its application to domain observation of $Pb(Mg1/3Nb_2/3)O_3$–$PbTiO_3$ single crystals. *Jpn. J. Appl. Phys.* 42 (Part 1, No. 9B):6209–6213.
104. Jesse, S., and S. V. Kalinin. 2011. Band excitation in scanning probe microscopy: Sines of change. *J. Phys. D Appl. Phys.* 44 (46):464006.
105. Miao, H., C. Tan, X. Zhou, X. Wei, and F. Li. 2014. More ferroelectrics discovered by switching spectroscopy piezoresponse force microscopy? *EPL (Europhys. Lett.)* 108 (2):27010.
106. Kim, B., D. Seol, S. Lee, H. N. Lee, and Y. Kim. 2016. Ferroelectric-like hysteresis loop originated from non-ferroelectric effects. *Appl. Phys. Lett.* 109 (10):102901.
107. Lee, K., H. Shin, W. Moon, J. U. Jeon, and Y. E. Pak. 1999. Detection mechanism of spontaneous polarization in ferroelectric thin films using electrostatic force microscopy. *Jpn. J. Appl. Phys.* 38 (Part 2, No. 3A):L264–L266.
108. Lushta, V., S. Bradler, B. Roling, and A. Schirmeisen. 2017. Correlation between drive amplitude and resonance frequency in electrochemical strain microscopy: Influence of electrostatic forces. *J. Appl. Phys.* 121 (22):224302.
109. Schön, N., R. Schierholz, S. Jesse, et al. 2021. Signal origin of electrochemical strain microscopy and link to local chemical distribution in solid state electrolytes. *Small Methods* 5:2001279.
110. Eliseev, E. A., A. N. Morozovska, A. V. Ievlev, et al. 2014. Electrostrictive and electrostatic responses in contact mode voltage modulated scanning probe microscopies. *Appl. Phys. Lett.* 104 (23):232901.
111. Kim, S., D. Seol, X. Lu, M. Alexe, and Y. Kim. 2017. Electrostatic-free piezoresponse force microscopy. *Sci. Rep.* 7:41657.
112. MacDonald, G. A., F. W. DelRio, and J. P. Killgore. 2018. Higher-eigenmode piezoresponse force microscopy: A path towards increased sensitivity and the elimination of electrostatic artifacts. *Nano Futures* 2 (1):015005.
113. Gomez, A., T. Puig, and X. Obradors. 2018. Diminish electrostatic in piezoresponse force microscopy through longer or ultra-stiff tips. *Appl. Surf. Sci.* 439:577–582.
114. Hong, S., H. Shin, J. Woo, and K. No. 2002. Effect of cantilever–sample interaction on piezoelectric force microscopy. *Appl. Phys. Lett.* 80 (8):1453–1455.
115. Huey, B. D., C. Ramanujan, M. Bobji, et al. 2004. The importance of distributed loading and cantilever angle in piezo-force microscopy. *J. Electroceram.* 13 (1):287–291.
116. Balke, N., S. Jesse, P. Yu, C. Ben, S. V. Kalinin, and A. Tselev. 2016. Quantification of surface displacements and electromechanical phenomena via dynamic atomic force microscopy. *Nanotechnology* 27 (42):425707.
117. Balke, N., S. Jesse, B. Carmichael, et al. 2017. Quantification of in-contact probe-sample electrostatic forces with dynamic atomic force microscopy. *Nanotechnology* 28 (6):065704.

118. Seal, K., S. Jesse, B. J. Rodriguez, A. P. Baddorf, and S. V. Kalinin. 2007. High frequency piezoresponse force microscopy in the 1–10 MHz regime. *Appl. Phys. Lett.* 91 (23):232904.
119. Bradler, S., A. Schirmeisen, and B. Roling. 2017. Amplitude quantification in contact-resonance-based voltage-modulated force spectroscopy. *J. Appl. Phys.* 122 (6):065106.
120. Zeng, Q., H. Wang, Z. Xiong, et al. 2021. Nanoscale ferroelectric characterization with heterodyne megasonic piezoresponse force microscopy. *Adv. Sci.* 8: 2003993.
121. Proksch, R. 2015. In-situ piezoresponse force microscopy cantilever mode shape profiling. *J. Appl. Phys.* 118 (7):072011.
122. Frederix, P. L. T. M., M. R. Gullo, T. Akiyama, et al. 2005. Assessment of insulated conductive cantilevers for biology and electrochemistry. *Nanotechnology* 16 (8):997–1005.
123. Kim, Y., C. Bae, K. Ryu, et al. 2009. Origin of surface potential change during ferroelectric switching in epitaxial $PbTiO_3$ thin films studied by scanning force microscopy. *Appl. Phys. Lett.* 94 (3):032907.
124. Son, J. Y., K. Kyhm, and J. H. Cho. 2006. Surface charge retention and enhanced polarization effect on ferroelectric thin films. *Appl. Phys. Lett.* 89 (9):092907.
125. Zhou, X., J. Fu, and F. Li. 2013. Contact resonance force microscopy for nanomechanical characterization: Accuracy and sensitivity. *J. Appl. Phys.* 114 (6):064301.
126. Kim, Y., A. Kumar, A. Tselev, et al. 2011. Nonlinear phenomena in multiferroic nanocapacitors: Joule heating and electromechanical effects. *ACS Nano* 5 (11):9104–9112.
127. Kalinin, S. V., S. Jesse, A. Tselev, A. P. Baddorf, and N. Balke. 2011. The role of electrochemical phenomena in scanning probe microscopy of ferroelectric thin films. *ACS Nano* 5 (7):5683–5691.
128. Minj, A., J. Serron, U. Celano, and K. Paredis. 2020. Surface contamination: A natural way toward high-resolution electric force microscopy in contact-resonant mode. *J. Phys. Chem. C* 124 (46):25331–25340.
129. Proksch, R. 2014. Electrochemical strain microscopy of silica glasses. *J. Appl. Phys.* 116 (6):066804.
130. Sekhon, J. S, L. Aggarwal, and G. Sheet. 2014. Voltage induced local hysteretic phase switching in silicon. *Appl. Phys. Lett.* 104 (16):162908.
131. Chen, Q. N., Y. Ou, F. Ma, and J. Li. 2014. Mechanisms of electromechanical coupling in strain based scanning probe microscopy. *Appl. Phys. Lett.* 104 (24):242907.
132. Seol, D., S. Park, O. V. Varenyk, et al. 2016. Determination of ferroelectric contributions to electromechanical response by frequency dependent piezoresponse force microscopy. *Sci. Rep.* 6: 30579.
133. Damjanovic, D. 1997. Stress and frequency dependence of the direct piezoelectric effect in ferroelectric ceramics. *J. Appl. Phys.* 82 (4):1788–1797.
134. Xu, F., S. Trolier-McKinstry, W. Ren, Baomin Xu, Z. L. Xie, and K. J. Hemker. 2001. Domain wall motion and its contribution to the dielectric and piezoelectric properties of lead zirconate titanate films. *J. Appl. Phys.* 89 (2):1336–1348.
135. Jesse, S., H. N. Lee, and S. V. Kalinin. 2006. Quantitative mapping of switching behavior in piezoresponse force microscopy. *Rev. Sci. Instrum.* 77 (7):073702.
136. Damjanovic, D. 1998. Ferroelectric, dielectric and piezoelectric properties of ferroelectric thin films and ceramics. *Rep. Prog. Phys.* 61 (9):1267–1324.
137. Li, F., L. Jin, Z. Xu, and S. Zhang. 2014. Electrostrictive effect in ferroelectrics: An alternative approach to improve piezoelectricity. *Appl. Phys. Rev.* 1 (1):011103.
138. Liu, Y., L. Collins, R. Proksch, et al. 2019. Reply to: On the ferroelectricity of $CH_3NH_3PbI_3$ perovskites. *Nat. Mater.* 18 (10):1051–1053.
139. Schulz, A. D., H. Röhm, T. Leonhard, S. Wagner, M. J. Hoffmann, and A. Colsmann. 2019. On the ferroelectricity of $CH_3NH_3PbI_3$ perovskites. *Nat. Mater.* 18 (10):1050.

140. Huang, B., G. Kong, E. N. Esfahani, et al. 2018. Ferroic domains regulate photocurrent in single-crystalline $CH_3NH_3PbI_3$ films self-grown on FTO/$TiO_2$ substrate. *NPJ Quantum Mater.* 3 (1):30.
141. Xia, G., B. Huang, Y. Zhang, et al. 2019. Nanoscale insights into photovoltaic hysteresis in triple-cation mixed-halide perovskite: Resolving the role of polarization and ionic migration. *Adv. Mater.* 31 (36):1902870.
142. Kutes, Y., L. Ye, Y. Zhou, S. Pang, B. D. Huey, and N. P. Padture. 2014. Direct observation of ferroelectric domains in solution-processed $CH_3NH_3PbI_3$ perovskite thin films. *J. Phys. Chem. Lett.* 5 (19):3335–3339.
143. Chen, B., J. Shi, X. Zheng, Y. Zhou, K. Zhu, and S. Priya. 2015. Ferroelectric solar cells based on inorganic-organic hybrid perovskites. *J. Mater. Chem. A* 3 (15):7699–7705.
144. Kim, D., J. S. Yun, P. Sharma, et al. 2019. Light-and bias-induced structural variations in metal halide perovskites. *Nat. Commun.* 10 (1):444.
145. Wang, P., J. Zhao, L. Wei, et al. 2017. Photo-induced ferroelectric switching in perovskite $CH_3NH_3PbI_3$ films. *Nanoscale* 9 (11):3806–3817.
146. Gómez, A., Q. Wang, A. R. Goñi, M. Campoy-Quiles, and A. Abate. 2019. Ferroelectricity-free lead halide perovskites. *Energy Environ. Sci* 12 (8):2537–2547.
147. Dong, Q., J. Song, Y. Fang, Y. Shao, S. Ducharme, and J. Huang. 2016. Lateral-structure single-crystal hybrid perovskite solar cells via piezoelectric poling. *Adv. Mater.* 28 (14):2816–2821.
148. Chen, B., T. Li, Q. Dong, et al. 2018. Large electrostrictive response in lead halide perovskites. *Nat. Mater.* 17 (11):1020–1026.
149. Liu, Y., N. Borodinov, M. Lorenz, et al. 2020. Hysteretic ion migration and remanent field in metal halide perovskites. *Adv. Sci.* 7: 2001176.
150. Giessibl, F. J. 2003. Advances in atomic force microscopy. *Rev. Mod. Phys.* 75 (3):949–983.
151. Hansma, P. K., and J. Tersoff. 1987. Scanning tunneling microscopy. *J. Appl. Phys.* 61 (2):R1–R24.
152. Kalinin, S. V., B. J. Rodriguez, S. Jesse, T. Thundat, and A. Gruverman. 2005. Electromechanical imaging of biological systems with sub-10nm resolution. *Appl. Phys. Lett.* 87 (5):053901.
153. Rodriguez, B. J., S. Jesse, A. P. Baddorf, and S. V. Kalinin. 2006. High resolution electromechanical imaging of ferroelectric materials in a liquid environment by piezoresponse force microscopy. *Phys. Rev. Lett.* 96 (23):237602.
154. Jungk, T., Á. Hoffmann, and E. Soergel. 2008. Impact of the tip radius on the lateral resolution in piezoresponse force microscopy. *New J. Phys.* 10 (1):013019.
155. Kalinin, S. V., S. Jesse, B. J. Rodriguez, et al. 2006. Spatial resolution, information limit, and contrast transfer in piezoresponse force microscopy. *Nanotechnology* 17 (14):3400–3411.
156. O'Donnell, J., E. U. Haq, C. Silien, T. Soulimane, D. Thompson, and S. A. M. Tofail. 2021. A practical approach for standardization of converse piezoelectric constants obtained from piezoresponse force microscopy. *J. Appl. Phys.* 129 (18):185104.
157. Yarajena, S. S., R. Biswas, V. Raghunathan, and A. K. Naik. 2021. Quantitative probe for in-plane piezoelectric coupling in 2D materials. *Sci. Rep.* 11 (1):7066.
158. Seol, D., S. Kang, C. Sun, and Y. Kim. 2019. Significance of electrostatic interactions due to surface potential in piezoresponse force microscopy. *Ultramicroscopy* 207:112839.
159. Kalinin, S. V., E. Karapetian, and M. Kachanov. 2004. Nanoelectromechanics of piezoresponse force microscopy. *Phys. Rev. B* 70 (18):184101.
160. Jesse, S., A. P. Baddorf, and S. V. Kalinin. 2006. Dynamic behaviour in piezoresponse force microscopy. *Nanotechnology* 17 (6):1615–1628.

161. Liu, Y., P. Trimby, L. Collins, et al. 2021. Correlating crystallographic orientation and ferroic properties of twin domains in metal halide perovskites. *ACS Nano* 15 (4):7139–7148.
162. Kolosov, O., A. Gruverman, J. Hatano, K. Takahashi, and H. Tokumoto. 1995. Nanoscale visualization and control of ferroelectric domains by atomic force microscopy. *Phys. Rev. Lett.* 74 (21):4309–4312.
163. Hong, J. W., K. H. Noh, Sang-il Park, S. I. Kwun, and Z. G. Khim. 1998. Surface charge density and evolution of domain structure in triglycine sulfate determined by electrostatic-force microscopy. *Phys. Rev. B* 58 (8):5078–5084.
164. Hong, J. W., Sang-il Park, and Z. G. Khim. 1999. Measurement of hardness, surface potential, and charge distribution with dynamic contact mode electrostatic force microscope. *Rev. Sci. Instrum.* 70 (3):1735–1739.
165. Jesse, S., S. V. Kalinin, R. Proksch, A. P. Baddorf, and B. J. Rodriguez. 2007. The band excitation method in scanning probe microscopy for rapid mapping of energy dissipation on the nanoscale. *Nanotechnology* 18 (43):435503.
166. Rodriguez, B. J., C. Callahan, S. V. Kalinin, and R. Proksch. 2007. Dual-frequency resonance-tracking atomic force microscopy. *Nanotechnology* 18 (47):475504.
167. Nath, R., Y.-H. Chu, N. A. Polomoff, R. Ramesh, and B. D. Huey. 2008. High speed piezoresponse force microscopy: <1 frame per second nanoscale imaging. *Appl. Phys. Lett.* 93 (7):072905.
168. Rodriguez, B. J., S. Jesse, S. Habelitz, R. Proksch, and S. V. Kalinin. 2009. Intermittent contact mode piezoresponse force microscopy in a liquid environment. *Nanotechnology* 20 (19):195701.
169. Shi, W., C. Lee, G. Pascual, J. P. Pineda, B. Kim, and K. Lee. PinPoint piezoelectric force microscopy (application note). Available from https://www.parksystems.com/images/media/appnote/App_Note-28_PinPoint-PFM_June_v1.2.pdf.
170. Zeng, K., and Q. Zeng. 2020. System and method for scanning probe microscopy. Singapore Patent Application No. 10202003740V.
171. Zeng, Q., Q. Huang, H. Wang, et al. 2021. Breaking the fundamental limitations of nanoscale ferroelectric characterization: Non-contact heterodyne electrostrain force microscopy. *Small Methods* 5: 2100639.
172. De Wolf, P., Z. Huang, B. Pittenger, et al. 2018. Functional imaging with higher-dimensional electrical data sets. *Microsc. Today* 26 (6):18–27.
173. Zeng, K., and Q. Zeng. 2021. Methods and apparatus for non-contact detection of surface strain on atomic force microscopy based technique. Singapore Patent Application No. 10202100483P.
174. Labardi, M., V. Likodimos, and M. Allegrini. 2000. Force-microscopy contrast mechanisms in ferroelectric domain imaging. *Phys. Rev. B* 61 (21):14390–14398.
175. Likodimos, V., M. Labardi, and M. Allegrini. 2002. Domain pattern formation and kinetics on ferroelectric surfaces under thermal cycling using scanning force microscopy. *Phys. Rev. B* 66 (2):024104.
176. Li, T., A. Lipatov, H. Lu, et al. 2018. Optical control of polarization in ferroelectric heterostructures. *Nat. Commun.* 9 (1):3344.
177. Yang, S., B. Yan, J. Wu, L. Lu, and K. Zeng. 2017. Temperature-dependent lithium-ion diffusion and activation energy of $Li_{1.2}Co_{0.13}Ni_{0.13}Mn_{0.54}O_2$ thin-film cathode at nanoscale by using electrochemical strain microscopy. *ACS Appl. Mater. Interfaces* 9 (16):13999–14005.
178. Young, T. J., M. A. Monclus, T. L. Burnett, W. R. Broughton, S. L. Ogin, and P. A. Smith. 2011. The use of the PeakForce™ quantitative nanomechanical mapping AFM-based method for high-resolution Young's modulus measurement of polymers. *Meas. Sci. Technol.* 22 (12):125703.

179. Asylum Research. Fast Force Mapping (FFM). Available from https://afm.oxinst.com/assets/uploads/products/asylum/documents/FFM_22APRIL2021.pdf.
180. Park Systems. Park PinPointTM Mode: PinPoint your sample via AFM. Available from https://parksystems.com/images/media/brochures/pinpoint/ParkPinPoint_201210.pdf.
181. Qiao, H., O. Kwon, and Y. Kim. 2020. Electrostatic effect on off-field ferroelectric hysteresis loop in piezoresponse force microscopy. *Appl. Phys. Lett.* 116 (17):172901.
182. Kumar, A., C. Chen, T. M. Arruda, S. Jesse, F. Ciucci, and S. V. Kalinin. 2013. Frequency spectroscopy of irreversible electrochemical nucleation kinetics on the nanoscale. *Nanoscale* 5 (23):11964–11970.
183. Jungk, T., Á. Hoffmann, and E. Soergel. 2006. Quantitative analysis of ferroelectric domain imaging with piezoresponse force microscopy. *Appl. Phys. Lett.* 89 (16):163507.
184. Anbusathaiah, V., S. Jesse, M. A. Arredondo, et al. 2010. Ferroelastic domain wall dynamics in ferroelectric bilayers. *Acta Mater.* 58 (16):5316–5325.
185. Collins, L., J. I. Kilpatrick, S. A. L. Weber, et al. 2013. Open loop Kelvin probe force microscopy with single and multi-frequency excitation. *Nanotechnology* 24 (47):475702.

# 4 Hybrid AFM Technique
## *Atomic Force Microscopy-Scanning Electrochemical Microscopy*

*Shuang Cao and Tong Sun*
Qingdao University

**CONTENTS**



## 4.1 INTRODUCTION

Numerous electrochemical characterization techniques, such as cyclic voltammetry [1], electrochemical impedance spectroscopy [2] and polarography [3], have been designed to investigate electrochemical activity on liquid/solid, liquid/gas and liquid/liquid interfaces and supply information about interfacial catalysis, corrosion and biological processes [4–7]. Nevertheless, the data obtained from traditional electrochemical measurements, which are based on potentiometric or voltammetric measurements, are usually at macro-level and cannot offer detailed information about local surface features at micro- or nano-level, which is significant to study the characteristics of heterogeneous systems, such as active sites, surface flaws, etc. Scanning Electrochemical Microscopy (SECM) developed by Bard et al. [8] employed

DOI: 10.1201/9781003174042-4

micro- or nano-scale electrode to scan sample surface and obtain spatially local electrochemical properties by analyzing the electrochemical response of the electrode to the sample surface [9–11]. Because of the advancements in nanoelectrode preparation technology, the spatial resolution for this technique has been increased up to 10 nm [12]. Detailed information about traditional SECM, such as working principle and main operation modes, will be discussed in the following section. However, the SECM signals are sensitive to interaction features, like the distance between electrode and sample surface. Generally, traditional SECM utilizes a constant height method, which simply leads to the convolution of gathered topographical and electrochemical data which has an impact on the data's correctness.

Various hybrid SECM techniques, including shear force positioning [13], scanning icon conductance microscopy [14], scanning tunneling microscopy [15] as well as atomic force microscope (AFM), have been developed to resolve the constraints of traditional SECM. Among these techniques, SECM incorporating AFM is considered to be more preferable due to its high-spatial-resolution imaging capabilities and the sharp cantilevered probe. High precision piezoelectric ceramics are used in the AFM system to control AFM tip position, which provides precise control of distance between the tip and sample surface. Due to the fact that the resolution of electrochemical or morphological scanning is limited by the probe size, a probe with a smaller radius is required to obtain a image with higher resolution. For the AFM-SECM technique, a sharp cantilevered AFM probe can be manufactured as an AFM-SECM probe with a small radius to achieve high spatial resolution. Furthermore, mechanical and electrical information can be obtained simultaneously in specific AFM working modes. Following Macpherson and Unwin's successful presentation about the AFM-SECM technique in 2000, significant progress has been made about the disciplines of designing and manufacturing for the AFM-SECM probe. Over the past two decades, various applications in chemical process, corrosion and biological material have been developed, which will be discussed in this chapter.

In this chapter, a hybrid AFM technique, atomic force microscopy-Scanning Electrochemical Microscopy (AFM-SECM), is presented. Working principles, modes and the key component for AFM-SECM probe are described in Section 4.2. And an up-to-date account of AFM-SECM technique applications is provided in Section 4.3, which includes some examples of AFM-SECM applications in three major research fields: electrocatalysis, corrosion research and life science. Among them, electrocatalysis will dominate the discussion, as it is inextricably linked to clean energy.

Since the main focus of this chapter is on the measurement related to electrochemistry, information about AFM topographical or mechanical measurement will not be given here, which can be found in other chapters presented in this book.

## 4.2 ATOMIC FORCE MICROSCOPY-SCANNING ELECTROCHEMICAL MICROSCOPY (AFM-SECM)

### 4.2.1 The Principles of AFM-SECM

Brief discussion about the SECM and AFM techniques will be presented in Sections 4.2.1.1 and 4.2.1.2 with their principles and major working modes. With these descriptions, the readers will get a better understanding about the combined AFM-SECM technique in the following section.

#### 4.2.1.1 SECM

SECM is a technique that uses an ultramicroelectrode (UME) to measure the electrochemical signals when it is held or scan across the substrate in a solution. In SECM system, UME is an electrode with a radius (*a*), which differs from few micrometers to 25 μm, and the substrate can be different types of solid surfaces (e.g., metal, glass, biological materials) or liquids. With the advancement of technology, the electrode with a radius of few nanometers, dubbed the nanoelectrode, can be built, which significantly enhances the resolution of SECM technology. The electrochemical signal of the redox mediator recorded by the UME can provide quantitative information regarding sample's morphological and electrochemical properties (e.g., reaction kinetics or catalytic efficiency) at micro/nanoscale levels [16,17].

As shown in Figure 4.1, there are three key components in the SECM system: (1) Positioning system. It includes the positioning elements, translation stages and controllers; (2) Data acquisition system. The bipotentiostat precisely controls the potentials and measures and amplifies the current between the tip and substrate current. The obtained data is stored and analyzed in computer; (3) SECM tip. It uses UME as working electrodes to scan across the substrate.

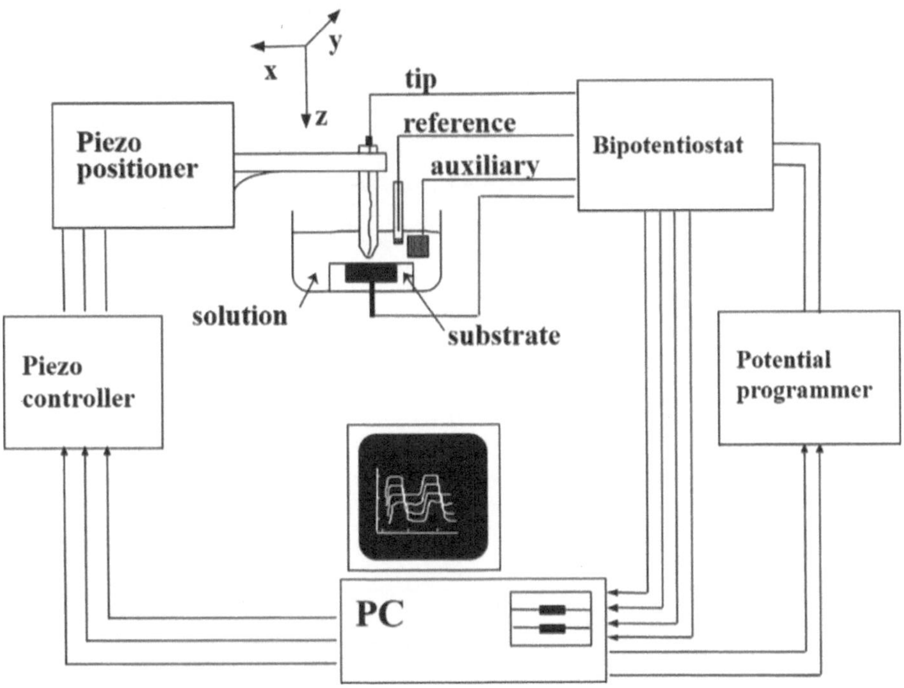


**FIGURE 4.1** Diagram of the SECM instrument.

The knowledge of electrochemistry at small electrode is crucial to understand the operation of SECM technique and achieve quantitative measurements for SECM system. What's more, the size of electrode or tip determines the spatial resolution of the SECM system [8]. Therefore, the chemical behaviors of UMEs in solution, including the UMEs being far from a substrate and close to a substrate, need to be considered carefully. The most common evaluation approach for the UMEs is voltammogram which is the diagram of current flow as a function of UME potential. The solution contains a species O (oxidized form of redox specie), with concentration (*c*), and supporting electrolyte is added in the solution to decrease the solution resistance. When potentiostat controlled voltage on UMEs, which against a stable reference electrode (e.g., reversible hydrogen electrode or silver/silver chloride electrode), is increasing, a reduction reaction $O+ne=R$ occurs at the UME, and the current will be measured by potentiostat. A typical voltammogram with an S-shaped curve can be obtained when the tip is far away from substrate as shown in Figure 4.2. In this case, the current will be limited by the mass transfer of $O$ from the bulk solution to the electrode surface at specific potentials (i.e., at potentials >0.2 V vs. Ag/AgCl in Figure 4.2), which we call it as steady-state diffusion controlled current.

The current for a conductive disk electrode can be expressed as

$$i_{T,\infty} = 4nFDca \tag{4.1}$$

where

$a$ is the radius of disk electrode
$c$ is the concentration of redox medium
$D$ is the diffusion coefficient of species
$F$ is the Faraday constant
$n$ is the number of electrons participating in the redox reaction

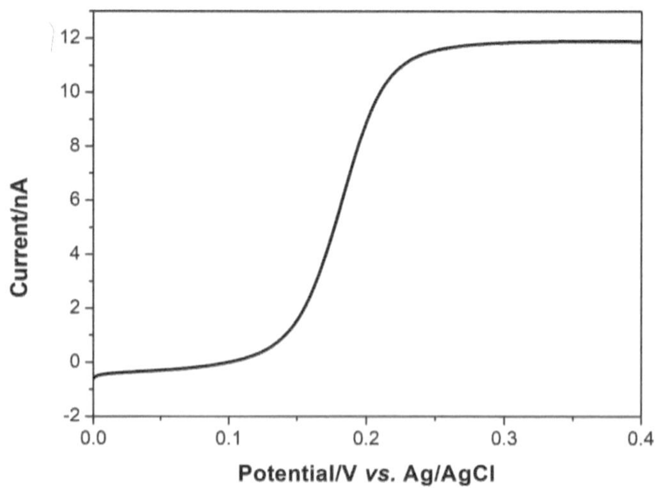


**FIGURE 4.2** Typical voltammogram for a UME, 1 mM FcMeOH and 0.1 M KCl were added in the solution.

Most of the traditional SECM experiments are carried out with disk-shaped electrodes. In comparison to disk-shaped electrodes, the side areas of electrode with other shape, such as the hemisphere or cone electrode, will limit the sensitivity of image. For example, a cone electrode whose apex has an area of ~ 50 nm in diameter and with ~200 nm tip height, the image resolution is around 120 nm [18]. What's more, even the steady-state diffusion controlled current for these other shape electrodes can be expressed in a similar way; the tip geometry still needs to be considered in some experiments, like current-distance curve.

Since the SECM was developed in 1989 by Bard et al. [8], several working modes have been developed, including generation-collection mode, feedback mode, transient mode, direct mode, redox competition mode and among others to accommodate more advanced applications. The major working modes are introduced below.

#### 4.2.1.1.1 Feedback Mode

As shown in equation (4.1), the UME tip current, $i_{T,\infty}$, is measured when it is far from substrate. It is necessary that the distance between tip and substrate is larger than a few tips' radius. In this situation, tip current is limited by the mass transfer by diffusion of $O$ from the bulk solution to the electrode surface. When the tip approaches an inert substrate, like glass or plastic, the tip current will decrease as the inert substrate suppresses the diffusion of $O$ to the tip. With the decrease of the distance between tip and substrate ($d$), the $i_T$ becomes smaller until both of them approach zero. It is called as negative feedback (Figure 4.3a). Alternatively, when the tip brought to an electrically conductive substrate, such as metal, where species $O$ can be reduced to

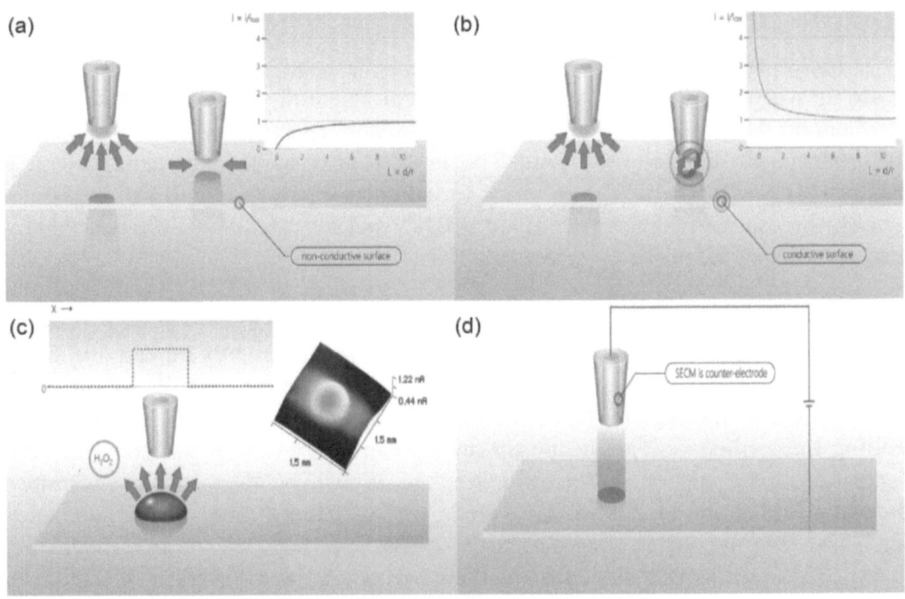


**FIGURE 4.3** Schematic diagram of SECM major working modes.

$R$, the tip current will increase compared to $i_{T,\infty}$ with the decrease in distance ($d$). We call it as positive feedback (Figure 4.3b). It should be noted that when $d$ is extremely small, like less than few nanometers, electron tunneling will occur and the tip current will switch from feedback current to tunneling current. In this situation, the value of current will rapidly increase. The diagram about the relationship between the probe-sample distance, $d$, and the tip current, $i_T$, is approach curve. As described above, the rate constant of electron transfer at the substrate species $O$ ($k_{b,s}$) is the main influencing factor for the trend of approach curve's change. Between limiting curves, $k_{b,s} \rightarrow 0$ (insulator) and $k_{b,s} \rightarrow \infty$ (conductor), different changing trends represent different $kb, s$ and the electron transfer constant of heterogeneous reaction at the substrate can be distinguished effectively.

#### 4.2.1.1.2 Generation-Collection Mode

For generation-collection mode, two modes are included: Substrate Generation-Tip Collection (SG-TC) mode and Tip Generation-Substrate Collection (TG-SC) mode. For SG-TC mode, different potentials are held at tip and substrate. As shown in Figure 4.3c, an electrode reaction occurs at the substrate and then generated product will be reacted on the tip surface and collected. In most cases, the substrate is much larger than the tip. We can use relatively small tip to scan the substrate in $x$, $y$ or $z$ direction to obtain the concentration profile of reaction product with high spatial resolution. For electrocatalysis and photocatalysis, we can scan over the surface and find the area with higher concentration of the catalytic product. Since the condition for whole substrate surface, such as applied voltage or light conditions, is same, we can determine the area with higher concentration of the catalytic product has higher catalytic efficiency. Beyond that, using the concentration profile of the catalytic product, quantitative information of the catalytic efficiency for complex catalysts can be provided.

The alternative mode is TG-SC mode. In this case, an electrode reaction occurs at the tip when it held a corresponding potential where the product of this reaction is generated. Meanwhile, different potentials are applied to the substrate to "collect" the products where it will be reacted. This mode is also useful for screening electrocatalysts. However, there is a limitation for complex catalysts due to the resolution. Briefly, the generation-collection mode can be used for detecting substances generated at the substrate or the tip selectively.

#### 4.2.1.1.3 Direct Mode

This mode can be used for modifying substrate surface, especially for enzyme deposition, semiconductor etching and micro-patterning measurements. When the tip is close to the surface, a certain electrochemical reaction will be carried out by controlling the applied voltage on the tip, which can lead to the metal etch or metal deposition on the substrate surface (Figure 4.3d). Localized electric field determines the spatial resolution of this working mode while feedback mode also can be used for micro- and nanopatterning. The major differences between the two modes are that the substrate is often unbiased and the UME is used as the working electrode in feedback mode. In direct mode, substrate is used as working electrode and UME acts as the auxiliary electrode.

#### 4.2.1.2 AFM

Atomic Force Microscopy (AFM) is considered as one of the most versatile scanned probe microscopes available to date. The history of development, instrumentation, theory and operation modes for this technique have been discussed in detail elsewhere in this book, so only a brief description and one special working mode, Lift mode, are provided here.

AFM technique was first introduced [19] as one powerful instrument in nano science and technology in 1986 by Binnig et al. It can provide nanoscale information from three dimensions with no vacuum, or contrast reagent is needed. The ultra-high spatial resolution is derived from the very sharp AFM tip attached to a force-sensing cantilever. In principle, when the AFM tip scans over sample surface, the interaction force between AFM tip and sample surface will change the deflection signal of the force-sensing cantilever. The piezoelectric elements in z direction will move up and down to keep the deflection of cantilever at a constant value which we call it "setpoint". By measuring the movement of piezoelectric element in $z$ direction, the sample surface topography information is detected. Or we can record the laser deflection signal to measure the interaction force between the tip and sample surface.

Over the past decades, several imaging modes and advanced operation modes, including nanomechanical modes, nanoelectrical modes and nanoelectrochemistry modes, have been widely used in various applications. The hybrid AFM technology introduced in this chapter, AFM-SECM, one of nanoelectrochemistry modes, is based on dual-pass function, or "Lift mode". As shown in Figure 4.4, scan twice over sample surface: surface topography is firstly measured and stored by AFM imaging mode; then the tip lifts away from sample surface with fixed height and scans again along with the sample surface fluctuation as stored in the first scan. Some properties of the sample, such as surface potential, magnetic properties and electrochemical activity, will be measured in second scan.

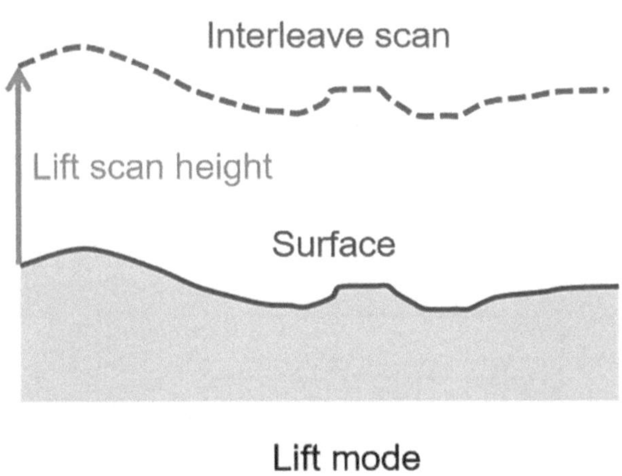


**FIGURE 4.4** Schematic diagram of lift mode.

#### 4.2.1.3 AFM-SECM

As described above, several types of sample properties can be obtained during AFM tip scanning: (1) surface topographical information can be acquired though piezoelectric element movement; (2) mechanical properties of sample can be obtained by analyzing the change of interaction force during tip contacting with sample surface; (3) electrical or electrochemistry properties, like electroconductivity, also can be measured through hybrid scanning electrical or electrochemical microscope.

For SECM technique, a robust method of distance determination is needed during measurement. As mentioned in Section 4.2.1.1, the distance between SECM tip and sample surface, *d*, will affect the diffusional flux of a specific electroactive moiety to the electrode, which leads to the change of tip current. Since the current that flows at the tip also will be affected by the chemical process occurring at the sample surface or in the space between sample and tip, we cannot interpret the investigated electrochemical activities correctly without obtaining the accurate *d*.

Given that AFM can control the tip position with sub-Å resolution in z direction and sub-nm resolution in *x*, *y* direction, combining SECM with AFM not only significantly extends the capability of detecting electrochemical properties for AFM but also provides an accurate electrochemical signal measurement for SECM. As shown in Figure 4.5, combining SECM with AFM is not complicated. For hardware part, an external bipotentiostat is integrated into the AFM system, which can control the tip and substrate potentials and measure and amplify the tip and substrate current. Beyond that, to confer dual functionality, integrating an electrochemical tip (electrode) into the AFM tip is required. The AFM-SECM tip should have a force-sensing cantilever with sharp tip to act as AFM tip, which can provide high spatial resolution topography imaging and precise tip-substrate distance control. Also, it can measure the current response occurring at the tip which can provide quantitative electrochemical properties information of the interface. Detailed discussion about different types of AFM-SECM probe and their application will be described in the following sections.

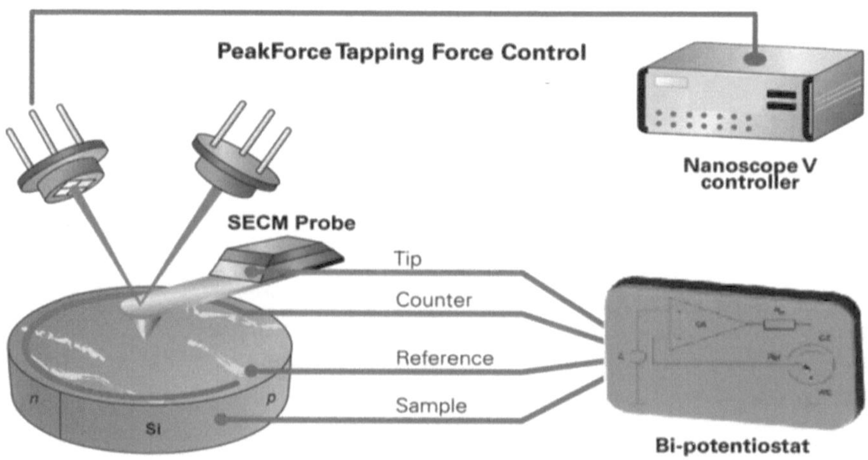


**FIGURE 4.5** Schematic illustration of an AFM-SECM system.

### 4.2.2 AFM-SECM Probe

The first documented AFM-SECM system is made by Unwin et al. [20] using Pt wire of 50 μm in diameter. The dual functional probes should have a well-defined electroactive area to work as working electrode for electrochemical data acquisition. At the same time, other conductive areas that are not involved in the measurement of electrochemical properties (e.g., the portion connecting the nanosized electroactive area to the external wire) should be insulated from solution to avoid unfavorable current leakage because the AFM-SECM characterization measurements are carried out in solution with redox mediators. In addition, the probes need to be etched to form a sharp AFM imaging tip to satisfy the AFM application needs of high precision and high resolution capability. The hand-made AFM-SECM probe fabricated by Unwin et al. followed these roles: Pt wire of 50 μm in diameter was etched, bent and flattened to form a typical AFM probe with force-sensing cantilever and sharp tip first. Then, an insulating film was attached to the Pt wire to insulate most of the conducting part except the very end of the tip. Uncovered Pt part acted as normal SECM tip to measure electrochemical activities; meanwhile, unfavorable current leakage was farthest avoided. An image of AFM-SECM probe produced by Unwin is shown in Figure 4.6. The major challenges for this method are the precise control of exposed Pt area size, insulating film and pinholes.

Another fabrication method developed by Abbou, J. et al. is using gold wire to get AFM-SECM probe [21]. A molded conical or spherical gold structure used controlled arc discharge and then coated in an electrodeposited insulating paint. The very end of the tip is well-defined gold structure in conical or spherical geometry made by arc discharge technique. The body of the gold wire is insulated by a layer of insulating paint. The image of this probe structure is shown in Figure 4.7. Except cones, the hand-made and fabrication AFM-SECM probes with other various geometries, such as needles [22], nanowires [23], pyramidal [24], and recessed frame [25], have been demonstrated in former reported works.

Over the past two decades, most of the AFM-SECM probes were self-designed and hand-fabricated because there was no high-quality commercial AFM-SECM that can be obtained until recently. As probes with reproducible and defined characteristics are crucial to carry out an AFM-SECM characterization experiment with high-quality, the need for development of commercial AFM-SECM with high availability, reproducibility and integrity is increasing. Kranz et al. first made the AFM-SECM probes with recessed electrode, and it is a significant progress during the development process of commercial probes. The strategies of designing commercial probes can be classified into two groups: (1) extended electrochemical functionality by modifying an existing commercial AFM probe. AFM-SECM probes with recessed electrode are typical examples; (2) bottom-up fabrication. Dual functionality probe structure was built from the beginning.

Kranz and his colleagues integrated a microelectrode with a conventional silicon nitride AFM tip using micromachining techniques [24]. In principle, to achieve a well-conducting effect, a standard AFM probe usually is metalized with a thick layer of Au film in 100~300 nm. Then, focused ion beam (FIB) technique was employed to remove the insulating layer including a part of the gold layer and parts of the original

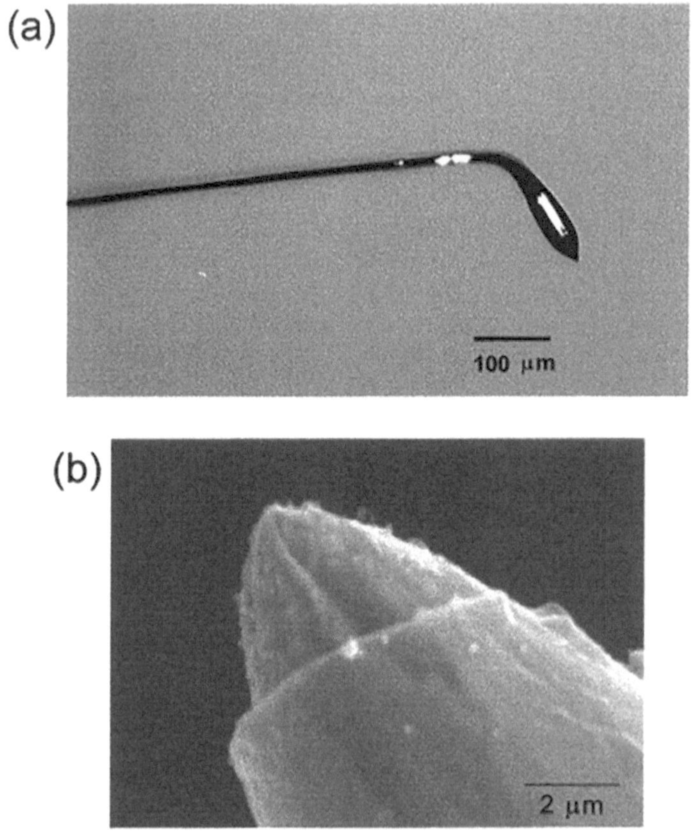


**FIGURE 4.6** Image of a coated AFM-SECM probe made by Unwin et al. [20]. (Reprinted with permission from Macpherson, Julie V., and Patrick R. Unwin, *Analytical Chemistry* 72 (2000): 276. Copyright 2000 American Chemical Society.)

$Si_3N_4$ tip. The obtained pillar-shaped tip was fabricated by FIB again to adjust the length and tip radius to define the distance between the microelectrode and the sample surface and increase the resolution of AFM imaging. At last, a gold integrated frame microelectrode was positioned around the base of the AFM tip as showed in Figure 4.8, which can be used as a dual functional AFM-SECM probe. This fabrication method was widely applied to make the commercial AFM-SECM probe with different geometries, like recessed disc [26] and ring-ring [27–28]. This design strategy keeps a fixed distance between the tip and sample surface during imaging by controlling pillar-shaped tip length. When the tip receives the topographical information from the sample in contact mode, the frame microelectrode works as SECM probe in a constant distance working mode avoiding short-circuit damaging and surface fouling of the electrode at the same time. When applied to soft or fragile sample scanning, this type of probe can also be operated in tapping mode to decrease frictional forces by employing very small oscillation amplitudes. However, the low

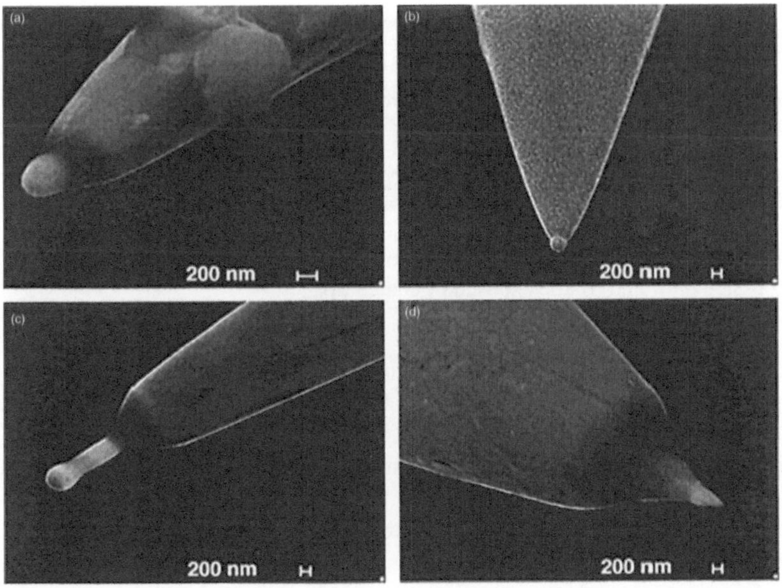


**FIGURE 4.7** Image of AFM-SECM probe with conical or spherical gold structure made by Abbou et al. [21]. (Reprinted with permission from Abbou, Jeremy, et al., *Analytical Chemistry* 74(24) (2002): 6355–6363. Copyright 2002 American Chemical Society.)

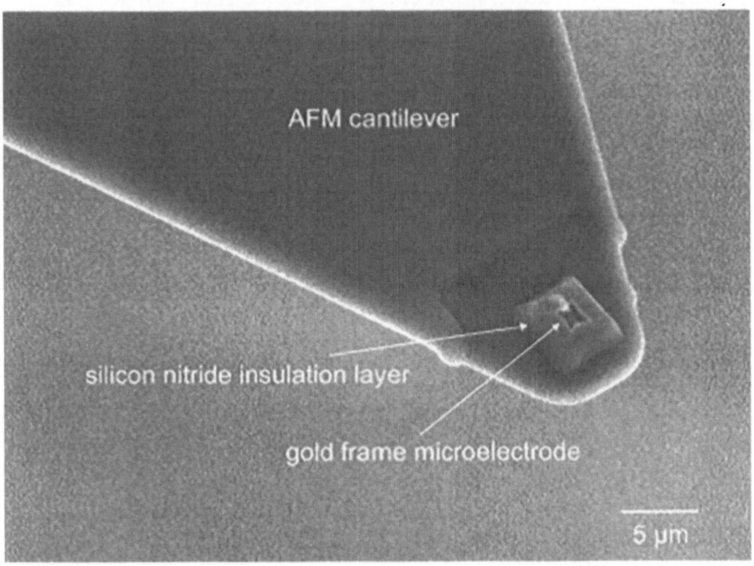


**FIGURE 4.8** Scanning electron micrograph of an integrated frame microelectrode made by Kranz et al. [25]. (Reprinted with permission from Kranz, Christine, et al., *Analytical Chemistry* 73(11) (2001): 2491–2500. Copyright 2001 American Chemical Society.)

**FIGURE 4.9** SEM images of PF-SECM probe from Bruker Corporation (Camarillo, CA, USA). (a) Side view and (b) the exposed Pt coated tip.

spatial resolution for SECM image is the insurmountable disadvantage for this probe. The spatial resolution for SECM depends on the electrode size and its distance from the surface. As the frame electrode part is positioned around the base of the AFM tip, it is difficult to achieve a higher spatial resolution (such as sub-100 nm) by decreasing its size. Similarly, the pillar-shaped AFM tip keeps the electrode part away from the surface, which limits the sensitivity of electrochemical signals.

Another type of commercial probe is AFM-SECM probe with protruding electrode, whose dual functional structure is built based on the first principles. As the electrode is located at the apex of the AFM tip, the spatial resolution of SECM image can be significantly improved by electrode size control. One typical example followed this design strategy is Peakforce-SECM (PF-SECM) probe [29], which is designed and developed by Bruker Corporation (Camarillo, CA, USA). For the characteristic tip of this probe, the dimension and height are ~50 and 200 nm, respectively (Figure.4.9b). It can provide electrochemical images with sub-100 nm resolution due to its unique design strategy. Microelectromechanical systems (MEMS) processing techniques are used to batch-fabricate this probe and the related components. For instance, proprietary chemically resistive epoxy and glass are used to seal and mount the nanoelectrode probe. The mounted probe is fully isolated with $SiO_2$ beside the exposed Pt coated conical tip, which is electrically connected with outside through a Pt conductive path of 11 μm in width on the top of $SiO_2$ coated cantilever. By this design strategy, the stray capacitance is reduced and the spatial resolution of SECM image is increased. Furthermore, topographical, mechanical, surface conductivity and electrochemical information can be simultaneously collected by using PF-SECM probe, paired Peakforce Tapping mode with interleaved scan mode.

It should be noted that no matter for hand-made or commercial AFM-SECM probes, the geometry of electrode part is not the disk shape. Therefore, theoretical simulation is needed to understand the mass transport near different types of probe surfaces. The finite element and the boundary element methods are the widely accepted numerical methods. Several research groups used these two methods to simulate diffusional mass transport to electrodes with different geometries in bulk solution and compared simulated limiting currents or feedback currents with experimentally measured currents [30–33]. Therefore, numerical methods, like finite difference method [34], finite element method [35] and boundary element method [36], are necessary for designing and fabricating a robust AFM-SECM probe.

### 4.2.3 AFM-SECM Working Modes

As mentioned above, this hybrid technique, AFM-SECM, is built based on AFM; its working modes are also based on a mentioned AFM working mode, Lift mode (as shown in Figure 4.4). In brief, AFM-SECM probe will work as AFM probe in the first scan to obtain the structure information of the sample surface, while a feedback loop keeps a fixed interaction force or cantilever deflection and stores it in software. In the second scan, the tip lifts from the sample surface slightly and scans again along the sample surface fluctuation as stored in the first scan. In this process, probe acts as SECM probe and the electrochemical activity is measured. Using lift mode, contact time between tip and surface can be decreased to avoid the damage of the sample surface by high shear force when it is fragile and soft and short circuit generated between the electrode and conductive sample surface.

Tip-substrate distance control is crucial to get high-quality SECM image since it will lead to the change of tip current. However, AFM-SECM generally relied on contact or tapping mode scanning to obtain topography information in the first scan. Contact mode is not suitable for soft or fragile samples and constant electrical short might damage the tip. Tapping mode is often unstable in liquid since it relies on the inherent mechanical resonance of the probe cantilever (more detailed discussion is described in another chapter). Additionally, convection caused by this unstable probe situation will affect the current value. For SECM technique, nearly all of the experiments are carried out in liquid environment. As a result, tapping mode is not a recommended mode of image. The Force Volume mode, one of traditional scanning mode of AFM, could avert these difficulties. However, the scanning speed limitation imposed by imaging principles is an issue for some SECM experiments. Now it seems that Peakforce tapping is the best option during the first scan. In this scanning mode, probe is modulated sinusoidally at an off-resonance, low frequency and contact with sample intermittently. Also, the feedback signal is the maximum force between the tip and surface to address unstable issue in liquid environment. The scan rate, normally from 0.1 to 80 Hz, can be adjusted to meet the needs of the experiment. Furthermore, quantitative imaging of mechanical properties of sample surface is measured from triggered force curve at every tapping cycle. In this way, topographic, electrochemical and quantitative nanomechanic properties are simultaneously mapped using Peakforce tapping combined SECM measurement.

## 4.3 APPLICATION AREAS OF AFM-SECM

AFM-SECM technique has been developed as a significant method for a variety of applications, such as clean energy, material fields, biology science, chemical science and so on in nearly two decades. In the following sections, various applications of this technique in several widely used study areas are briefly discussed.

### 4.3.1 Application in Electrocatalysis

Development of renewable, environmental friendly and highly efficient technologies for energy conversion and storage is essential for solving the issues of detrimental environmental pollution and increasing energy crisis [37]. Over the past decades, some promising techniques, such as fuel cell and photo-electrocatalysis system, have been developed to address the worldwide need for zero-emission and sustainable energy sources. In the investigation of electrocatalysts, conventional SECM can be used as an invaluable tool since it can provide valuable information, such as sample characteristic for the investigation of new materials (electrocatalysts), quantitative electrochemical data including the rate of catalytic reaction and catalytic activity of different sites. For example, Bard et al. [38] and Mirkin et al. [39] have investigated the relationship between the structure and catalytic activities for metal nanoparticles (NPs) (Pt and Au). In these works, the geometric properties were obtained by employing feedback mode, and the catalytic activity of individual NPs was obtained by generation-collection mode. SECM technique achieved the information about the relationships between the structure and activity, which provides valuable data to design electrocatalytic devices with higher efficiencies. However, it should be noted that the structure or topographical information of the studied materials comes from the electrochemical image, normally obtained by feedback mode. The electrical properties of different components, such as conductivity, have an effect on feedback current. It means that when the material contains complex components, the accuracy of the topographical image will be reduced. Additionally, since the catalytic activity is highly relying on the local atomic structures, particularly the less coordinated sites at edges and corners, the SECM with high spatial resolution and sensitivity is demanded. For this point, Mirkin et al. [40] employed the SECM with 15-nm spatial resolution and high sensitivity to map out the oxygen evolution activity of a semi-2D nickel oxide nanosheet and distinguished that the nickel oxide nanosheet edges possessed the higher catalytic activities. However, the state-of-the-art electron tomography technique is still needed to obtain the topographical information and reconstruct the 3D structures at the edges, because the feedback current is not only influenced by surface morphology but also related to component activity.

Different from SECM, AFM uses piezoelectric elements to record the topographical information. It has been proved as a surface analysis technique which offers both qualitative and quantitative information of morphology. Hence, the technique combines SECM with AFM providing a unique tool to study the structure-activity relationships for complex nanostructured catalysts. Recently, Gao et al. introduced AFM-SECM technique to investigate the catalytic activity of a nickel foam-based monolithic electrode, which shows high performance for both hydrogen and oxygen evolution reactions in neutral media [41]. After a convenient synergic chemical and electrochemical etching (nickel foam was applied 0.3V vs. RHE for 5 minutes in 0.5M $H_2SO_4$ solution), nickel foam-based monolithic electrode with typical karst landform features was obtained, and it consists of metallic Ni in the valley areas and Ni/a-Ni$(OH)_2$ heterostructures on the towers. As shown in Figure 4.10, AFM-SECM was employed to obtain the correlation between the surface topography and the electrocatalytic activity. From AFM image (Figure 4.10b), topographical information of the electrode surface was captured for the valley areas (dark part) and tower areas (bright part). SECM current profiles of hydrogen evolution reaction (HER) and oxygen evolution reaction (OER) were measured by different applied voltages

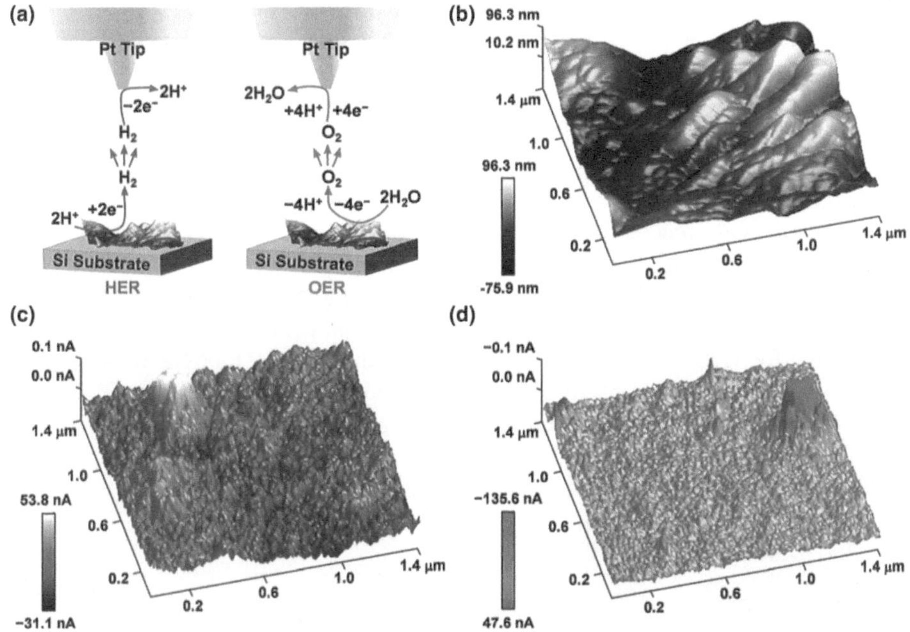


**FIGURE 4.10** SECM studies of the karst Nickle foam for both HER and OER by Gao et al. [41]. (Reprinted with permission from Gao, Xueqing, et al., *Energy & Environmental Science* 13(1) (2020): 174–182. Copyright 2019 The Royal Society of Chemistry.)

(Figure 4.10c and d). From the SECM current profiles, HER current was mainly observed in valley area, and OER current was mainly observed on the tower areas. As AFM-SECM is an *in-site* image and analysis technique (the topographical and electrochemical data can be collected simultaneously), the topographical information obtained by AFM and the activity information measured by SECM at same location can be connected, and the importance of different reaction sites for HER and OER was examined and interpreted.

Photoelectrochemical (PEC) water-splitting systems generate hydrogen from water by sunlight and electric energy. In these systems, semiconductors are usually used as light-absorber and generate separate charges [42,43]. Further investigation of the properties of the separated charges plays a significant role in the mechanism exploration. Brunschwig and coworkers use AFM-SECM to investigate the topographical, mechanical and electrochemical properties of Pt NPs deposited on p-type Si surfaces [44]. In this work, Peakforce-SECM mode from Bruker was used to measure and correlate these properties. Based on the contact current, the charge transfer pathway from the light absorber of Si to the catalyst of Pt can be obtained intuitively. At the same time, by analyzing these data, understanding of interfaces between individual Pt NPs and Si substrates is achieved and provides a method to make a comparison between different Pt NPs. Unlike normal macroscopic measurements, AFM-SECM technique provides valuable information about the electrical and electrochemical properties of individual NPs.

Metal NPs, especially noble metal NPs, have attracted increasing research interests since their extensive applications in electrocatalysis and sensing [45,46]. Not only studying of NP shape, size and orientation is important to achieve higher electrocatalytic efficiencies, but the surface property, like functionalized groups on the NP surface and capping agents, can also provide valuable information to study their electrocatalytic behavior [47,48]. AFM-SECM technique can be used to address the challenges of studying the factors at the individual nanoparticle level and provide the topographical, mechanical and electrochemical information simultaneously. For example, Schechter et al. obtained the topographical and electrochemical characteristics of individual Pt NP at the same time [49]. AFM-SECM technique was employed to study active ORR catalytic sites of Pt NPs with different sizes. Different active sites for ORR were identified by using current map, topography was measured using AFM, and the intermediate of peroxide generated during the ORR on the same Pt NPs was probed using SG-TC mode as shown in Figure 4.11. Furthermore,

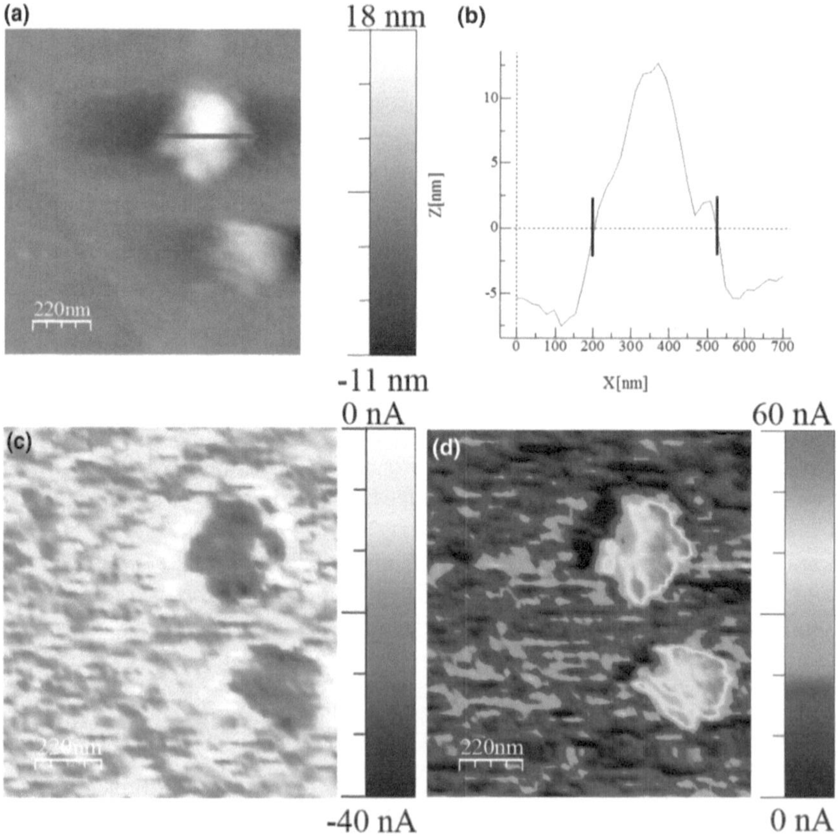


**FIGURE 4.11** AFM-SECM image of Pt NPs on HOPG. (a) Surface topography, (b) Z-axis height profile, (c) oxygen reduction current mapping, (d) peroxide oxidation current mapping. (Reprinted with permission from Kolagatla, Srikanth, et al. *Applied Catalysis B: Environmental* 256 (2019): 117843. Copyright 2019 Elsevier B.V. [49].)

the relationship between particle size and distribution of active sites was studied in this work. However, it is controversial to apply the 16 nm-spatial resolution to achieve these current mappings. Note that the spatial resolution of the SECM image with disk-shaped probe is defined by the probe radius (which is the exposed Pt part). Due to the larger exposed Pt area for the protruding electrode, the spatial resolution should be lower than the radius of the protruding electrode used. The AFM-SECM probe used in this work is 200 nm in protrusion length and 50 nm in radius for the Pt part; therefore, it is a great challenge to obtain such high-resolution current map in this situation.

Demaille et al. achieved the size and electrochemical behavior measurement on an individual redox PEGylated Au NPs by using AFM-SECM technique, which provides required nanometer resolution to probe Au NPs [50]. Figure 4.12 shows the AFM image of individual NPs. At the same time, the position, size and electrochemical activity derived from the SECM image for the grafted redox-PEG chains were obtained (Figure 4.12b). Moreover, same group also explored electrochemical properties of several hundreds of individual nanodots in a single image. It shows

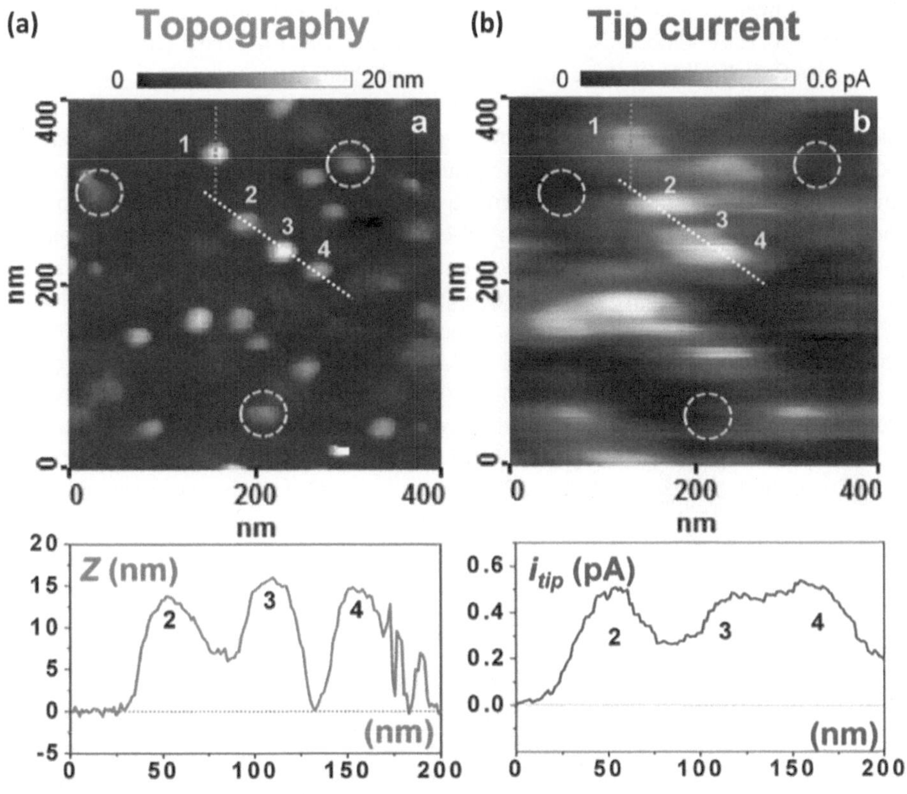


**FIGURE 4.12** AMF-SECM images of a gold surface bearing a high-density random array of ~20 nm Fc-PEGylated gold NPs. Topographical information (a) and electrochemical activity image (b). (Reprinted with permission from Huang, Kai, et al., *ACS Nano* 7(5) (2013): 4151–4163. Copyright 2013 American Chemical Society [50].)

the possibility of achieving high-throughput reading of dense nanoarrays with high sensitivity and a read-out speed for AFM-SECM technique [51].

### 4.3.2 Application in Corrosion Research

The corrosion of a metal or metal alloy can be considered as chemical or electrochemical reaction between metal and its environment. Monitoring topographical changes for metal surface and electrochemical reaction processes occurring at the metal/liquid interface will provide fundamental insight into corrosion processes. Application of AFM-SECM technique about corrosion on bare metal or metal alloy substrates will be described in this section. More details about the theory have been discussed elsewhere [52].

The oxide layers formed on metallic materials surface, such as the widely industrial used copper and iron, act as a barrier to ionic migration and show excellent resistance towards corrosion. Nevertheless, the oxide layer will undergo local degradation in some environment, like acidic chloride solution, which will reduce its corrosion resistance. Using AFM-SECM technique, the topographical changes and electrochemical reaction pathway can be monitored, which can provide key information to study the corrosion process and decrease degradation of the metallic materials. Kranz et al. have explored AFM-SECM to record the topographical changes and cooper ions generation process from a pure copper substrate in acidic chloride solution at the initial stage of pit formation [53,54]. The results show the potential of AFM-SECM to investigate localized corrosion on otherwise passive surfaces. Kranz and coworkers fabricated a suitable AFM-SECM probe by sputtering a gold layer on a commercial silicon nitride cantilever. Contact mode was used for AFM image to motor the topographical changes caused by corrosion process; generation-collection mode was used for SECM image to record the release of $Cu^{2+}$ ions simultaneously as shown in Figure 4.13. In addition to copper, Kranz and Souto *in-situ* monitored the generation of single corrosion and growth of pit on iron surface in 0.5 M NaCl solution with AFM-SECM technique [55].

Understanding of the influence of intermetallic particles in the multi-component alloys is essential for studying the mechanism of localized corrosion of alloys. Norgren et al. applied the integrated AFM/SECM as a valuable technique to *in-situ* investigate the localized corrosion of Al alloys caused by the intermetallic particles in NaCl solution [56]. Concurrent topographic images and electrochemical current maps using $I^-/I_3^-$ as redox mediator obtained from same scanned area were recorded though the dual-mode probe. Detailed information about the localized dissolution associated with different kinds of intermetallic particles was obtained from these images, and it can be concluded that the localized dissolution derived from intermetallic particles may occur well below the breakdown potential.

Another typical example of AFM-SECM application for alloy corrosion was also studied by Norgren et al. The influence of intermetallic particles on localized corrosion of EN AW-3003 alloy in chloride solutions with/without acetic acid was examined and interpreted using both conventional electrochemical measurements and *in-situ* AFM-SECM technique [57]. SECM imaging is used to visualize local electrochemical activity with controlled potential. Meanwhile, the changes in the surface topography caused by ongoing electrochemical reactions were quantified using an AFM image. In this

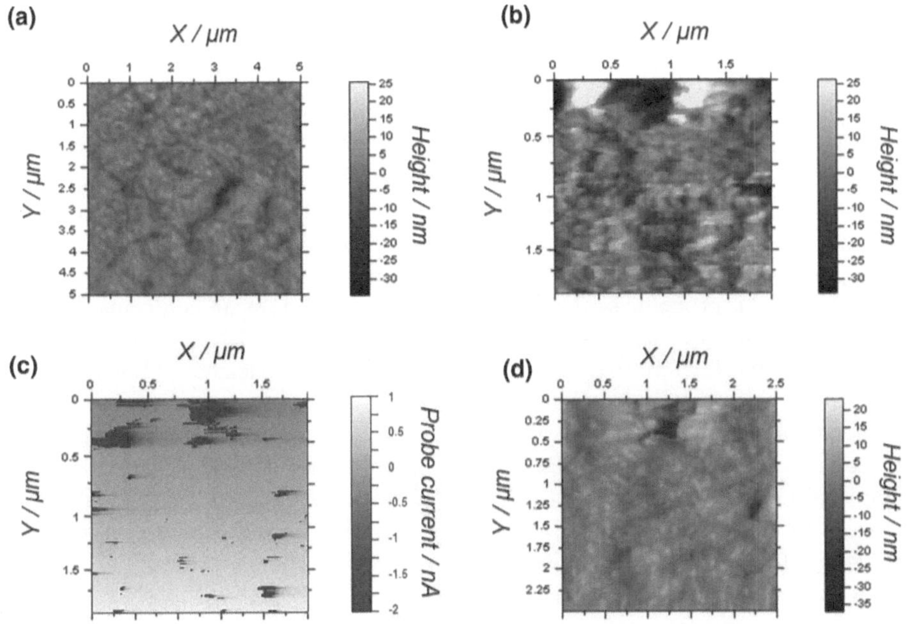


**FIGURE 4.13** Topographical information recorded by AFM (a, b, d) and (c) current mapping recorded by SECM while scanning a random area using an AFM-SECM probe and −0.45 V was applied. (Reprinted with permission from Izquierdo, Javier, et al., *Electrochimica Acta* 247 (2017): 588–599. Copyright 2017 Elsevier Ltd [54].)

study, the alloy was recorded using AFM-SECM technique in a 10 mM NaCl + 2 mM $K_4Fe(CN)_6$ solution with a 200 mV cathodic potential supplied on it. Large intermetallic particles are harder than the matrix in the AFM image, and some of these particles exhibit greater electrochemical current in the SECM image. These results indicated large intermetallic particles can enhance cathodic activities on the surface, but not all of them will exhibit this effect. Additionally, AFM and SECM data were measured simultaneously when the alloy sample was exposed to 10 mM NaCl + 5 mM KI solution with 200 mV applied anodic potential, which will lead to the passivity breakdown. SECM images are used to determine the local electrochemical activity at this applied potential, and AFM images are used to determine the topographical changes on the surface where early pits occurred. By examining these results, it is possible to deduce that the local electrochemical active sites are related to intermetallic particles and that large intermetallic particles can accelerate anodic dissolution. Beyond that, continuous SECM and AFM pictures can be used to examine localized corrosion processes on the alloy surface, such as pit development and corrosion product deposition.

### 4.3.3 Application in Life Science

Almost from its inception, SECM has been applied to investigate biochemically relevant problem [58]. SECM possesses particular advantages in life science studies because (1) it can be carried out in liquid environment which is key factor to keep biological samples alive and with less interferences than other scanning probe microscopy (SPM) techniques; (2) biological processes or redox chemistry can be detected by SECM which are often associated to process life sciences; (3) the size of SECM probe can be made small enough to insert into a living cell and detect biological processes inside cell without killing it. Therefore, conventional SECM technique has been popularly used to investigate complex biological samples (such as cells, DNA, antibodies tissues and bacteria) [59–63]. Undoubtedly, AFM-SECM can be used as a valuable characterization tool for life science.

AFM-SECM probe has been applied to detect the local conformation and motional dynamics of single-stranded and double-stranded DNA with oligonucleotides by Demaille et al. [64]. The DNA samples were fabricated with low-density monolayer of single-stranded oligonucleotides $(dT)_{20}$, alkyl-ferrocene has been borne to 3′-end and this modified DNA sample was attached to the gold surface as shown in Figure 4.14a. When AFM-SECM probe with a very smooth spherical tip-end approaches the DNA sample, the deflection curves can be recorded and the distance between tip and sample can be obtained. Meanwhile, current signal caused by Fc oxidized when the tip was applied an enough positive potential in pH 7 buffered

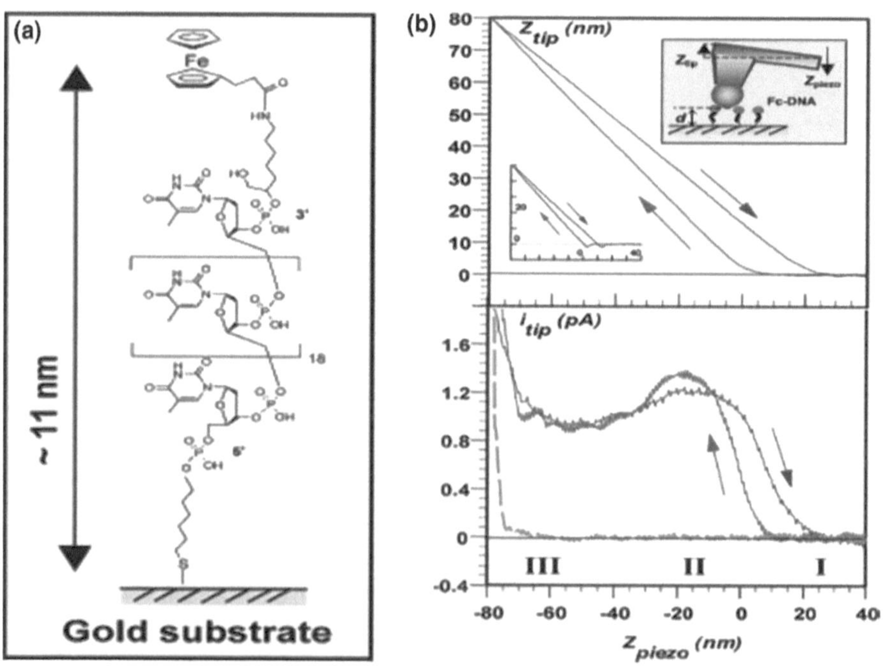


**FIGURE 4.14** Chemical structure of the Fc-$(dT)_{20}$ oligonucleotide 5′-$C_6$-thiol end-grafted onto a gold substrate (a), AFM approach curve (b) and current approach recorded simultaneously when AFM-SECM probe approaches the sample. (Reprinted with permission from Wang, Kang, et al. *The Journal of Physical Chemistry B* 111(21) (2007): 6051–6058. Copyright 2007 American Chemical Society [64].)

solution with 1 M $NaClO_4$ will be measured by SECM part as shown in Figure 4.14b. From the deflection and current curves, motional dynamics of the DNA chain can be analyzed and can be used for the design of electrochemical DNA sensors.

Similarly, Demaille and coworkers used redox ferrocene (Fc)-PEG chains to label the lettuce mosaic virus (LMV) which immobilized on a gold substrate and characterize the virus [65]. Viruses, as natural nanomachines, can be used for several applications, such as drug delivery to targeted location [66], providing nanocontainer for enzymatic catalysis [67,68], etc. Traditional techniques for virus mapping, like TEM or X-ray diffraction, can only provide *ex-situ* image and lead to structure alteration. In Demaille's work, the molecule touching mode AFM-SECM probe was carried out to *in-situ* map the distribution of proteins on individual virus particles. As the LMV was labeled with Fc-PEG chains, the local tip current was measured when it approached the LMV at applied potential of 0.3 V vs. SCE in 10 mM pH 7.4 phosphate buffer solution. In this situation, the Fc was alternatively oxidized at the tip and reduced at the substrate. From point to point, the current map was formed and the distribution of proteins on individual virus particles was obtained as shown in Figure 4.15. Moreover, they also image the redox-immunomarked proteins and make proteins electrochemically "visible" with molecule touching mode AFM-SECM [69].

The activity studies on enzyme surfaces have also been extended with the development of AFM-SECM technique. For example, Mizaikoff et al. used FIB technique to integrate electrodes with different geometries into AFM tip and introduced the AFM-SECM tip to simultaneously obtain the images of the peroxidase immobilized enzyme activity and contact AFM on a protein gel spots [70]. Without $H_2O_2$, negligible current can be detected in the solution at applied potential of 0.05 V vs. Ag/AgCl as shown in Figure 4.16b, which meant no enzymatic reaction occurs in this situation. The simultaneously obtained topographical information (Figure 4.16a) from contact mode corresponded well with the surface. Figures 4.16c and d show the topographical information and current obtained with the same AFM-SECM probe and applied voltage when $H_2O_2$ was added into the solution, respectively. The electrochemical responses represent the enzymatic activity of the peroxidase and correspond well with the topography properties obtained from AFM image. Another example for enzyme activity image by AFM-SECM was also studied by Mizaikoff using glucose oxidase as the sample loading in a soft polymer matrix [71]. Similar result was obtained: there is no recorded current in the absence of glucose; enzymatic activity occurred and current was recorded when glucose existed in the solution. According to the above results, the electrochemical reaction processes that occurred on the surface of the soft biological samples can be characterized by AFM-SECM technique.

AFM-SECM technique can also be used for the application of cells, membranes and proteins [72–75]. For example, Mizaikoff et al. applied AFM-SECM as amperometric microbiosensor to obtain high-resolution imaging of membrane transport [76]. One interesting research conducted by Nishizawa et al. is *in-situ* manipulation of the environment surrounding a single cultured cell and controlling the morphological shape of the cell by AFM-SECM probe [77]. The key components for fundamental cell biology study and cell-based diagnostic are the surface micropatterning techniques, which can be used for manipulation of the adhesion of living cells. After grounding mechanically, a commercially Pt-coated AFM probe with Parylene C

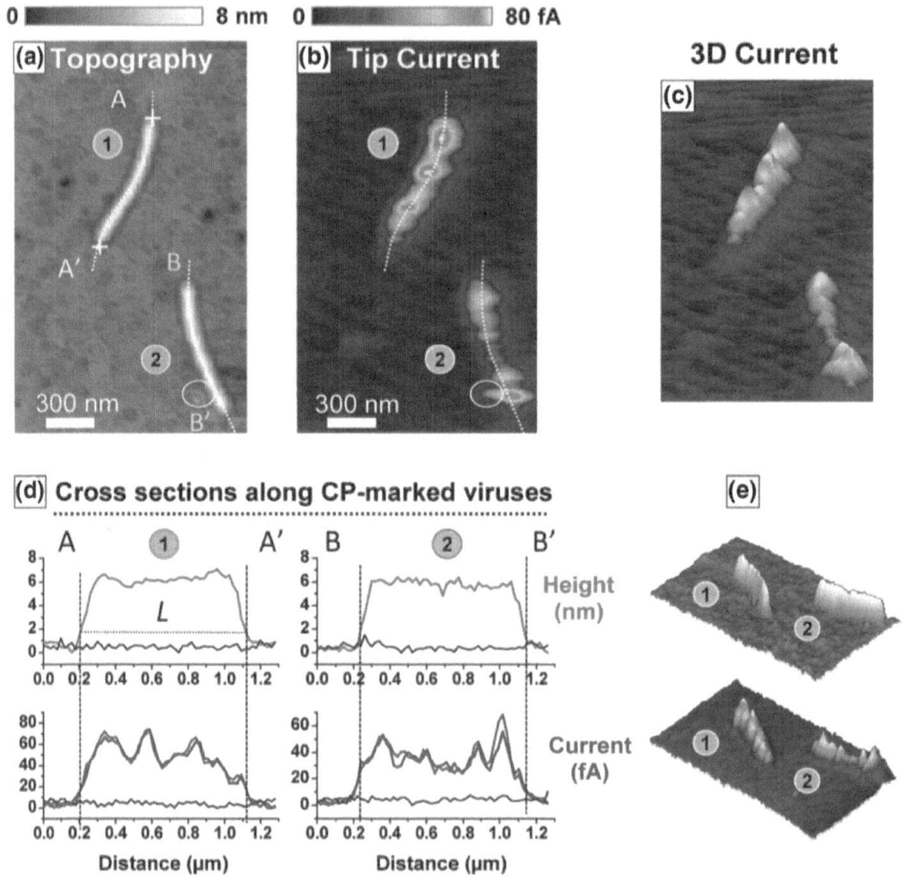


**FIGURE 4.15** *In-situ* Mt/AFM-SECM imaging of CP-marked LMV particles immobilized on a gold substrate. (a) Topography, (b) tip current, (c) tip current image in a 3D format, (d) cross section of the topography and current images and (e) 3D tilted views of the topography and current images. (Reprinted with permission from Nault, Laurent, et al., *ACS Nano* 9(5) (2015): 4911–4924. Copyright 2015 American Chemical Society [65].)

insulator can be converted into electrode to accommodate electrochemical-based biolithography (ECBL). Then electrical pulse was applied on the electrode to generate the oxidant HBrO, and the cytophobic material, such as heparin or albumin, layered adjacent to a living NIH-3T3 mouse fibroblast, was locally etched by the HBrO (Figure 4.17a). The 2–3 nm deep etch area can be used as protein-adhesive sites to confine the NIH-3T3 fibroblast in a patterned area and modify the ambient environment of a single cell (Figure 4.17b and c).

## 4.4 PERSPECTIVE

This hybrid AFM technique, AFM-SECM, which can obtain electrochemical and topographical imaging simultaneously, has successfully delivered on its initial

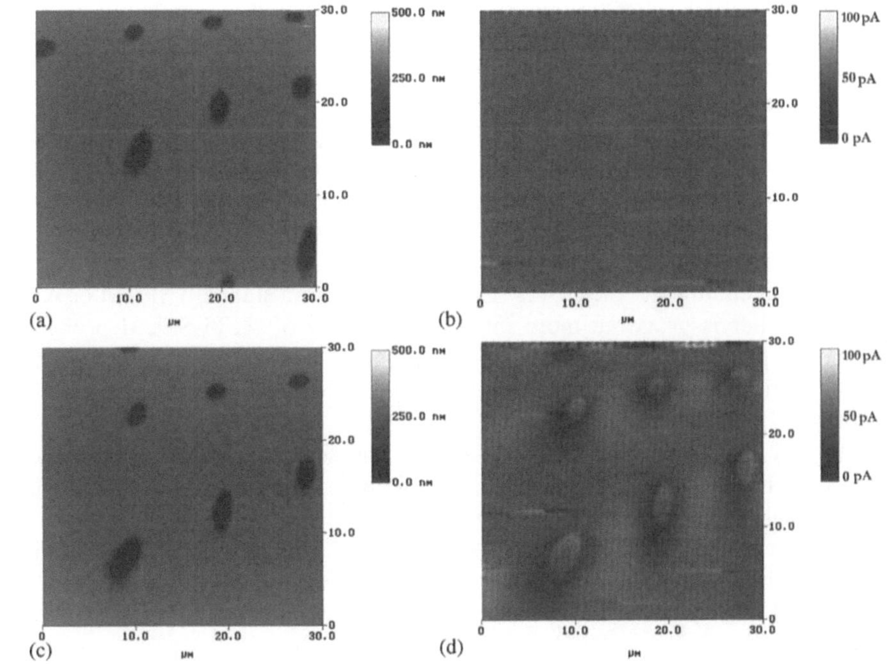


**FIGURE 4.16** Simultaneously topography and current images of peroxidase activity in AFM contact mode. (a) AFM image and (b) current image without the enzymatic substrate $H_2O_2$. (c) AFM image and (d) current image with $H_2O_2$. (Reprinted with permission from Kranz, C., et al., *Ultramicroscopy* 100(3–4) (2004): 127–134. Copyright 2004 Elsevier B.V. [70].)

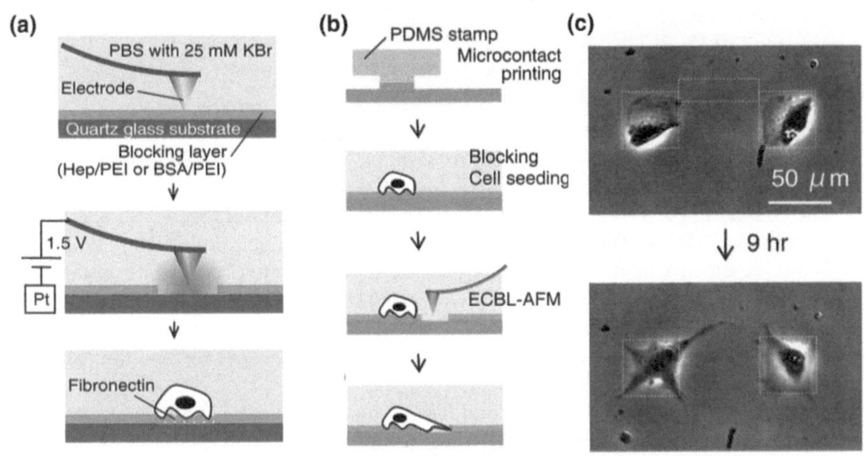


**FIGURE 4.17** Illustration of AFM-SECM as electrochemical-based biolithography technique to *in-situ* lithography adjacent to a single, cultured cell, and manipulate the morphological shape of the cell. (Reprinted with permission from Sekine, Soichiro, et al., *Electrochemistry Communications* 11(9) (2009): 1781–1784. Copyright 2009 Elsevier B.V. [77].)

promise and continues to showcase its extraordinary potential in numerous applications over the previous two decades. However, there are still numerous key difficulties that need to be overcome in order to achieve wide application and fully realize its value.

The first challenge is probe design and preparation. The combined probe is essentially an AFM probe with an ultramicroelectrode integrated at or near the tip region; nevertheless, the fabrication technique is challenging. The image quality with self-designed and produced probes is limited by their dependability and endurance. Although certain commercial probes have been produced with good reliability and endurance, the cost is still much more than standard SECM or AFM probes. Another issue is the more intricate geometry of AFM-SECM probes in contrast to typical SECM disk-shaped probes, which necessitate theoretical simulation to explain mass transport at various types of probe surfaces, as discussed in Section 4.2.2.

Other key challenges are spatial resolution and scan rate for AFM-SECM. In order to reach the ultimate single molecule resolution or information, AFM-SECM spatial resolution will have to be upgraded (compared to the 10 nm resolution that can be attained with traditional SECM). In addition, the scan rate may reach as high as 600 Hz with high-speed AFM and up to 100 Hz with normal AFM, the scan rate for AFM-SECM is often <5 Hz and needs to be further enhanced.

## REFERENCES

1. Elgrishi, Noémie, Kelley J Rountree, Brian D McCarthy, Eric S Rountree, Thomas T Eisenhart, and Jillian L Dempsey. 2018. A practical beginner's guide to cyclic voltammetry. *Journal of Chemical Education* 95 (2):197–206.
2. Bellezze, Tiziano, Giampaolo Giuliani, Annamaria Viceré, and Gabriella Roventi. 2018. Study of stainless steels corrosion in a strong acid mixture. Part 2: Anodic selective dissolution, weight loss and electrochemical impedance spectroscopy tests. *Corrosion Science* 130:12–21.
3. Heyrovský, Jaroslav, and Jaroslav Kůta. 2013. Principles of Polarography. Amsterdam: Elsevier.
4. O'Connell, Michael A, and Andrew J Wain. 2015. Combined electrochemical-topographical imaging: A critical review. *Analytical Methods* 7 (17):6983–6999.
5. Sun, Peng, Francois O Laforge, and Michael V Mirkin. 2007. Scanning electrochemical microscopy in the 21st century. *Physical Chemistry Chemical Physics* 9 (7):802–823.
6. Mirkin, Michael V, Wojciech Nogala, Jeyavel Velmurugan, and Yixian Wang. 2011. Scanning electrochemical microscopy in the 21st century. Update 1: Five years after. *Physical Chemistry Chemical Physics* 13 (48):21196–21212.
7. Barker, Anna L, Marylou Gonsalves, Julie V Macpherson, Christopher J Slevin, and Patrick R Unwin. 1999. Scanning electrochemical microscopy: Beyond the solid/liquid interface. *Analytica Chimica Acta* 385 (1–3):223–240.
8. Bard, Allen J, Fu Ren F Fan, Juhyoun Kwak, and Ovadia Lev. 1989. Scanning electrochemical microscopy. Introduction and principles. *Analytical Chemistry* 61 (2):132–138.
9. Amemiya, Shigeru, Allen J Bard, Fu-Ren F Fan, Michael V Mirkin, and Patrick R Unwin. 2008. Scanning electrochemical microscopy. *Annual Review of Analytical Chemistry* 1:95–131.
10. Wain, Andrew J. 2014. Scanning electrochemical microscopy for combinatorial screening applications: A mini-review. *Electrochemistry Communications* 46:9–12.

11. Bergner, Stefan, Preety Vatsyayan, and Frank-Michael Matysik. 2013. Recent advances in high resolution scanning electrochemical microscopy of living cells–a review. *Analytica Chimica Acta* 775:1–13.
12. Sun, Tong, Dengchao Wang, and Michael V Mirkin. 2018. Tunneling mode of scanning electrochemical microscopy: Probing electrochemical processes at single nanoparticles. *Angewandte Chemie International Edition* 57 (25):7463–7467.
13. Ludwig, Markus, Christine Kranz, Wolfgang Schuhmann, and Hermann E Gaub. 1995. Topography feedback mechanism for the scanning electrochemical microscope based on hydrodynamic forces between tip and sample. *Review of Scientific Instruments* 66 (4):2857–2860.
14. Morris, Celeste A, Chiao-Chen Chen, and Lane A Baker. 2012. Transport of redox probes through single pores measured by scanning electrochemical-scanning ion conductance microscopy (SECM-SICM). *Analyst* 137 (13):2933–2938.
15. Meier, Jan, Andreas Friedrich, and Ulrich Stimming. 2002. Novel method for the investigation of single nanoparticle reactivity. *Faraday Discussions* 121:365–372.
16. Fernández, José L, Manjula Wijesinghe, and Cynthia G Zoski. 2015. Theory and experiments for voltammetric and SECM investigations and application to ORR electrocatalysis at nanoelectrode ensembles of ultramicroelectrode dimensions. *Analytical Chemistry* 87 (2):1066–1074.
17. Polcari, David, Philippe Dauphin-Ducharme, and Janine Mauzeroll. 2016. Scanning electrochemical microscopy: A comprehensive review of experimental parameters from 1989 to 2015. *Chemical Reviews* 116 (22):13234–13278.
18. Huang, Zhuangqun, Peter De Wolf, Rakesh Poddar, et al. 2016. PeakForce scanning electrochemical microscopy with nanoelectrode probes. *Microscopy Today* 24 (6):18–25.
19. Binnig, Gerd, Calvin F Quate, and Ch Gerber. 1986. Atomic force microscope. *Physical Review Letters* 56 (9):930.
20. Macpherson, Julie V, and Patrick R Unwin. 2000. Combined scanning electrochemical: Atomic force microscopy. *Analytical Chemistry* 72 (2):276–285.
21. Abbou, Jeremy, Christophe Demaille, Michel Druet, and Jacques Moiroux. 2002. Fabrication of submicrometer-sized gold electrodes of controlled geometry for scanning electrochemical-atomic force microscopy. *Analytical Chemistry* 74 (24):6355–6363.
22. Comstock, David J, Jeffrey W Elam, Michael J Pellin, and Mark C Hersam. 2012. High aspect ratio nanoneedle probes with an integrated electrode at the tip apex. *Review of Scientific Instruments* 83 (11):113704.
23. Burt, David P, Neil R Wilson, John MR Weaver, Phillip S Dobson, and Julie V Macpherson. 2005. Nanowire probes for high resolution combined scanning electrochemical microscopy: Atomic force microscopy. *Nano Letters* 5 (4):639–643.
24. Pust, Sascha E, Marc Salomo, Egbert Oesterschulze, and Gunther Wittstock. 2010. Influence of electrode size and geometry on electrochemical experiments with combined SECM–SFM probes. *Nanotechnology* 21 (10):105709.
25. Kranz, Christine, Gernot Friedbacher, Boris Mizaikoff, Alois Lugstein, Jürgen Smoliner, and Emmerich Bertagnolli. 2001. Integrating an ultramicroelectrode in an AFM cantilever: Combined technology for enhanced information. *Analytical Chemistry* 73 (11):2491–2500.
26. Davoodi, Ali, Jinshan Pan, Christofer Leygraf, and Stefan Norgren. 2005. In situ investigation of localized corrosion of aluminum alloys in chloride solution using integrated EC-AFM/SECM techniques. *Electrochemical and Solid State Letters* 8 (6):B21–B24.
27. Smirnov, Waldemar, Armin Kriele, René Hoffmann, et al. 2011. Diamond-modified AFM probes: From diamond nanowires to atomic force microscopy-integrated boron-doped diamond electrodes. *Analytical Chemistry* 83 (12):4936–4941.

28. Shin, Heungjoo, Peter J Hesketh, Boris Mizaikoff, and Christine Kranz. 2007. Batch fabrication of atomic force microscopy probes with recessed integrated ring microelectrodes at a wafer level. *Analytical Chemistry* 79 (13):4769–4777.
29. Nellist, Michael R, Yikai Chen, Andreas Mark, et al. 2017. Atomic force microscopy with nanoelectrode tips for high resolution electrochemical, nanoadhesion and nanoelectrical imaging. *Nanotechnology* 28 (9):095711.
30. Dobson, Phillip S, John MR Weaver, Mark N Holder, Patrick R Unwin, and Julie V Macpherson. 2005. Characterization of batch-microfabricated scanning electrochemical-atomic force microscopy probes. *Analytical Chemistry* 77 (2):424–434.
31. Leonhardt, Kelly, Amra Avdic, Alois Lugstein, et al. 2011. Atomic force microscopy-scanning electrochemical microscopy: Influence of tip geometry and insulation defects on diffusion controlled currents at conical electrodes. *Analytical Chemistry* 83 (8):2971–2977.
32. Sklyar, Oleg, Angelika Kueng, Christine Kranz, et al. 2005. Numerical simulation of scanning electrochemical microscopy experiments with frame-shaped integrated atomic force microscopy– SECM probes using the boundary element method. *Analytical Chemistry* 77 (3):764–771.
33. Leonhardt, Kelly, Amra Avdic, Alois Lugstein, et al. 2013. Scanning electrochemical microscopy: Diffusion controlled approach curves for conical AFM-SECM tips. *Electrochemistry Communications* 27:29–33.
34. Selzer, Yoram, and Daniel Mandler. 2000. Scanning electrochemical microscopy. Theory of the feedback mode for hemispherical ultramicroelectrodes: Steady-state and transient behavior. *Analytical Chemistry* 72 (11):2383–2390.
35. Nann, Thomas, and Jürgen Heinze. 2003. Simulation in electrochemistry using the finite element method part 2: Scanning electrochemical microscopy. *Electrochimica Acta* 48 (27):3975–3980.
36. Fulian, Qiu, Adrian C. Fisher, and Guy Denuault. 1999. Applications of the boundary element method in electrochemistry: Scanning electrochemical microscopy. *The Journal of Physical Chemistry B* 103 (21):4387–4392.
37. Omer, Abdeen Mustafa. 2008. Energy, environment and sustainable development. *Renewable and Sustainable Energy Reviews* 12 (9):2265–2300.
38. Kim, Jiyeon, Christophe Renault, Nikoloz Nioradze, Netzahualcóyotl Arroyo-Currás, Kevin C Leonard, and Allen J Bard. 2016. Electrocatalytic activity of individual Pt nanoparticles studied by nanoscale scanning electrochemical microscopy. *Journal of the American Chemical Society* 138 (27):8560–8568.
39. Sun, Tong, Yun Yu, Brian J Zacher, and Michael V Mirkin. 2014. Scanning electrochemical microscopy of individual catalytic nanoparticles. *Angewandte Chemie International Edition* 53 (51):14120–14123.
40. Sun, Tong, Dengchao Wang, Michael V Mirkin, et al. 2019. Direct high-resolution mapping of electrocatalytic activity of semi-two-dimensional catalysts with single-edge sensitivity. *Proceedings of the National Academy of Sciences* 116 (24):11618–11623.
41. Gao, Xueqing, Yingdong Chen, Tong Sun, et al. 2020. Karst landform-featured monolithic electrode for water electrolysis in neutral media. *Energy & Environmental Science* 13 (1):174–182.
42. Lewis, Nathan S. 2016. Research opportunities to advance solar energy utilization. *Science* 351:6271.
43. Walter, Michael G, Emily L Warren, James R McKone, et al. 2010. Solar water splitting cells. *Chemical Reviews* 110 (11):6446–6473.
44. Jiang, Jingjing, Zhuangqun Huang, Chengxiang Xiang, et al. 2017. Nanoelectrical and nanoelectrochemical imaging of Pt/p-Si and Pt/p+-Si electrodes. *ChemSusChem* 10 (22):4657–4663.

45. Zamborini, Francis P, Lanlan Bao, and Radhika Dasari. 2012. Nanoparticles in measurement science. *Analytical Chemistry* 84 (2):541–576.
46. Kleijn, Steven EF, Stanley CS Lai, Marc TM Koper, and Patrick R Unwin. 2014. Electrochemistry of nanoparticles. *Angewandte Chemie International Edition* 53 (14):3558–3586.
47. Wolfe, Rebecca L, Ramjee Balasubramanian, Joseph B Tracy, and Royce W Murray. 2007. Fully ferrocenated hexanethiolate monolayer-protected gold clusters. *Langmuir* 23 (4):2247–2254.
48. Murray, Royce W. 2008. Nanoelectrochemistry: Metal nanoparticles, nanoelectrodes, and nanopores. *Chemical Reviews* 108 (7):2688–2720.
49. Kolagatla, Srikanth, Palaniappan Subramanian, and Alex Schechter. 2019. Catalytic current mapping of oxygen reduction on isolated Pt particles by atomic force microscopy-scanning electrochemical microscopy. *Applied Catalysis B: Environmental* 256:117843.
50. Huang, Kai, Agnes Anne, Mohamed Ali Bahri, and Christophe Demaille. 2013. Probing individual redox PEGylated gold nanoparticles by electrochemical–atomic force microscopy. *ACS Nano* 7 (5):4151–4163.
51. Chennit, Khalil, Jorge Trasobares, Agnès Anne, et al. 2017. Electrochemical imaging of dense molecular nanoarrays. *Analytical Chemistry* 89 (20):11061–11069.
52. Bard, Allen J, and Michael V Mirkin. 2012. *Scanning Electrochemical Microscopy*. Milton Park,⋅:Abingdon: Taylor & Francis.
53. Izquierdo, Javier, Bibiana Maria Fernandez-Perez, Alexander Eifert, Ricardo M Souto, and Christine Kranz. 2016. Simultaneous atomic force: Scanning electrochemical microscopy (AFM-SECM) imaging of copper dissolution. *Electrochimica Acta* 201:320–332.
54. Izquierdo, Javier, Alexander Eifert, Christine Kranz, and Ricardo M Souto. 2017. In situ investigation of copper corrosion in acidic chloride solution using atomic force: Scanning electrochemical microscopy. *Electrochimica Acta* 247:588–599.
55. Izquierdo, Javier, Alexander Eifert, Christine Kranz, and Ricardo M Souto. 2015. In situ monitoring of pit nucleation and growth at an iron passive oxide layer by using combined atomic force and scanning electrochemical microscopy. *ChemElectroChem* 2 (11):1847–1856.
56. Davoodi, Ali, Jinshan Pan, Christofer Leygraf, and Stefan Norgren. 2005. In situ investigation of localized corrosion of aluminum alloys in chloride solution using integrated EC-AFM/SECM techniques. *Electrochemical and Solid State Letters* 8 (6):B21.
57. Davoodi, Ali, Jinshan Pan, Christofer Leygraf, and Stefan Norgren. 2008. Multianalytical and in situ studies of localized corrosion of EN AW-3003 alloy: Influence of intermetallic particles. *Journal of the Electrochemical Society* 155 (4):C138.
58. Wang, Joseph, Li-Heuy Wu, and Ruiliang Li. 1989. Scanning electrochemical microscopic monitoring of biological processes. *Journal of Electroanalytical Chemistry and Interfacial Electrochemistry* 272 (1–2):285–292.
59. Hu, Keke, Yun Li, Susan A Rotenberg, Christian Amatore, and Michael V Mirkin. 2019. Electrochemical measurements of reactive oxygen and nitrogen species inside single phagolysosomes of living macrophages. *Journal of the American Chemical Society* 141 (11):4564–4568.
60. Holt, Katherine B, and Allen J Bard. 2005. Interaction of silver (I) ions with the respiratory chain of Escherichia coli: An electrochemical and scanning electrochemical microscopy study of the antimicrobial mechanism of micromolar Ag+. *Biochemistry* 44 (39):13214–13223.
61. Zhu, Renkang, Sheila M Macfie, and Zhifeng Ding. 2005. Cadmium-induced plant stress investigated by scanning electrochemical microscopy. *Journal of Experimental Botany* 56 (421):2831–2838.

62. Pierce, David T, and Allen J Bard. 1993. Scanning electrochemical microscopy. 23. Retention localization of artificially patterned and tissue-bound enzymes. *Analytical Chemistry* 65 (24):3598–3604.
63. Hengstenberg, Andreas, Andrea Blöchl, Irmgard D Dietzel, and Wolfgang Schuhmann. 2001. Spatially resolved detection of neurotransmitter secretion from individual cells by means of scanning electrochemical microscopy. *Angewandte Chemie International Edition* 40 (5):905–908.
64. Wang, Kang, Cédric Goyer, Agnès Anne, and Christophe Demaille. 2007. Exploring the motional dynamics of end-grafted DNA oligonucleotides by in situ electrochemical atomic force microscopy. *The Journal of Physical Chemistry B* 111 (21):6051–6058.
65. Nault, Laurent, Cécilia Taofifenua, Agnes Anne, et al. 2015. Electrochemical atomic force microscopy imaging of redox-immunomarked proteins on native potyviruses: From subparticle to single-protein resolution. *ACS Nano* 9 (5):4911–4924.
66. Steinmetz, Nicole F. 2010. Viral nanoparticles as platforms for next-generation therapeutics and imaging devices. *Nanomedicine: Nanotechnology, Biology and Medicine* 6 (5):634–641.
67. Carette, Noëlle, Hans Engelkamp, Eric Akpa, et al. 2007. A virus-based biocatalyst. *Nature Nanotechnology* 2 (4):226–229.
68. Cardinale, Daniela, Noëlle Carette, and Thierry Michon. 2012. Virus scaffolds as enzyme nano-carriers. *Trends in Biotechnology* 30 (7):369–376.
69. Anne, Agnes, Arnaud Chovin, Christophe Demaille, and Manon Lafouresse. 2011. High-resolution mapping of redox-immunomarked proteins using electrochemical–atomic force microscopy in molecule touching mode. *Analytical Chemistry* 83 (20):7924–7932.
70. Kranz, Christine, Angelika Kueng, Alois Lugstein, Emmerich Bertagnolli, and Boris Mizaikoff. 2004. Mapping of enzyme activity by detection of enzymatic products during AFM imaging with integrated SECM–AFM probes. *Ultramicroscopy* 100 (3–4):127–134.
71. Kueng, Angelika, Christine Kranz, Alois Lugstein, Emmerich Bertagnolli, and Boris Mizaikoff. 2003. Integrated AFM–SECM in tapping mode: Simultaneous topographical and electrochemical imaging of enzyme activity. *Angewandte Chemie International Edition* 42 (28):3238–3240.
72. Bergner, Stefan, Preety Vatsyayan, and Frank-Michael Matysik. 2013. Recent advances in high resolution scanning electrochemical microscopy of living cells–a review. *Analytica Chimica Acta* 775:1–13.
73. Knittel, Peter, Hao Zhang, Christine Kranz, Gordon G Wallace, and Michael J Higgins. 2016. Probing the PEDOT: PSS/cell interface with conductive colloidal probe AFM-SECM. *Nanoscale* 8 (8):4475–448.
74. Frederix, Patrick LTM, Patrick D Bosshart, Terunobu Akiyama, et al. 2008. Conductive supports for combined AFM–SECM on biological membranes. *Nanotechnology* 19 (38):384004.
75. Sekine, Soichiro, Hirokazu Kaji, and Matsuhiko Nishizawa. 2008. Integration of an electrochemical-based biolithography technique into an AFM system. *Analytical and Bioanalytical Chemistry* 391 (8):2711–2716.
76. Kueng, Angelika, Christine Kranz, Alois Lugstein, Emmerich Bertagnolli, and Boris Mizaikoff. 2005. AFM-Tip-integrated amperometric microbiosensors: High-resolution imaging of membrane transport. *Angewandte Chemie* 117 (22):3485–3488.
77. Sekine, Soichiro, Hirokazu Kaji, and Matsuhiko Nishizawa. 2009. Spatiotemporal subcellular biopatterning using an AFM-assisted electrochemical system. *Electrochemistry Communications* 11 (9):1781–1784.

# 5 Scanning Microwave Impedance Microscopy

*Yongliang Yang, Nicholas Antoniou, and Ravi Chintala*
PrimeNano Inc.

**CONTENTS**



DOI: 10.1201/9781003174042-5

## 5.1 INTRODUCTION

Electrical property characterization plays an important role in materials research and semiconductor manufacturing. For relatively large devices, traditional methods such as four-point probe can be used to measure the conductivity of materials. With the development of advanced fabrication technologies and requirements for high performance and high energy efficiency, the devices are getting smaller and smaller with the critical features reaching down to a few nano meters. It has been increasingly important to be able to characterize the electrical properties of devices at the nanoscale. Atomic force microscopy (AFM), using sharp tips to interact with the sample, has the capability to characterize material properties down to the atomic level. As demonstrated in previous chapters, the AFM and AFM based characterization methods have been driving the nanoscience and technology, advanced materials, and frontier physics research in recent decades.

In this chapter, we introduce a recently developed nanoscale electrical characterization technology, scanning microwave impedance microscopy (sMIM) [1]. sMIM is based on near field microwave images and measures the complex impedance between a sharp conductive tip and sample materials at microwave frequencies. It characterizes local conductivity and permittivity of materials at nanoscale resolution. sMIM technology was developed by the ZX Shen group at Stanford University and commercialized by PrimeNano Inc. sMIM is compatible with multiple industrial AFMs at ambient and low temperature. Several types of low temperature sMIM systems were recently released to the market to meet requirements of frontier physics and advanced materials research.

## 5.2 sMIM WORKING PRINCIPLE

sMIM measures the reflected microwave signals from the tip-sample interaction. As shown in Figure 5.1, microwave signals are conducted to a customized fabricated microwave cantilever probe via an impedance matching network for high efficiency microwave transmission. At the tip apex, high frequency AC electromagnetic fields are generated and interact with the sample material beneath the tip apex. The tip-sample impedance affects the reflected microwave signals, which are, in turn, collected and conducted to microwave electronics along the same path. The microwave

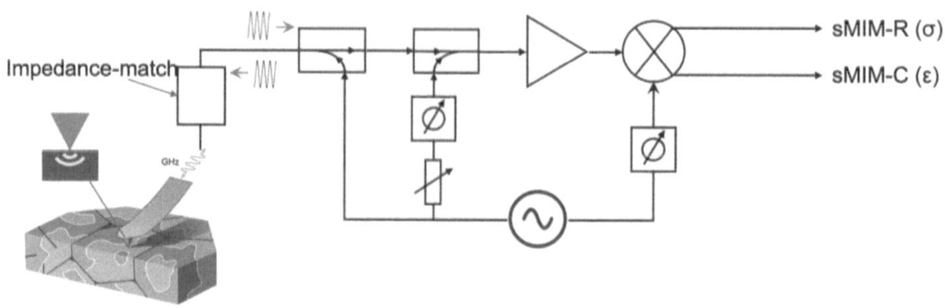


**FIGURE 5.1** Schematic of sMIM working principle.

electronics process the reflected microwave signals and output two channels (sMIM-C and sMIM-R) related to the imaginary and real components of tip-sample impedance/admittance. Note that the generated electromagnetic fields are focused near the tip apex and as a result the sMIM outputs depend only on the materials beneath the tip apex in a small volume of interaction comparable in size to the tip apex, generally from a couple of nm to 10s of nm depending on the sample material properties. As the tip scans the sample surface, driven by an AFM platform, sample electrical information is obtained simultaneously with topographic information.

A typical sMIM system includes the following components:

1. sMIM probe
2. Probe interface module
3. Microwave electronics
4. Scanning platform

We will discuss these components in detail in Sections 5.2.1–5.2.4.

### 5.2.1 sMIM Probe

An sMIM probe transmits microwave signals to the tip apex and senses the reflected microwave signals. It is an essential component for sMIM performance because the probe directly interacts with the sample mechanically and electrically. The microwave signals conducted to the tip apex generate strong AC electromagnetic fields near the tip apex and interact with the sample materials beneath the tip. The resulting tip-sample impedance, indicative of the sample material electrical properties, is measured by sMIM electronics via the reflected microwave signals.

To achieve functional and high quality sMIM imaging, an sMIM probe should meet the following requirements:

a. Sharp tip apex. The size of the tip apex mainly determines the spatial resolution of sMIM. It is important to have a sharp tip apex and robust tip that can maintain the sharpness during scanning.
b. Shielded conductive path. The microwave signals are applied to the electrode on the chip and conducted to the tip apex via a conductive path. It is important to fully shield the conductive path except for the tip apex. Otherwise, a larger, poorly controlled capacitance will be in parallel with the tip-sample impedance, resulting in unstable sMIM signals and difficulty in data interpretation. The shielded conductive path also minimizes external environmental noise pick up for high quality imaging.
c. Small chip impedance (resistance and capacitance). sMIM measures sample induced impedance change over the background impedance from the chip. Small chip impedance minimizes microwave loss and maximize sensitivity.
d. Straight cantilever that remains straight with temperature change. The shielded conductive path is part of a multi-layered cantilever structure. It is important to balance the internal stresses to form a straight cantilever,

ensuring stable laser reflection during AFM scanning. For applications at different temperatures, it is also important to keep the cantilever straight when temperature changes.

The sMIM probes were developed based on MEMS fabrication. [2] As shown in Figure 5.2, the sMIM probes have a similar shape and dimensions as commonly used

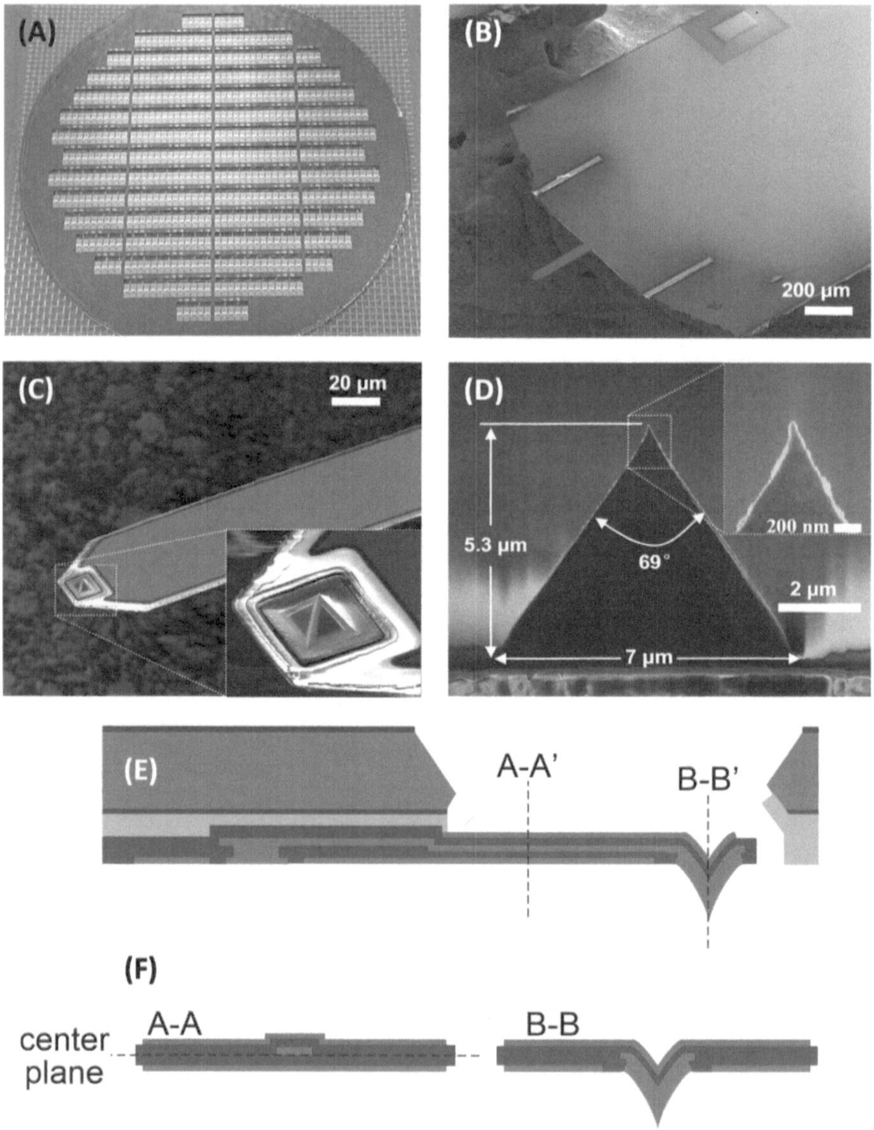


**FIGURE 5.2** sMIM probe. (a) Batched fabricated wafer with hundreds of sMIM probes. SEM images of sMIM probe (b), cantilever (c) and pyramid tip (d). Schematic of sMIM probe (e) and cross-section of cantilever A-A' and tip B-B' (f).

AFM probes and are fully compatible with major AFM platforms. A pyramid shaped metal tip with sharp apex is integrated on a micro cantilever near the free end. The tip is fabricated by depositing metal into an etched sharp pyramid shape and has a tip apex diameter less than 50 nm. The tip is gold encapsulated TiW. Combining the good electrical conductivity of gold and the mechanical stiffness of TiW, the sMIM tip maintains its sharpness during scanning without compromising the microwave transmission efficiency. A metallic center conductive path connects the tip and electrode on the chip for microwave transmission. Shielding metal layers are fabricated on the top and bottom side of the cantilever and insulated from the conductive path by dielectric layers. Gold is used as the main body of the metal layers for lower resistivity. The width and shape of the conductive path and dielectrics are optimized for small series resistance ($< 5\ \Omega$) and small background capacitance (~1 pF). The shielding ensures that only the tip apex interacts with the sample and blocks any environmental noise from coupling to the conductive path. $Si_3N_4$ layers are used not only as insulation between the conductive path and shielding but also the cantilever main body, ensuring good cantilever mechanical properties. Cantilevers of different length can be fabricated to be suitable for AFM scanning in different modes. Long cantilevers (spring constant of ~1 N/m) and short cantilevers (spring constant of ~8 N/m) can be used for applications ranging from soft organic materials to inorganic semiconductors.

The schematic of the cantilever cross-sectional view in Figure 5.2f, shows that the cantilever is symmetric about the center plane except for a small region within the center conductor. The top half and bottom half of the cantilever have the same internal stresses and thermal expansion coefficient. Both stress-induced and temperature-induced bending moments are well balanced. Thus, the cantilever is straight and will remain so with temperature changes.

### 5.2.2 Probe Interface Module

A probe interface module has the function of mechanically holding an sMIM probe and electrically matching it to the characteristic impedance of 50 Ω. The probe interface modules have different designs to be compatible with different AFM platforms. Figure 5.3 shows an example of probe interface modules for major AFM platforms.

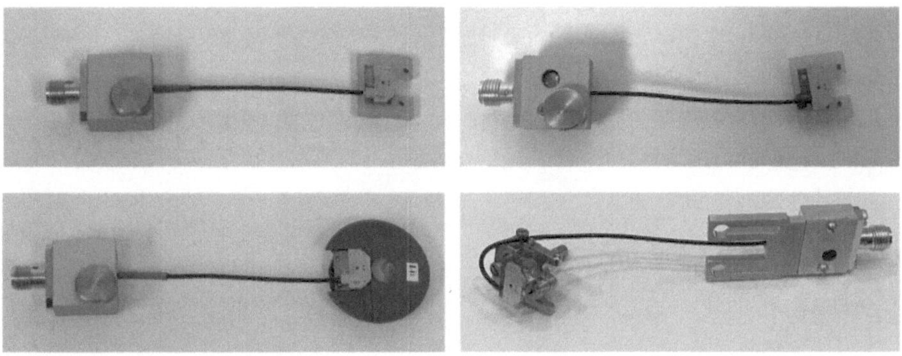

**FIGURE 5.3** Probe interface modules for different AFMs.

The probe interface module has a base compatible with the AFM scanning head. A slot was fabricated between the base and a thin printed circuit board. The sMIM probe is loaded into the slot and pushed from the backside by a spring for holding the probe stable during scanning. In the meantime, the electrodes and grounding on the probe chip contact the corresponding electrodes on the PCB for microwave signal transmission. Custom designed circuits with electronic components are integrated to match the probe interface module to be of 50 Ω impedance at a frequency within the operational range. As each probe has slight variations in impedance, a trimmer was integrated on the probe interface module to adjust the electronic component value to match different probes.

### 5.2.3 Microwave Electronics

The sMIM microwave electronics send microwave signals to the probe interface module and receive the reflected microwave signals from the probe interface module. Microwave signals are processed and converted into analog sMIM outputs that are related to the sample electrica and dielectric properties. A bias T is integrated so that a DC and AC bias can be applied to the tip to modulate material properties. The stability of sMIM electronics is extremely important as changes in electronic component behavior can easily overwhelm the small variation in reflected microwave signals. Therefore, the microwave electronics are powered by a custom power supply with ultra-stable voltages to ensure stable electronics performance. An USB interface is integrated for controlling the sMIM microwave electronics on a computer locally or remotely. Figure 5.4 shows the pictures of sMIM Microwave electronics box and the power supply box.

### 5.2.4 Scanning Platform

A scanning platform is an AFM that the sMIM probe interface module can be mounted on and performs the scanning on a sample with high spatial resolution and stability. The AFM should have a topographic feedback mechanism compatible with the sMIM probe. The controller should also be able to record sMIM outputs and synchronize them with the topographic information. With customized probe interface modules, sMIM is compatible with major room temperature AFM's and selected low temperature AFM's.

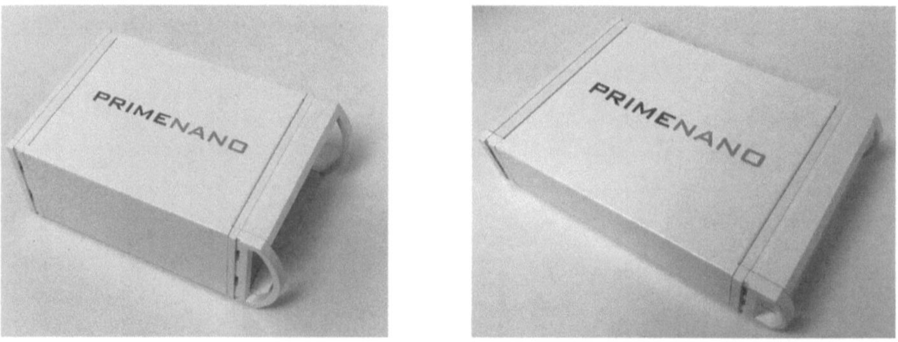


**FIGURE 5.4** sMIM microwave electronics and power supply.

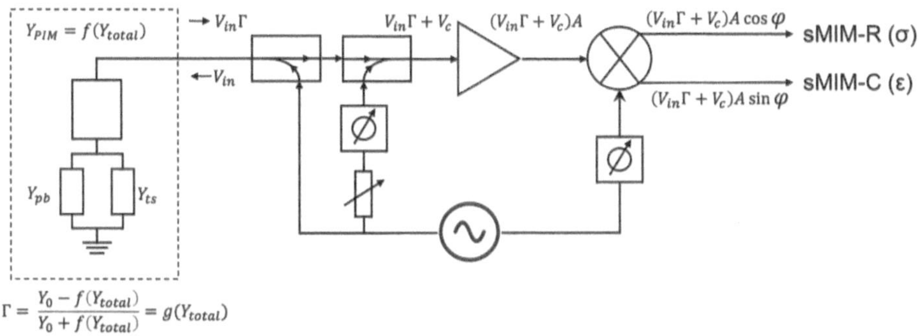


**FIGURE 5.5** Equivalent circuit diagram of sMIM.

### 5.2.5 sMIM Image Mechanism

An equivalent circuit diagram of sMIM is shown in Figure 5.5. Microwave signals are transmitted to the probe via impedance matching. The reflected signals are cancelled to remove background signal then amplified and finally demodulated into two orthogonal sMIM outputs, sMIM-R and sMIM-C.

The relevant length scale of the probe and tip to sample interaction is between about 1 mm and 10s-100s nm, which is very small compared with the microwave wavelength, 10 cm for 3GHz. The tip-sample interaction can be described by using lumped elements, impedance Z or admittance Y. Here we use admittance Y for simpler analysis. The electromagnetic fields generated at tip apex interact with the sample material beneath it in a region surrounding the tip apex with the scale comparable with the tip apex, inducing a tip-sample admittance $Y_{ts}$. As we discussed previously, the probe has a co-axial structure. The admittance between the center conductor and the grounding, defined as $Y_{pb}$, is only related to the probe and can be treated as constant during the scanning. $Y_{ts}$ and $Y_{pb}$ are parallel and we define $Y_{\text{total}} = Y_{ts} + Y_{pb}$

The impedance matching in the probe interface module transfers $Y_{\text{total}}$ into an admittance value $Y_{\text{PIM}}$ that is close to the characteristic impedance of $Y_0 = 1/50$ S so that most of the microwave energy can be delivered to the probe. The circuit for the impedance matching is complicated but here we simplify the relationship between $Y_{\text{PIM}}$ and $Y_{\text{total}}$ as function:

$$Y_{\text{PIM}} = f(Y_{\text{total}}) \tag{5.1}$$

The microwave reflection coefficient from the probe interface module can be expressed as:

$$\Gamma = \frac{Y_0 - Y_{\text{PIM}}}{Y_0 + Y_{\text{PIM}}} = \frac{Y_0 - f(Y_{\text{total}})}{Y_0 + f(Y_{\text{total}})} = g(Y_{\text{total}}) \tag{5.2}$$

With $V_{\text{in}}$ sent to the probe interface module, the reflected microwave signals $V_{\text{in}}\Gamma$ are nulled by adding a cancellation signal $V_c$, which has the same amplitude and reversed phase to the background signal, in order to remove the background signals before amplifying by a gain of A. Finally, the microwave signals are demodulated at a reference phase $\phi$ with respect to the incident signal:

$$\begin{aligned} \text{sMIM} &= (V_{\text{in}}\Gamma + V_c)A\ e^{i\varphi} \\ &= A\ V_{\text{in}}e^{i\varphi}g(Y_{\text{total}}) + AV_c \end{aligned} \quad (5.3)$$

Note the cancellation signal $V_c$ is set during the setup and stays constant during scanning.

In typical testing, the value of $Y_{ts}$ is much smaller than $Y_{pb}$, $|Y_{ts}| \ll |Y_{pb}|$, so the $Y_{ts}$ can be treated as a small perturbation to $Y_{ts}$. A Taylor expansion of the sMIM signal at $Y = Y_{pb}$ :

$$\text{sMIM} \approx A\ V_{\text{in}}e^{i\varphi}g(Y_{pb}) + A\ V_{\text{in}}e^{i\varphi}g'(Y_{pb})Y_{ts} + AV_c\ e^{i\varphi} \quad (5.4)$$

where $g'(Y)$ is the first derivative of $g(Y)$ with respect to Y. The sMIM signal can be further simplified as:

$$\text{sMIM} \approx C + k \quad (5.5)$$

where $C = A\ e^{i\varphi}\left[V_{\text{in}}g(Y_{pb}) + V_c\right]$ and $k = A\ V_{\text{in}}e^{i\varphi}g'(Y_{pb})$ are complex constants.

During the sMIM measurements, we can adjust the demodulation phase $\phi$ to make $k$ to be a real number. This can be done when only the tip-sample capacitance changes, for example, scanning on a purely capacitive feature or doing an approach curve when tunning the demodulation phase. So we can get:

$$\text{sMIM} \approx r\ Y_{ts} + C \quad (5.6)$$

or

$$\text{sMIM\_R} + i\ \text{sMIM\_C} \approx r\left[\text{Re}(Y_{ts}) + i\ \text{Im}(Y_{ts})\right] + C \quad (5.7)$$

where $r$ is a real constant and $C$ is a complex constant.

In sMIM measurements, the sample electrical properties variation induced sMIM signals changes are the contrasts we are interested in. Therefore, the sMIM contrasts are:

$$\Delta(\text{sMIM\_R}) = r\ \text{Re}(\Delta Y_{ts}), \quad (5.8)$$

$$\Delta(\text{sMIM\_C}) = r\ \text{Im}(\Delta Y_{ts}), \quad (5.9)$$

From the analysis above, the sMIM electronics can be simply treated as a lock-in amplifier operating at microwave frequencies. The electronics demodulate the reflected microwave signals into the real and imaginary components and output sMIM-R and sMIM-C. Once the sMIM channels are aligned, the sMIM-R and sMIM-C signals are proportional to the real and imaginary components of the tip-sample admittance respectively.

Here it is worth to discussing the cancellation. One can try to adjust the matching to have $Y_{PIM} = f(Y_{total})$ very close to $Y_0$, but the matching is never perfect so there is always a finite reflection background. The tip sample induced change due to $Y_{ts}$ is even smaller than the background. In the electronics, the cancellation signal $V_c$, is adjusted in amplitude and phase to reduce the background signal to the same level as the tip-sample induced reflection signal. Thus, the signal after cancellation can be further amplified without worrying about saturation. One should note the cancellation was done before the scanning and keeps its settings during the scanning. However, the reflection background and the cancellation signals change with environment changes, resulting in long term drift in the sMIM signal offset.

### 5.2.6 sMIM Operational Modes

sMIM can be installed on major AFM platforms and is compatible with main AFM scanning modes, such as contact mode, tapping mode, lift mode, and so on. Combined with the function of AFM controllers and other equipment, sMIM can be operated at different modes to meet many sample characterization requirements.

#### 5.2.6.1 Direct sMIM

The direct sMIM is the most common and straightforward mode. The sMIM signals are directly recorded during the scanning and synchronized with the topographic information. As we discussed, the drift in the sMIM offset generally induces a slowly changing background in sMIM images. Line-leveling to a known region (e.g., substrate) where the materials have the same electrical properties is usually used to flatten the drift. The sMIM signals obtained in direct sMIM mode are only relative values.

#### 5.2.6.2 Lift Mode

In lift mode, the sMIM probe scans two paths at each line. During the first path the tip is in contact with the sample, and during the second path the tip follows the topography from the first path but is lifted by a certain distance (100s nm). The lift mode sMIM signals are calculated by subtracting the second path scanning sMIM signals from first path scanning sMIM signals at each data point. It measures the differences in sMIM signals between tip contact with the sample and tip away from sample, without worrying about background drift. The lift sMIM mode can be used for quantitative sMIM measurements.

#### 5.2.6.3 sMIM C-V Curve with DC Bias Sweep

For certain kinds of materials, the electric property variations under electric fields are important as they reveal the carrier properties or polarization in the material. For large devices, C-V curves can be obtained on a probe station for macro scale characterization. With the benefit of sMIM, the C-V curves can be realized at the nanoscale. In sMIM based C-V curve testing, the tip is parked at point of interest. DC bias is applied either on the tip or sample to create electrical fields at the sample materials beneath the tip. By sweeping the DC bias, the electrical fields modulate the sample properties and sMIM signal changes accordingly. The measured sMIM-C versus bias

V is from nano-scale volumes beneath the tip apex and can be used to characterize the carrier type and concentration, defects, traps, and more.

#### 5.2.6.4 sMIM dC/dV with AC Bias Modulation

sMIM based dC/dV measures the sMIM-C variations due to the AC modulation of the sample materials. In sMIM dC/dV, an AC modulation up to hundreds of kHz is applied between the tip and sample, and a lock-in amplifier is used to extract the amplitude and phase of sMIM-C signal variation referenced to the bias. The sMIM dC/dV achieves all the SCM functions, the dC/dV amplitude relates to the doping level and the dC/dV phase shows the carrier type. What is more, the sMIM dC/dV can scan directly on semiconducting materials without any insulating layer, while SCM requires an insulation layer between the tip and sample to for a metal-insulator-semiconductor structure.

## 5.3 sMIM FEATURES

### 5.3.1 Sub-aF Electrical Resolution

sMIM has sub aF ($< 10^{-18}$ Farad) electrical resolution, which allows this technology to measure very small electrical variations on a sample. The sMIM electrical resolution is tested by scanning on a capacitance standard sample developed by NIST and manufactured by $MC^2$ technology [3] as shown in Figure 5.6a. The measurements and analysis follow a method published in the literature, [4] and as such, validate the sMIM sensitivity on a known commercially available standard using an objective procedure. The samples are circular gold pads on a silicon substrate with $SiO_2$ between. The silicon substrate and the gold metal pads form a series of capacitors. The capacitance values depend only on the dimensions of the pads and thickness of the $SiO_2$ and are not affected by other sensitive factors in AFM scanning such as tip size and surface oxidation. Following the fundamental physics principles, the capacitance values can be calculated. The measured sMIM signal in Figure 5.6b can be converted into capacitance and Figure 5.6b shows the line profile of sMIM in $\Delta C$ over the smallest capacitor. The sMIM electrical resolution of $3 \times 10^{-19}$ Farad is extracted by the noise level of sMIM in Farads. Note the sMIM electrical resolution

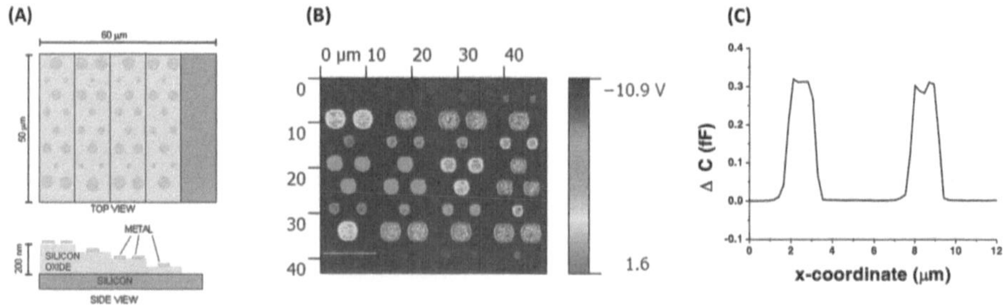


**FIGURE 5.6** sMIM scanning on capacitance standard sample. (a) Schematic of capacitance calibration sample. (b) sMIM image on the sample. (c) Line profile of sMIM in $\Delta C$ over the smallest capacitor.

tested was using an older version sMIM system. It was reported that the current version of sMIM system shows more than 10X in signal-to-noise ratio improvement [5], leading us to expect an electrical resolution below $10^{-20}$ Farad.

### 5.3.2 Nano-Meter Spatial Resolution

sMIM is a near filed measurement. The spatial resolution is defined by the size of the electric fields near the tip and affected by the tip apex size and sample electrical properties. The sharper the tip, the higher the spatial resolution. The electric field concentration drops faster on a conductive sample than on an insulating sample, thus sMIM shows higher resolution on more conductive samples. Generally, a fresh sMIM probe has a tip apex size less than 50 nm and sub 20 nm spatial resolution can be routinely achieved.

Figure 5.7 shows the sMIM image of a Moiré pattern of graphene on boron nitride. Single layer graphene is superimposed on a thin layer of boron nitride (h-BN) which has a different periodicity from graphene. The periodicity of Moiré patterns formed depend on the twist angle between the graphene and the h-BN. While there is no contrast in the topography, the contrast in the sMIM signal shows the honeycomb pattern with a period of 14 nm in Figure 5.7a and b. A line profile in Figure 5.7c indicates a sharp transition in the honeycomb pattern with ~4 nm lateral spatial resolution (full width at half maximum).

### 5.3.3 Linear Response to Dielectric Constant

It is proven that sMIM-C signals have a linear relationship to the logarithmic dielectric constant by finite element modeling and experimental testing [6]. Figure 5.8 shows the sMIM-C results on a variety of bulk dielectric samples with dielectric constant between 2.5 and 300. The tests were done in lift mode and the sMIM data are derived from measurements with the tip in contact with and raised 1 um above the sample. Finite element modeling was used to simulate tip-sample impedance with the tip in contact with and raised 1um above the sample. The modeling data were scaled to

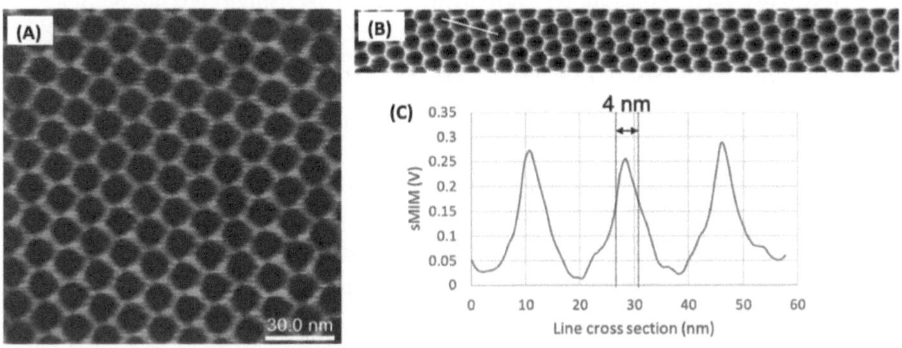


**FIGURE 5.7** sMIM image of Moiré pattern of graphene on boron nitride. (a) and (b) sMIM image. (c) Line profile in (b).

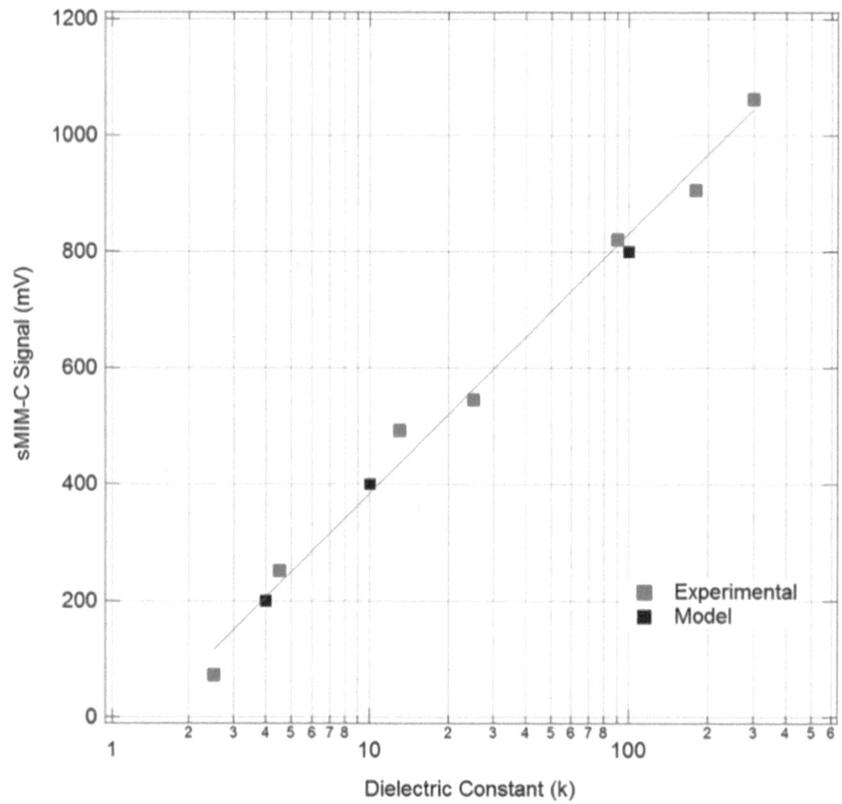


**FIGURE 5.8** Response of sMIM-C signal to dielectric constant (k).

compensate the difference between real tip size and tip in the model. The experimental data and modeling data are plotted together in Figure 5.8 and they match very well. The sMIM-C signal response is linear to the logarithm of dielectric constant, enabling quantitative dielectric constant measurement with sMIM using reference samples.

### 5.3.4 Linear Response to Doping Concentration

In semiconductor manufacturing, accurate control and characterization of doping levels in a device is critical for the chip performance. sMIM signal responds in a predictable way to dopant concentration in semiconductors as shown in Figure 5.9 [7]. For most of the range of dopant concentrations encountered in semiconductor manufacturing, the response of the sMIM-C signal is linear. For very low concentrations $\left(<10^{16}/\text{cm}^3\right)$ and very high concentrations $\left(>10^{20}/\text{cm}^3\right)$ the response is not linear, but it is known and can be easily calculated. With sMIM, one can therefore create images of dopants with nanoscale lateral resolution. Since the dopant level is predictable it is possible to quantify the sMIM signal into dopant concentration in atoms per cubic centimeter in order to provide a quantified map of dopant distribution in a sample.

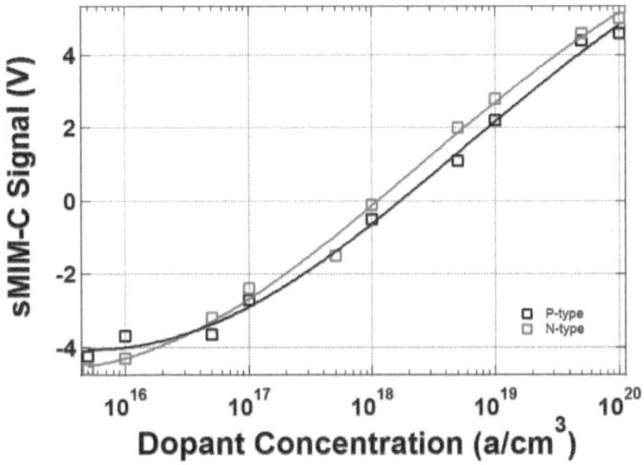


**FIGURE 5.9** Response of sMIM-C signal to dopant concentration for N and P type.

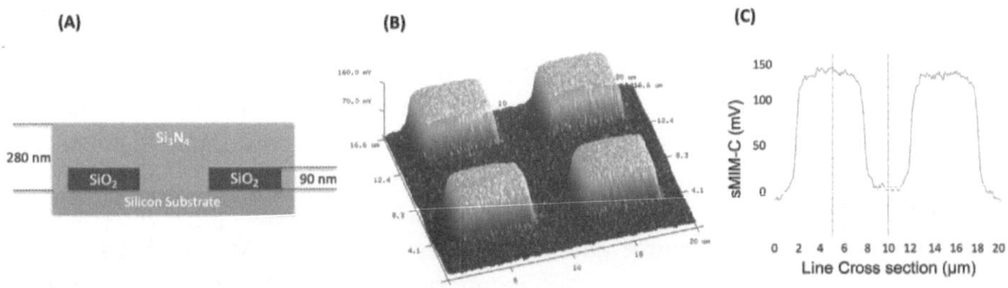


**FIGURE 5.10** sMIM senses islands of $SiO_2$ buried through 190nm of $Si_3N_4$. (a) sample structure. (b) sMIM images. (c) sMIM line profile.

### 5.3.5 Subsurface Sensing

The high frequency electromagnetic fields at the sMIM probe apex can penetrate into materials with depths up to a few hundred nanometers depending on sample properties. So sMIM has the unique advantage of sensing structures beneath the surface. Figure 5.10 shows the sMIM images of $SiO_2$ buried in $Si_3N_4$. $SiO_2$ islands were deposited on a silicon substrate and coved with $Si_3N_4$. The sample surface was polished to be flat to avoid topography induced artifacts. The sMIM can clearly image the buried structures due to the permittivity differences between the oxide and nitride surrounding it.

## 5.4 sMIM APPLICATIONS AT ROOM TEMPERATURE

sMIM installed on a commercial room temperature AFM platform can simultaneously measure sample topography and electrical properties. sMIM uses microwave to measure the tip-sample impedance and it does not require complicated sample preparation such as a counter electrode. The sMIM signals are dominated by the sample properties in a small volume beneath the tip, typical in the order of tip apex size, 10s nm. These unique features make sMIM a versatile technique to characterize microscopic conductivity variations in a wide variety of systems, such as semiconductors, materials research, and 1D/2D materials. In this section, we present a few sMIM application examples to demonstrate sMIM capabilities and features.

### 5.4.1 Semiconductors

Understanding the doping profile in the nano scale is critical to the semiconductor device performance. Generally, scanning capacitance microscopy (SCM), measuring the capacitance change with applied AC modulation, is used to extract the dopant information. However, the SCM's response to the doping level is not monotonic with peaks at medium doping level and SCM is only sensitive to semiconductors, not dielectrics [8]. sMIM measures the conductivity and permittivity properties and is sensitive to a variety of materials (semiconductors, dielectrics, and other) and has a linear response to doping levels. More and more reports and publications demonstrate the advantage of sMIM in semiconductors [9][10].

Vertical Insulated Gate Bipolar Transistors (IGBTs) are one of the most important types of discrete power transistors. These devices were developed to have more efficient operation and faster switching. Mapping the dopant levels at different locations in a device provides guidance to device design and fabrication. Figure 5.11 shows an example of sMIM characterization of an IGBT cross-sectional sample together with scanning electron microscope (SEM) and SCM images. The SEM image reveals the different device components, such as metal contacts, gate oxide, poly-Si trench gate and single crystal silicon source and body. Comparing the SCM and sMIM images at the similar region, sMIM has more dynamic range than SCM and shows more features about the device. Even though both sMIM and SCM shown the doping in emitter region, the sMIM has a clearer image with fine structures related to the small doping variations. sMIM also shows the gate oxide between n-substrate and trench gates while the gate oxide and substrate merged together in SCM. Defects in trench gate, gate-gate contract, metal can also be clearly seen in sMIM image, but not in SCM.

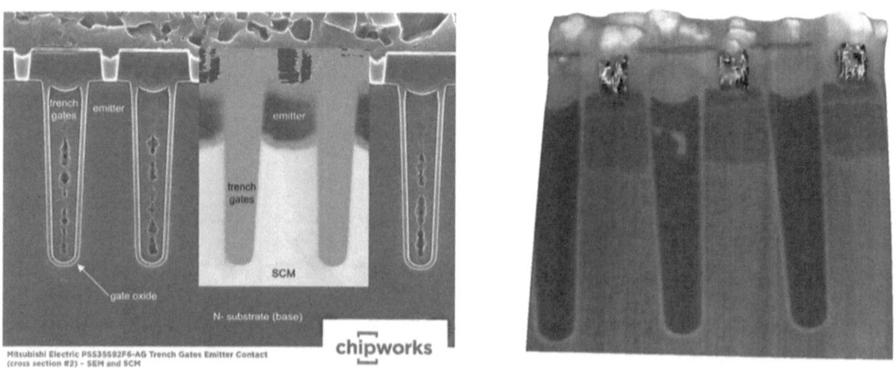


**FIGURE 5.11** IGBT sample characterization with SEM, SCM and sMIM. SEM and SCM images courtesy of Chipworks.

For failure analysis, SCM, SEM and other conventional imaging techniques have limitations in studying the various aspects of specific devices. In the case of SEM, complicated sample preparation such as etching are needed to preferably etch doped regions for adequate contrast. SCM is generally applied to highlight the varied doping regions. However, SCM is only sensitive to doped semiconductor regions or nonlinear regions, and does not reveal any contrast in dielectrics, metals or regions with uneven oxide thickness. sMIM, on the contrary, requires no special sample preparation to image the doped regions, and can even distinguish metals, oxides, and semiconductors. Furthermore, sMIM dC/dV, which is equivalent to SCM and provides dopant information, can be simultaneously obtained.

A front-side illuminated global-shutter Complementary metal-oxide semiconductor (CMOS) image sensor was employed to demonstrate sMIM's ultra-high sensitivity to various doping concentrations, and the contrast from those linear regions as shallow trench separation, dielectrics, and polysilicon [11] [12]. As shown in Figure 5.12, the sample includes 3 μm pitch pixels. In Figure 5.12b, the sMIM-C channel resolves all of the important features, including n-well photocathode diffusion, n-well storage diffusion, shallow trench isolation, p-type substrate surrounding cathode, and contact. At the same time, the sMIM-R signal is captured as well (Figure 5.12c). When compared to the sMIM-C, the sMIM-R response is weaker which can be probably

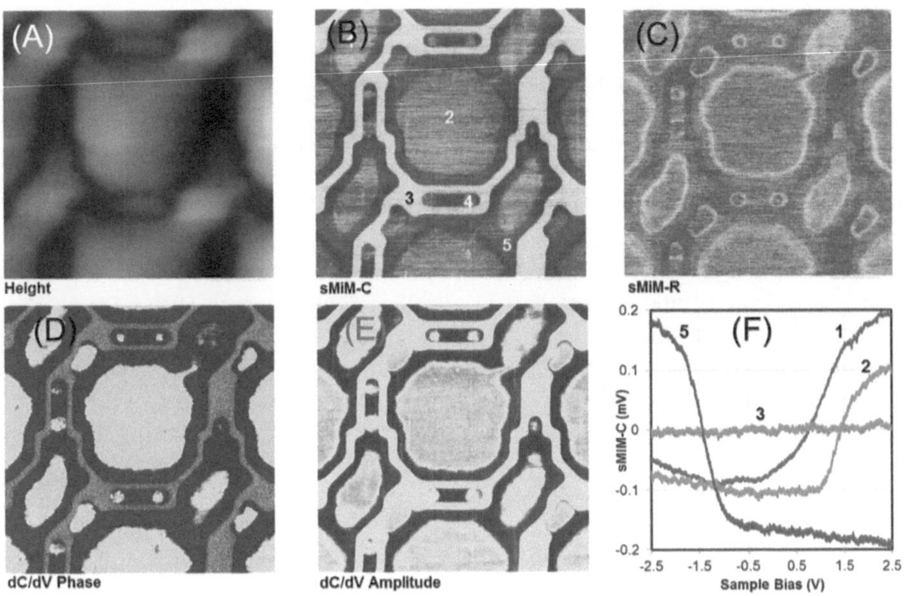


**FIGURE 5.12** sMIM images of a front-side illuminated global-shutter CMOS image sensor from Chipworks. The sample features 3 μm pitch pixels. (a) Topography; (b) sMIM-C; (c) sMIM-R; (d) dC/dV phase, N-type and P-type regions have dC/dV phase difference of 180 degrees; (e) dC/dV amplitude; and (F) site-specific capacitance-voltage curves at locations labelled in (b). The numbering in (B) features: (1) n-well storage diffusion; (2) n-well photocathode diffusion; (3) shallow trench isolation; (4) contact; and (5) p-type substrate surrounding cathode. These images were captured with a Dimension Edge AFM at Chipworks.

attributed to the presence of the surface oxide, and also due to the fact that domains are highly insulating or conductive.

The dC/dV phase emerges as predicted and analogous to traditional SCM imaging (Figure 5.12d). P-type and N-type regions have clearly defined dC/dV phase value with a difference of 180 degrees and the rest of the dielectric/insulating regions have a noisy dC/dV phase close to 0 degrees. In Figure 5.12e, it can be seen that the dC/dV amplitude has varied doping densities with respect to non-linear material, for example Si, in this sample. The dopant sensitivity spans from intrinsic silicon to $10^{20}$ /cm$^3$ for sMIM. The dC/dV amplitude channel appears to be analogous to the sMIM-C; however, the sMIM-C channel showed a wider dynamic range. For instance, when locations #4 (highly doped Si) and #3 (oxide) in the sMIM-C channel are compared with the dC/dV amplitude channel, the SMIM-C signal in #3 is found to be different from #4, while dC/dV amplitude displays no contrast between #3 and #4. While the sMIM-C signal remains monotonic with dopant concentration, the variation in the signal can be attributed to dC/dV amplitude being peaked at intermediate dopant values.

sMIM is also capable of conducting capacitance-voltage (C-V) spectroscopy utilizing the sMIM-C signal. For semiconductor devices, site-specific C-V spectroscopy can be utilized as a failure analysis tool. Plot #1 and #2 substantiate n-type regions (n-well photocathode diffusion and n-well storage diffusion) plot #3 is in parallel to the oxide of shallow trench isolation, and plot #5 shows the p-type substrate enclosing the cathode (Figure 5.12f).

An important attribute of sMIM is that the sMIM signal varies in a monotonic way with the dopant concentration. This opens up opportunities to compare doping levels of semiconductors without introducing unknown integration constants that are associated with differential techniques like SCM. The conductivity of semiconductors is affected by the amount of mobile hole and electron charges in the semiconductor. Both n and p doped samples exhibit an increase of sMIM signal with increased doping concentration. This is illustrated in Figure 5.13 which is a semiconductor test sample having a range of different doping concentrations fabricated by ion implantation at a spacing of approximately 2 μm. In the middle regions of the doping concentration, the sMIM signal depends approximately logarithmically on the doping concentration and then the rate of change of the sMIM signal falls off at the very lowest and highest doping concentrations. The sensitivity of sMIM to detect the low and higher doping concentrations has recently been improved by redesigning the system to generate higher signals with reduced noise, opening up opportunities for sMIM to characterize semiconductor devices with very high and low doping concentration implementations.

The linear relationship between sMIM-C and doping concentration make quantitative doping level measurements possible. By scanning on a reference sample with known doping levels, one can extract the transfer function between sMIM signal and doping level. The sMIM-C signal on an "unknown" sample can be converted into doping concentration.

Figure 5.14 shows the cross-sectional view of the p-Channel Metal-Oxide-Semiconductor (PMOS) power transistors found on the Linear Technologies LTC3612 die, which is used as the "unknown" sample [9]. Figure 5.14a shows a cross sectional SEM image of the PMOS transistor structure, while Figures 5.14b

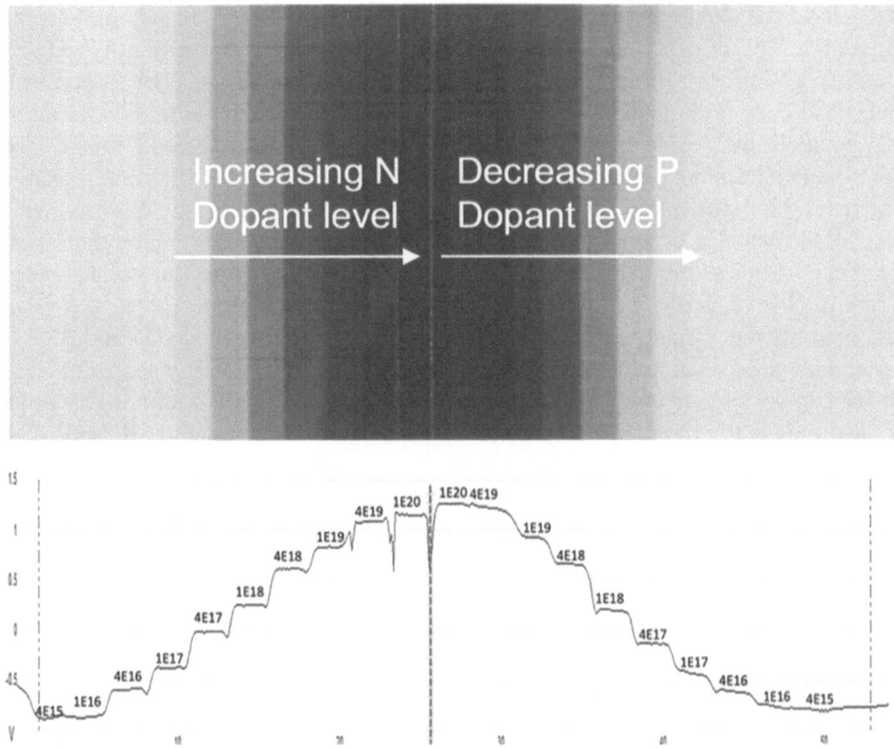


**FIGURE 5.13** Dopant Reference Standard with a staircase of P and N doped regions mapped using sMIM.

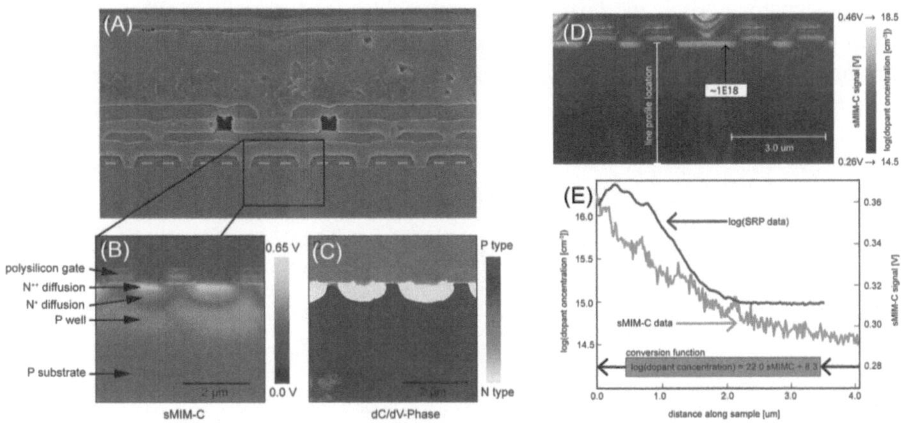


**FIGURE 5.14** Cross-sectional analysis of the NMOS power transistors in the Linear Technology LTC3612. A SEM image in (a) with details of the area in sMIM-C (b) and dC/dV phase image in (v). Line profile of sMIM-C (d) is converted into log(dopant concentration) and plotted in (e) together with SRP data obtained on the same device.

and c show the corresponding sMIM-C and dC/dV phase images respectively of the same area. The dC/dV phase image provides information of the dopant type, with the N-type source/drain diffusions yielding a negative signal and the P-type well and substrate yielding a positive signal. The sMIM-C image shows additional information about the device structure, which are not present in the dC/dV image. The polysilicon gate is well defined. The gradient from N++ to N+ diffusion region is clear as is the P well and the substrate region material features. Figure 5.14d presents the sMIM image in units of log (doping concentration) using calibration data from reference samples. Figure 5.14e is a line profile extracted from the scaled image where the data is presented in log of dopant concentration with a comparison to an SRP measurement made on the same area. Agreement between sMIM and SRP is better than an order of magnitude. Since the dopant concentration in a semiconductor device can vary by up to six or more orders of magnitude, this is useful information at the dramatically better spatial resolution for sMIM as compared to SRP [13]. The dopant concentration in the N+ source/drains was also estimated from the sMIM data, and a value of $10^{18}/cm^3$ was found.

### 5.4.2 Subsurface Sensing

sMIM is based on the electromagnetic interaction between tip and sample. Since the generated concentrated electromagnetic field can penetrate into dielectric materials, the sMIM has the capability to sense structures beneath the sample surface.

Figure 5. 15 illustrates sMIM image narrow buried dopants below surface [14]. Highly conductive two-dimensional sheets of ultra-dense 10 nm wide phosphorus dopant nanostructures spaced with various gaps were created by scanning tunneling microscopy (STM) lithography and then buried under a protective cap of 30 nm epi silicon. The sample surface is flat without any obvious features related to the doping (Figure 5.15a). The sMIM capacitive image is shown in 5.15b where the metallic nanostructures show as dark patterns. It is clear from this image that individual 10 nm lines with pitches above 88 nm can be resolved. An example of an unresolved region (line 1 in Figure 5.15b) is shown in Figure 5.15c, which resolves no discrete

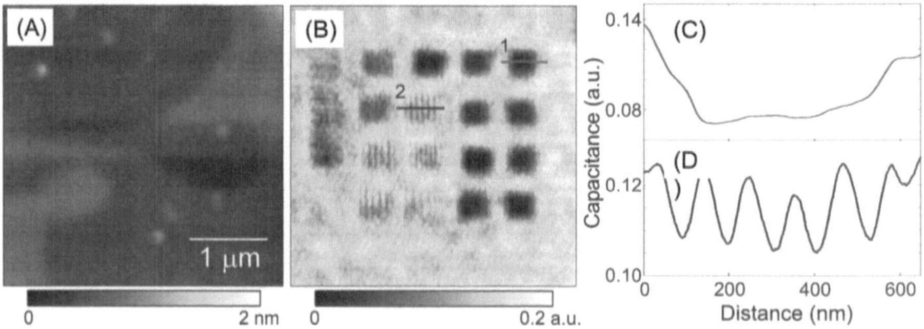


**FIGURE 5.15** (a) sMIM topography image and corresponding (b) sMIM capacitive image of buried resolution test structure. An averaged line cut across (c) line 1 (32 nm grating) showing no resolved lines and (d) line 2 (100 nm line spacing) showing resolved lines from (b).

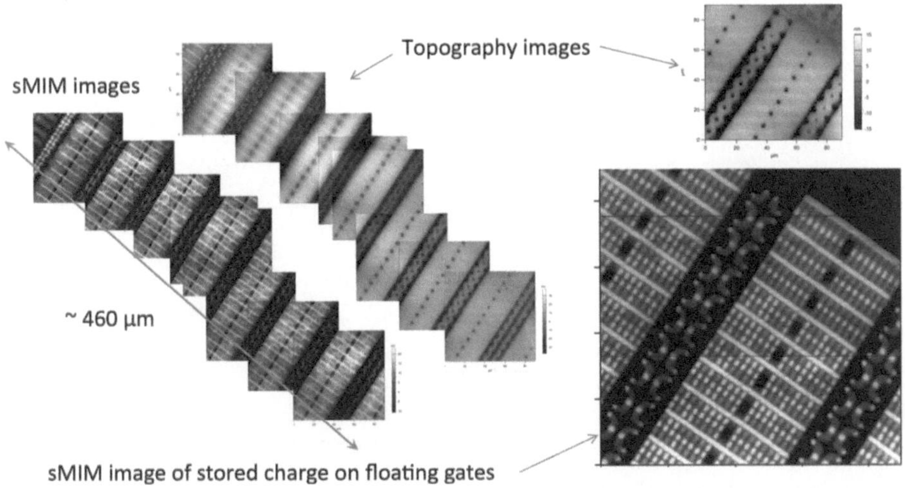


**FIGURE 5.16** sMIM image of Flash memory device after back-side polishing. A pattern of 1's and 0's had been programmed in the device before it was processed. sMIM was able to detect the programmed charge through the residual backside silicon.

features on the 32 nm pitch and instead shows a uniform contrast. Figure 5.15d shows line 2 in Figure 5.15b, which crosses 10 nm lines with 100 nm pitch. The sMIM successfully resolved the buried structures.

sMIM can be used to detect charge. Figure 5.16 shows a flash memory device which was pre-programmed. The device was polished from the back side to remove much of the material below the layer where the charges were stored. Some residual silicon was left so that the charges would not bleed away. sMIM was used to scan the area where the charge was stored. Both sMIM and topography of the same area are shown in Figure 5.16. On the right side of the figure is a zoomed in image. The orange and purple regions correspond to regions of different charge state in the flash, corresponding to the programming of the 1's and 0's.

### 5.4.3 2D Materials

The ability to measure different electrical properties is helpful to understand the behavior of advanced materials, such as 2D materials. An example is shown in Figure 5.17 where the electrical behavior of layer-structured III-VI $In_2Se_3$ nanoribbons was explored [15]. There was an unexpectedly large resistance when a voltage was applied between some of the adjacent contacts. Examination by SEM and AFM did not reveal any differences that could explain the different electrical behaviors. However, sMIM was able to clearly discriminate that some regions of the material had higher resistance. This provided guidance on where to make further studies and it was found that structural phases of the $In_2Se_3$ caused higher resistance in certain regions.

2D semiconducting materials, such as $MoS_2$, are in the limelight of current materials research. These layered materials can be exfoliated into atomically thin 2D sheets

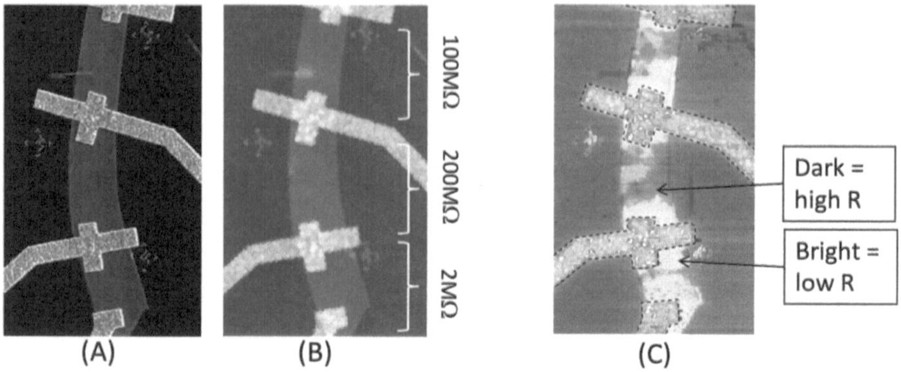


**FIGURE 5.17** (a) SEM of nanoribbon material. (b) AFM and resistance measured between contacts of nanoribbon material. (c) sMIM image. High resistance regions are darker orange.

with unique electrical and optical properties, which are attractive for nanoelectronics and optoelectronics. However, electrical characterization of these materials is challenging for traditional methods due to its sub-nanometer thickness, generally small pieces randomly distributed on insulating substrate. Figure 5.18 present the electrical characterizaiton of $MoS_2$ with sMIM [16]. The sample was made of a few layers of $MoS_2$ flakes in the field-effect transistor configuration and the device is covered by 15 nm thick $Al_2O_3$ to avoid direct contact between the metallic tip and the 2D sheet. Selected sMIM images within the channel regions at different backgate voltages are displayed. As backgate voltage increased, the conductance signal emerged initially at the edges and then in the interior, with appreciable spatial nonuniformity. The results suggest that the contribution of defect induced edge states to the total conductance is significant in the subthreshold regime but negligible once the bulk becomes conductive. The observation of conductance inhomogeneity also provides a guideline for future improvement of the device performance.

### 5.4.4 Ferroelectrics

Many ferroelectrics are wide-band gap semiconductors and commonly there is a Schottky barrier formed between the metal tip and the ferroelectric. A large voltage on the probe is required to overcome the Schottky barrier in traditional AC or DC characterization, such as conductive AFM. A high voltage on the probe may result in a stong electrical field that may change the sample properties, for example, alter the local electric polarization. sMIM overcomes these difficulties as the measurement is based on high frequency interaction. A small fraction of the voltage required at DC or AC detections can measure the ferroelectric properties because the interface barrier is effectively a shunt due to its relatively larger capacitance. Furthermore, the capabilities of applying DC/AC bias on sMIM probes enables the possibilities of controlling the ferroelectric domains at the nano-scale. Figures 5.19 shows the charcterization of the ferroelectrics with sMIM. The sample is 100 nm epitaxial thin film on lead zirconate titanate [$Pb(Zr_{0.2}Ti_{0.8})O_3$ 100 (PZTO)] grown on a single crystal strontium titanate [SrTiO3 (STO)] with a 50 nm layer of metal strontium ruthenate

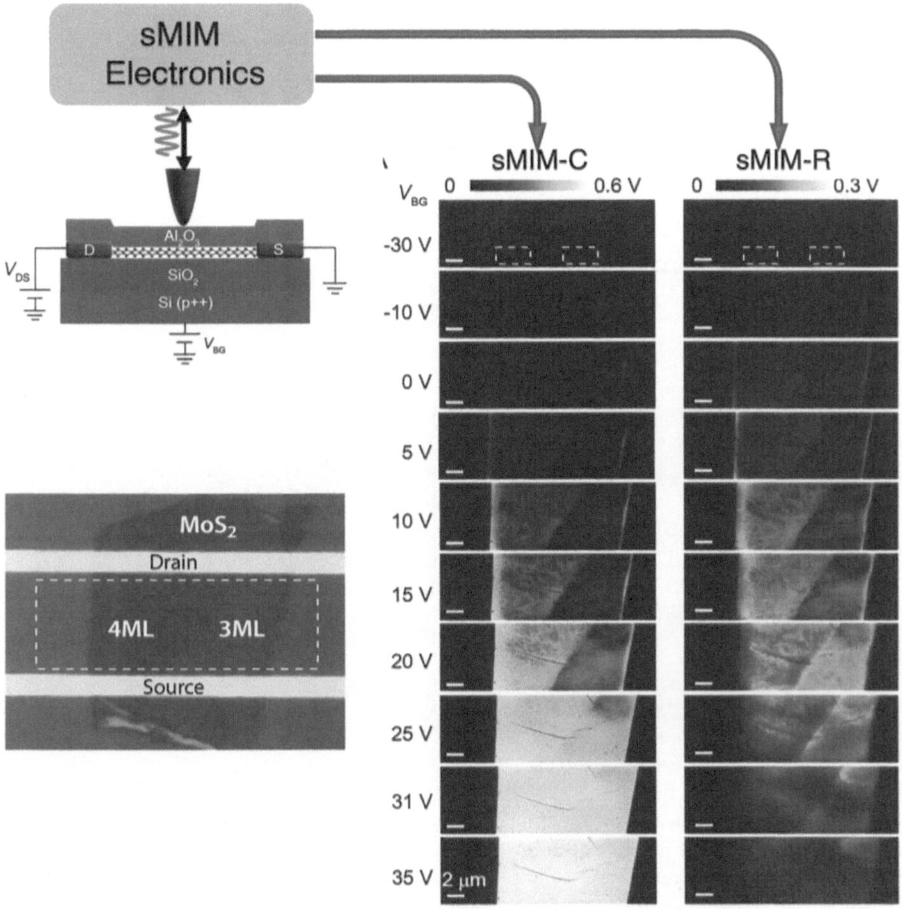


**FIGURE 5.18** sMIM scanning on $MoS_2$ sample with different gate biases.

[$SrRuO_3$ (SRO)] as a bottom electrode. As shown in Figure 5.19b, the spontaneous domain walls distributed on the sample and sMIM shows that the domain walls are conductive. After the scanning, a DC bias was applied on the sMIM tip to write the artificial ferroelectrics domain structures in nano-scale. Here a box-in-box domain structure was created. Spontaneous domains and domain walls were erased inside the structure and artificial domain structures are created with conducting domain walls. Note the artificial ferroelectrics can be created and erased with sMIM and in-situ characterized.

### 5.4.5 C-V Curve

The dielectric film quality is one of the most important factors that will greatly impact device performance and reliability. With device getting smaller and smaller, the traditional capacitance-voltage (C-V) measurement based on a probe station will

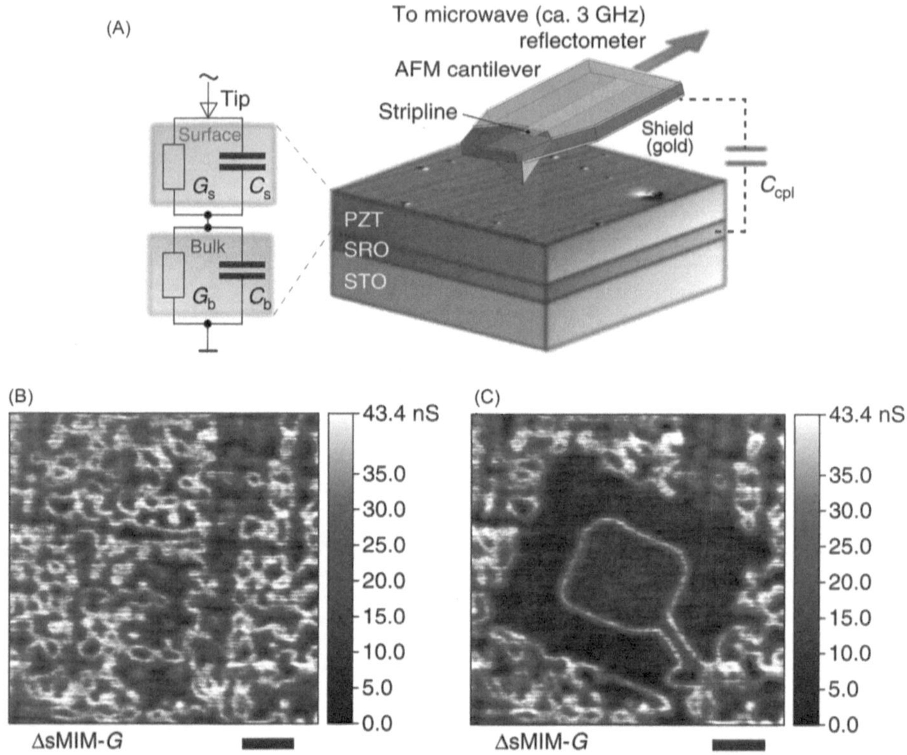


**FIGURE 5.19** (a) A schematic of the sample and sMIM. (b) A sMIM image of the PZT film, where the conducting walls of spontaneous domains of a few hundred-nanometer size are clear seen. (c) A sMIM image of the box-in-box domain structure written by the probe after the image in (b) was taken.

not have the capability to characterize nano-scale devices. sMIM, based on nano-scale probes, can be used to measure the C-V response of these small devices, and provide nano-scale material properties. Figure 5.20 shows sMIM based C-V measurements on a Metal-Oxide-Semiconductor (MOS) array [18]. The sMIM C-V curve and dC/dV bias curve clearly show the typical MOS accumulation, depletion and inversion characteristics of semiconductor behavior. The point nano C-V curves show excellent signal to noise ratio, providing high quality device C-V curve analysis and electrical information for dielectric quality analysis. The C-V curve and dC/dV bias from a good device show no hysteresis, while for a bad device, the C-V curve and dC/dV bias curve shows hysteresis, which is from the poor interface state in the dielectric oxide.

## 5.5 sMIM AT LOW TEMPERATURE

A great deal of cutting-edge physical investigations concerning fundamental mechanisms will frequently involve electrical measurements at low temperatures and high magnetic fields. The measurement of conductivity of novel material systems at

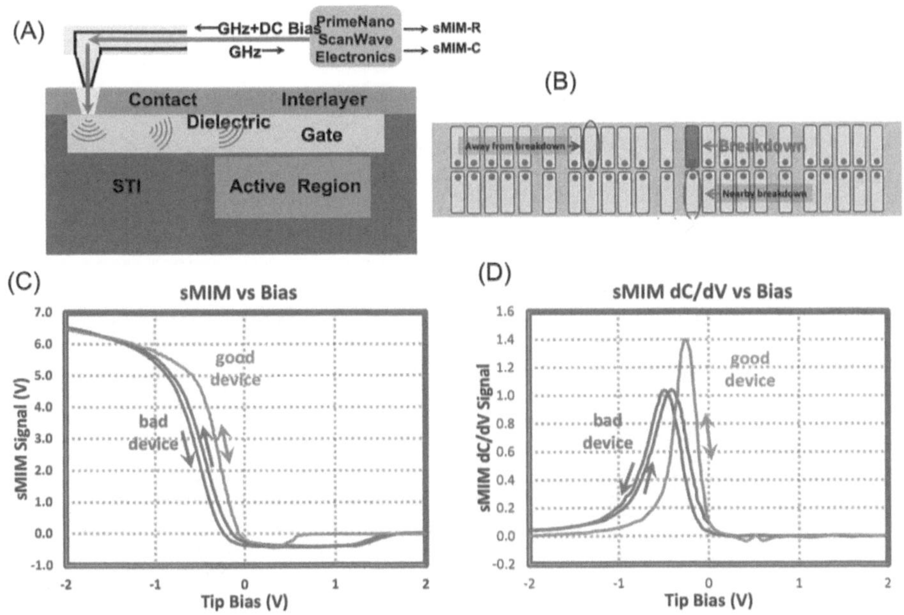


**FIGURE 5.20** (a) Schematic of the sMIM shielded probe and sample measurement setup. (b) Plane view drawing of the device test array. (c) and (d) sMIM nano C-V and dC/dV sweep curve of the good and bad devices.

varying temperatures was at the core of many Nobel Prize winning discoveries over the last 30 years, ranging from high temperature superconductivity (1987), quantum hall effect (1985), fractional quantum hall effect (1998), giant magnetoresistance (2007), and graphene (2010). More recently, being able to image electrical properties at the nano- and meso-scale has emerged as a fertile ground for discoveries. sMIM has been playing an important role in condensed matter physics and material research at low temperature and high magnetic field.

Figure 5.21 shows the schematic of an LT sMIM system. A low temperature AFM with sMIM probe interface module is integrated in a cryostat for reaching low temperature. A laser interferometer is implemented to detect the cantilever deformation for topographic feedback. A pre-amplifier is installed in the insert to amplify the microwave signal for better signal to noise ratio. The microwave electronics connected to the low temperature system control unit to provide electrical information synchronized with sample topography during the scanning.

There are three types of LT sMIM systems commercially available for different application requirements:

1. 2K ScanWave™: sMIM operated in a dry or liquid cryostat with helium exchange gas sample environment. The base temperature can reach below 2 K.
2. mK ScanWave™: sMIM operated in a dilution refrigerator. The base temperature is below 100 mK.

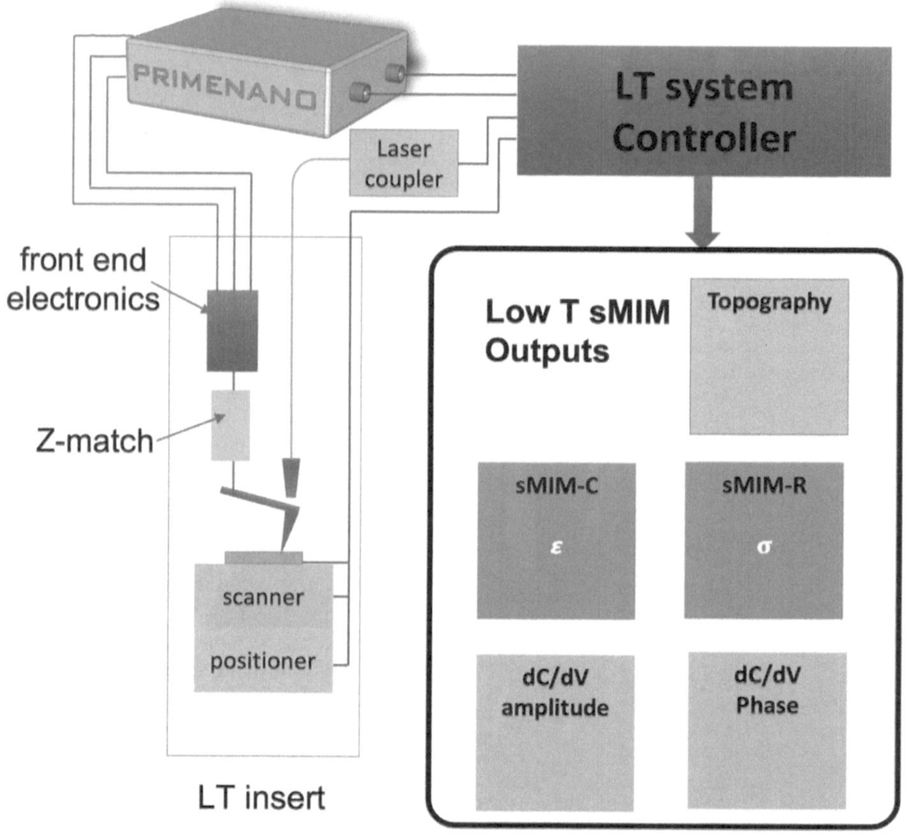


**FIGURE 5.21** Schematic of a low temperature sMIM system.

3. UHV ScanWave™: sMIM operated in a low temperature, UHV environment. The base temperature is below 500 mK and the sample is in UHV condition with in-situ sample transfer capability.

All the LT sMIM systems come with an option to install solenoid magnets or vector magnets for testing samples under magnetic fields.

With the unique capabilities, sMIM has been widely used in advanced physics research and played an important role in the most advanced research. A couple of selected examples will be discussed in this section.

### 5.5.1 Quantum Effect

sMIM is a useful technique to understand novel physics, such as quantum effect in topological insulators [19]. Figure 5.22a shows the schematic of a quantum hall device made of mercury telluride (HgTe). The sMIM measurement results in Figure 5.22b and c revealed that at the edge of the quantum hall device the electrical conductivity was high when measured at low temperatures without a magnetic field being applied.

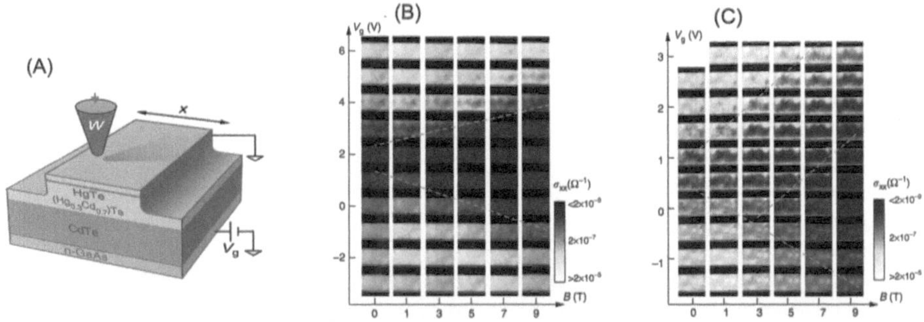


**FIGURE 5.22** (a) Schematic of quantum well sample. (b) sMIM results on 5.5 nm quantum well and (c) sMIM results on 7.5 nm quantum well.

This was expected based on theoretical quantum modeling of the device. But the modelling predicted that the edge conductivity would disappear once the magnetic field reached 3.8 Tesla. Condensed matter physicists use magnetic fields to turn off what is called time-reversal symmetry. Time-reversal symmetry means that the physics remains the same whether one goes forward or backward in time in a model. Time-reversal symmetry is consistent with the presence of conservation of entropy or randomness. Application of a strong magnetic field is a powerful technique to change the time-reversal physics behavior of materials and these changes can be modeled. However, a surprising result showed up when sMIM measurements were made with high magnetic fields. The edge conductivity unexpectedly continued to persist even up to applied fields above 9 Tesla. The measurements showed that the edges did not become electrically insulating until a very high magnetic field and that the modeling did not capture all the effects that were applicable to the device. This provides an example of how sMIM measurements in conjunction with carefully designed experiments and modeling provides guidance to scientists on where to focus their efforts to improve the models and how to design other experiments to test other theories.

Figure 5.23 shows sMIM direct imaging of a conductivity transition in single layer graphene triggered by the filling of bulk landau levels in the quantum Hall effect regime [20]. Two monolayers of graphene are sandwiched between two 20 nm layers of boron nitride and sit on a 300 nm oxide on a doped Si wafer. As the bias on the Si substrate is changed, carriers are driven in or out of the graphene, changing the occupancy of Landau states. When a Landau level becomes fully occupied or fully empty in the bulk, the local bulk resistivity is high. Figure 5.23d shows the sMIM signal recorded during repeated scanning along a single line across a device as the gate voltage is tuned from −40 to 40 V. A series of high-resistivity states are observed with sMIM, matching the Landau Level structure. Note, this conductivity change is being observed through a 20 nm thick insulating BN layer and is an excellent example of sMIM's ability to directly observe fundamental electrical properties even through insulating layers. It is important to note that measuring these phenomena could only occur through performing sMIM measurements in a cryogenic environment with a strong magnetic field, showing that the LT ScanWave sMIM system is a uniquely capable tool.

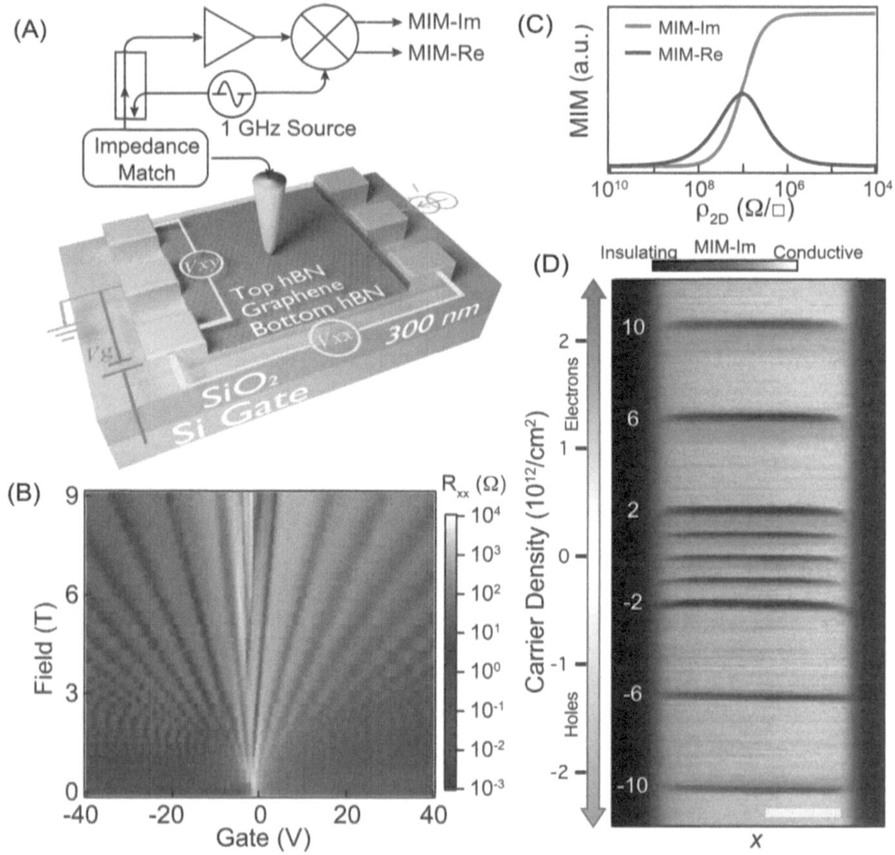


**FIGURE 5.23** sMIM testing on graphene device. (a) Schematic of device structure. (b) Landau fan diagram of (c) Typical response curves of sMIM as function of 2D resistivity (d) MIM signals of repeated scans along the same line.

## 5.6 SUMMARY

In summary, sMIM uses microwaves to measure dielectric and conductivity information of materials at the nanoscale. It also measures doping type and doping level for semiconductor materials. The high sensitivity electronics and shielded cantilever probe enable its high signal to noise ratio for ultra-high electrical resolution and spatial resolution. The capability to measure the sample surface and subsurface without counter electrodes on the sample makes it an idea technique for electrical characterization. Compatible with major commercial AFM's at room temperature and low temperature, sMIM has an important role in research and industry.

Currently the sMIM images are generally qualitative results, showing the difference in electrical properties over an area. It should be noted that the sMIM has a linear response to dielectric constant and doping levels. sMIM has the potential to provide quantitative results by using a reference sample. Future development of

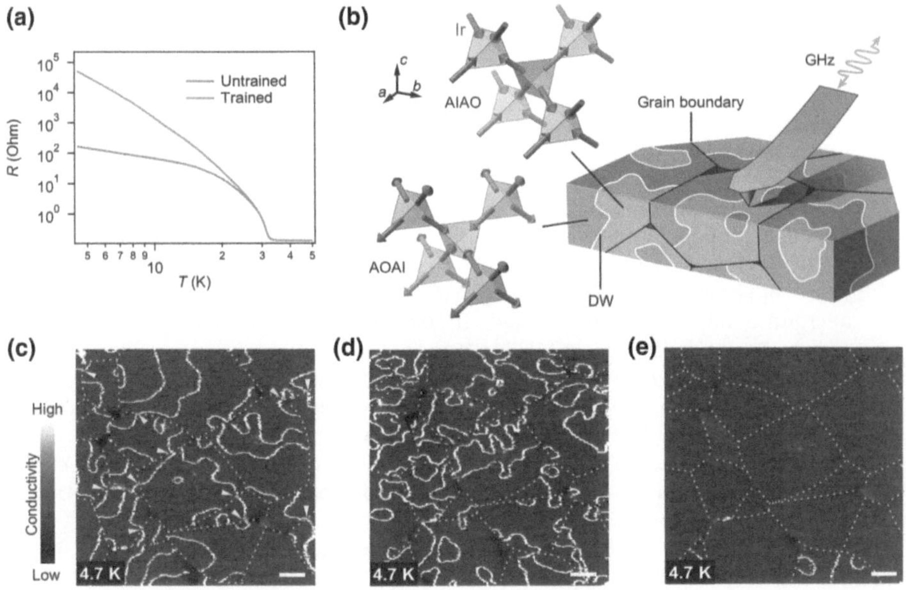


quantitative measurement solutions could make sMIM an even more powerful tool that can benefit the semiconductor industry and materials research.

## BIBLIOGRAPHY


1. K. Lai, W. Kundhikanjana, M. Kelly and Z.-X. Shen, "Modeling and characterization of a cantilever-based near-field scanning microwave impedance microscope," *Review of Scientific Instruments*, vol. 79, p. 063703, 2008.
2. Y. Yang, K. Lai, Q. Tang, W. Kundhikanjana, M. A. Kelly, K. Zhang, Z.-X. Shen and X. Li, "Batch-fabricated cantilever probes with electrical shielding for nanoscale dielectric and conductivity imaging," *Journal of Micromechanics and Microengineering*, vol. 22, p. 115040, 2012.
3. Available: https://www.mc2-technologies.com/smm-calibration-kit/, [Online].
4. H. P. Huber, M. Moertelmaier, T. M. Wallis, C. J. Chiang, M. Hochleitner, A. Imtiaz, Y. J. Oh, K. Schilcher, M. Dieudonne, J. Smoliner, P. Hinterdorfer, S. J. Rosner, H. Tanbkuchi, P. Kabos and F. Kienberger, "Calibrated nanoscale capacitance measurements using a scanning microwave microscope," *Review of Scientific Instruments*, vol. 81, p. 113701, 2010.
5. https://www.primenanoinc.com/scanwave™-pro.html, [Online].
6. S. Friedman, O. Amster and Y. Yang, "Recent advances in scanning Microwave Impedance Microscopy (sMIM) for nano-scale measurements and industrial applications," In *Proceedings of the SPIE 9173, Instrumentation, Metrology, and Standards for Nanomanufacturing, Optics, and Semiconductors VIII*, 2014.
7. N. Antoniou, R. Chintala and Y. Yang, "Scanning microwave impedance microscopy for materials metrology," In *Proceedings of the SPIE 11611, Metrology, Inspection, and Process Control for Semiconductor Manufacturing XXXV*, 2021.
8. J. J. Kopanski, "Scanning capacitance microscopy for electrical characterization of semiconductors and dielectrics," In Kalinin, S., Gruverman, A. (eds.) *Scanning Probe Microscopy*, Springer, New York, 2007.

9. O. Amster, F. Stanke, S. Friedman, Y. Yang, S. Dixon-Warren and B. Drevniok, "Practical quantitative scanning microwave impedance microscopy," *Microelectronics Reliability*, Vols. 76–77, pp. 214–217, 2017.
10. R. C. Germanicus, P. D. Wolf, F. Lallemand, C. Bunel, S. Bardy, H. Murray and U. Luders, "Mapping of integrated PIN diodes with a 3D architecture by scanning microwave impedance microscopy and dynamic spectroscopy," *Journal of Nanotechnology*, vol. 11, pp. 1764–1775, 2020.
11. CMOS (Image Sensors). Available: https://www.azonano.com/article.aspx?ArticleID=4207, [Online].
12. B. Drevnoik, S. J. Dixon-Warren, O. Amster, S. L. Friedman and Y. Yang, "Extending electrical scanning probe microscopy measurements of semiconductor devices using microwave impedance microscopy," In *International Symposium for Testing and Failure Analysis*, Portland, OR, USA, 2015.
13. "How Big a Pattern Do We Need for Spreading Resistance Analysis?," Available: http://www.solecon.com/pdf/how_big_a_pattern_do_we_need_for_sra.pdf, [Online].
14. D. Scrymgeour, A. B. Fishgrab, R. J. Simonson, M. Marshall, E. Bussmann, C. Y. Nakakura, M. Anderson and S. Misra, "Determining the resolution of scanning microwave impedance microscopy using atomic-precision buried donor structures," *Applied Surface Science*, vol. 423, pp. 1097–1102, 2017.
15. K. Lai, H. Peng, W. Kundhikanjana, D. T. Schoen, C. Xie, S. Meister, Y. Cui, M. A. Kelly and Z.-X. Shen, "Nanoscale electronic inhomogeneity in $In_2Se_3$ nanoribbons revealed by microwave impedance microscopy," *Nano Letter*, vol. 9, no. 3, pp. 1265–1269, 2009.
16. D. Wu, X. Li, L. Luan, X. Wu, W. Li, M. N. Yogeesh, R. Ghosh, Z. Chu, D. Akinwande, Q. Niu and K. Lai, "Uncovering edge states and electrical inhomogeneity in $MoS_2$ field-effect transistors," *Proceedings of the National Academy of Sciences*, vol. 113, no. 31, pp. 8583–8588, 2016.
17. A. Tselev, P. Yu, Y. Cao, L. R. Dedon, L. W. Martin, S. V. Kalinin and P. Maksymovych, "Microwave a.c. conductivity of domain walls in ferroelectric thin films Alexander," *Nature Communications*, vol. 7, p. 11630, 2016.
18. W.-S. Hu, J.-H. Lee, M.-H. Kao, H.-W. Yang, P. D. Wolf and O. Amster, "Device dielectric quality analysis and fault isolation at the contact level by scanning microwave impedance microscopy," In *ISTFA 2016: Conference Proceedings from the 42nd International Symposium for Testing and Failure Analysis*, Fort Worth, TX, USA, 2016.
19. E. Y. Ma, M. R. Calvo, J. Wang, B. Lian, M. Muhlbauer, C. Brune, Y.-T. Cui, K. Lai, W. Kundhikanijana, Y. Yang, M. Baenninger, M. Konig, C. Ames, H. Buhmann, P. Leubner and L. W. Molenkamp, "Unexpected edge conduction in mercury telluride quantum wells under broken time-reversal symmetry," *Nature Communications*, vol. 6, p. 7252, 2016.
20. Y.-T. Cui, B. Wen, E. Y. Ma, G. Diankov, Z. Han, F. Amet, T. Taniguchi, K. Watanabe, D. Goldhaber-Gordon, C. R. Dean and Z.-X. Shen, "Unconventional correlation between quantum hall transport quantization and bulk state filling in gated graphene devices," *Physical Review Letters*, vol. 117, p. 186601, 2016.

# 6 Atomic Force Microscopy-Based Infrared Microscopy for Chemical Nano-Imaging and Spectroscopy

*Xiaoji G. Xu*
Lehigh University

## CONTENTS



Mid-infrared radiation directly couples to the dipole-allowed vibrational modes of molecules at characteristic frequencies. Infrared (IR) spectroscopy—typically through the detection of IR absorption—allows for convenient identifications of chemicals based on the presence of their functional groups. However, the traditional IR spectroscopy, when implemented in a microscopy configuration, suffers from low spatial resolutions because of Abbe's diffraction limit [1], on the order of several microns. Consequently, many meaningful samples with spatial features at the nanoscale cannot be spatially resolved by regular IR microscopy. In the research of energetic materials, such heterogeneous samples include and are not limited to heterojunction organic photovoltaics, multi-composition perovskite, and hydrocarbon-containing oil shale.

The integration of atomic force microscopy (AFM) with IR radiation provides a reliable route to bypass the diffraction limit to achieve spatial resolution spectroscopic imaging at a 10-nm scale. As of 2021, two main categories of AFM-based IR microscopy exist based on their detection principles: mechanical detection on the tip-enhanced photothermal response of the sample; optical detection through light scattering from the near field of the AFM tip. This chapter on AFM-based IR

DOI: 10.1201/9781003174042-6 

microscopy will separately describe the working principles of these two routes, followed by some of their applications in the characterization of energetic materials.

## 6.1 PHOTOTHERMAL AFM-IR MICROSCOPY

The hallmarks of AFM are ultrahigh sensitivity on force detection through the deflection of a microscale cantilever [2] and the nanoscopically sharp apex of the AFM tip to deliver excellent spatial resolution. The AFM-IR microscopy leverages these two aspects of AFM to probe the frequency-dependent IR photothermal response of the sample. AFM-IR microscopy also utilizes the field enhancement by metal-coated nanoscopic AFM tip through the lightning rod effect: the IR field intensity is greatly amplified and spatially confined to an excitation volume comparable to the radius of curvature of the tip. The highly confined excitation volume, small force detection area, and ultra-sensitive force detection enable the mechanical measurement of local IR absorption through the photothermal effect. No optical detection of infrared photons is involved in the operational principle of AFM-IR microscopy, which means the restriction of spatial resolution from Abbe's diffraction limit is bypassed.

The operational principles of photothermal AFM-IR have evolved in several stages. In the early work by Pollock and coworkers in 1999, a small resistive thermometer was used as an AFM probe. Photothermally induced temperature fluctuations were detected at the sample surface when illuminated by infrared light from a Michelson interferometer [3]. In this implementation, the specially designed temperature-sensing AFM probe was required, limiting its applications. The relatively large sizes of the temperature-sensing probe also limit the spatial resolution through this temperature measurement approach.

In 2000, Mark Anderson demonstrated a novel approach without temperature-sensing AFM probes. Instead, the photothermal-induced deformations of the sample surface were measured through the mechanical deflection of the AFM cantilever, where the AFM tip and the sample were in constant contact [4]. Figure 6.1a displays the schematics of Anderson's apparatus. Modulated IR radiations from a Fourier transform infrared (FTIR) spectrometer were guided to the tip-sample region of the AFM. The measurement of the frequency-dependent photothermal deformation enabled the collection of IR spectrum through mechanical detection with AFM. Despite its conceptual novelty, this seminal AFM-IR approach has a critical drawback: detecting the photothermal deformation from the low-speed cantilever deflection occupies the same signal channel as the AFM topography feedback. In the perspective of the contact mode AFM, the photothermal deformation of a sample is indistinguishable from the topography change of the sample. In Anderson's measurement, the contact mode AFM feedback was practically switched off to avoid the vertical piezo stage to compensate for the photothermal deformation. However, an AFM topography feedback mechanism is always required to maintain proper tip-sample contact or distances. Without the AFM feedback, small external vibrations from the environment would cause the AFM tip to crash into the sample surface, causing irreversible damages to both the tip and the sample. This drawback severely limits practical applications of the initial design of photothermal AFM-IR.

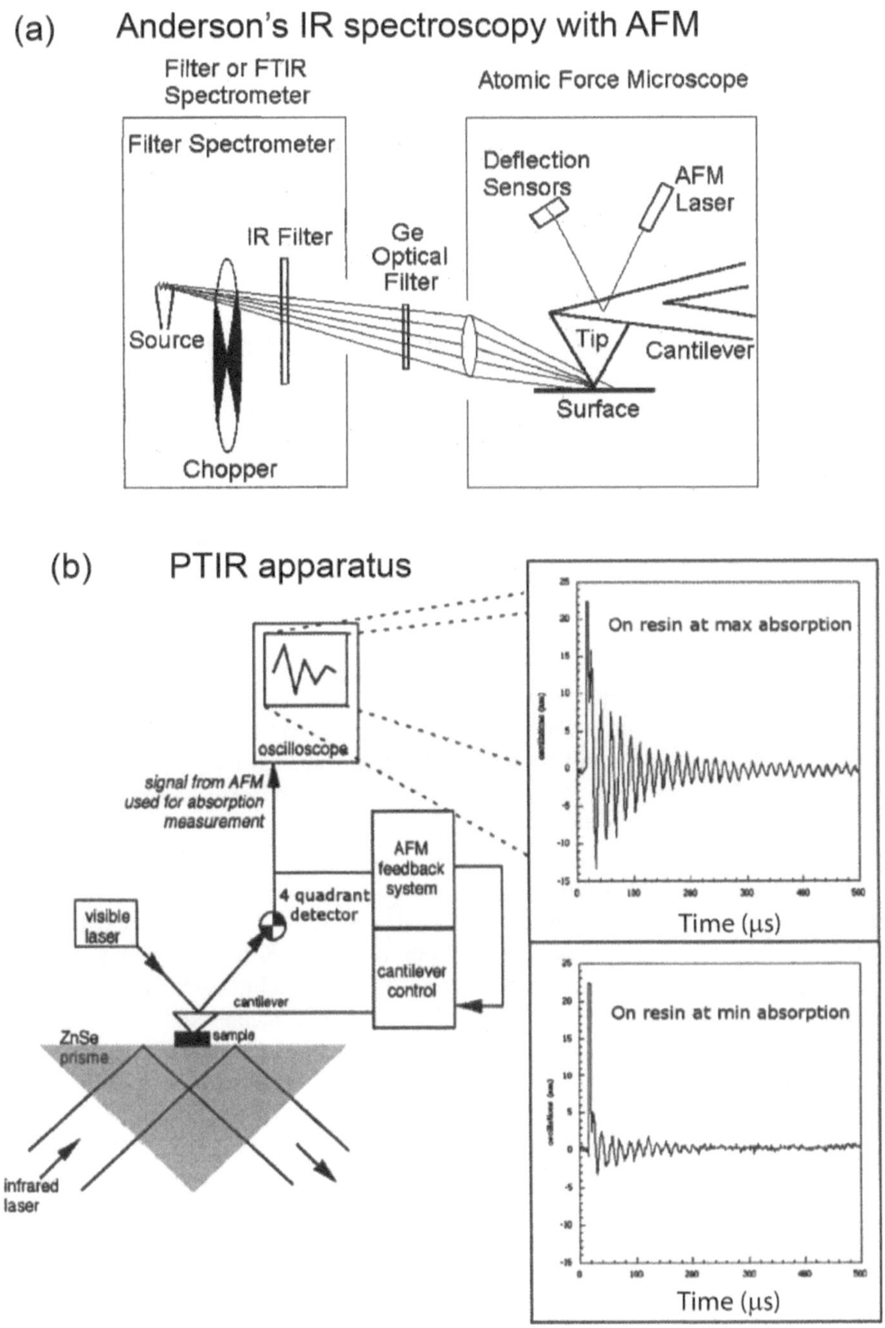


**FIGURE 6.1** (a) The schematics of infrared spectroscopy with AFM with mechanical detection by Mark Anderson. (Reproduced from Ref. [4] with permission.) (b) The operational schematics of AFM-IR through PTIR by Alexandre Dazzi and colleagues. (Reproduced from Ref. [5] with permission.)

A milestone in AFM-IR microscopy is the invention of the photothermal induced resonance (PTIR) methods by Alexandre Dazzi and colleagues in 2005 [5]. In PTIR, photothermal expansion of the sample from absorption of short-duration IR pulses causes high-frequency cantilever oscillations, which are detected as the IR absorption signal. Figure 6.1b illustrates the schematics of the first PTIR apparatus and the cantilever oscillations with on-resonance excitation and off-resonance excitation. IR excitation on the vibrational resonance of the sample causes much stronger cantilever oscillations than the IR is off-resonant with the sample. In PTIR, the contact mode AFM feedback is always on, and the tip effectively follows the sample topography. In PTIR, the cantilever deflection signal for the AFM feedback is different from the cantilever oscillation signal for IR readout through the difference in the time scale. The response time of the AFM feedback that relies on cantilever deflection is much slower than cantilever oscillatory motions (typically in the tens of kHz frequency for PTIR). Consequently, the AFM feedback does not act on the cantilever oscillation through extension or contraction of the vertical piezo stage. The contact mode feedback follows the variations of sample height in the lateral scans because the cantilever deflections caused by height variations are relatively slow and within the feedback loop bandwidth. The photothermal measurement of infrared absorption and the AFM topography feedback are simultaneously performed. Effective AFM topography feedback greatly alleviates the vulnerability of the tip crashing into the sample surface during the photothermal measurement. It is a significant improvement of the AFM-IR compared with Anderson's original design. Moreover, the cantilever oscillations from the photothermal excitation of the sample are detected through Fourier transform at the high frequencies, outside the low-frequency noises from the surroundings, which improves signal quality. The photothermal induced cantilever oscillations signals from PTIR have been demonstrated to be proportional to the IR absorption of the sample, delivering spectra similar to the regular FTIR spectra [6].

The initial PTIR was demonstrated with IR radiations from a free-electron laser [5]. Later, frequency tunable nanosecond duration mid-IR sources based on optical parametric oscillator (OPO) became commercially available, such as NT270 series from Ekspla, Lithuania. The tabletop OPO IR source accelerated the adoption of PTIR as a nano-IR technique. The advantage of PTIR with a frequency tunable IR source is that both imaging and point spectroscopy can be achieved with a single tabletop mid-IR source. The spectroscopy is performed by placing the AFM tip on the location of interest under contact mode feedback, sweeping the frequency of the IR source, and registering the PTIR signal from Fourier transform at one of the cantilever oscillation frequencies. The IR imaging is achieved by fixing the frequency of the mid-IR light source, spatially scanning the AFM tip under the contact mode feedback, and registering the PTIR signal. The ability to acquire IR image and broadband spectroscopy in an AFM coupled to a tabletop IR source is advantageous for applications in research labs. Also, the PTIR instrument is commercially available in the NanoIR product lines of Anasys Instruments (acquired by Bruker in 2018). PTIR has been widely used in the chemical identifications of soft matters, including polymers and biological samples, with applications spanning academic research and industry [6].

Further development of PTIR in the contact mode AFM was achieved with the resonance enhancement. In the initial commercially available PTIR instrument, the repetition rate of the mid-IR laser pulses is limited to the availability of the light source. For example, the NT270 series OPO from Ekspla has a repetition rate of 1 kHz, much lower than the cantilever mechanical resonant frequency. The invention of quantum cascade laser (QCL) as a mid-infrared light source enables tabletop infrared pulses with a tunable repetition rate up to MHz. In 2014, Mikhail Belkin and coworkers introduced the resonantly enhanced infrared photoexpansion nanospectroscopy (REINS) technique into the AFM-IR family [7]. REINS is also colloquially known as the resonant enhanced PTIR or resonant AFM-IR. In REINS, the AFM operates in the contact mode, and the repetition rate of a QCL is tuned to match a mechanical resonance of the AFM cantilever. The mechanical mode of the cantilever is resonantly driven by periodical photothermal expansions of the sample, which gains the oscillation amplitude according to the quality factor of the AFM cantilever. Monolayer sensitivity is achieved with ~25 nm spatial resolution by REINS. In comparison, the typical spatial resolution of non-resonance enhanced regular PTIR is between 50 and 100 nm [6]. Furthermore, REINS and resonance enhanced PTIR have been applied to nano-imaging in the fluid [8,9]. Resonance enhanced PTIR delivers a significant signal improvement over the early PTIR imaging in the aqueous phase [10].

On the other hand, the drawbacks of AFM contact mode also affect the PTIR applications. The AFM tip has to be always in contact with the sample surface under a load. Lateral scans of the AFM tip in the contact mode exert lateral forces to the sample, resulting in scratching the sample surface. In addition, rough sample surfaces with abrupt height changes are difficult for the tip to follow, often causing both tip wear and surface damage. In AFM-IR, IR absorptions occasionally soften the sample surface and cause photothermal deformations, aggravating lateral scratches by the AFM tip.

AFM-IR methods beyond the contact mode are needed to preserve sample integrity. In 2016, Photo-induced Force Microscopy (PiFM), invented by H. Kumar Wickramasinghe, was introduced into the AFM-IR community [11,12]. PiFM operates in the non-contact mode of AFM, or equivalently the attractive regime of the tapping mode, which preserves the integrity of the sample surface. The schematics of a typical PiFM setup are illustrated in Figure 6.2a. A frequency tunable IR laser with variable repetition rate, i.e., a QCL, is used as the IR source. The IR beam is steered and focused into the tip-sample region of a tapping mode AFM. A parabolic mirror is commonly used in PiFM to focus the beam without frequency dispersion. The cantilever deflection signal from the quadrant photodiode of the AFM is routed into a lock-in amplifier for signal demodulation.

PiFM utilizes two mechanical eigenmodes of an AFM cantilever. The pair of first two mechanical eigenmodes for a cantilever has the relationship of $\Omega_2 \cong 6.3\,\Omega_1$, with $\Omega_1$ and $\Omega_2$ representing the first and second modes. The operation procedure of a typical IR PiFM is as follows:

1. The tapping mode feedback operates on driving one of the two modes (e.g., $\Omega_2$) in the attractive regime of intermolecular force;

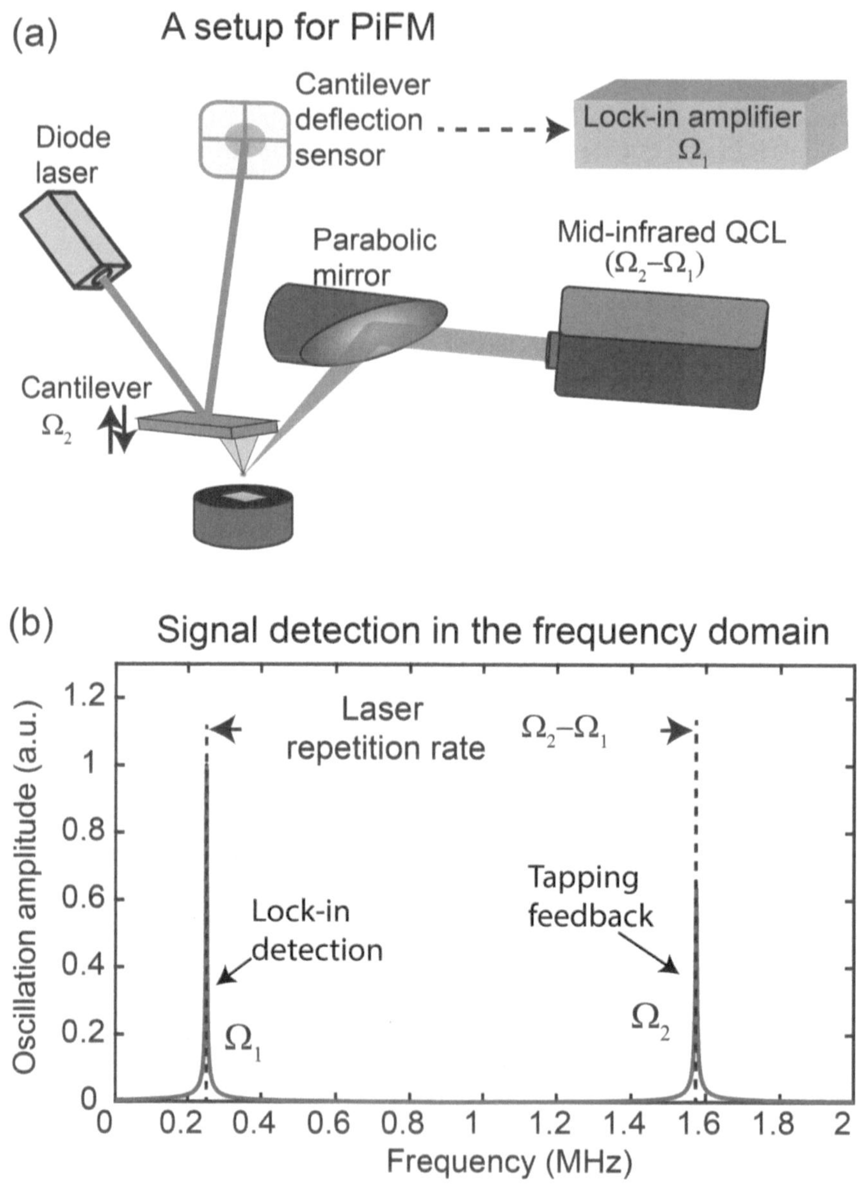


**FIGURE 6.2** (a) A schematic of PiFM. (b) Signal detection scheme of PiFM in the frequency domain.

2. The repetition rate of the infrared laser, i.e., from a QCL, matches the frequency difference $(\Omega_2 - \Omega_1)$ between the two mechanical modes;
3. The cantilever deflection waveform from the quadrant photodiode is routed to a lock-in amplifier to detect the oscillations from the other mechanical eigenmode.

PiFM employs a signal detection scheme of force heterodyne [13]. A mechanical signal is generated at $\Omega_1$ through heterodyne between the cantilever oscillation frequency at $\Omega_2$ and the modulated photo-induced force at the laser repetition rate at $(\Omega_2 - \Omega_1)$, as illustrated in Figure 6.2b. This innovative design of the operational principle is that the mechanical resonance of the cantilever amplifies the heterodyned signal of the same frequency. A high-quality factor of the AFM cantilever retains the heterodyne signal at $\Omega_1$ and amplifies it. Unwanted background signals, primarily through the periodical photothermal responses of the AFM cantilever by direct IR illumination at the laser repetition rate $(\Omega_2 - \Omega_1)$, do not match the mechanical resonance of the cantilever and are not amplified. IR PiFM has similar operation modes as the PTIR microscopy. IR imaging is performed by scanning the tip over the sample with tapping mode and co-registering the lock-in demodulation signal. Broadband spectroscopy at a location of interest is achieved by recording the lock-in demodulation while sweeping the IR frequency. IR PiFM has been demonstrated to reach < 10 nm spatial resolution over a range of samples [12,14–16]. The detection sensitivity of IR PiFM has been demonstrated to reach monolayer [17].

An intriguing aspect of PiFM is the nature of the photo-induced force. An attractive force is generated under light illumination when the tip and the sample are close but not in direct contact [18]. The initial model of PiFM attributed the photo-induced force to dipole-dipole interaction [11,19]. However, this dipole-dipole model would predict the dispersive line shape of PiFM spectra across a resonance [20] that contracts with the absorptive line shape of IR PiFM spectra. Later, the photothermal expansion-mediated van der Waals force model [18,21] and the optomechanical damping model [22] were respectively introduced to interpret the nature of the photo-induced force. It is possible that both induced dipole and photothermal expansion are present in PiFM [23]. In the IR regime, the signal contribution from the photothermal expansion is more dominant than the induced dipole contribution, particularly on soft matters [21].

Inspired by PiFM, a variant of tapping mode AFM-IR is also developed to achieve similar spatial resolution and performance. The tapping mode AFM-IR operates on the repulsive regime of the tapping mode with intermittent contact between the tip and the sample. A similar force heterodyne scheme to PiFM is used in tapping mode AFM-IR, with the laser repetition rate tuned to match the frequency difference between two mechanical resonance modes of the cantilever. The tapping mode AFM-IR directly measures the photothermal expansion from the sample through direct tip-sample contact, similar to that of the PTIR. Although the experimental hardware and signal detection scheme of tapping mode AFM-IR is essentially the same as PiFM, there is a small difference. PiFM operates in the attractive regime of the intermolecular force in the tapping mode; the tapping mode AFM-IR operates in the repulsive regime of the intermolecular force. This subtle difference means different operational mechanisms. The force gradient in the repulsive regime of the tapping mode is stronger than that in the attractive regime. Because the presence of the force gradient shifts the mechanical resonance of the cantilever, the cantilever resonance is shifted much more in the repulsive regime than in the attractive regime. The intermittent contact in the repulsive regime also means that the mechanical properties of the local area (modulus and adhesion) contribute to the shift of the effective cantilever resonant frequency. On the other hand, effective force heterodyne requires the repetition rate of the laser to match the exact frequency difference between the two mechanical resonant modes: one of which is driven under external fixed frequency for tapping mode feedback; another one shifts on different regions of the sample. A frequency tracking mechanism is often used in the tapping mode AFM-IR to follow the effective cantilever resonance and dynamically adjusts the repetition rate of the IR laser. This extra requirement results in additional complexity for the tapping mode AFM-IR than PiFM. In comparison, the shift of effective cantilever resonance frequency is negligible in the attractive regime. The frequency-tracking mechanism is not needed for PiFM operated in the attractive regime of intermolecular force.

Peak force tapping (PFT) mode [25], also known as the pulsed force mode [26], is an emerging AFM feedback mode with versatile applications. In PFT mode, the distance between the AFM tip and the sample surface is modulated at a low frequency (e.g., 2 kHz), much lower than the cantilever resonance. At the turning point of the modulation, the AFM tip indents into the sample surface to a maximal force. A feedback loop is employed to regulate the maximal indentation force as a process variable with a user-defined set point by adjusting the vertical piezo stage. Compared with the regular tapping mode, the PFT mode provides a deterministic temporal region of tip-sample contact. The PFT mode also avoids the building up of high lateral force between the tip and sample during the lateral scan, as the AFM tip is detached from the sample surface in every oscillation cycle. Thus, the surface deformation from the lateral force is effectively avoided.

The PFT mode provides another route to achieve high spatial resolution AFM-IR that preserves surface integrity. The resulting method is the peak force infrared (PFIR) microscopy, developed by Xiaoji Xu and colleagues in 2017 [24]. Figure 6.3a illustrates the operational principle of PFIR microscopy. A PFIR microscope consists of an AFM operating in the PFT mode, an IR light source (e.g., QCL) that allows pulse emission by an external trigger, and a data acquisition (DAQ) device. In PFIR, a voltage waveform at the frequency of the PFT is routed from the AFM controller or the vertical piezo stage. The waveform is routed into a phase lock loop (PLL) to generate a transistor-transistor logic (TTL) waveform that is phase synchronized to the PFT oscillation. The TTL waveform is used to trigger the IR source to emit one or multiple IR pulses [27]. By adjusting the timing of the laser trigger from the PLL, the emission of the IR pulse is synchronized to the temporal regime of tip-sample contact in a PFT cycle. The exact moment of IR illumination is typically chosen to be right after the maximal cantilever deflection. So the process variable for the PFT feedback is not perturbed by subsequent IR photothermal responses. The cantilever deflection detects the photothermal expansion of the sample under the tip-enhanced IR illumination through the direct force from tip-sample contact (Figure 6.3b). The photothermal expansion has two effects on the cantilever deflections: first, rapid photothermal expansion effectively excites the mechanical resonance of the cantilever, causing PTIR-type oscillations; second, long-term photothermal volume expansion causes an offset in the cantilever deflection. The extracted photothermal effects are displayed in Figure 6.3c. In PFIR, the cantilever deflection waveform after the

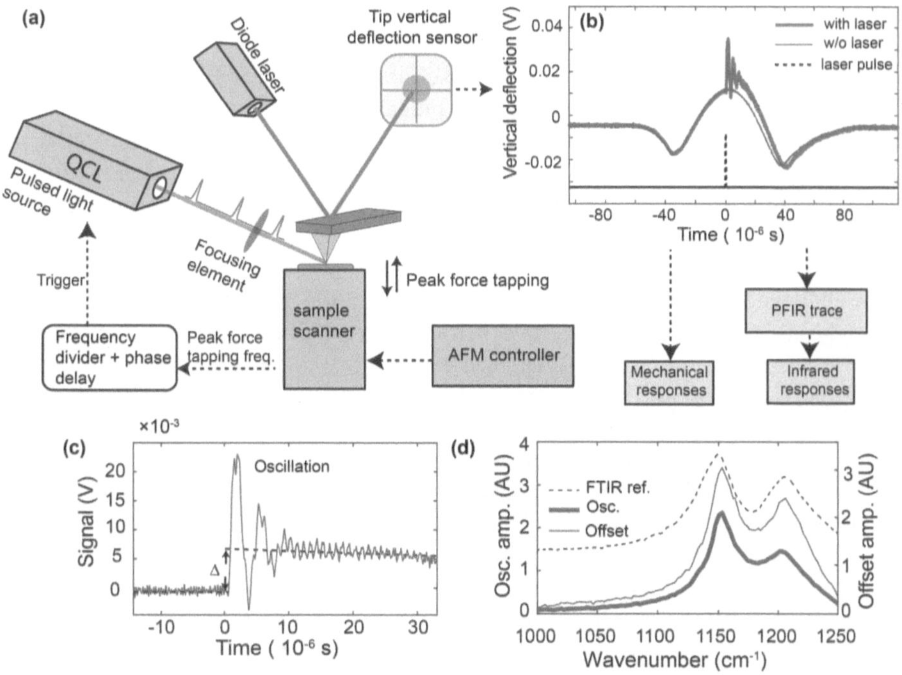


**FIGURE 6.3** (a) Schematics of a PFIR microscope. (b) Cantilever deflection waveform with IR laser illumination (thick curve), without illumination (thin curve), and the timing of the laser pulse (vertical dashed line). (c) A PFIR trace is obtained from the subtraction of cantilever deflection curves. (d) PFIR spectra of PTFE polymer obtained through frequency-dependent cantilever oscillation amplitude (thick curve) and magnitude of baseline offset (thin curve). The FTIR reference from the bulk polymer sample is included as a dashed line. (Figure reproduced from Ref. [24] with permission.)

laser illumination is extracted with a digitalization card (e.g., PXI-5122 National Instruments). The amplitude of the fast cantilever oscillations is extracted with a fast Fourier transform. The integration window in the frequency domain from the fast Fourier transform is chosen to be wide enough to accommodate a slight shift of the cantilever oscillation frequency on the heterogeneous sample surface. The integrated amplitude of the cantilever oscillations in the frequency domain is used as the PFIR signal. Similar to PTIR and PiFM, PFIR microscopy delivers both chemical imaging and broadband spectroscopy through scanning the sample at fixed IR frequency or sweeping the IR frequency at a fixed spatial location.

If the frequency of the phase-synchronized TTL waveform is divided by two and used to trigger the IR source, half of the cantilever deflection waveforms carry the photothermal response; the other half of deflection waveforms do not. The two types of cantilever deflection waveforms can be subtracted to obtain the PFIR trace—the cantilever response to the photothermal response of the sample, without the slow curvature from the cantilever deflection of PFT feedback. Fourier transform of the

PFIR trace reveals the photothermal induced oscillations, similar to the readout of PTIR. The long-term photothermal volume expansion, which is absent from contact mode PTIR, can be retrieved from the baseline offset of the PFIR trace before and after illumination from the IR pulse. The magnitude of the baseline offset can also be used as a quantity to represent the IR absorption signal. Figure 6.3d shows the PFIR spectra collected from the oscillation channel and the offset channel for the PTFE polymer, which show good agreement with the FTIR measurement of the bulk sample. Since its invention, PFIR has been applied to many samples, from soft matters, polaritonic materials to photovoltaics [27–31].

A challenge for infrared microscopy is the operation in the aqueous phase. Water strongly attenuates lights in the mid-infrared range. The attenuated total internal reflection (ATR) geometry FTIR can access the infrared absorption from the liquid/solid interface through the frequency-dependent attenuation of the IR evanescent wave. However, regular micro-ATR-FTIR spectroscopy suffers from low spatial resolution [32]. In comparison, AFM-IR represents an alternative approach to the optical detection of ATR-FTIR microscopy. In 2008, a proof-of-principle experiment was done by Alexandre Dazzi and colleagues to measure a biological cell in aqueous condition with PTIR [10]. In 2017, Mikhail Belkin and coworkers utilized REINS in the total internal reflection geometry to image polymer film in heavy water [8]. In 2018, Andrea Centrone and coworkers used the resonantly enhanced PTIR to measure the secondary structures of amyloid fibrils in the aqueous condition [9]. The PFIR microscopy with total internal reflection geometry was also developed [33]. Its extension in fluid allows for nanoscopy of soft matters and polaritonic materials [34,35]. A common challenge for fluid-phase AFM-IR operation is the hydraulic drag to the AFM cantilever motions. Cantilever oscillations are quickly damped, including the oscillation caused by IR photothermal responses. Comparably speaking, the PFIR microscopy suffers the least from the hydraulic drag among existing AFM-IR methods. The mechanical detection of the cantilever deflection is performed right after the IR illumination. The AFM cantilever does not need to maintain long-lasting oscillations. In the liquid phase PFIR, the sample was also excited with multiple IR pulses per PFT cycle to increase the total detectable photothermal responses.

Besides the IR measurement, the AFM-IR microscopy may also provide complimentary mechanical information. In the PFIR microscopy, the PFT mode allows simultaneous collection of the force-distance curve, which are used to directly extract mechanical properties of modulus and adhesion, through the built-in PeakForce QNM™ functionality from Bruker [25]. In contact mode PTIR, the shift of the contact resonance frequencies of the AFM cantilever allows for the extraction of relative mechanical properties of the sample: harder surface exhibits an increase of the contact resonance frequency; softer surface sees a reduction. Similar indirect mechanical property measurement is also possible by tracking the shift of the effective resonant frequency of the cantilever in the tapping mode AFM-IR, in which the tip intermittently touches the sample surface.

As a summary of AFM-IR methods, comparisons of PTIR, resonance enhanced PTIR, PiFM, and PFIR are summarized in Table 6.1.

**TABLE 6.1**
**Comparisons of AFM-IR Techniques**

| | PTIR | REINS/Resonance Enhanced PTIR | PiFM/Tapping Mode AFM-IR | PFIR |
|---|---|---|---|---|
| Parent AFM mode | Contact mode | Contact mode | Non-contact/ tapping mode | PFT mode |
| Signal origin | Photothermal | Photothermal | Photo-induced force/ photothermal | Photothermal |
| Sample damage due to tip scratch | Possible | Possible | Low | Low |
| Typical spatial resolution | 50 ~ 100 nm | 25 ~ 50 nm | ~10 nm | ~10 nm |
| Complimentary mechanical property | Mechanical through contact resonance | Mechanical through contact resonance | Not available for PiFM, possible for tapping mode AFM-IR | Quantitative and direct by PeakForce QNM |
| IR source | OPO, QCL | QCL | QCL | QCL, OPO |
| The repetition rate of IR source | ~$10^3$ Hz | ~$10^5$ Hz | $10^5$ ~ $10^6$ Hz | ~$10^3$ Hz |

## 6.2 APPLICATION OF AFM-IR IN ENERGETIC MATERIALS

The AFM-IR techniques through mechanical detection of photothermal response from IR absorption are suitable for the measurement of samples with high thermal expansion coefficients. Within the topics of energetic materials, PiFM and PFIR have been demonstrated for measurement on polymer-based OPV, perovskite photovoltaics, as well as organic matters in oil shale source rock.

The nanoscale chemical imaging capability of AFM-IR is particularly suitable for studying heterogeneous materials, in particular, in samples that nanoscale spatial features affect their performance, such as polymer-based OPV. PiFM has been demonstrated to reveal the chemical heterogeneity of OPV. A joint research work by Sung Park and Zhenan Bao and colleagues reveal the donor and acceptor domains of the PII-2T-PS:PPDI-T all-polymer OPV [36]. Chemical nanoscopy with PiFM proves that the addition of 1-chloronaphthalene modifies the domain sizes of OPV. The presentative images are displayed in Figure 6.4a–b. PiFM also works on perovskite photovoltaic samples based on the chemical contrast between degraded and undegraded photovoltaic regions. Figure 6.4c–e illustrates one of the representative results. PFIR microscopy has also been utilized for IR imaging of photovoltaic materials. A demonstration of PFIR imaging CsFAMA perovskite film is shown in Figure 6.4f–h.

On fossil energy sources, AFM-IR techniques have been applied to the oil shale source rock. Contact mode AFM-IR has been utilized to image the organic matters extracted from oil shale source rock to reveal the chemical species [38]. Recently, PFIR microscopy was used to image the kerogens from the nanopores of

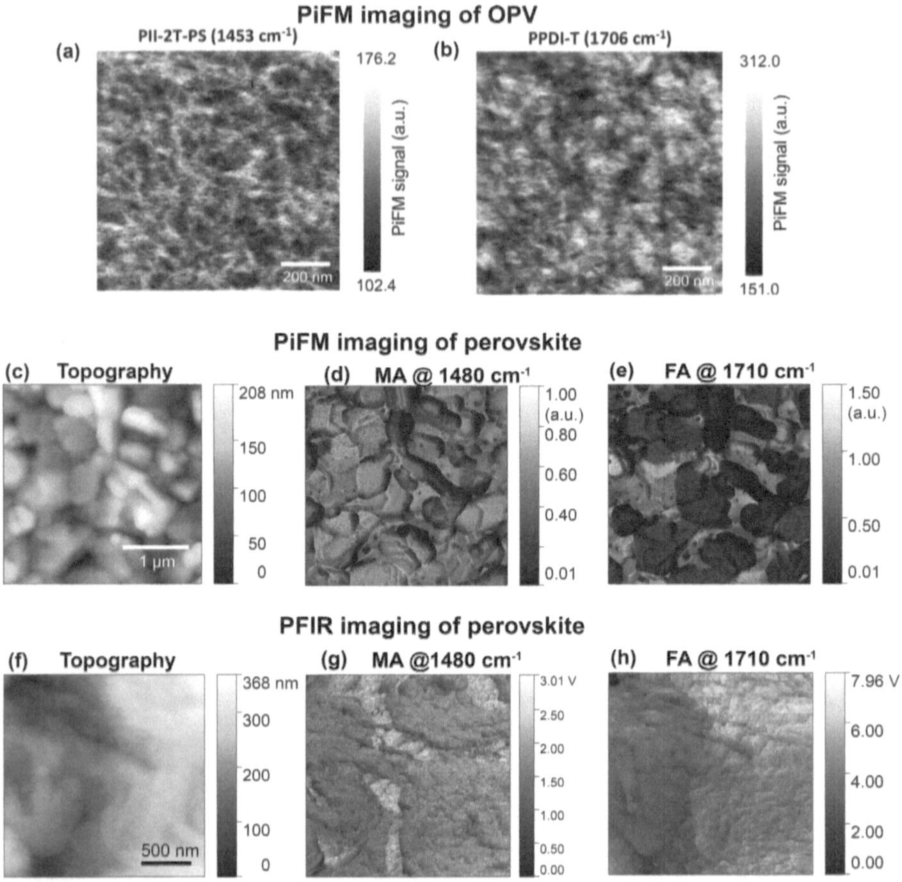


**FIGURE 6.4** (a–b) PiFM imaging of nanoscale donor/acceptor domains of OPV at characteristic IR absorption frequencies. (Reproduced from Ref. [36] with permission.) (c–e) Topography and PiFM images of CsFAMA perovskite film. (f–h) Topography and PFIR images of another CsFAMA perovskite film. MA-rich domains are revealed at 1,680 cm$^{-1}$ and FA-rich domains are revealed at 1,710 cm$^{-1}$. Both PiFM and PFIR images were collected by Devon Jakob and used with permission.

the oil shale source rock with both chemical imaging and mechanical mapping [37]. Representative PFIR measurements of oil shale source rock with both IR imaging and spectroscopy are included in Figure 6.5.

## 6.3 SCATTERING-TYPE SCANNING NEAR-FIELD OPTICAL MICROSCOPY

Complimentary to AFM-IR's mechanical detection of the photothermal responses, the optical detection of the near-field scattered light by AFM tip represents another feasible route to bypass Abbe's diffraction limit. The apex of a sharp metallic tip can act as an optical antenna that locally enhances the electromagnetic field, with a spatial

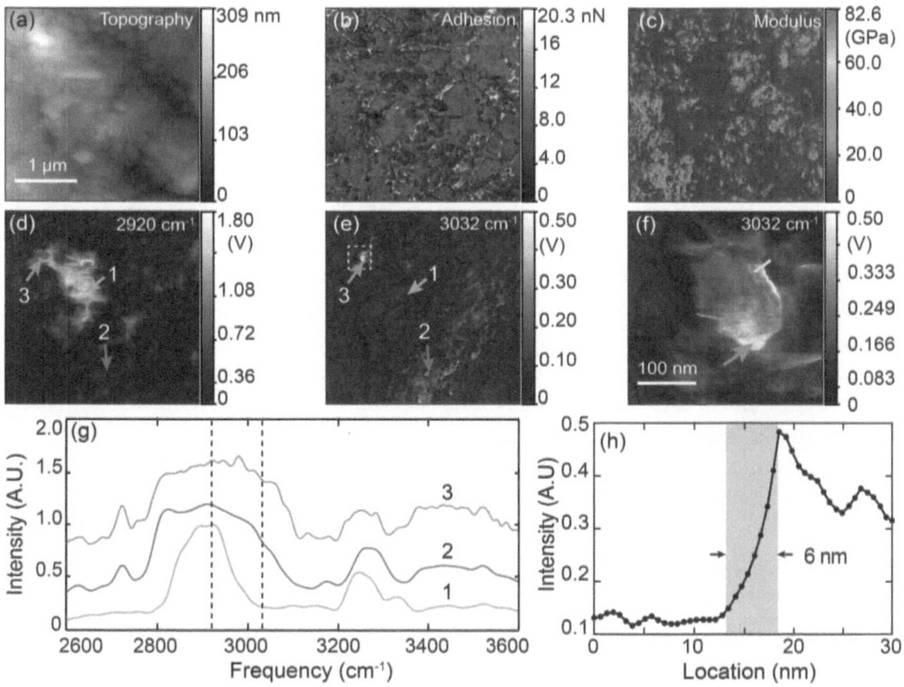


**FIGURE 6.5** PFIR-acquired images and point spectra of an immature 3 μm×3 μm Eagle Ford source rock sample. (a) Topography of the source rock. The scale bar is 1 μm. (b) Adhesion between the AFM tip and the sample. (c) Modulus map of the source rock sample surface. (d) A PFIR image taken at 2,920 $cm^{-1}$, indicating the presence of saturated hydrocarbon compounds. (e) A PFIR image taken at 3,032 $cm^{-1}$, characteristic of unsaturated compounds. (f) A 330 nm×330 nm PFIR image taken at 3,032 $cm^{-1}$ positioned within the location of the white box in panel e. (g) Point spectra taken from chosen areas on the surface as indicated by the colored arrows in the accompanying PFIR images. (h) A spatial resolution of 6 nm is observed from a cross-section in (f), indicated by the yellow line. (The figure is reproduced from Ref. [37] with permission.)

scale much smaller than the half wavelength of the light [39]. The same antenna also scatters the nearfield of electromagnetic waves with high spatial frequencies not propagating through far-field radiation. If a sample contains a resonance that matches the light frequency, elastic scattering from the tip-sample junction carries the spectroscopic signatures of the resonance. Thus, detection of the scattered light reveals the presence of resonance in the sample. The AFM-based imaging methods through the detection of elastically scattered light form the basis of the Scattering-Type Scanning Near-Field Optical Microscopy (s-SNOM). s-SNOM is also known as apertureless near-field scanning optical microscopy, because it emerged to prominence after the fiber-based near-field scanning optical microscopy that has a resolution limit due to the finite size of the fiber aperture. Modern s-SNOM microscopy routinely delivers IR nanoscopy at sub-20 nm spatial resolution, which is described in this section.

The early concept of scattering-type near-field microscope was conceived by an Irish scientist Edward Synge and documented in correspondence to Albert Einstein in 1928 [40]. Synge proposed the scheme of spatially manipulating the position of a tiny metal particle and detecting its scattered light as a route to overcome the diffraction limit of optical imaging. The idea was well ahead of its time and never realized in Synge's lifetime. The modern realization of s-SNOM with IR light was first achieved by Claude Boccara and coworkers in 1995 [41]. s-SNOM has gained wide popularity after a series of key technology developments by Fritz Keilmann, Rainer Hillenbrand, Renaud Bachelot, Thomas Taubner, Dmitri Basov, Gilbert Walker, and others [42,43–51]. By the early 2020s, s-SNOM has become one of the most successful AFM-based infrared microscopy techniques for investigations of low-dimensional materials, polariton structures, semiconductors, as well as chemical imaging of soft matter. Successful and early commercialization of the s-SNOM instrument by Neaspec facilitates its popularity.

The theoretical treatment of near-field light scattering in s-SNOM is based on modifying the effective polarizability of a metallic tip due to the presence of the sample underneath. A qualitative but informational model is the image dipole model. The image dipole model treats the half sphere of the tip as a dipole and causes the redistribution of charges at the sample surface, schematically illustrated in Figure 6.6a. Equation (6.1) mathematically describes the image dipole model [44,52]:

$$\alpha_{\text{eff}}(\omega) = \alpha\left(1 - \frac{\alpha\beta}{16\pi(r+d)^3}\right)^{-1} \tag{6.1}$$

where $r$ is the tip apex radius and $d$ is the tip-sample distance. $\alpha$ is the polarizability of the tip in the free space without sample underneath, which is equal to $4\pi r^3 \frac{(\varepsilon_t - 1)}{(\varepsilon_t + 2)}$. $\beta = \frac{\varepsilon(\omega)-1}{\varepsilon(\omega)+1}$ is a coefficient to satisfy the boundary condition required by solving Maxwell equations. $\varepsilon_t$ is the complex dielectric function of the AFM tip, and $\varepsilon(\omega)$ is the frequency $\omega$-dependent complex dielectric function of the sample. The essence of this model is that the resonance-dependent dielectric function of the sample affects the effective polarizability of the tip $\alpha_{\text{eff}}(\omega)$, which determines light scattering. When the sample has resonances, the complex $\varepsilon(\omega)$ contains the imaginary part, which leads to the presence of both real and imaginary parts in the effective polarizability $\alpha_{\text{eff}}(\omega)$. Numerical plot of the real and imaginary parts of $\alpha_{\text{eff}}(\omega)$ are calculated from Equation (6.1) and shown in Figure 6.6b and c, respectively. The details of the simulation parameters can be found in Ref. [53]. An interesting condition for the image dipole model is for samples with the real part of $\varepsilon(\omega)$ equaling to $-1$, often found in samples with strong phonons. In this case, the value of $\beta$ is significantly increased, leading to a dramatic enhancement of the near-field scattering signal [46]. Note that the image dipole model underestimates the interaction range between the tip and the sample and leads to a discrepancy between the experimental data and numerical simulation. More elaborate models have been developed to improve accuracy [54,55].

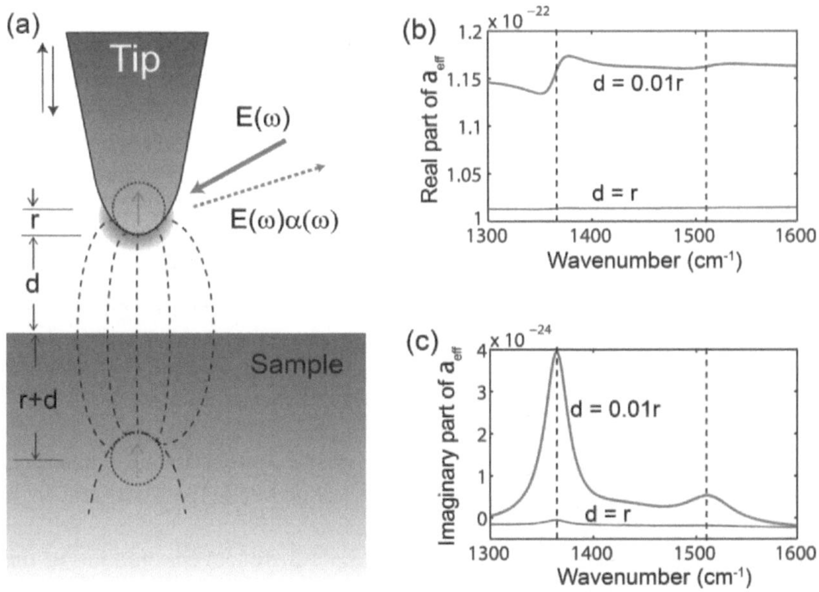


**FIGURE 6.6** (a) A schematic illustration of the image dipole model for near-field scattering. (b) The frequency-dependent real part of the effective tip polarizability, calculated from the image dipole model with $d=r$ (thin curve) and $d=0.01\ r$ (thick curve). (c) The frequency-dependent imaginary part of the effective tip polarizability, calculated from the image dipole model. The imaginary part has the absorption profile, peaked at IR resonances, marked by vertical dashed lines. A close tip-sample distance increases both the real and the imaginary part of the polarizability, while the real part remains the dominant one.

The sample IR resonance is encoded in its frequency-dependent dielectric function $\varepsilon(\omega)$ to modify the effective polarizability $\alpha_{\text{eff}}(\omega)$ of the tip. As a result, the amplitude and phase of the scattered light acquire the frequency dependency of the sample resonance. However, the intensity spectrum of the tip scattered light assumes a dispersive profile across the sample resonance, which is dominated by the contribution from the real part of $\alpha_{\text{eff}}(\omega)$. In contrast, regular FTIR spectroscopy and AFM-IR measure the dissipation of IR light by the sample, which originates from the imaginary part of $\varepsilon(\omega)$, and has a similar profile in the imaginary part of $\alpha_{\text{eff}}(\omega)$. The directly measured spectrum of tip-scattered light gives a different profile from the familiar FTIR absorption spectra. In chemical identification, the lack of direct spectral correspondence to FTIR absorption spectra represents the first hurdle for s-SNOM for IR nanoscopy.

Moreover, there is a signal background from the scattered light directly from the tip cone region outside the tip-sample junction. The focused spot size of IR laser by a parabolic mirror is often tens of microns due to the diffraction limit. The reflections and scattering from the relatively large tip cone contribute to and often dominate the IR signals at the optical detector, which are not from the tip-sample junction. This type of unwanted background signal is colloquially referred to as the far-field background. Such a far-field background does not carry the information of the sample resonance underneath the AFM tip; thus, it has to be filtered out. Additional optical and signal processing techniques are used to obtain meaningful near-field scattering signals from the tip-sample junction. An s-SNOM instrument provides such a capability.

The core elements of an s-SNOM instrument are illustrated in Figure 6.7. It consists of an AFM, an asymmetric Michelson interferometer, an IR source, and an IR detector attached to a lock-in amplifier or digitalization device. The AFM is typically operated in the tapping mode; IR emission from the source is guided and focused to the tip-sample region of the AFM; the scattered light is interferometrically amplified and guided into the IR detector. An MCT detector (HgCdTe) is typically used for the mid-IR radiation.

The tapping mode AFM allows the lateral scan of the tip position and the vertical modulation of the tip-sample distance. Vertical modulation of the AFM tip is needed because of the elastic scattering nature of s-SNOM: the photon frequency does not change, and the near-field signal from the tip-sample junction has the same frequency as the far-field background. On the other hand, the near-field signal is strongly and nonlinearly dependent on the tip-sample distance, whereas the far-field background is linearly dependent on the tip-sample distance. Modulation of the tip-sample distance allows for the extraction of near-field scattering signals from the far-field background, based on their dependence on the tip-sample distance. In s-SNOM, the cantilever of the tapping mode AFM is driven at its mechanical resonance, typically the first mechanical resonance of the cantilever. The tapping frequency is used to provide a reference frequency to the lock-in demodulation. The key practice of s-SNOM of lock-in demodulation is to utilize a non-fundamental harmonic of the

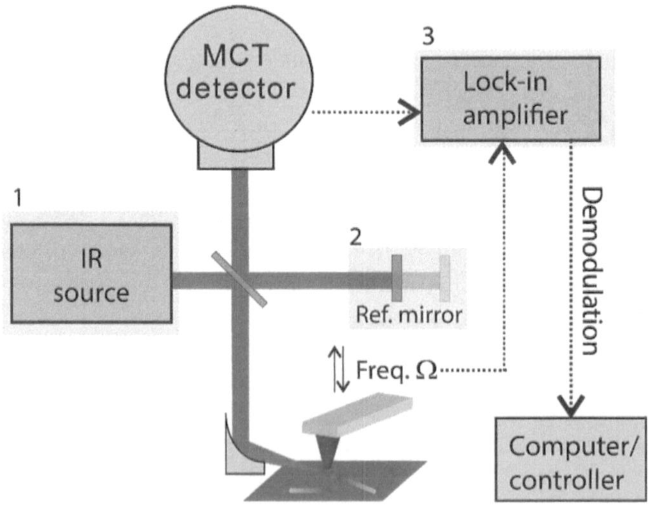


**FIGURE 6.7** (a) A typical s-SNOM instrument. Depending on specific types of s-SNOM methods, combinations of the light source (highlighted as region 1), reference modulation scheme (highlighted as region 2), and lock-in detection frequencies (highlighted as region 3) are used.

reference frequency [43,44,56]. For example, if the cantilever oscillates at ~250 kHz (reference frequency), a demodulation frequency at 500 kHz (the second harmonic) or 750 kHz (the third harmonic) is typically used. Why is the non-fundamental harmonic demodulation used for s-SNOM? The far-field background signal is approximately linearly dependent on the tip-sample distance. The harmonic oscillation of the AFM tip results in a harmonic modulation of the far-field background signal, which is almost absent in the non-fundamental lock-in demodulation. On the other hand, the near-field signal has a strong nonlinear dependence on the tip-sample distance. Harmonic oscillation of the AFM tip over the sample causes appearance of non-fundamental harmonic in the lock-in demodulation. Therefore, non-fundamental harmonic demodulation extracts a part of the near-field signal without the far-field background. Here, noted that the near-field signal also co-exists with the far-field background in the first harmonic demodulation that is typically discarded. If all amplitude and phase from lock-in harmonics are collected, the Fourier synthesis method of vertical reconstruction of near-field interactions can recover the near-field scattering signal vs. tip-sample distance relationship in s-SNOM [57].

Besides AFM, the asymmetric Michelson interferometer is another central component of s-SNOM. Interferometric detections are used to recover the IR absorption from near-field scattering and to amplify the near-field signals. The asymmetric Michelson interferometer with the IR detector is constructed as follows:

1. The radiation from the IR source is split into two replicas by a beam splitter;
2. One beam is guided and focused into the tip-sample region by a parabolic mirror or an IR lens;
3. The scattered light from the tip-sample region of the AFM is collected by the same focusing element and guided back to the beam splitter;
4. The other replica of the beam travels through a reference arm and is retro-reflected by a reference mirror on a position controlling stage (piezo stage or motorized stage);
5. The replica of reflected light is collinearly combined with the scattered light from the tip-sample region;
6. The combined beam reaches the IR detector, and the intensity of the IR beam is converted to an electrical signal and routed to a lock-in amplifier for demodulation.

The way how the asymmetric interferometer is used distinguishes different s-SNOM methods. Homodyne detection, pseudoheterodyne detection [42], and nano-FTIR [48,49] are three widely used s-SNOM methods. The homodyne and pseudoheterodyne detections are used for spectroscopic imaging, i.e., obtain near-field images at fixed IR frequency; nano-FTIR is used to collect broadband spectrum comparable to the traditional FTIR absorption.

Homodyne detection is straightforward to implement. The replica of the IR beam from the reference arm amplifies the near-field scattered light from the AFM tip-sample junction after they are combined and reach the IR detector [44,58,59]. The near-field scattered light is selectively amplified by the strong reference field depending on their phase difference, which can be adjusted by moving the position of the reference mirror with a piezo stage. In the frequency domain, a spectral phase shift across the sample resonance exists in the near-field scattered light from the tip-sample junction. In a nutshell, sample resonance causes near-field scattered light by the tip to gain an additional phase. The total amplitude of the near-field scattered light can be obtained with quadrature phase-shifted homodyne [47,60]. The reference mirror is modulated back and forth in two positions with a length difference of $\lambda/8$, in which $\lambda$ is the wavelength of the narrowband IR radiation. This modulation corresponds to quadrature phase difference ($\pi/2$ in optical phase) after retroreflection by the light. Two lock-in demodulation signals (e.g., $S_1$ and $S_2$) from non-fundamental harmonic demodulation at these two quadrature positions are collected, and a total amplitude $S$ is calculated as $S = \sqrt{S_1^2 + S_2^2}$. This total amplitude is independent of the relative optical path difference between the two interferometer arms.

A notable special homodyne condition is the $\pi/2$ homodyne condition, in which the optical path difference between the two interferometer arms is the odd integer multiple of $\lambda/4$ or equivalently $\pi/2$ in phase. The $\pi/2$ homodyne condition selectively amplifies the near-field scattered light from the imaginary part of the effective polarizability of the tip-sample junction. The underlying reason for the $\pi/2$ homodyne is as follows. A metallic tip does not typically contain resonances. The imaginary part of the effective polarizability originates from the sample resonances. The sample resonance causes phase retardation of $\pi/2$ in the near-field scattered light. Since the imaginary part of the effective polarizability has the same absorptive profile as the regular FTIR spectra, the s-SNOM response obtained from $\pi/2$ homodyne condition corresponds to the IR absorption response. How to set the homodyne phase to $\pi/2$ homodyne condition? A simple way is to use a non-resonant region of the substrate as a reference and then adjust and maintain the position of the reference mirror to minimize the lock-in demodulation signal. This reference mirror position satisfies the $\pi/2$ homodyne condition [53]. If the optical path of the asymmetric Michelson interferometer is balanced, i.e., the optical paths are the same for both arms, then the $\pi/2$ homodyne condition can be maintained across a range of IR frequencies through small adjustment of the piezo stage to account for the wavelength difference. The direct homodyne configuration of s-SNOM requires extremely high stability of the interferometers, which is challenging for a regular home-built apparatus. An active stabilization mechanism for homodyne detection s-SNOM was developed at additional instrument complexity [50].

Pseudoheterodyne detection is an elegant and widely adopted technique for narrowband s-SNOM imaging, developed by Rainer Hillenbrand and coworkers [42]. Figure 6.8a describes its implementation in s-SNOM. In pseudoheterodyne, the position of the reference mirror is modulated sinusoidally at a low-frequency M, while the AFM tip oscillates under the tapping frequency of $\Omega$. As a result, the sideband frequency components of $n\Omega \pm m\mathrm{M}$ emerge in the time-varying signal from the IR detector, with $m$ and $n$ being integers equal or larger than 1 and 2, respectively. These sidebands are schematically illustrated in Figure 6.8b. These beating frequencies are the result of short-range near-field interaction between the tip and sample and the optical interference between the near-field signal and phase-modulated reference field. The unwanted far-field background is removed from demodulation from one of $n\Omega \pm m\mathrm{M}$ frequencies. Moreover, the pseudoheterodyne technique allows for the

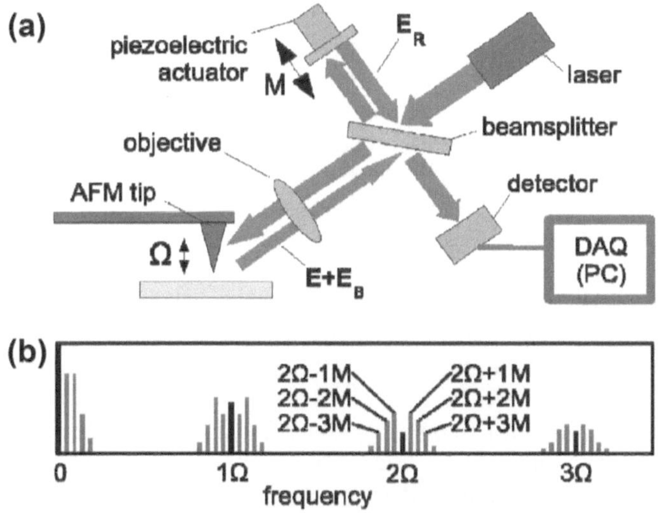


**FIGURE 6.8** (a) Schematics of a pseudoheterodyne signal detection of s-SNOM. (b) A frequency domain spectrum of the detector signal. (Reproduced from Ref. [42] with permission.)

readout of amplitude and phase of the lock-in demodulated signal. If the depth of the modulation of the reference mirror is set to a specific value of $2.63\lambda/4\pi$, then the demodulations at $n\Omega \pm 1\text{M}$ and $n\Omega \pm 2\text{M}$ frequency are proportional to the imaginary and real part of the near-field signal at the *n*th harmonic. The total amplitude and phase can be calculated through the modulus and argument from the real and imaginary parts of the near-field signal. The details on the derivation of this modulation depth are described in Ref. [42]. In practice, how to set this "magic" modulation depth? An empirical way is to apply a slow DC voltage ramp to the reference mirror while watching the calculated amplitude of the near-field from the two sidebands of the pseudoheterodyne detection. If the modulation depth is set correctly to $2.63\lambda/4\pi$, the calculated amplitude should remain constant even if a DC voltage ramp is applied to the piezostage of the reference mirror. A minor limitation of the pseudoheterodyne detection for s-SNOM is its relatively low signal strength from the sideband frequency detection. As a result, longer signal acquisition time is often required for pseudoheterodyne than for homodyne detection.

When the IR radiation is broadband, the asymmetric Michelson interferometer is used analogously to a Fourier transformation spectroscopy, i.e., FTIR. The optical path of the reference arm is scanned over a distance of several centimeters while collecting the near-field signal from non-fundamental lock-in demodulation. During the collection, the lateral position of the AFM tip is fixed, and the AFM is under the tapping mode feedback. The near-field signal versus the optical path of the reference arm forms an *asymmetric* interferogram if the sample contains resonance. Figure 6.9a displays two interferograms from a boron nitride nanotube and a Si substrate [61]. Fourier transform of the interferogram reveals a complex spectrum with both amplitude and phase—or equivalently the real and imaginary parts of the spectrum. Figure 6.9b displays the amplitude and phase of the BNNT IR response from nano-FTIR measurement. The resonances in the sample excited by the IR field remain excited within their lifetime, causing the interferogram to exhibit asymmetric elongation. After Fourier transforms, the asymmetric interferogram leads to a presence of the imaginary component, which is extracted to represent the IR absorption due to sample resonances (Figure 6.9c). The imaginary part of the Fourier transform is equivalent to that of the regular FTIR spectrum. Because an FTIR-equivalent spectrum is collected underneath a nanoscopic AFM tip, the method is termed nano-FTIR. It has become a standard practice for s-SNOM with broadband infrared sources to acquire IR spectra.

How an s-SNOM is operated depends on the type of light sources. Homodyne and pseudoheterodyne require narrowband and Continuous Wave (CW) IR source; nano-FTIR needs a broadband light source. In both cases, the IR source is preferred to be CW or have a high repetition rate. The repetition rate of the source has to exceed the requirement of Nyquist frequency—determined by twice the non-fundamental harmonic of the lock-in reference frequency.

The further development of s-SNOM is accompanied by the development of frequency tunable or broadband IR sources. In the early years of s-SNOM, IR radiations from gas lasers (e.g., $CO_2$ or CO) were used [41,46,58], which lack continuous frequency tunability. The development of QCL in the mid-IR provides a better alternative to the gas lasers with wide-range frequency tunability. Extra-cavity CW QCLs become the standard component for the s-SNOM instrument. High repetition rate narrowband IR from difference frequency generation (DFG) of an OPO can also provide mid-IR laser radiations for s-SNOM, with an even wider frequency coverage

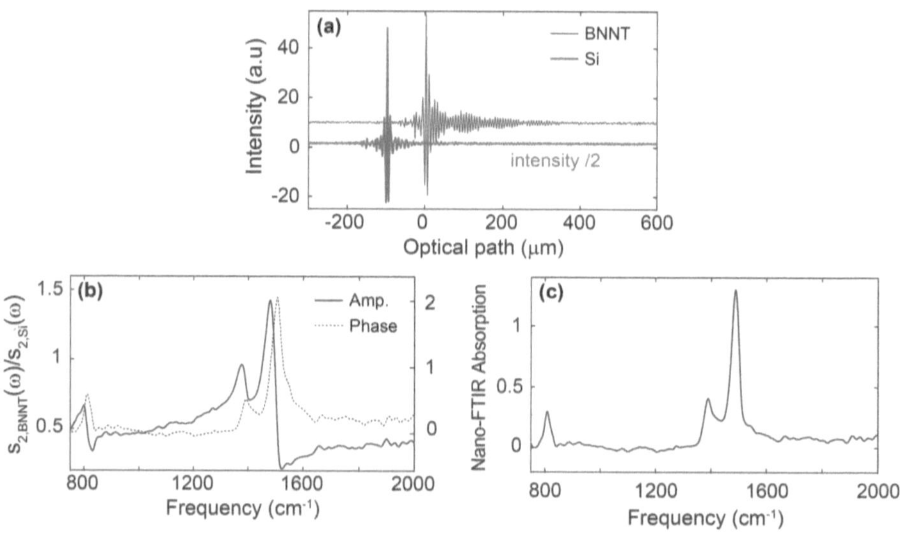


**FIGURE 6.9** Interferograms that was collected by nano-FTIR on a boron nitride nanotube (thin curve) and on a Si substrate as reference (thick curve). (b) The retrieved amplitude and spectral phase of the BNNT from the Fourier transform of interferograms. (c) Extracted absorption spectrum from the nano-FTIR. (The figure is reproduced from Ref. [61] with permission.)

**TABLE 6.2**
**Types of Interferometric Detection Methods for s-SNOM**

| Interferometric Detection Method | Homodyne | Pseudoheterodyne | Nano-FTIR |
|---|---|---|---|
| Imaging modality | Single-frequency imaging | Single-frequency Imaging | Broadband spectroscopy |
| Reference mirror operation | Fixed position or position shifting between two quadrature | Sinusoidal modulation with "magic" depth of $2.63\lambda/4\pi$ | Scan over a long distance of millimeters to centimeters |
| Lock-in detection frequency | $n\Omega$ with $n \geq 2$ | $n\Omega \pm m$M with $n \geq 2$, $m = 1, 2, \ldots$ | $n\Omega$ with $n \geq 2$ |
| Suitable IR sources | QCL, gas laser | QCL, gas laser | DFG, synchrotron, blackbody/plasma source |

than QCLs [62]. These widely tunable infrared sources facilitate the application of s-SNOM for IR nano-imaging. On the other hand, the collection of broadband spectra with nano-FTIR requires broadband radiations to create interferograms. The popular broadband IR source is often from the DFG between broadband femtosecond laser radiations of high repetition rate, either from two synchronized fiber lasers [49] or from the signal and idler outputs of a femtosecond OPO [63]. One limitation of DFG light source is its finite bandwidth of several hundreds of wavenumbers, limited by the phase-matching conditions of nonlinear optical crystal. Several nano-FTIR spectra are stitched together to obtain a broad coverage within the fingerprint region [64]. An alternative to DFG for nano-FTIR is the high brilliance incoherent light from synchrotron radiation of beamlines. The synchrotron radiation comes with an ultrawide bandwidth spanning from near IR to far IR, providing full coverage for nano-FTIR [65–67]. On the other hand, the synchrotron facilities are not available in a typical laboratory setting. Efforts to provide ultra-broadband IR sources have been explored on the blackbody radiation, in which heated bodies emit a wide range of frequency components—like the IR source of a regular FTIR spectrometer. Globar sources have been explored for nano-FTIR [68]. However, the brilliance of regular globar is much below the DFG lasers or synchrotrons. Plasma from the electrical arc was also explored as a source for s-SNOM [69]. The drawback of electrical arc plasma is the high noise and low stability of the plasma. Laser-driven plasma from a compressed gas chamber was also developed to provide improved photon fluence of ultra-broadband blackbody light source [61].

As a summary of s-SNOM methods, Table 6.2 lists the characteristics of homodyne, pseudoheterodyne, and nano-FTIR methods in s-SNOM.

## 6.4 APPLICATIONS OF s-SNOM IN ENERGY MATERIALS

s-SNOM has been widely adopted in condensed phase matter physics for applications where spatial variations of dielectric functions are studies, such as Mott transition of $VO_2$ [70], as well as the measurement of polaritons in low dimensional materials [71–76]. s-SNOM has also been applied to characterize energetic material. For example, the contrast of dielectric function between the donor and acceptor of a certain type of OPV was revealed by s-SNOM imaging [77]. In particular, s-SNOM coupled with the synchrotron radiation with ultrabroad bandwidth allows for characterization of almost any IR active materials, including energetic materials. Ana Nogueira and colleagues have utilized s-SNOM imaging and nano-FTIR to reveal spatial heterogeneity in CsFAMA perovskite [78]. Figure 6.10a displays the representative measurement of perovskite with synchrotron s-SNOM. Bing-Wei Mao and colleagues have utilized s-SNOM to image and supported with models to reveal the increase of carrier at the boundary of the perovskite domain boundaries [79].

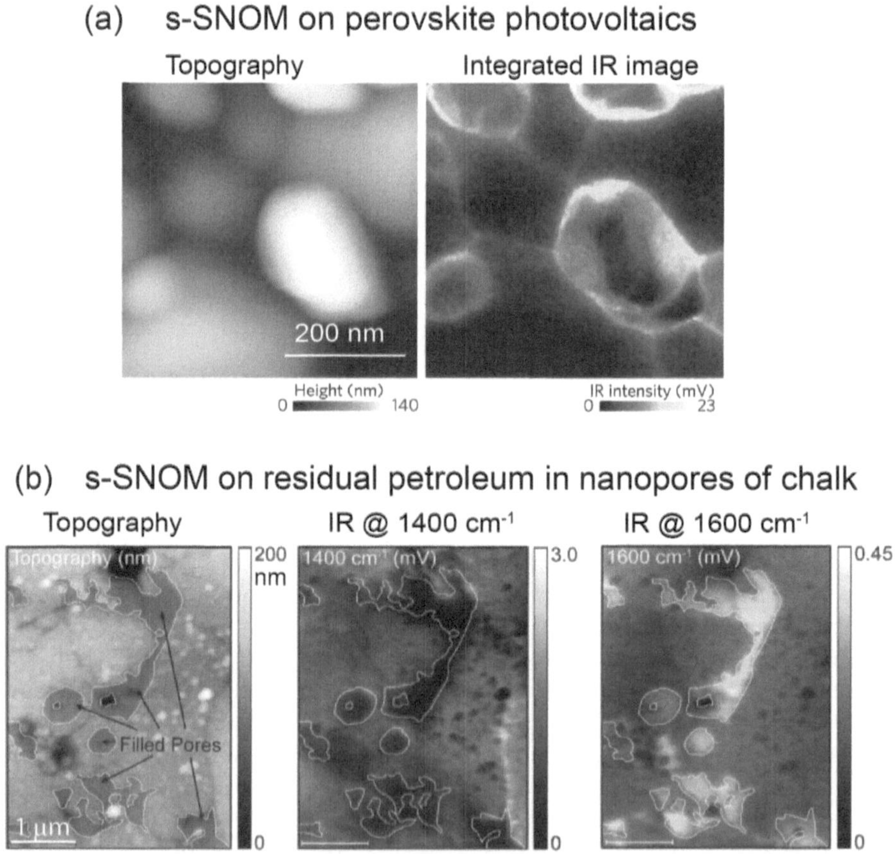


**FIGURE 6.10** (a) s-SNOM measurement of CsFAMA perovskite with a synchrotron source. The left image is topography. The right image is the integrated s-SNOM response from broadband IR frequency. (The figure is reproduced from Ref. [78] with permission.) (b) s-SNOM measurement on residual petroleum in nanopores of chalk. Left image: Topography of a chalk sample with residual petroleum. Middle image: s-SNOM IR image at 1,400 $cm^{-1}$ on *calcite*. Right image: s-SNOM IR image at 1,600 $cm^{-1}$ revealing residual petroleum in nanopores. (The figure is reproduced from Ref. [81] with permission.)

On the energetic materials of fossil fuel, s-SNOM with synchrotron light source has been used to image the inorganic compositions and organic matters in oil shale source rock [80]. David Budd and colleagues have used s-SNOM nano-imaging with a tabletop OPO IR source to reveal the petroleum in nanoporous rock. Figure 6.10b displays the representative result for s-SNOM of migrated residual petroleum in nanopores of Niobrara chalk samples [81].

## 6.5 SUMMARY AND COMPARISON BETWEEN AFM-IR AND IR s-SNOM

Both AFM-IR and s-SNOM deliver the much-needed spatial resolution for nanoscale chemical imaging and spectroscopy for the material research community. Overall, they provide a complementary set of infrared imaging methods with high reliability. If available resources permit, having both techniques in the laboratory should satisfy the needs for nano-IR characterization for heterogeneous IR active materials. On the other hand, one may wonder about the criteria for choosing a better technique for a specific sample between AFM-IR and s-SNOM. The photothermal-based AFM-IR is preferred over s-SNOM if the samples have large thermal expansion coefficients—more mechanical response per given photothermal temperature increase. These samples include soft matters and organic materials. The optical detection s-SNOM excels in conditions where local enhanced IR fields are present, most notably, if the real part of the sample dielectric function equals -1 or the sample supports polaritons. A typical set of materials suitable for s-SNOM is the two-dimensional materials, graphene, hexagonal boron nitride, and transition metal dichalcogenide [72,74,76].

AFM-IR instruments have a lower complexity than s-SNOM that requires an interferometer and an optical detector. A single-frequency tunable pulsed IR source (e.g., QCL) can provide both spectrum collection and chemical imaging for AFM-IR. The AFM-IR spectra usually resemble the IR absorption spectrum, which provides quick identification of sample composition. s-SNOM usually requires both broadband IR source and frequency tunable narrowband IR source for nano-FTIR and spectroscopic near-field imaging, respectively. On the other hand, the signal generation principle of s-SNOM is well understood and analytically modellable with a few parameters and even available through machine learning analysis [82]. In contrast, modeling of AFM-IR signal requires the local information thermal expansion coefficients, local heat diffusion rate, and mechanical properties that are not known a priori. Given their respective strength and limitation, both types of AFM-based IR methods should be used complimentarily for material characterizations.

## REFERENCES


1. E. Abbe, Beiträge zur Theorie des Mikroskops und der mikroskopischen Wahrnehmung, *Archiv für mikroskopische Anatomie*, 9 (1873) 413–468.
2. S. Alexander, L. Hellemans, O. Marti, J. Schneir, V. Elings, P.K. Hansma, M. Longmire, J. Gurley, An atomic-resolution atomic-force microscope implemented using an optical lever, *Journal of Applied Physics*, 65 (1989) 164–167.

3. A. Hammiche, H.M. Pollock, M. Reading, M. Claybourn, P.H. Turner, K. Jewkes, Photothermal FT-IR spectroscopy: A step towards FT-IR microscopy at a resolution better than the diffraction limit, *Applied Spectroscopy*, 53 (1999) 810–815.
4. M.S. Anderson, Infrared spectroscopy with an atomic force microscope, *Applied Spectroscopy*, 54 (2000) 349–352.
5. A. Dazzi, R. Prazeres, F. Glotin, J.M. Ortega, Local infrared microspectroscopy with subwavelength spatial resolution with an atomic force microscope tip used as a photothermal sensor, *Optics Letters*, 30 (2005) 2388–2390.
6. A. Dazzi, C.B. Prater, Q. Hu, D.B. Chase, J.F. Rabolt, C. Marcott, AFM–IR: Combining atomic force microscopy and infrared spectroscopy for nanoscale chemical characterization, *Applied Spectroscopy*, 66 (2012) 1365–1384.
7. F. Lu, M. Jin, M.A. Belkin, Tip-enhanced infrared nanospectroscopy via molecular expansion force detection, *Nature Photonics*, 8 (2014) 307–312.
8. M. Jin, F. Lu, M.A. Belkin, High-sensitivity infrared vibrational nanospectroscopy in water, *Light: Science & Applications*, 6 (2017) e17096–e17096.
9. G. Ramer, F.S. Ruggeri, A. Levin, T.P.J. Knowles, A. Centrone, Determination of polypeptide conformation with nanoscale resolution in water, *ACS Nano*, 12 (2018) 6612–6619.
10. C. Mayet, A. Dazzi, R. Prazeres, F. Allot, F. Glotin, J.M. Ortega, Sub-100nm IR spectromicroscopy of living cells, *Optics Letters*, 33 (2008) 1611–1613.
11. I. Rajapaksa, K. Uenal, H.K. Wickramasinghe, Image force microscopy of molecular resonance: A microscope principle, *Applied Physics Letters*, 97 (2010) 073121.
12. D. Nowak, W. Morrison, H.K. Wickramasinghe, J. Jahng, E. Potma, L. Wan, R. Ruiz, T.R. Albrecht, K. Schmidt, J. Frommer, Nanoscale chemical imaging by photoinduced force microscopy, *Science Advances*, 2 (2016) e1501571.
13. M.T. Cuberes, H.E. Assender, G.A.D. Briggs, O.V. Kolosov, Heterodyne force microscopy of PMMA/rubber nanocomposites: Nanomapping of viscoelastic response at ultrasonic frequencies, *Journal of Physics D: Applied Physics*, 33 (2000) 2347.
14. G. Delen, M. Monai, F. Meirer, B.M. Weckhuysen, In situ nanoscale infrared spectroscopy of water adsorption on nanoislands of surface-anchored metal-organic frameworks, *Angewandte Chemie International Edition*, 133 (2021) 1644–1648.
15. L. Wang, D.S. Jakob, H. Wang, A. Apostolos, M.M. Pires, X.G. Xu, Generalized heterodyne configurations for photoinduced force microscopy, *Analytical Chemistry*, 91 (2019) 13251–13259.
16. Z. Zheng, N. Xu, S.L. Oscurato, M. Tamagnone, F. Sun, Y. Jiang, Y. Ke, J. Chen, W. Huang, W.L. Wilson, A mid-infrared biaxial hyperbolic van der Waals crystal, *Science Advances*, 5 (2019) eaav8690.
17. J. Li, J. Jahng, J. Pang, W. Morrison, J. Li, E.S. Lee, J.-J. Xu, H.-Y. Chen, X.-H. Xia, Tip-enhanced infrared imaging with sub-10nm resolution and hypersensitivity, *The Journal of Physical Chemistry Letters*, 11 (2020) 1697–1701.
18. L. Wang, H. Wang, D. Vezenov, X.G. Xu, Direct measurement of photoinduced force for nanoscale infrared spectroscopy and chemical-sensitive imaging, *The Journal of Physical Chemistry C*, 122 (2018) 23808–23813.
19. J. Jahng, D.A. Fishman, S. Park, D.B. Nowak, W.A. Morrison, H.K. Wickramasinghe, E.O. Potma, Linear and nonlinear optical spectroscopy at the nanoscale with photoinduced force microscopy, *Accounts of Chemical Research*, 48 (2015) 2671–2679.
20. U. Yang, Markus B. Raschke. Resonant optical gradient force interaction for nano-imaging and -spectroscopy. *New Journal of Physics*, 18 (5) 053042.
21. J. Jahng, E.O. Potma, E.S. Lee, Tip-enhanced thermal expansion force for nanoscale chemical imaging and spectroscopy in photoinduced force microscopy, *Analytical Chemistry*, 90 (2018) 11054–11061.

22. M.A. Almajhadi, S.M.A. Uddin, H.K. Wickramasinghe, Observation of nanoscale opto-mechanical molecular damping as the origin of spectroscopic contrast in photo induced force microscopy, *Nature Communications*, 11 (2020) 1–9.
23. J. Jahng, E.O. Potma, E.S. Lee, Nanoscale spectroscopic origins of photoinduced tip–sample force in the midinfrared, *Proceedings of the National Academy of Sciences*, 116 (2019) 26359.
24. B. Pittenger, N. Erina, C. Su, Quantitative mechanical property mapping at the nanoscale with PeakForce QNM, *Application Note Veeco Instruments Inc*, 1 (2010) 1–11.
25. L. Wang, H. Wang, M. Wagner, Y. Yan, D.S. Jakob, X.G. Xu, Nanoscale simultaneous chemical and mechanical imaging via peak force infrared microscopy, *Science Advances*, 3 (2017) e1700255.
26. A. Rosa-Zeiser, E. Weilandt, S. Hild, O. Marti, The simultaneous measurement of elastic, electrostatic and adhesive properties by scanning force microscopy: Pulsed-force mode operation, *Measurement Science and Technology*, 8 (1997) 1333.
27. L. Wang, M. Wagner, H. Wang, S. Pau-Sanchez, J. Li, J.H. Edgar, X.G. Xu, Revealing phonon polaritons in hexagonal boron nitride by multipulse peak force infrared microscopy, *Advanced Optical Materials*, 8 (2020) 1901084.
28. L. Wang, D. Huang, C.K. Chan, Y.J. Li, X.G. Xu, Nanoscale spectroscopic and mechanical characterization of individual aerosol particles using peak force infrared microscopy, *Chemical Communications*, 53 (2017) 7397–7400.
29. W. Li, H. Wang, X.G. Xu, Y. Yu, Simultaneous nanoscale imaging of chemical and architectural heterogeneity on yeast cell wall particles, *Langmuir*, 36 (2020) 6169–6177.
30. C. Gusenbauer, D.S. Jakob, X.G. Xu, D.V. Vezenov, É. Cabane, J. Konnerth, Nanoscale chemical features of the natural fibrous material wood, *Biomacromolecules*, 21 (2020) 4244–4252.
31. N. Li, X. Niu, L. Li, H. Wang, Z. Huang, Y. Zhang, Y. Chen, X. Zhang, C. Zhu, H. Zai, Y. Bai, S. Ma, H. Liu, X. Liu, Z. Guo, G. Liu, R. Fan, H. Chen, J. Wang, Y. Lun, X. Wang, J. Hong, H. Xie, S. Jakob Devon, G. Xu Xiaoji, Q. Chen, H. Zhou, Liquid medium annealing for fabricating durable perovskite solar cells with improved reproducibility, *Science*, 373 (2021) 561–567.
32. K.L.A. Chan, S.G. Kazarian, New opportunities in micro-and macro-attenuated total reflection infrared spectroscopic imaging: Spatial resolution and sampling versatility, *Applied Spectroscopy*, 57 (2003) 381–389.
33. H. Wang, L. Wang, E. Janzen, J.H. Edgar, X.G. Xu, Total internal reflection peak force infrared microscopy, *Analytical Chemistry*, 93 (2021) 731–736.
34. H. Wang, J.M. González-Fialkowski, W. Li, Q. Xie, Y. Yu, X.G. Xu, Liquid-phase peak force infrared microscopy for chemical nanoimaging and spectroscopy, *Analytical Chemistry*, 93 (2021) 3567–3575.
35. H. Wang, E. Janzen, L. Wang, J.H. Edgar, X.G. Xu, Probing mid-infrared phonon polaritons in the aqueous phase, *Nano Letters*, 20 (2020) 3986–3991.
36. K.L. Gu, Y. Zhou, W.A. Morrison, K. Park, S. Park, Z. Bao, Nanoscale domain imaging of all-polymer organic solar cells by photo-induced force microscopy, *ACS Nano*, 12 (2018) 1473–1481.
37. D.S. Jakob, L. Wang, H. Wang, X.G. Xu, Spectro-mechanical characterizations of kerogen heterogeneity and mechanical properties of source rocks at 6 nm spatial resolution, *Analytical Chemistry*, 91 (2019) 8883–8890.
38. J. Yang, J. Hatcherian, P.C. Hackley, A.E. Pomerantz, Nanoscale geochemical and geomechanical characterization of organic matter in shale, *Nature Communications*, 8 (2017) 1–9.
39. L. Novotny, S.J. Stranick, Near-field optical microscopy and spectroscopy with pointed probes, *Annual Review of Physical Chemistry*, 57 (2006) 303–331.

40. L. Novotny, From near-field optics to optical antennas, *Physics Today*, 64 (2011) 47–52.
41. R. Bachelot, P. Gleyzes, A.C. Boccara, Near-field optical microscope based on local perturbation of a diffraction spot, *Optics Letters*, 20 (1995) 1924–1926.
42. N. Ocelic, A. Huber, R. Hillenbrand, Pseudoheterodyne detection for background-free near-field spectroscopy, *Applied Physics Letters*, 89 (2006) 101124.
43. G. Wurtz, R. Bachelot, P. Royer, Imaging a GaAlAs laser diode in operation using apertureless scanning near-field optical microscopy, *European Physical Journal Applied Physics*, 5 (1999) 269–275.
44. B. Knoll, F. Keilmann, Enhanced dielectric contrast in scattering-type scanning near-field optical microscopy, *Optics Communications*, 182 (2000) 321–328.
45. R. Hillenbrand, F. Keilmann, Complex optical constants on a subwavelength scale, *Physical Review Letters*, 85 (2000) 3029–3032.
46. R. Hillenbrand, T. Taubner, F. Keilmann, Phonon-enhanced light–matter interaction at the nanometre scale, *Nature*, 418 (2002) 159–162.
47. T. Taubner, R. Hillenbrand, F. Keilmann, Performance of visible and mid-infrared scattering-type near-field optical microscopes, *Journal of Microscopy*, 210 (2003) 311–314.
48. S. Amarie, P. Zaslansky, Y. Kajihara, E. Griesshaber, W.W. Schmahl, F. Keilmann, Nano-FTIR chemical mapping of minerals in biological materials, *Beilstein Journal of Nanotechnology*, 3 (2012) 312–323.
49. F. Huth, A. Govyadinov, S. Amarie, W. Nuansing, F. Keilmann, R. Hillenbrand, Nano-FTIR absorption spectroscopy of molecular fingerprints at 20 nm spatial resolution, *Nano Letters*, 12 (2012) 3973–3978.
50. X.G. Xu, L. Gilburd, G.C. Walker, Phase stabilized homodyne of infrared scattering type scanning near-field optical microscopy, *Applied Physics Letters*, 105 (2014) 263104.
51. A.J. Sternbach, J. Hinton, T. Slusar, A.S. McLeod, M.K. Liu, A. Frenzel, M. Wagner, R. Iraheta, F. Keilmann, A. Leitenstorfer, M. Fogler, H.T. Kim, R.D. Averitt, D.N. Basov, Artifact free time resolved near-field spectroscopy, *Optics Express*, 25 (2017) 28589–28611.
52. M.B. Raschke, C. Lienau, Apertureless near-field optical microscopy: Tip–sample coupling in elastic light scattering, *Applied Physics Letters*, 83 (2003) 5089–5091.
53. X.G. Xu, A.E. Tanur, G.C. Walker, Phase controlled homodyne infrared near-field microscopy and spectroscopy reveal inhomogeneity within and among individual boron nitride nanotubes, *The Journal of Physical Chemistry A*, 117 (2013) 3348–3354.
54. A. Cvitkovic, N. Ocelic, R. Hillenbrand, Analytical model for quantitative prediction of material contrasts in scattering-type near-field optical microscopy, *Optics Express*, 15 (2007) 8550–8565.
55. A.S. McLeod, P. Kelly, M.D. Goldflam, Z. Gainsforth, A.J. Westphal, G. Dominguez, M.H. Thiemens, M.M. Fogler, D.N. Basov, Model for quantitative tip-enhanced spectroscopy and the extraction of nanoscale-resolved optical constants, *Physical Review B*, 90 (2014) 085136.
56. M. Labardi, S. Patanè, M. Allegrini, Artifact-free near-field optical imaging by apertureless microscopy, *Applied Physics Letters*, 77 (2000) 621–623.
57. L. Wang, X.G. Xu, Scattering-type scanning near-field optical microscopy with reconstruction of vertical interaction, *Nature Communications*, 6 (2015) 8973.
58. B.B. Akhremitchev, Y. Sun, L. Stebounova, G.C. Walker, Monolayer-sensitive infrared imaging of DNA stripes using apertureless near-field microscopy, *Langmuir*, 18 (2002) 5325–5328.
59. L. Stebounova, B.B. Akhremitchev, G.C. Walker, Enhancement of the weak scattered signal in apertureless near-field scanning infrared microscopy, *Review of Scientific Instruments*, 74 (2003) 3670–3674.

60. Z.H. Kim, B. Liu, S.R. Leone, Nanometer-scale optical imaging of epitaxially grown GaN and InN islands using apertureless near-field microscopy, *The Journal of Physical Chemistry B*, 109 (2005) 8503–8508.
61. M. Wagner, D.S. Jakob, S. Horne, H. Mittel, S. Osechinskiy, C. Phillips, G.C. Walker, C. Su, X.G. Xu, Ultrabroadband nanospectroscopy with a laser-driven plasma source, *ACS Photonics*, 5 (2018) 1467–1475.
62. Q. Hu, H. Yang, Nanoscale IR spectroscopy and imaging with a versatile broad band IR laser source, In: Proceedings of the SPIE, 2021.
63. X.G. Xu, M. Rang, I.M. Craig, M.B. Raschke, Pushing the sample-size limit of infrared vibrational nanospectroscopy: From monolayer toward single molecule sensitivity, *The Journal of Physical Chemistry Letters*, 3 (2012) 1836–1841.
64. I. Amenabar, S. Poly, M. Goikoetxea, W. Nuansing, P. Lasch, R. Hillenbrand, Hyperspectral infrared nanoimaging of organic samples based on Fourier transform infrared nanospectroscopy, *Nature Communications*, 8 (2017) 14402.
65. P. Hermann, A. Hoehl, P. Patoka, F. Huth, E. Rühl, G. Ulm, Near-field imaging and nano-Fourier-transform infrared spectroscopy using broadband synchrotron radiation, *Optics Express*, 21 (2013) 2913–2919.
66. H.A. Bechtel, E.A. Muller, R.L. Olmon, M.C. Martin, M.B. Raschke, Ultrabroadband infrared nanospectroscopic imaging, *Proceedings of the National Academy of Sciences*, 111 (2014) 7191–7196.
67. R.O. Freitas, C. Deneke, F.C.B. Maia, H.G. Medeiros, T. Moreno, P. Dumas, Y. Petroff, H. Westfahl, Low-aberration beamline optics for synchrotron infrared nanospectroscopy, *Optics Express*, 26 (2018) 11238–11249.
68. F. Huth, M. Schnell, J. Wittborn, N. Ocelic, R. Hillenbrand, Infrared-spectroscopic nanoimaging with a thermal source, *Nature Materials*, 10 (2011) 352–356.
69. D.J. Lahneman, T.J. Huffman, P. Xu, S.L. Wang, T. Grogan, M.M. Qazilbash, Broadband near-field infrared spectroscopy with a high temperature plasma light source, *Optics Express*, 25 (2017) 20421–20430.
70. M.M. Qazilbash, M. Brehm, B.-G. Chae, P.C. Ho, G.O. Andreev, B.-J. Kim, S.J. Yun, A.V. Balatsky, M.B. Maple, F. Keilmann, Mott transition in $VO_2$ revealed by infrared spectroscopy and nano-imaging, *Science*, 318 (2007) 1750–1753.
71. A. Huber, N. Ocelic, D. Kazantsev, R. Hillenbrand, Near-field imaging of mid-infrared surface phonon polariton propagation, *Applied Physics Letters*, 87 (2005) 081103.
72. Z. Fei, A.S. Rodin, G.O. Andreev, W. Bao, A.S. McLeod, M. Wagner, L.M. Zhang, Z. Zhao, M. Thiemens, G. Dominguez, M.M. Fogler, A.H.C. Neto, C.N. Lau, F. Keilmann, D.N. Basov, Gate-tuning of graphene plasmons revealed by infrared nano-imaging, *Nature*, 487 (2012) 82–85.
73. J. Chen, M. Badioli, P. Alonso-González, S. Thongrattanasiri, F. Huth, J. Osmond, M. Spasenović, A. Centeno, A. Pesquera, P. Godignon, Optical nano-imaging of gate-tunable graphene plasmons, *Nature*, 487 (2012) 77–81.
74. S. Dai, Z. Fei, Q. Ma, A.S. Rodin, M. Wagner, A.S. McLeod, M.K. Liu, W. Gannett, W. Regan, K. Watanabe, Tunable phonon polaritons in atomically thin van der Waals crystals of boron nitride, *Science*, 343 (2014) 1125–1129.
75. X.G. Xu, B.G. Ghamsari, J.-H. Jiang, L. Gilburd, G.O. Andreev, C. Zhi, Y. Bando, D. Golberg, P. Berini, G.C. Walker, One-dimensional surface phonon polaritons in boron nitride nanotubes, *Nature Communications*, 5 (2014) 1–6.
76. F. Hu, Y. Luan, M.E. Scott, J. Yan, D.G. Mandrus, X. Xu, Z. Fei, Imaging exciton–polariton transport in $MoSe_2$ waveguides, *Nature Photonics*, 11 (2017) 356–360.
77. P. Li, J. Fang, Y. Wang, S. Manzhos, L. Cai, Z. Song, Y. Li, T. Song, X. Wang, X. Guo, M. Zhang, D. Ma, B. Sun, Synergistic effect of dielectric property and energy transfer on charge separation in non-fullerene-based solar cells, *Angewandte Chemie International Edition*, 60 (2021) 15054–15062.

78. R. Szostak, J.C. Silva, S.H. Turren-Cruz, M.M. Soares, R.O. Freitas, A. Hagfeldt, H.C.N. Tolentino, A.F. Nogueira, Nanoscale mapping of chemical composition in organic-inorganic hybrid perovskite films, *Science Advances*, 5 (2019) eaaw6619.
79. T.-X. Qin, E.-M. You, M.-X. Zhang, P. Zheng, X.-F. Huang, S.-Y. Ding, B.-W. Mao, Z.-Q. Tian, Quantification of electron accumulation at grain boundaries in perovskite polycrystalline films by correlative infrared-spectroscopic nanoimaging and Kelvin probe force microscopy, *Light: Science & Applications*, 10 (2021) 84.
80. Z. Hao, H.A. Bechtel, T. Kneafsey, B. Gilbert, P.S. Nico, Cross-scale molecular analysis of chemical heterogeneity in shale rocks, *Scientific Reports*, 8 (2018) 2552.
81. R.E. Simon, S.C. Johnson, O. Khatib, M.B. Raschke, D.A. Budd, Correlative nano-spectroscopic imaging of heterogeneity in migrated petroleum in unconventional reservoir pores, *Fuel*, 300 (2021) 120836.
82. X. Chen, Z. Yao, S. Xu, A.S. McLeod, S.N. Gilbert Corder, Y. Zhao, M. Tsuneto, H.A. Bechtel, M.C. Martin, G.L. Carr, M.M. Fogler, S.G. Stanciu, D.N. Basov, M. Liu, Hybrid machine learning for scanning near-field optical spectroscopy, *ACS Photonics*, 8 (2021) 2987–2996.

# 7 Application of AFM in Lithium Batteries Research

*Shuang-Yan Lang, Zhen-Zhen Shen, Jing Wan, and Rui Wen*
Chinese Academy of Sciences
University of Chinese Academy of Sciences

## CONTENTS



DOI: 10.1201/9781003174042-7

## 7.1 INTRODUCTION

Advanced energy storage becomes the driving force of modern society. Electrochemical batteries, such as lead–acid, nickel–cadmium, nickel–metal hydride, and lithium-ion batteries (LIBs), have been proposed for the commercial market for their low cost.[1] These electrical energy systems especially LIBs enable the rise of portable electronics and electric vehicles. In the early 1980s, a layered cathode material capable of reversibly extracting lithium ions was successfully prepared by Goodenough et al. This type of material has gradually developed in subsequent research and has now become a widely used cathode material.[2,3] In 1990, the concept of "lithium-ion battery" was first proposed.[4] This lithium-ion secondary battery composed of carbon as the negative electrode and lithium cobalt oxide as the positive electrode attracted the attention of the majority of battery researchers, and the technology of lithium-ion battery became the world's research hotspots. In the second year, Sony launched a commercial lithium-ion battery with graphite as the negative electrode and lithium cobalt oxide as the positive electrode and quickly occupied the market of portable consumer electronic products, which promoted the development of small electronic devices.[5,6] Commonly, LIBs consist of Li transition cathode, separator with organic electrolyte and graphite anode. Performance parameters such as battery output voltage, energy density, power density, service life and safety are important to measure battery quality. In order to increase the energy density of the battery, many optimizations have been made to the electrode material and assembly structure of the battery in the subsequent development and research. The traditional lithium cobalt oxide cathode and graphite anode are still used. At present, the energy density of lithium-ion batteries is about 240–250 Wh/kg, which is gradually approaching the theory of graphite anode. Thus, there is an urgent need to develop anode materials to meet the requirements for high-energy-density LIBs.[7,8]

To satisfy the ever-increasing demands for energy use, Li metal anode has been intensively investigated in recent years, due to its highest theoretical specific capacity (3,860 mAh/g) and lowest electrochemical potential (–3.040 V vs. standard hydrogen electrode). Two typical Li metal battery systems are lithium-sulfur (Li-S) and lithium-oxygen (Li-$O_2$) batteries. For Li-S batteries, the insulating properties of sulfur have largely hindered the utilization of active materials. Numerous efforts have been devoted in recent years to maximizing the cathode capacity, including cathode structural design and the use of redox mediators.[9,10] The interfacial performance at Li metal anode is equally important for the commercialization of Li-S batteries. Targeted information such as the effects of the polysulfides on the Li performance could provide useful strategies for Li protection in Li-S batteries. Deep insights into the interfacial evolution at both S and Li/electrolyte interfaces in Li-S batteries are important for our fundamental understanding of Li-S electrochemistry as well as further optimization of Li-S batteries.

In recent decades, Li-$O_2$ batteries have excited huge research attention due to their ultra-high theoretical specific energy of ~3,500 Wh/kg, which is far beyond that of the conventional Li-ion battery. The operation of Li-$O_2$ batteries relies on the reversible conversion chemistry between $O_2$ and $Li^+$ on the electrode. However, the passivation and degradation of the $O_2$ electrode and the occurrence of side reactions make Li-$O_2$ batteries suffer from large over-potential and poor cycle life. To address these challenges facing Li-$O_2$ batteries and further optimize the Li-$O_2$ batteries, a fundamental investigation of the Li-$O_2$ electrochemistry/chemistry on the surface of the electrode must come first. An in-depth understanding of the morphological evolution, structural transformation and reaction kinetics of Li-$O_2$ reactive products under complex electrochemical conditions is a prerequisite for taking a massive leap forward of Li-$O_2$ batteries.

In brief, for all of the electrochemical systems, the interfacial reactions play key roles on the cell performance. With nanoscale resolution, atomic force microscopy (AFM) can measure the interfacial morphology at the solid/liquid or solid/gas interfaces, which is very useful to track the interfacial processes of Li batteries systems, such as Li-ion, Li-S, and Li-$O_2$ batteries as mentioned above. In previous chapters, we have systematically introduced the powerful functions of AFM, basically focused on the fundamental mechanism of AFM technique and its combination with other characterization methods. In this chapter, we mainly introduce the applications of AFM in Li battery research, including Li-ion and Li metal batteries. In an effort to track the real-time evolution of interfacial reactions, electrochemical AFM (EC-AFM) is highly recommended to investigate the interfacial processes during cell operation. For Li metal batteries, we first focus its applications on the cathode/electrolyte interfaces in Li-S and Li-$O_2$ batteries. Then we emphasize some recent works on the Li metal anode using AFM, as well as the state-of-the-art solid-state systems. In the end, we provide a brief summary and outlook of the applications of AFM in Li battery research.

## 7.2 *IN SITU* VISUALIZATION OF ON-SITE FORMATION OF CEI AND SEI IN LITHIUM-ION BATTERIES

### 7.2.1 Introduction: Interfacial Electrochemistry in Li-ion Batteries

LIBs are commonly used for portable electronics and electric vehicles in recent years. They are rechargeable batteries using intercalated lithium compound as cathode materials and typically graphite as the anode, where Li ions move from anode to cathode during the discharge process, and back during the recharging. To realize long cycle life as well as high capacity of the Li-ion batteries, it is essential to understand and optimize the interfacial reactions at both the cathode/electrolyte interfaces and the anode/electrolyte interfaces, or electrode/solid-electrolyte interfaces.

The transition metal oxides with layered structures are popular cathode materials which allow Li ions to intercalate into or extract from the host materials reversibly. Typical examples are $LiFePO_4$, $LiCoO_2$ and $LiNi_xCo_yMn_zO_2$ (NMC) that have been widely investigated and commercially available.[11,12] The fast degradation during the cycling is one of the most significant challenges hindering their further deployment.

The solid electrolyte interphase (SEI) at the anode/electrolyte interface is considered to be a passivation layer generated from the reduction of electrolyte components during the charging process.[13–15] Ideally, the SEI film has the characteristics of ionic conduction but electronic insulation, compactness, uniformity and certain mechanical strength, which can achieve effective lithium ion transportation at the interface and adapt to the volume expansion of the electrode during charge and discharge.[16–20] Its formation mechanism and physicochemical properties have a crucial impact on the ion transport at the interface of electrode/electrolyte and battery performance. Therefore, a deep understanding of the nanostructure and dynamic evolution of the SEI film is of far-reaching significance for the optimal design of battery systems. The structure and composition of the SEI film are complex, and it is sensitive to air and moisture. AFM can perform non-destructive and *in situ* high-resolution imaging on them in an anhydrous and oxygen-free environment. Therefore, *in situ* electrochemical AFM is widely used in the research field of the SEI film.[21–26]

### 7.2.2 *In situ* AFM Imaging of the Evolution of the CEI Film

Layered $LiCoO_2$ is one of the most widely employed cathode materials in Li-ion battery systems. Understanding the interfacial evolution and properties of its CEI is vital for cell performance. Using EC-AFM, Lu et al.[27] successfully investigated the formation of CEI at the basal plane and edge plane of $LiCoO_2$ crystal (Figure 7.1). Since the $Li^+$ intercalation/extraction processes are highly dependent on the orientation of the crystals, no evident CEI is observed on the basal planes as shown in Figure 7.1A. While on the edge plane of $LiCoO_2$ crystals, CEI film with loose fibrillary structures

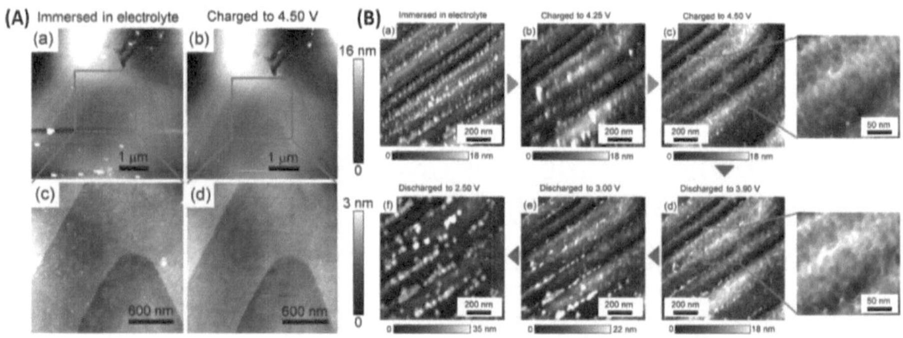


**FIGURE 7.1** AFM images of *in situ* monitoring of the interfacial evolution of (A) basal plane and (B) edge plane of $LiCoO_2$ crystal.[27]

is detected under high voltage, which is unstable and can be decomposed at low voltage (Figure 7.1B). These results show the importance of controlling the surface composition for high stability $LiCoO_2$, which are nice examples that represent the powerful functions of AFM for the CEI investigation towards Li-ion batteries. More excellent research focused on kinds of CEI investigation are provided in recommended papers.[12,28,29]

### 7.2.3 SEI Live Formation at the Anode/Electrolyte Interfaces in Classical Liquid Electrolytes

In the typical carbonate electrolyte system (1M lithium hexafluorophosphate ($LiPF_6$) dissolved in equimolar ethylene carbonate (EC)/dimethyl carbonate (DMC)), *in situ* AFM monitors the real-time dynamic growth of the SEI film during the lithiation/delithiation process of solvated lithium ions on the highly oriented pyrolytic graphite (HOPG) (Figure 7.2A). When charged to 2.0V and continued dropping to 0.91V, irregularly distributed tiny island-like structures appear on the surface of HOPG electrode. Subsequently, when the potential continues to move negatively to 0.36V, the SEI film mainly grows along the edge of HOPG instead of its steps (Figure 7.2a). It can be considered that the formation of SEI precedes or starts through the co-intercalation of solvated $Li^+$, which can only occur at the edge of the electrode. With the negative shift of the potential, the thickness of the deposits at the edge of the electrode continues to enhance, and the height also increases. During the charging process from 0.74 to 0.62V, a large number of nanoparticles (NPs) on the electrode surface gradually evolve into wider and higher convex structures. It can be considered that the formation of the initial SEI film occurs between 0.74 and 0.62V, and the height increasing below 0.62V is due to the reduction of more solvated $Li^+$

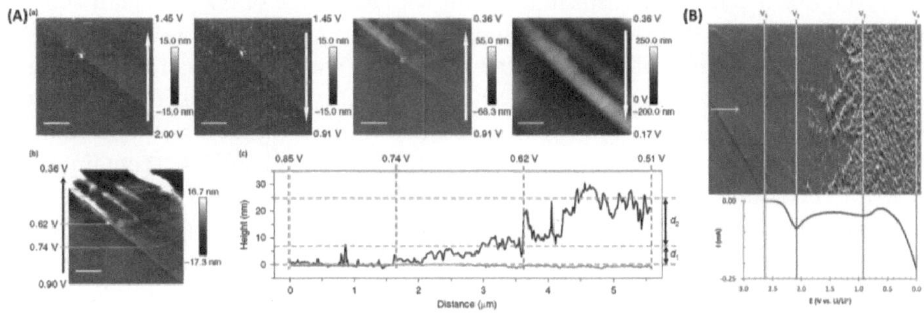

FIGURE 7.2 (A) (a) Morphological evolution of the SEI film via *in situ* and operando AFM measurements of the HOPG electrode during the first lithiation process. (b) Correlation of morphology and potential during the initial cathodic scan from 0.90 to 0.36V. (c) Height distribution (black curve) of the SEI on the stage labeled by the blue line in (b). The scale bars are 1 µm in (a, b).[30] (B) SEI live formation on a HOPG surface in 1.5M LiTFSI dissolved in EC (control electrolyte) imaged (top) during electrochemical potential sweep (bottom) from open circuit voltage (OCV) at $V_1$ to 0V at $V_4$, at a rate of 5mV/s. Vertical lines correspond to key features described in the text. The arrow indicates the slow scan direction. Image parameters: $3.5\times4.5\,\mu m^2$.[31]

and co-intercalation into the HOPG layer. In this process, there are still new SEI films continuously growing.[30] Moreover, SEI live formation on the HOPG surface in 1.5 M LiTFSI dissolved in EC (control electrolyte) was also monitored via *in situ* AFM. It displays that the surface morphology of HOPG changes significantly below 1.5 V, where pronounced accumulation of interphasial species occurred along edge sites, corresponding to the initial intercalation of solvated $Li^+$ into the top layers of HOPG, and the subsequent reduction of EC molecules coordinated to intercalated $Li^+$. Then, surface species engulfed the basal planes around 1.0 V, correlating to the typical reduction of carbonate species such as EC. Finally, the extensive presence of surface species below 0.6 V obscured the view of both basal planes and edge sites (Figure 7.2B), revealing the whole SEI live-formation process.[31]

### 7.2.4 Regulation Strategies for SEI Films

Utilizing organic liquid electrolytes with different components, such as adjusting the concentration of lithium salt,[32,33] changing the type of lithium salt,[34] and adding additives,[35] can effectively regulate the structural morphology of SEI films and therefore improve the electrochemical performance.

In the electrolyte system with a conventional lithium salt concentration (1 mol/L), at the initial stage of SEI formation, the main components of SEI film grow along the edge sites rather than on the basal planes of HOPG; meanwhile, the co-intercalation of solvated Li-ions tends to occur at the edge sites. With the cell potential negatively shifted, the step heights of HOPG continuously increase caused by the simultaneous co-intercalation of more solvated Li into HOPG layers along with their reduction. When a large number of solvated Li-ions and their decomposed products embed into the graphite layers, the HOPG will be exfoliated and further inducing the destruction of electrode structure. When the concentration is increased to 3.37 mol/L, it is observed that a dense and uniform SEI film with the shape of NPs gradually covers the whole electrode surface along the edge and plane sites of HOPG (Figure 7.3A). Compared with what is observed in the electrolyte with lithium salt concentration of 1 mol/L, a more stable and homogeneous SEI film is beneficial for the protection of HOPG layers from being exfoliated due to the continuous co-intercalation of solvated Li-ions.[32] In a high-salt concentrated aqueous electrolyte (containing 21 mol/L LiTFSI salt), *in situ* electrochemical AFM is used to reveal the morphological evolution and mechanical properties of the SEI film and indicate the chemistry composition of the SEI film combined with the *ex situ* XPS characterization. The research results show that in the high-salt-concentration aqueous electrolyte system, the SEI film with a thickness of about 4–6 nm is unevenly distributed on the surface of the HOPG electrode. The SEI film can be divided into a region with a modulus of $30 \pm 10$ GPa and a part with a modulus below 1–2 GPa, and its main inorganic components are $LiCO_3$ and LiF.[33]

Lithium salt is an important component of the electrolyte, and changing the type of lithium salts will have a significant effect on the SEI film at interface.[34] With the LiFSI salt involved, the SEI film tends to nucleate as NPs on the edge of the HOPG firstly and the platform of HOPG later, and then diffuses in a two-dimensional plane to form a thin SEI film (Figure 7.3B). In the LiTFSI-based electrolyte, loose SEI

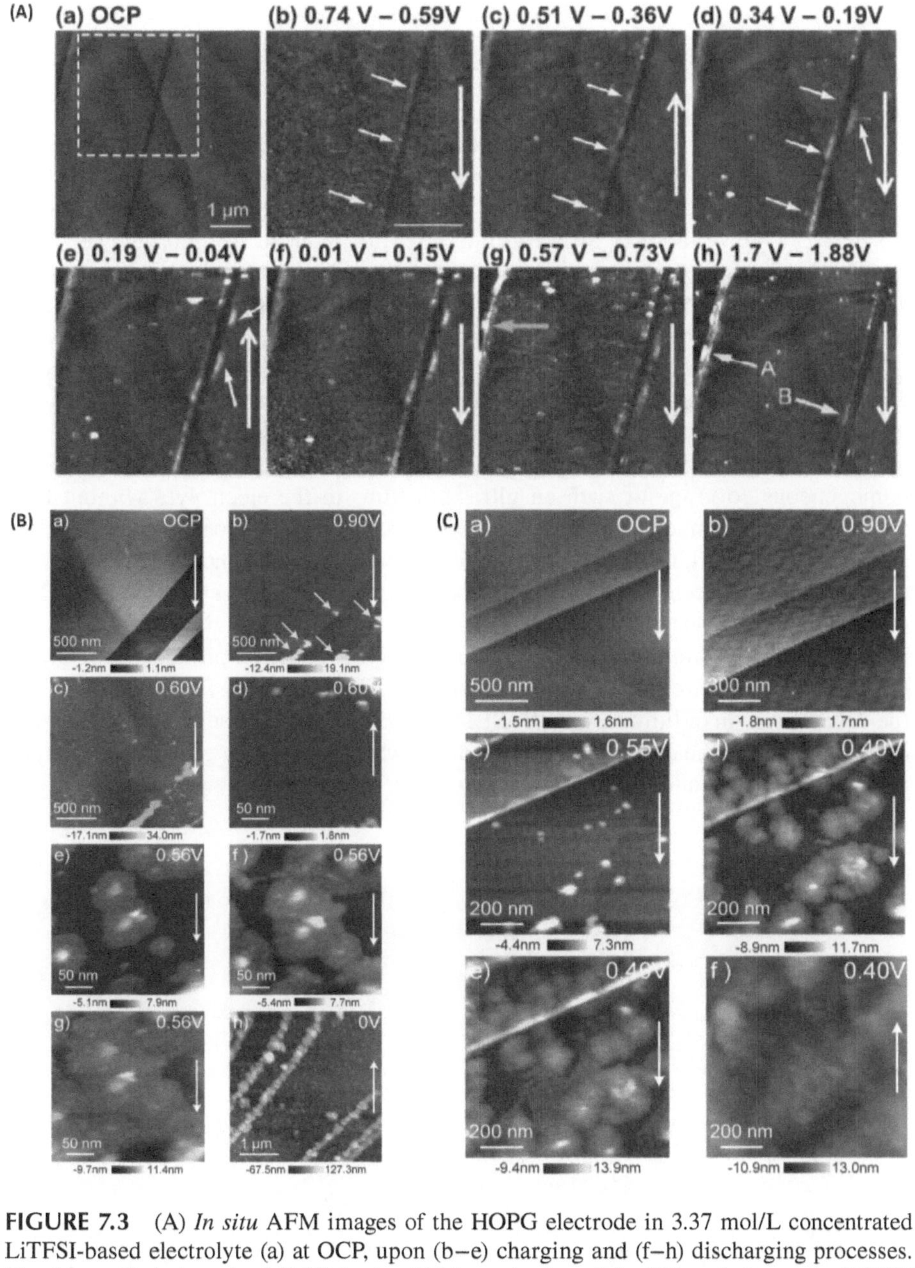


**FIGURE 7.3** (A) *In situ* AFM images of the HOPG electrode in 3.37 mol/L concentrated LiTFSI-based electrolyte (a) at OCP, upon (b–e) charging and (f–h) discharging processes. The data scale is 4 μm (c–i).[32] (B) *In situ* AFM monitoring of the SEI evolution at the HOPG/electrolyte interface at (a) OCP and (b–h) charging process in LiFSI-based electrolyte.[34] (C) *In situ* AFM monitoring of the SEI evolution at the HOPG/electrolyte interface at (a) OCP and (b–f) charging process in LiTFSI-based electrolyte.[34]

films that grow in stacks are easily formed on the electrode surface, causing the co-intercalation of solvated ions and other cations, resulting in instability and destruction of the electrode structure during the cycles (Figure 7.3C). LiFSI salt can also induce a dense and uniform granular SEI film on the surface of the high-specific-energy silicon anode, which is beneficial to protect the stability of the electrode.[36]

Relevant research results display that adding additives to the electrolyte can effectively form a stable SEI film at the electrode/electrolyte interface. *In situ* electrochemical AFM captures an ultra-thin SEI film induced by fluoroethylene carbonate (FEC) on the surface of ultra-flat monolayer molybdenum disulfide ($MoS_2$) electrode,[35] real-time monitoring its initial nucleation, growth and formation process (Figure 7.4A). An atomically flat platform of $MoS_2$ without any impurities is exhibited at OCP. Upon charging to 1.77 V, a brush-shaped film is initially generated on the $MoS_2$ surface. Interestingly, the next scanned plot shows that this initial start to lateral growth develops into an intensive and uniform film covering the majority of the electrode. Finally, the whole $MoS_2$ surface was wrapped by the dense and homogeneous coverage of such an ultra-thin film. In the electrolyte without FEC additive, bright NP nuclei appear on the triangular monolayer $MoS_2$ electrode surface at cathodic 1.88 V. The average heights of the NPs gradually increase, manifesting their growth and accumulation at 1.39 V on both basal planes and edge sites. This process could be ascribed to the initial nucleation and further development of SEI film, considered as an NP-shaped SEI, consisting of reduction products of $[FSI]^-$ in the electrolyte (Figure 7.4B). Compared with the NP-shaped SEI film generated from the decomposition of lithium salt, the SEI film induced by FEC additives densely and uniformly covers the entire electrode surface, which provides a guarantee for stabilizing the interface and improving battery performance during the cycles.

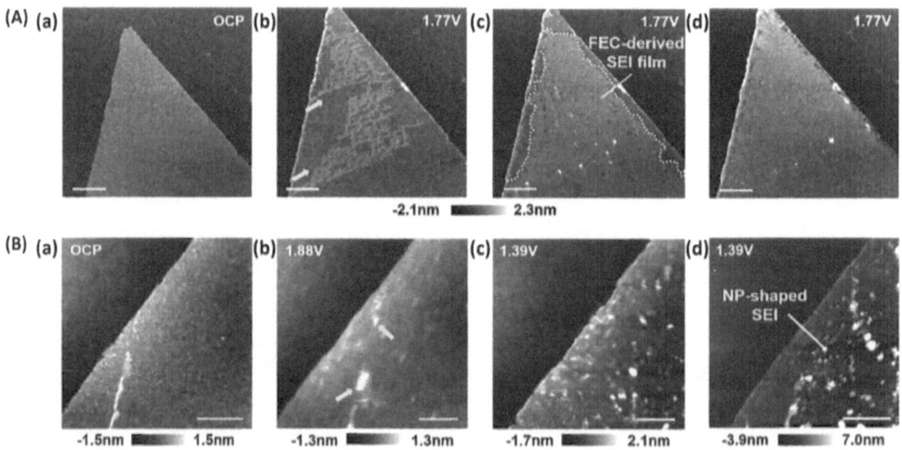


**FIGURE 7.4** (A) *In situ* AFM images of the monolayer $MoS_2$ electrode (a) at OCP and (b–d) upon charging process in FEC-containing electrolyte. The scale bars are 600 nm.[35] (B) *In situ* AFM images of the interfacial evolution (a) at OCP and (b–d) upon charging process in LiFSI-based electrolyte without FEC. The scale bars are 500 nm in (b, c) and 1 μm in (a, d).[35]

## 7.3 INTERFACIAL EVOLUTION IN LITHIUM-SULFUR BATTERIES

### 7.3.1 Lithium-Sulfur Batteries: Introduction and Interfacial Electrochemistry

To meet the ever-increasing energy demands of modern consumer electronics, lithium-sulfur (Li-S), as one of the most promising post-lithium-ion technologies, are under intense study globally because of their high theoretical energy density (2,600 Wh/kg) and natural abundance of sulfur.[8] Tremendous efforts have been made to increase the overall performance of Li-S batteries, via electrode design, electrode/electrolyte interface protection and/or electrolyte regulation.[9,10] These efforts have largely promoted the improvement of the reversibility and stability of the Li-S batteries in recent times. Great industrial opportunities of Li-S batteries lie in a common development of engineering techniques of advanced materials and fundamental understanding of Li-S electrochemistry, in general, and multistep, complex Li-S redox processes, in particular.

Li-S redox processes involve multistep evolution of intermediate polysulfides and phase transitions. During the discharge process, sulfur would be first reduced to soluble long-chain polysulfides ($S_x^{2-}$, 8<x<4). They are subsequently reduced to intermediate polysulfides ($S_x^{2-}$, 4<x<2) and, finally, short-chain solid $Li_2S_2$ and $Li_2S$. A schematic impression of the stepwise reduction process from sulfur to $Li_2S$ is shown in Figure 7.5.[37] The recharge processes undergo the reverse processes of the discharge, upon the solid-liquid-solid transition, where the $Li_2S_2$/$Li_2S$ is oxidized to polysulfides and then to sulfur. The interfacial evolution of different sulfide species has been regarded as a critical issue that hinders the full utilization of active materials. Long-chain polysulfides can dissolve into the electrolyte, shuttle back and forth between the electrodes, known as "shuttle effect", causing severe capacity fade and passivation on both electrodes. In addition, the electrodeposition of insulating $Li_2S_2$ and $Li_2S$ blocks further electron transfer at the interface, which limits the discharge capacity and aggravates the interfacial inhomogeneity.

Recent advances in *in situ/operando* techniques enable the real-time investigation of Li-S redox processes, such as electron microscopy (EM), synchrotron-based method, Raman and Fourier-transform infrared spectroscopy (FTIR), which provide us basic evidence of morphological changes and/or redox pathways at the interfaces.[38–40] For in-depth insights into the Li-S redox processes, including the effects of different reaction environments, more powerful, accurate characterization and comprehensive analysis methods are required, in an effort to reveal mechanistic principles of Li-S redox in detail and guide further optimization in battery design. AFM is undoubtedly an advanced science and technology, as introduced in the previous chapters, which presents three-dimensional topological information in a high spatial resolution from nano- to micrometer. Although the use of AFM in battery research is in its early stages, preliminary results do show important and unique contributions to our fundamental understanding of Li-S electrochemistry. We believe that further research using AFM along with related advanced techniques will forge the way forward, substantially reinforce our understanding of Li-S degradation mechanisms and potentially enable further scientific breakthroughs for Li-S and other state-of-the-art energy storage technologies.

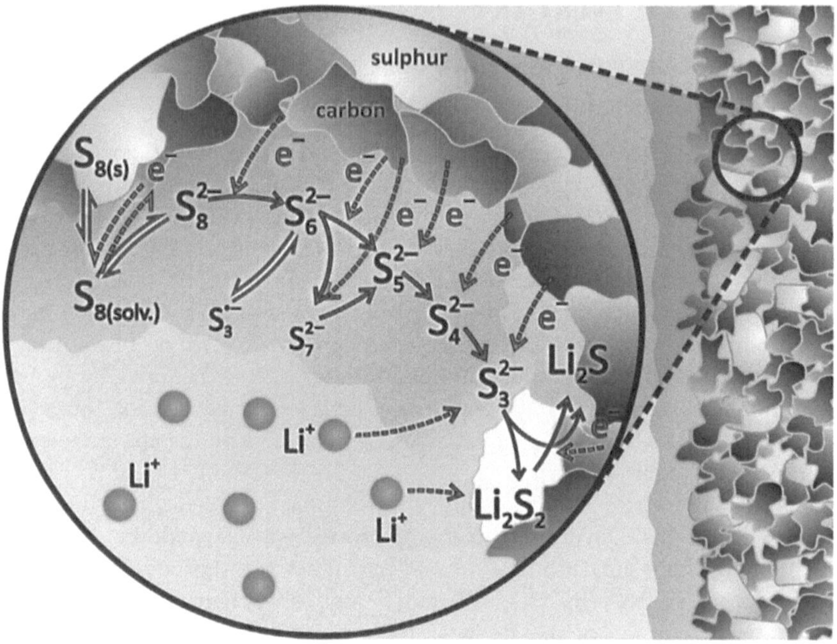


**FIGURE 7.5** Schematic illustration of the stepwise reduction process of sulfur.[37]

In this section, we would like to introduce some recent cases of AFM applications focused on the cathode/electrolyte interface, and anode/electrolyte interface in Li-S batteries, as well as the application of other AFM-based techniques. We hope our description in this section could provide the authors a brief overview of how to utilize AFM in Li-S field and arouse more interest and thinking in its deeper exploration.

### 7.3.2 Dynamic Evolution at the Cathode/Electrolyte Interfaces in Lithium-Sulfur Batteries

In general, sulfur powders, mixed with the carbon and the conductive binder, are employed as cathode materials of Li-S batteries. This mixture is a homogeneous slurry that makes it difficult to pinpoint the surfaces of active sulfur and track their evolution during Li-S redox. In addition, these surfaces are usually buried under the thick, rough slurry, with curved, irregular shapes, which gives rise to additional challenges for their characterization using AFM. To overcome these issues, Wen and Wan et al.[41–43] creatively employed HOPG as the electrode substrate, $Li_2S_8$ additive in the electrolyte as the active material, which converts the buried, curved active interface inside the cathode into one atomic-flat electrode/electrolyte interfaces, enabling the nanoscale visualization using AFM. A schematic illustration of the electrochemical cell is shown in Figure 7.6.

When coupled with a potentiostat or galvanostat, AFM can perform simultaneous scanning and bias-induced electrochemical measurements, known as electrochemical

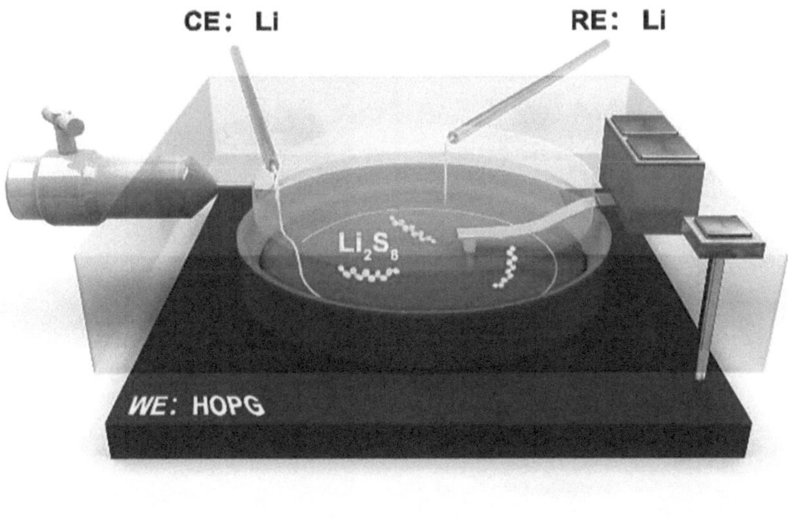


**FIGURE 7.6** Schematic illustration of the Li-S cell using $Li_2S_8$ as the catholyte for EC-AFM study.[41]

AFM (EC-AFM), enabling *in situ* visualization of the active interfaces during Li-S redox process and thus revealing the fundamental mechanisms in detail. On the basis of this design, systematic investigations, including the effects of temperature and Li salts, are provided, which expand our understanding of Li-S reaction, especially the electrodeposition/dissolution processes of solid short-chain $Li_2S_2$ and $Li_2S$, and, in turn, correlate interfacial structure to overall cell function.

Figure 7.7 shows the electrodeposition process of $Li_2S_2$ and $Li_2S$ during the discharge in a typical electrolyte of Li-S cells with 0.5 M lithium bis(trifluoromethanesulfonyl) imide (LiTFSI) in 1,3-dioxolane (DOL)/1,2-dimethoxyethane (DME) (1:1).[41] As shown in Figure 7.7a, the cyclic voltammetry (CV) curve performed in AFM cell presents evident cathodic peaks at 2.3, 2.0 and 1.86 V, related to the formation of intermediate polysulfides, short-chain $Li_2S_2$ and $Li_2S$, respectively, which consists with typical ones performed in Li-S coin cells. During the recharge process, reverse peaks at 2.55–2.6 V reveal the subsequent oxidation of short-chain $Li_2S_2$ and $Li_2S$ into long-chain polysulfides. Correspondingly, evident morphological changes can be observed at the electrode/electrolyte interfaces during the redox processes. Figure 7.7b-f provide *in situ* images captured during the discharge.

Since the long-chain polysulfides are soluble, the interfaces remain clean before 2.0 V (Figure 7.7b). At 1.97 V, NPs, with the average height of 6.5 nm, deposited onto both the step edges and terraces of the HOPG surface, which increased and accumulated as the potential was further swept to negative direction (Figure 7.7c and d). When the potential approached 1.83 V, as shown in Figure 7.7e and f, micro-sized lamellar deposits were observed at the interfaces, which grew rapidly and covered the whole interfaces that can be visualized. After the recharge and subsequent cycling processes, NPs would accumulate while lamella could be dissolved completely. In the combination with *ex situ* Raman and X-ray photoelectron spectroscopy (XPS),

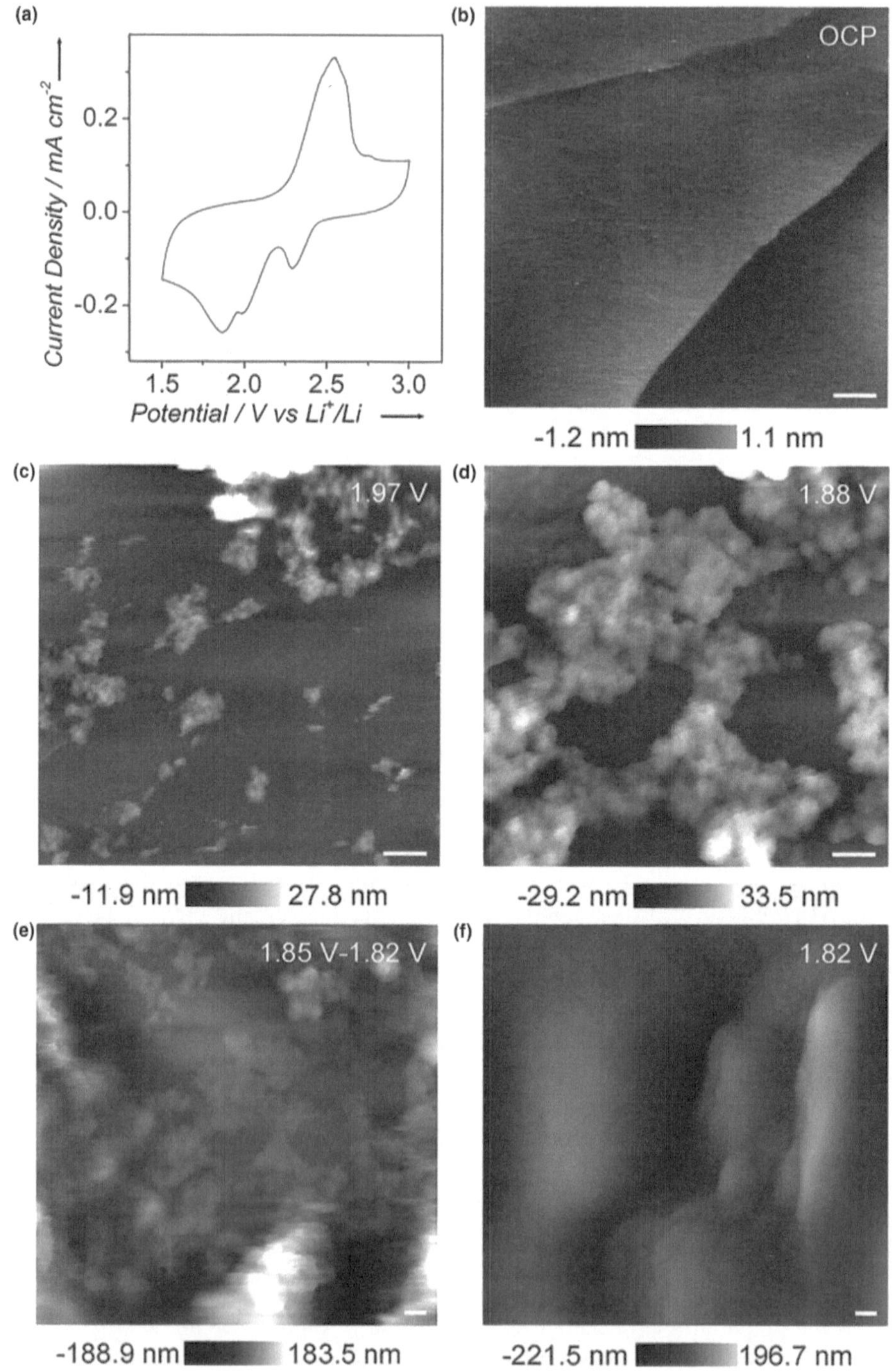


**FIGURE 7.7** EC-AFM investigation of $Li_2S_2$ and $Li_2S$ electrodeposition during the reduction process of $Li_2S_8$ in 0.5 M LiTFSI in DOL/DME. The scale bars are 100 nm.[41]

the NPs were recognized as $Li_2S_2$, and micro-sized lamella as $Li_2S$. This piece of work, for the first time, visualizes the electrodeposition and dissolution processes of short-chain $Li_2S_2$ and $Li_2S$ at the nanoscale using EC-AFM and clearly distinguishes $Li_2S_2$ with $Li_2S$ in terms of interfacial morphology, reversibility upon redox processes, and correlation with cell failure, which indicates the great application prospects of AFM in Li-S field, and is of significant importance to unveil mechanistic details and promote advances of Li-S batteries.

In the follow-up works, using EC-AFM, Wen and Wan et al.[42,43] further investigated the effects of high temperature and Li salts, which would mediate the electrochemical environments of sulfide species, on the electrodeposition of $Li_2S_2$ and $Li_2S$. Figure 7.8 provides the chemical structures of LiTFSI and lithium bis(fluorosulfonyl) imide (LiFSI). The only difference between them is the terminal group that connects to the sulfonyl group, which is trifluoromethyl (-$CF_3$) in LiTFSI (Figure 7.8a), but fluorine (-F) in LiFSI (Figure 7.8b). This difference can actually lead to evident changes in the electrolyte properties, including electronic conductivity and viscosity, as well as their interplay with different sulfide species, giving rise to the unique interfacial evolution of $Li_2S_2$ and $Li_2S$ during Li-S redox processes.

In the electrolyte of 0.5 M LiFSI in DOL/DME, $Li_2S$ would deposit as spheres, which tightly packed at the interfaces, instead of lamellas in LiTFSI. Figure 7.9 provides the interfacial evolution during the discharge process at the high-temperature of 60°C.[42] With the potential decreasing, NPs deposited at 2.15 V and sphere-like products deposited at 2.02 V (Figure 7.9a–c), which were ascribed to the electrodeposition of $Li_2S_2$ and $Li_2S$, respectively. Interestingly, different from the case at room-temperature, an amorphous film can be observed at the interfaces when the potential was further scanned to 1.5 V (Figure 7.9d–f). Note that the LiF peak from XPS shows much higher signal than the one at room-temperature (Figure 7.9g). This film was then ascribed to a net of LiF, which was deposited from the high-temperature degradation of LiFSI and served as a functional film that was able to confine the polysulfides onto the electrode/electrolyte interfaces, suppressing the loss of active materials (Figure 7.9h).

Furthermore, the effects of Li salts were investigated using EC-AFM in the electrolyte with both LiTFSI and LiFSI.[43] Both kinds of the $Li_2S$ deposits, lamellas and spheres, can be observed during the discharge process. An evident difference of their morphological evolution was detected during the recharge process, as shown in Figure 7.10. Lamellas dissolved from edges to the center (Figure 7.10a–f);

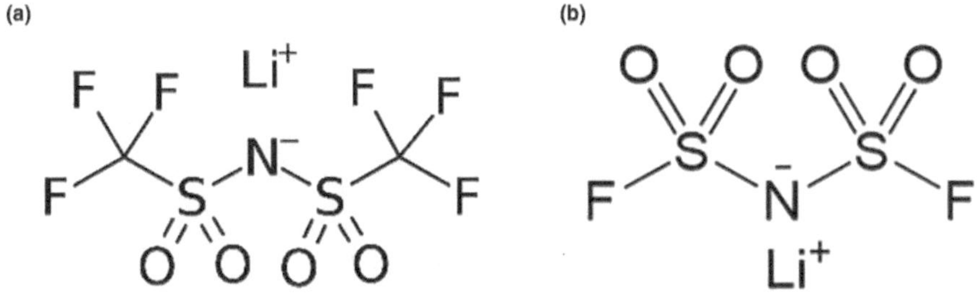


**FIGURE 7.8** Chemical structure of (a) LiTFSI and (b) LiFSI.

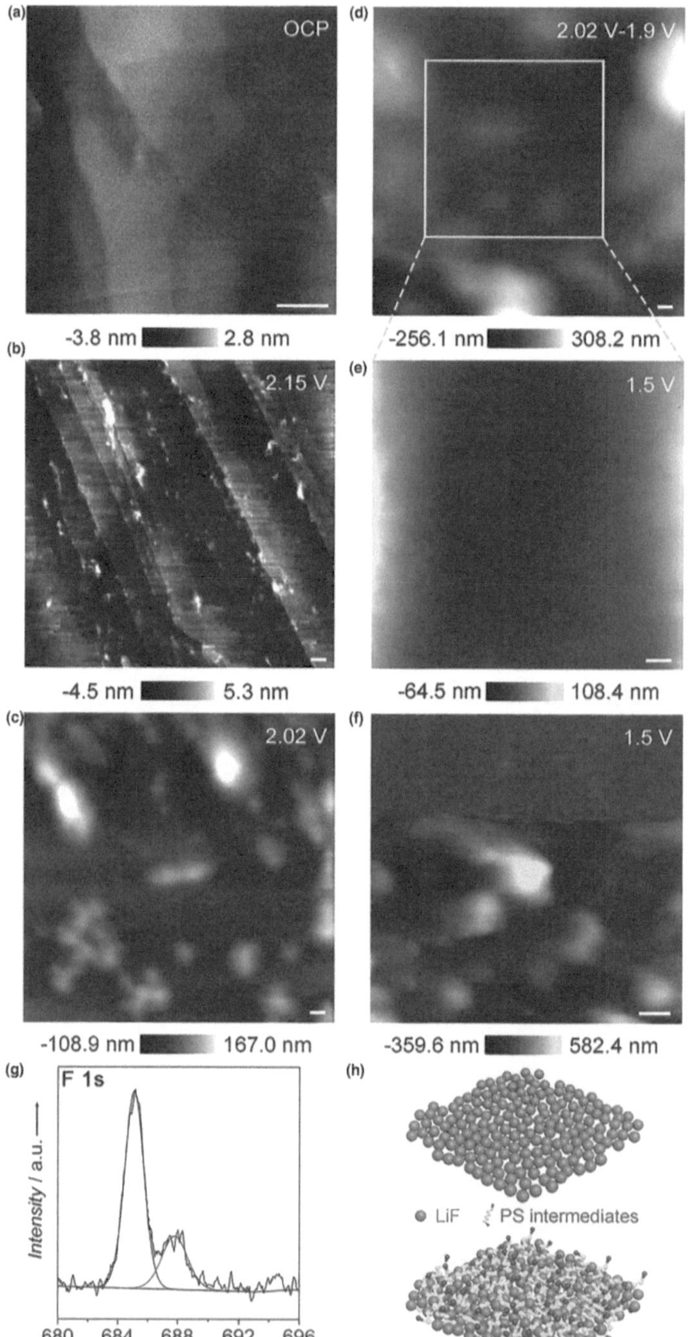


**FIGURE 7.9** EC-AFM investigation during the reduction process of $Li_2S_8$ in 0.5 M LiFSI in DOL/DME at 60°C. The scale bars are 250 nm in (a–e) and 1 μm in (f).[42]

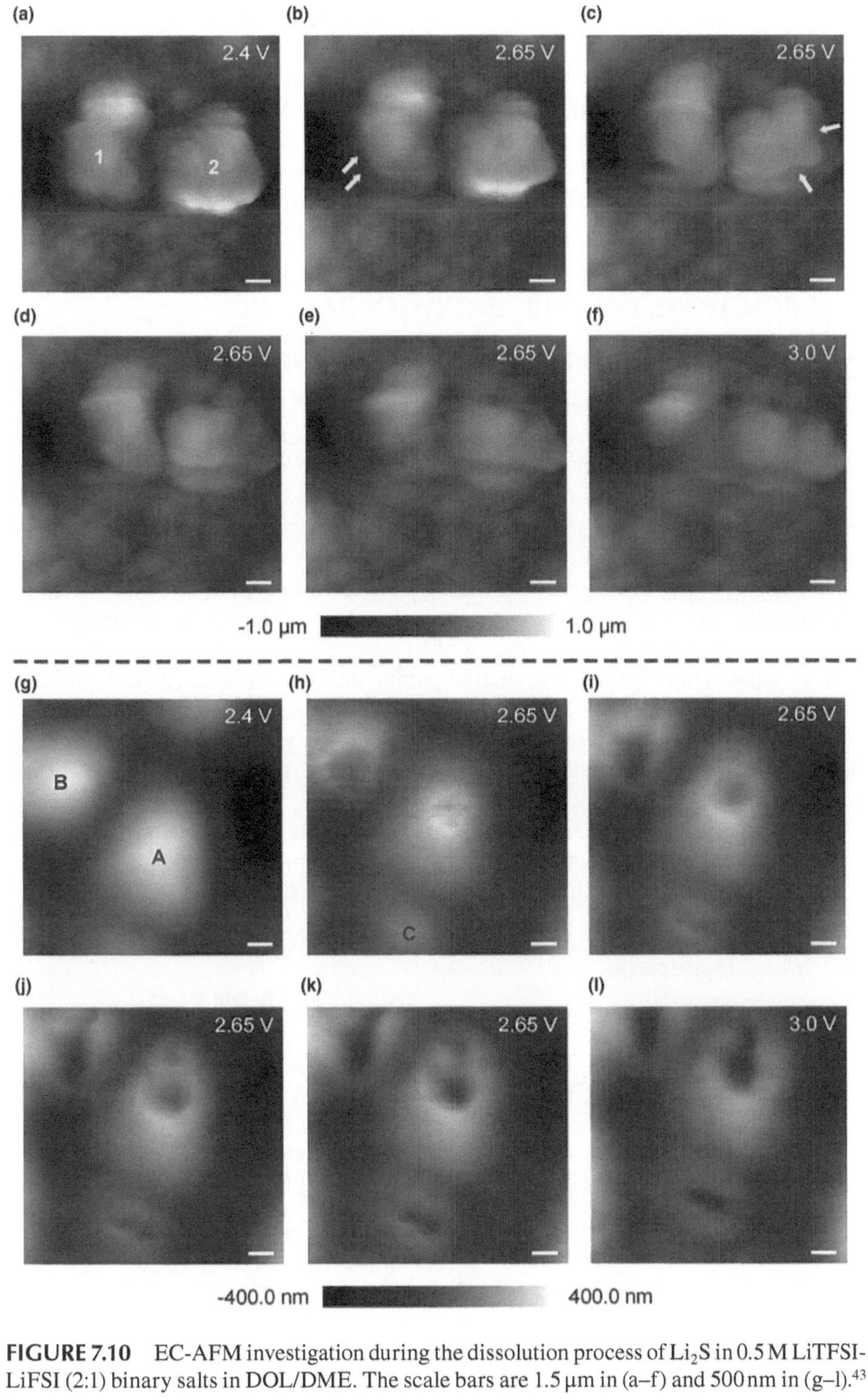


**FIGURE 7.10** EC-AFM investigation during the dissolution process of $Li_2S$ in 0.5 M LiTFSI-LiFSI (2:1) binary salts in DOL/DME. The scale bars are 1.5 μm in (a–f) and 500 nm in (g–l).[43]

however, the spheres started to dissolve from the center with the hollow formation (Figure 7.10g–l). This provides direct evidence of the precise control of Li salts over the reaction pathways of electrodeposited $Li_2S$, which, in turn, affects the overall cell performance. These results emphasize the unique role of EC-AFM on the nanoscale visualization of cathode/electrolyte interfaces in Li-S batteries, paving the way to the in-depth understanding of Li-S electrochemistry and improvements in battery performance.

## 7.4 CORRELATING THE CATALYTIC EFFECT AND INTERFACIAL REACTIONS IN LITHIUM-OXYGEN BATTERIES

### 7.4.1 Interfacial Electrochemistry in Lithium-Oxygen Batteries

Due to the ultrahigh theoretical specific energy of ~3,500 Wh/kg, Li–$O_2$ battery has been paid great research attention in the past several decades.[8,44,45] As shown in Figure 7.11a, the operation of Li–$O_2$ battery is based on the reduction of $O_2$ upon the discharge and its reverse release upon the charge. Compared with the intercalation reaction conducted in the conventional Li–ion battery, the conversion reaction between $O_2$ and $Li^+$ in Li–$O_2$ battery stores more charge per unit mass. In addition, $O_2$ as the positive electrode active material is low-cost and environmentally friendly, which makes Li–$O_2$ chemistry more practical and promising for future energy storage devices. However, the development of Li–$O_2$ batteries is facing huge challenges. The accumulation of insoluble discharge product lithium peroxide ($Li_2O_2$) with poor ion and electric conductivity leads to passivation of the electrode and hinders the subsequent $O_2$ reduction, limiting the cell discharge capacity. Besides that, the discharge product is difficult to reversibly decompose during charging, causing the poor cycle stability and a large over-potential of the battery.[46]

The discharge/charge reactions of the battery mainly occur at the electrode/electrolyte interface. Besides that, the morphology, size, deposition pathway, decomposition behavior and oxidation kinetics of the discharge product on the electrode surface

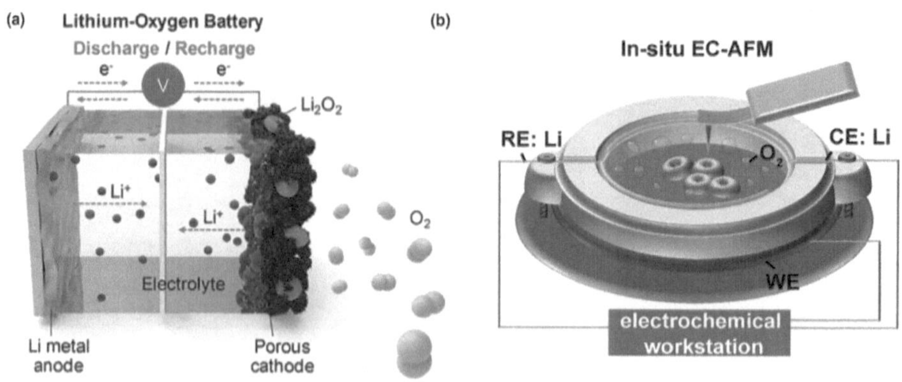


**FIGURE 7.11** Schematic presentations of (a) Li–$O_2$ battery[45] and (b) *in situ* Li–$O_2$ model cell.[47]

are directly related to the performance of the battery. Therefore, directly visualizing the interfacial evolution, including the morphology evolution, structural transition and kinetic behavior, in an operation battery could provide a deep understanding of the correlation between interfacial reaction and battery performance, which is the prerequisite for addressing the challenges facing Li–$O_2$ batteries. *In situ* AFM with high spatial resolution and low invasiveness could realize the real-time tracking of the interfacial reaction occurring at electrode/electrolyte/oxygen three-phase interface (Figure 7.11b).[47] Up to date, *in situ* AFM has been applied in several studies and has made a significant contribution to clarifying the discharging/charging mechanism in Li–$O_2$ batteries.

Before discussing the application of *in situ* AFM for Li–$O_2$ batteries, it is necessary to discuss electrochemistry in Li–$O_2$ batteries first. It is generally considered that upon the discharge, $O_2$ undergoes a one-electron reaction to form the reactive intermediate lithium superoxide ($LiO_2$) (equation 7.1). Then, $LiO_2$ absorbed on the electrode transforms to the discharge product $Li_2O_2$ via a surface electrochemical pathway (equation 7.2) and $LiO_2$ dissolved in the electrolyte generates $Li_2O_2$ through a solution-mediated (electrochemical) pathway (equation 7.3). Film-like $Li_2O_2$ with thickness <10 nm generated via surface-mediated pathway passivates the electrode severely and hinders the subsequent reduction of $O_2$, while toroidal $Li_2O_2$ with large size formed by solution-mediated pathway is well-known to enhance the discharge capacity of the battery. Compared with the toroidal $Li_2O_2$, film-like $Li_2O_2$ has a stronger interaction with the electrode and is easier to decompose upon the charge (equation 7.4).[48–51]

$$O_2 + Li^+ + e^- \rightarrow LiO_2 \tag{7.1}$$

$$LiO_2 + Li^+ + e^- \rightarrow Li_2O_2 \tag{7.2}$$

$$2LiO_2 \rightarrow Li_2O_2 + O_2 \tag{7.3}$$

$$Li_2O_2 \rightarrow O_2 + 2Li^+ + 2e^- \tag{7.4}$$

It has been widely reported that the electrode reaction can be promoted by modifying the electrolyte or constructing catalytic electrodes, thereby improving the performance of the battery.[52] Therefore, in this section, we focus on the current state of AFM application in exploring electrolyte, solid catalyst and soluble catalyst effect in Li-$O_2$ batteries, aiming at gaining a deeper understanding of Li-$O_2$ batteries and providing ideas for using AFM to investigate the solid/liquid/gas three-phase interfacial issues.

### 7.4.2 *In Situ* AFM Observation of the Electrolyte Effect

The physico-chemical property of the electrolyte is one of the most critical factors that determine the electrochemical performance of the Li–$O_2$ battery. Electrolyte solvents with a low polarity, such as tetraethylene glycol dimethyl ether (TEGDME) and dimethyl ether (DME), have a high $Li^+$–ion transport efficiency and $O_2$ solubility.

But the low donor number (DN) values of such solvents limit their ability to stabilize the solvated $LiO_2$ intermediates. In contrast, solvents with high DN values, such as dimethyl sulfoxide (DMSO), exhibit a good ability to dissolve $LiO_2$ intermediates. But here, their low $O_2$ solubility and instability towards Li–metal become the key challenges during cell operation.[44,53] Understanding the Li–$O_2$ reaction mechanism mediated by electrolytes is the key to developing suitable electrolytes for Li–$O_2$ batteries. Using *in situ* AFM, Wen et al. visualized Li–$O_2$ electrochemical reaction in the TEG–based electrolyte that shows good stability and is widely used. At the beginning of the discharge, most of the NPs nucleate along the step edge of HOPG. Then the NPs develop into stable nanoplates and eventually form a $Li_2O_2$ film (thickness of ~5 nm) covering the electrode (Figure 7.12A). Upon the charge, the decomposition of this $Li_2O_2$ film occurs at the $Li_2O_2$/HOPG interface, and the thick $Li_2O_2$ film delays the complete decomposition due to the low electronic conductivity. These *in situ* AFM images give the first visualization of dynamic Li–$O_2$ interfacial reaction and morphological information of $Li_2O_2$ in ether–based electrolyte.[54] Further, Hong et al. monitored the decomposition processes of conformal $Li_2O_2$ films using *in situ* AFM and combined gas analysis and electrochemical methods to explore the preferential oxygen evolution reaction (OER) route within $Li_2O_2$ (Figure 7.12B). *In situ* views show that at low potentials of 3.2–3.7 V (vs. $Li^+$/Li), the decomposition is exclusive to the thinner part of the $Li_2O_2$ films, leading to the evolution of $O_2$ and subsequent release of superoxide species. At a higher potential (>3.7 V), the oxidation initiates at the sidewalls of the thicker $Li_2O_2$ film, and the succeeding lateral decomposition of $Li_2O_2$ film suggests the preferential ion and electron transport occurring at the sidewalls of the thick $Li_2O_2$ film.[55]

Using *in situ* EC–AFM, Shen et al. observed the nucleation, growth and oxidation process of $Li_2O_2$ on HOPG surface in a Li–$O_2$ model cell with DMSO–based electrolyte (Figure 7.12C). The authors found that upon the discharge, $O_2$ is reduced with the deposition of NPs, followed by the formation of toroidal $Li_2O_2$. High-resolution AFM image reveal that numerous nanosheets are arranged spirally to form the toroidal structure with a diameter of 300–400 nm and a thickness of ~100 nm. During charging, the decomposition of the toroid proceeded slowly before 3.88 V. Holding the potential at 3.88 V, the toroidal $Li_2O_2$ was quickly depleted. The author considered that the decomposition of toroidal $Li_2O_2$ mainly take a bottom-up way. After the bottom layer is completely decomposed, the toroid is directly desorbed from the electrode surface, causing irreversible capacity degradation.[56]

The solution-mediated growth pathway of $Li_2O_2$ with large size during the discharge can be boosted by introducing electrolyte additives, such as $H_2O$ and $LiNO_3$, improving discharge capacity significantly.[46,52,57] *In situ* EC-AFM is employed to explore the correlation between the morphological changes in the reaction deposits mediated by electrolyte additives and battery performance. Mediated by $H_2O$ additive, the origin of the large structure features upon the oxygen reduction reaction (ORR) and the decomposition behavior of discharge product upon the OER are revealed by *in situ* AFM probes. At the water content of ~6,000 ppm, during the discharge, $Li_2O_2$ grows at the electrolyte/$Li_2O_2$ interface, which involves lateral growth to form nanoplates followed by vertical growth mainly along the peripheral area to form a toroidal structure (Figure 7.13A). In contrast, at a high $H_2O$ content of

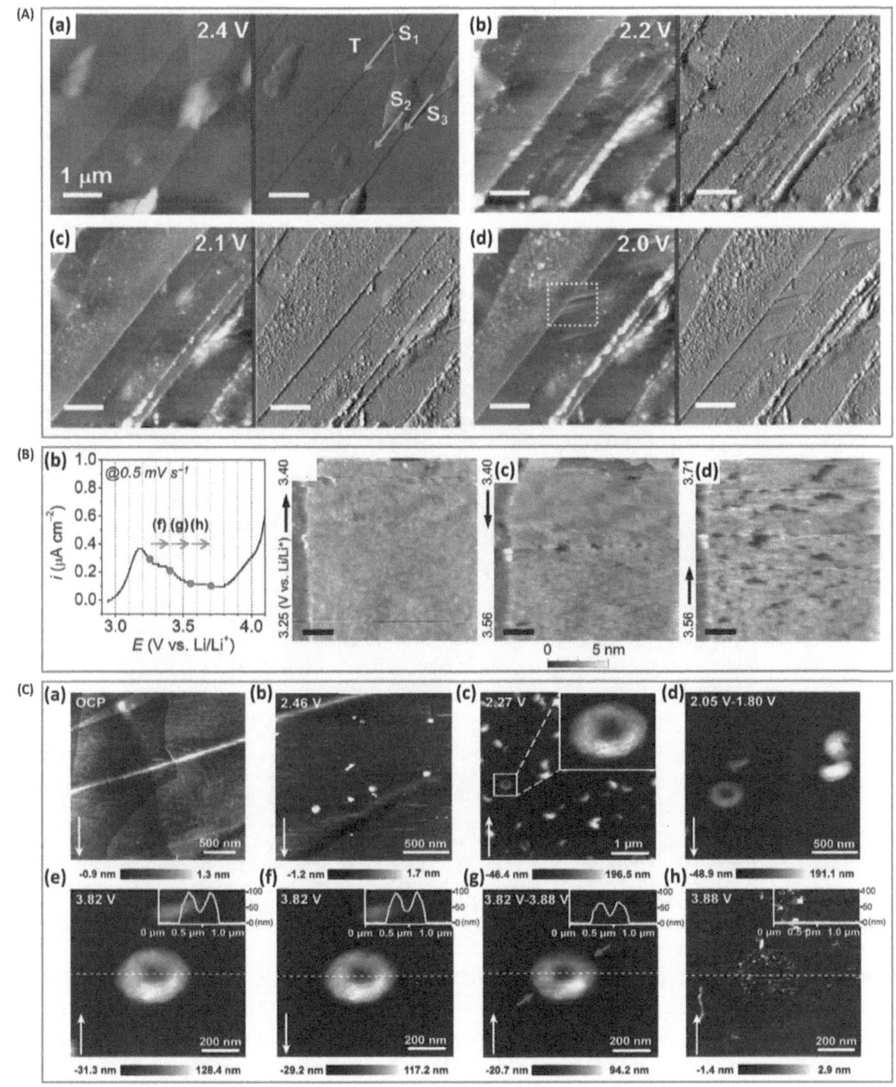

**FIGURE 7.12** (A) *In situ* AFM image of film-like $Li_2O_2$ formation processes on HOPG in 0.5 M LiTFSI/TEGDME. The scale bars in (a–d) are 1 μm.[54] (B) (a) Anodic potential sweep at 0.5 mV/s in 0.5 M LiTFSI/TEGDME and (b–c) the corresponding *in situ* AFM images of the decomposition process of $Li_2O_2$ film before charging to ~3.70 V. The scale bars in (b–d) denote 500 nm. The black arrows in (b–d) indicate the scanning direction.[55] (C) *In situ* AFM image of toroidal $Li_2O_2$ (a–d) formation and (e–h) decomposition processes on HOPG in 0.5 M LiTFSI/DMSO. The white arrows in (a–h) indicate the scanning direction.[56]

~20,000 ppm, lamellar $Li_2O_2$ deposits along the step edge of HOPG surface upon discharging (Figure 7.13B). During the charge, the lamellar $Li_2O_2$ decomposes ahead of the toroid, showing higher electrochemical reversibility (Figure 7.13C). By reducing

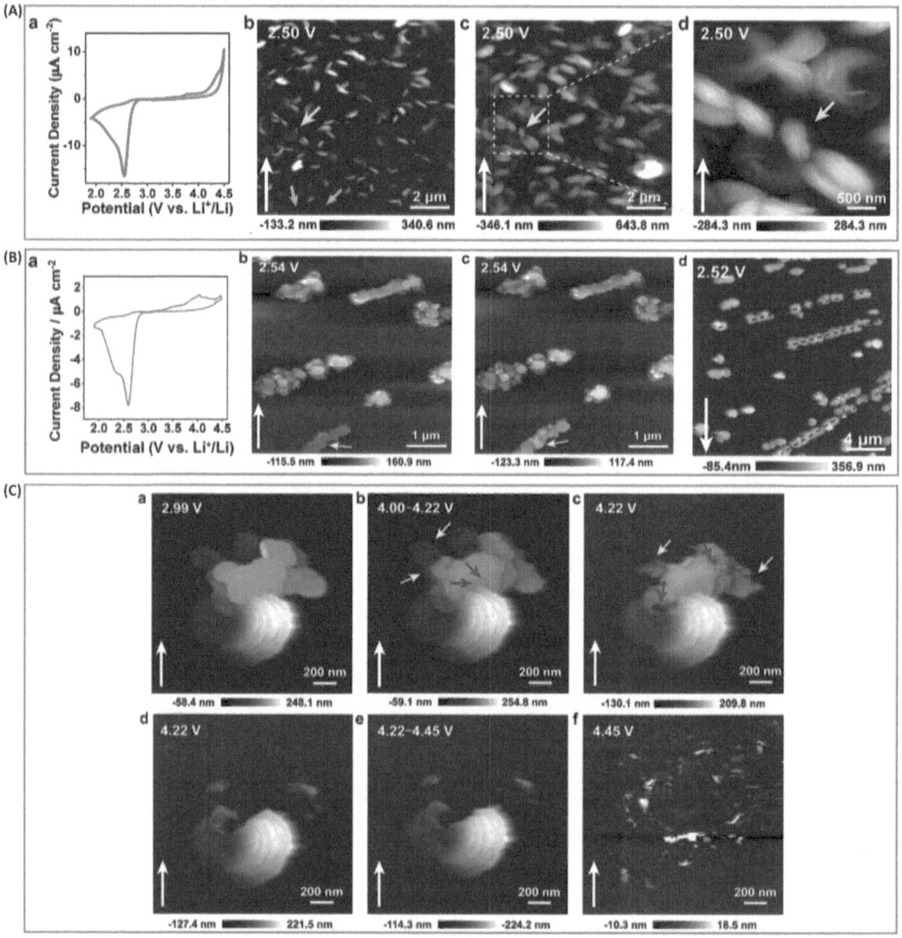


**FIGURE 7.13** (A) (a) Cyclic voltammetry (CV) curve taken in the *in situ* Li–$O_2$ model cell at a sweep rate of 1 mV/s in 0.5 M LiTFSI/DMSO containing ~6,000 ppm $H_2O$. (b–d) *In situ* AFM images of HOPG upon discharge.[58] (B) (a) CV curve taken in the *in situ* Li–$O_2$ model cell at a sweep rate of 1 mV/s in 0.5 M LiTFSI/DMSO containing ~20,000 ppm $H_2O$. (b–d) *In situ* AFM images of HOPG upon discharge.[58] (C) *In situ* AFM image of lamellar and toroidal $Li_2O_2$ decomposition on HOPG. The white arrows in the images indicate the scanning direction.[58]

the discharge rate, the discharge product can be transformed from a toroidal shape to a lamellar one, thereby reducing the battery overpotential.[58]

### 7.4.3 *In Situ* AFM Monitoring the Catalytic Effect of Solid Catalysts

Li–$O_2$ interfacial reaction can be enhanced by constructing an efficient catalytic electrode, which is crucial to the development of Li–$O_2$ batteries.[59,60] Clarifying the exact role of electrocatalysts for ORR during discharge and OER during charge is of fundamental significance for the design of high-performance catalyst. In response, *in situ* AFM was applied to explore the precise mechanism and details of $Li_2O_2$ growth/decomposition on catalytic electrodes as well as the structure-activity relationship of the catalytic cathode.

Because of its excellent catalytic activity, Au has more applications in Li–$O_2$ batteries. Besides that, the nanostructure of Au is directly related to its catalytic activity.[61,62] Figure 7.14A shows the *in situ* AFM images of $Li_2O_2$ nucleating, growing and decomposing on the nanoporous gold (NPG) upon the Li–$O_2$ electrochemical

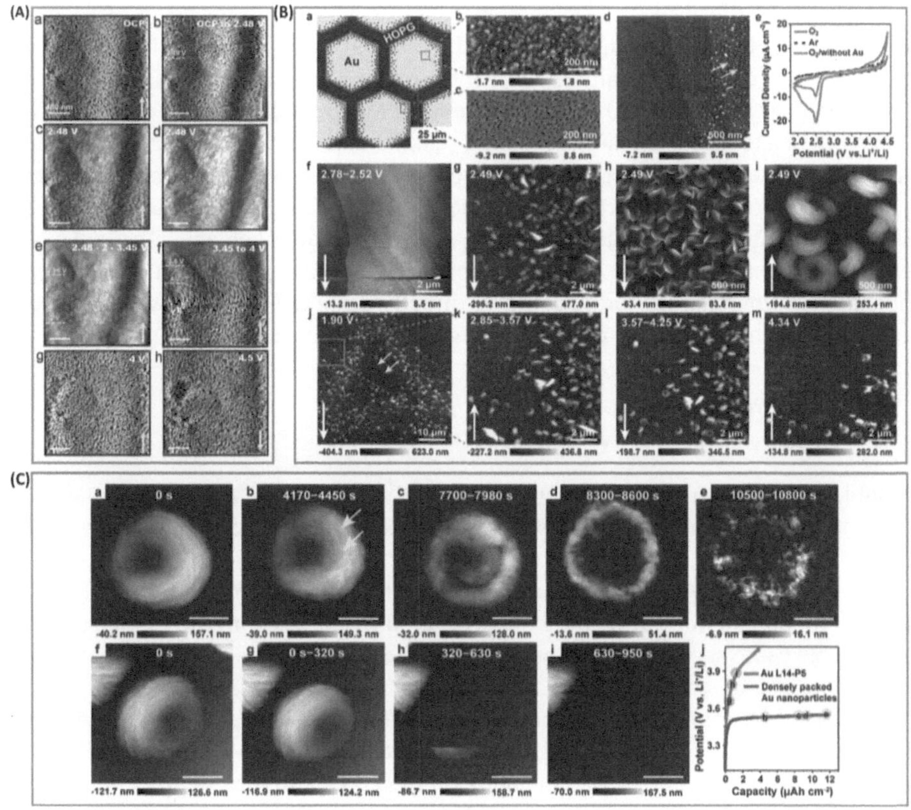


**FIGURE 7.14** (A) *In situ* AFM images of the NPG electrode (a–d) upon ORR and (e–h) OER in 0.5 M LiTFSI/tetraglyme. The arrows in the AFM images indicate the scanning direction.[63] (B) (a) Schematic illustration of the Au-based electrode with various nanostructures. AFM image of (b) Au nanoparticles domain and (c) nanoporous Au. (d) AFM image of the border between the nanoporous Au and Au nanoparticles domain. (e) CV curve taken in the Li–$O_2$ model cell at a scan rate of 1 mV/s. The electrolyte is 0.5 M LiTFSI/DMSO. *In situ* AFM images of the Au–based electrode (f–j) upon discharging and (k–m) charging. The white arrows in the AFM images indicate the scanning direction.[64] (C) Time-dependent AFM images of the decomposition processes of toroidal $Li_2O_2$ on (a–e) Au nanoparticles electrode and (f–i) nanoporous Au under galvanostatic control at a charge rate of 4 μA/cm$^2$. (j) The corresponding charge curves at a charge rate of 4 μA/cm$^2$. The scale bars in (a–i) are 200 nm.[64]

reaction in the ether-based electrolyte. As discharging to 2.48 V, the NP-shaped products gradually grow along the ligaments of the NPG over time, resulting that $Li_2O_2$ film covers the NPG surface. During the charging, the size of $Li_2O_2$ decreases after 2.96 V and eliminates from the NPS over 3.8–4.0 V. The rapid growth and decomposition of the $Li_2O_2$ at low overpotentials demonstrate that the NPG/$Li_2O_2$ interface where the ORR/OER occurs is beneficial to the electron transfer due to the high electric conductivity of the NPG.[63]

Further, *in situ* EC-AFM was applied to explore the Li–$O_2$ interfacial reaction occurring at the composite electrode with various Au nanostructures, as shown in Figure 7.14B. Figure 7.14Ba–j display the discharge processes on the electrode which contains closely packed Au NPs, nanoporous Au with pore size range of 5–15 nm and bare HOPG. Discharging to 2.52 V, the toroidal $Li_2O_2$ with small size (~50 nm in thickness, ~250 nm in diameter) distribute on the densely packed Au NPs and the toroid with larger size (~250 nm in thickness, ~600 nm in diameter) tend to accumulate on the nanoporous domain, revealing that the nanoporous structure could effectively boost the growth of $Li_2O_2$ with large size. In sharp contrast to the Au domain, the amounts of toroidal products on the bare HOPG are significantly reduced. Figure 7.14C shows that the toroidal $Li_2O_2$ are in continuous contact with the Au NPs electrode during the charge and $Li_2O_2$ could be fully decomposed at a low charge potential. In contrast, the toroid could be directly desorbed from the Au nanoporous electrode at a higher charge potential, causing irreversible capacity degradation.[64]

### 7.4.4 *In Situ* AFM Visualization of the Surface Effect of Soluble Catalysts

Soluble catalyst is a class of soluble redox-active molecules which can potentially promote the growth/oxidation of solid and immobile $Li_2O_2$ occurring at the $Li_2O_2$/electrolyte interface, avoiding the catalyst being passivated by insulating $Li_2O_2$ and only catalyzing $Li_2O_2$ that is in direct contact with the electrode.[65,66] *In situ* AFM was used to investigate the Li–$O_2$ interfacial reaction dynamics with the participation of the soluble catalyst, which could provide visual evidence for in-depth understanding of the catalytic mechanism.

Figure 7.15A shows the decomposition process of film-like $Li_2O_2$ catalyzed by the representative redox mediator of 2,2,6,6–tetramethylpiperidin–1–yl)oxyl (TEMPO). As the charging progresses, the nanopits on the $Li_2O_2$ film are significantly broadened. As the potential rises, the accessibility of $TEMPO^+$ to the $Li_2O_2$/electrolyte interfaces is increased by the extending $TEMPO^+$ concentration, which reduces the height of the $Li_2O_2$ film until it is eliminated. These results visualized the advantage of soluble catalyst to overcome the sluggish charge transport in the bulk $Li_2O_2$.[55]

Due to the high catalytic performance, soluble catalyst of DBBQ has been used in Li–$O_2$ batteries.[67,68] Figure 7.15B illustrated the interfacial evolution processes with the addition of DBBQ in DMSO-based electrolyte. Upon the discharge, NPs with a height of ~60 nm deposit and grow into micron-sized flower-like $Li_2O_2$ with distribution along the step edges of HOPG. Upon the charge, many nanopits randomly appear on the outer layer of the flower-like $Li_2O_2$, and then the breaking layer with a thickness of around 60 nm peels off, exposing the inner structure. This outside-in

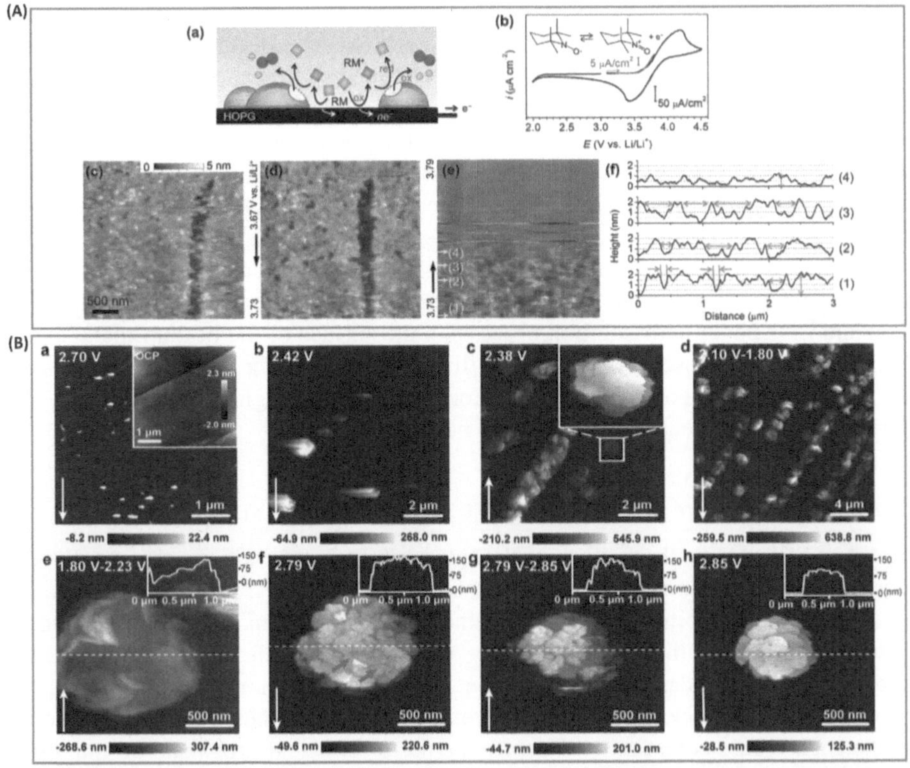


**FIGURE 7.15** (A) (a) Illustration of $Li_2O_2$ decomposition process catalyzed by redox mediator. (b) CV curve with 10 mM TEMPO in $O_2$-free (black curve, 5 mV/s) and $O_2$-containing electrolyte (red curve, 0.2 mV/s). *In situ* AFM images (c) at the beginning and (d–e) along the anodic LSV. (f) Section profiles of the lines indicated by grey arrows in (e).[55] (B) *In situ* AFM images of flower-like $Li_2O_2$ (a–d) formation and (e–h) decomposition on HOPG in 0.5 M LiTFSI/DMSO with the 5 mM DBBQ redox mediator.[56]

pathway of the decomposition in a three-dimensional direction greatly promotes the oxidation process and enhances the reversibility of $Li_2O_2$. These results provide a reliable basis for understanding DBBQ to improve the performance of Li–$O_2$ batteries at the nanoscale.[56]

## 7.5 SEI EVOLUTION AND LI PLATING/STRIPPING PROCESSES ON LI METAL ANODE

### 7.5.1 Introduction

With the increasing energy consumption, materials and devices with high energy density are urgently demanded. The long-lasting pursuit for high-energy-density storage systems has stimulated the development of lithium metal anode for its high specific capacity (3,860 mA h/g) and the lowest electrochemical potential (−3.04 V vs standard hydrogen electrode). Nevertheless, the unstable SEI formation, uneven Li deposition and further Li dendrite penetration hinder the development and deployment of the Li metal batteries. Therefore, it is of great significance to study the evolution process of the SEI on the surface of lithium metal anode and the kinetic behavior of lithium deposition/dissolution, and to reveal the reaction mechanism and capacity degradation principle of lithium metal batteries. However, due to the chemical activity of lithium metal and its instability to detector such as electron beams, techniques that are friendly and non-destructive to lithium metal anode are required, which can reveal the reaction mechanism of lithium batteries on nanoscale. EC-AFM is able to meet the above requirements. Therefore, using *in situ* AFM to explore the SEI formation and Li deposition/dissolution process on the surface of the lithium metal anode during the charge/discharge process will help revealing the relationship between them and provide theoretical guidance for the optimal design of materials and performance improvement of lithium metal batteries.

### 7.5.2 The SEI Film at Lithium Metal Anode/Electrolyte Interface

An ideal SEI film on the surface of the lithium metal electrode has the characteristics of compactness and uniformity, smoothness and flatness, and the combination of rigidity and flexibility. It can help induce uniform lithium deposition/dissolution behavior. As is often the case, the rigidity of SEI film is beneficial for effectively suppressing the lithium dendrites penetration, and the interphasial property of flexibility protects the lithium electrode from uneven deformation. When the batteries operate under high C-rate conditions, Li deposits tend to grow and propagate on the top of SEI film. In that case, precipitation behavior and dendritic penetration of lithium have little correlation with the interphasial properties of SEI film. The mechanical properties of the as-formed SEI layers after Li deposition at different current densities were elucidated by force-distance curves and Young's modulus maps.[69] It is investigated that the compact and thinner SEI formed at low current density was mainly composed of inorganic products with high mechanical strength, while the SEI formed at a high current density contained organic–inorganic mixed layers which were relatively thicker and fragile. The mechanical property of SEI film at the electrode/electrolyte interface could be distinctly revealed and understood by AFM, thus further realizing the precise regulation of the SEI film. Employing the electrochemical polishing method based on the potentiostatic and galvanostatic control strategy, the formation of the ultra-smooth and thin SEI film can be obtained on the surface of the alkali metal anode, and the control of electrochemical behavior can be achieved.[70] Results indicate that the polished metal lithium electrode has a large-scale flat surface with a uniform smooth SEI film (Figure 7.16A). The AFM force curve and the XPS characterization with depth analysis revealed that the SEI film consists of soft organic cross-linking polymer with stiff inorganic component, which is with an adjustable multilayer film structure, and significantly improved ionic conductivity (Figure 7.16B). This microscopically flat organic–inorganic composite multilayer SEI film with both rigidity and flexibility has a good effect on inhibiting the growth and puncture of lithium dendrites, and is beneficial to increase the coulombic efficiency and cycle stability during the deposition/dissolution process of lithium metal

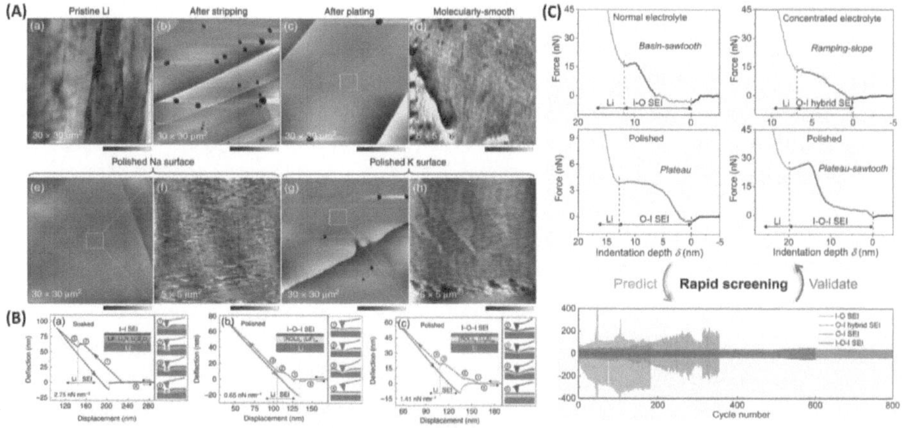


**FIGURE 7.16** (A) (a–d) AFM images of Li surface (a) before, (b) after stripping and (c, d) plating processes. AFM images of (e, f) Na and (g, h) K surfaces after polishing.[70] (B) Typical force-displacement curves of (a) soaked Li surface, polished Li surfaces with (b) single potential step in the ESP process and (c) multiple potential steps in the ESP process (I–O–I structured SEI).[70] (C) Mechanical properties of the representative SEIs on Li-metal and Li-free anodes are obtained by AFM nanoindentation features for rapidly assessing the qualities of unknown SEIs and the electrochemical performancs.[71]

electrode. The surface polishing method of alkali metal based on electrochemistry, *in situ* construction of SEI film, and multi-scale and multi-dimensional joint characterization provide new solutions to the problems of large volume expansion and uncontrollable lithium dendritic growth on alkali metal anodes. The idea and guidance of the strategy also provide high-quality alkali metal surface for basic research of surface science.

Adopting the method of AFM nano-indentation, the SEI film on the lithium-containing and lithium-free anodes can be evaluated for the prediction of the corresponding electrochemical performance (Figure 7.16C). Wang et al.[71] constructed a series of single-layer and multilayer SEI films with known compositions and structures in a targeted manner. Three basic single-layer SEIs, each with mechanical stiffness, medium stiffness and softness, were established with a certain chemical composition and spatial arrangement. More complex multilayer SEI structures are composed of layers along the surface normal. The AFM nano-indentation test was performed on the above-mentioned SEI films, and it was used as a standard result to measure the thickness, smoothness and Young's modulus of the unknown SEI film. These characteristics are further related to the electrochemical cycle performance of the corresponding lithium metal anode and can be used as a standard for quickly evaluating the mechanical and electrochemical properties of the unknown SEI film and its Young's modulus, smoothness and thickness. The combination of AFM and chemical composition characterization further reveals the correlation between the SEI composition on the surface of the lithium metal anode and the nucleation of lithium dendrites.[72]

In addition, the spontaneous formation of the SEI at Li metal anode is due to the decomposition of electrolyte components.[73–75] The quality of SEIs is highly affected by the additives of electrolytes, which can be visualized at nanoscale by AFM, providing fundamental understanding and important guidance for Li protection. For example, lithium nitrate ($LiNO_3$) has been highly recommended for a high-quality SEI formation at Li metal anode.[76–79] Figure 7.17 shows the AFM images of Li metal anode immersed in 0.5 M $LiNO_3$ in DOL/DME for different lengths of time.[76] In 10 minutes, a smooth surface is observed which is similar to the one immersed in solvents (Figure 7.17a). Large particles deposited onto the surface after immersing 1 hour,

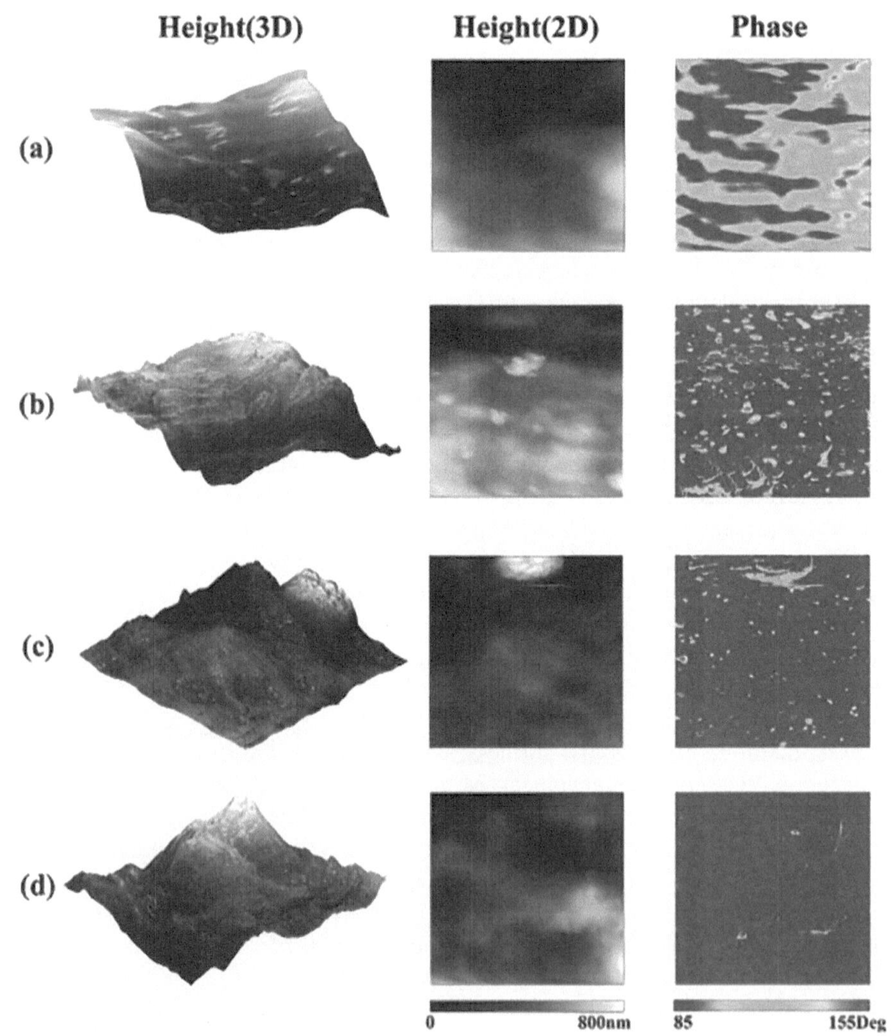


**FIGURE 7.17** 10 μm × 10 μm AFM images of Li metal anode immersed in 0.5 M $LiNO_3$ in DOL/DME for (a) 10 min, (b) 1 h, (c) 3 h and (d) 5 h.[76]

which gradually transform into a homogeneous flat film in 5 hours (Figure 7.17b–d). This can be attributed to the continuous decomposition of electrolyte components, revealing their different reaction rates and correlating structure to function.

In addition, using EC-AFM, a recent work elucidated the effect of $LiNO_3$ on SEI formation in detail.[80] Figure 7.18 presents the real-time morphological changes during the pristine SEI formation (from large particles into homogeneous film, Figure 7.18a–e) and its evolution upon Li deposition/dissolution (Figure 7.18g–i) in 5.0 wt% $LiNO_3$. Quantitative mapping of the elastic modulus can be produced and processed typically through the Derjaguin–Muller–Toporov (DMT) model.[81] With the help of DMT Modulus information, a soft-stiff amorphous-NP bilayer structure of SEI can be recognized (Figure 7.18f), which ensures the formation of a dendrite-free and stable interface during the Li deposition and dissolution processes. AFM has

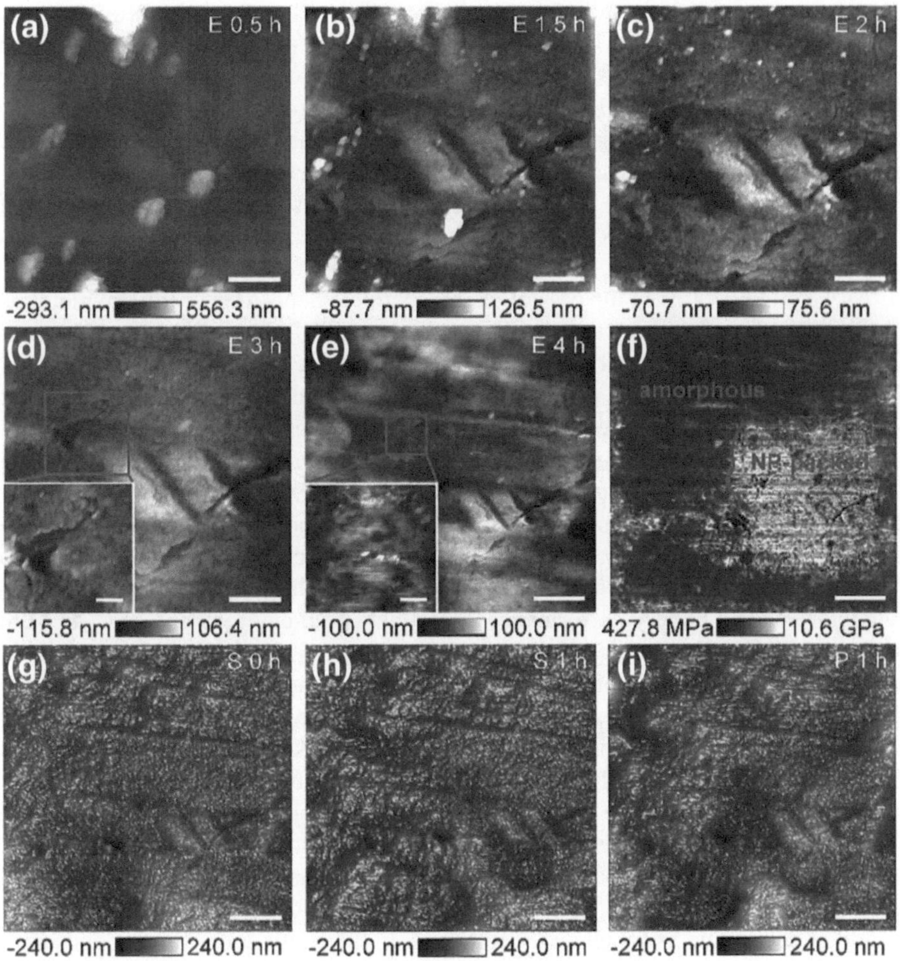


**FIGURE 7.18** EC-AFM investigation of the formation and evolution of the bilayer SEI in the electrolyte of 0.5 M LiTFSI, 5 mM $Li_2S_8$ in DOL/DME with 5.0 wt% $LiNO_3$.[80]

great advantages in terms of the visualization of nanoscale structures, key features as well as local physicochemical properties of SEI. More attention can be paid to the interfacial exploration of the anode/electrolyte interface using AFM.

### 7.5.3 Dynamic Evolution and Artificial Regulation of Li Precipitation Behaviors

Regarding of the dynamic deposition/dissolution behavior of lithium metal, Kitta et al.[25] observed the initial lithium deposition on the surface of lithium metal and the corresponding process of protrusion growth by *in situ* AFM. They also studied the mechanical properties of the SEI film by analyzing its adhesion to the surface of lithium metal. The results found that the protruding lithium metal deposits are related to its weak adhesion, and the uniformly distributed SEI film can induce the deposition of lithium to proceed evenly on the surface of the lithium metal (Figure 7.19A). Shen et al.[26] used EC-AFM to study the first lithium deposition of EC-based and FEC-based electrolytes on graphite anodes in real time, and the results showed that the SEI formed by FEC electrolytes contains more LiF inorganic salts. Because LiF has better hardness and stability, the SEI formed by FEC electrolyte is harder and denser than that of EC electrolyte. It possesses better mechanical properties and larger resistance and slows down the intercalation of lithium ions. Therefore, it can effectively inhibit the growth of lithium dendrites (Figure 7.19B).

As we all know that in the case of electrochemical deposition of lithium metal on the surface of anode during the charging process, an SEI layer will also on-site form on the surface of deposited Li, which can directly affect the deposition morphology, Li growth behavior and cell performance. Different from the SEI film formed on the surface of the electrode, the SEI layer grown on–site on the surface of the deposited lithium will produce corresponding continuous morphological changes during the process of charging, discharging and cycling along with the dynamic process of lithium deposition and dissolution. The *in situ* detection and on-site analysis of the SEI-like layer are extremely challenging. Therefore, tracking the dynamic evolution process of this type of SEI layer is the key to clarify its impact on the anode/electrolyte interface and battery performance. Shi et al.[82] used environmental EC-AFM to study the lithium deposition/dissolution behavior at the anode/electrolyte interface, and successfully *in situ* monitored the dynamic evolution process of the SEI shell growth on the surface of the spherical metallic lithium particles. They also found that during the first cycle of lithium deposition, spherical-like metallic lithium particles nucleated on the surface of the copper electrode and further grew. During the lithium dissolution process, with the continuous dissolution of metallic lithium, the SEI shell formed on the surface of the deposited Li shrinks and collapses significantly. In the subsequent cycles, the deposited metal lithium particles preferentially nucleate and grow at the position where there is no SEI shell remaining, which reflects the passivation effect of the SEI shell on the interface. In the continuous lithium deposition/dissolution process, the repeating formation and collapse of the SEI shell layer causes the continuous consumption of the electrolyte; at the same time, the continuous

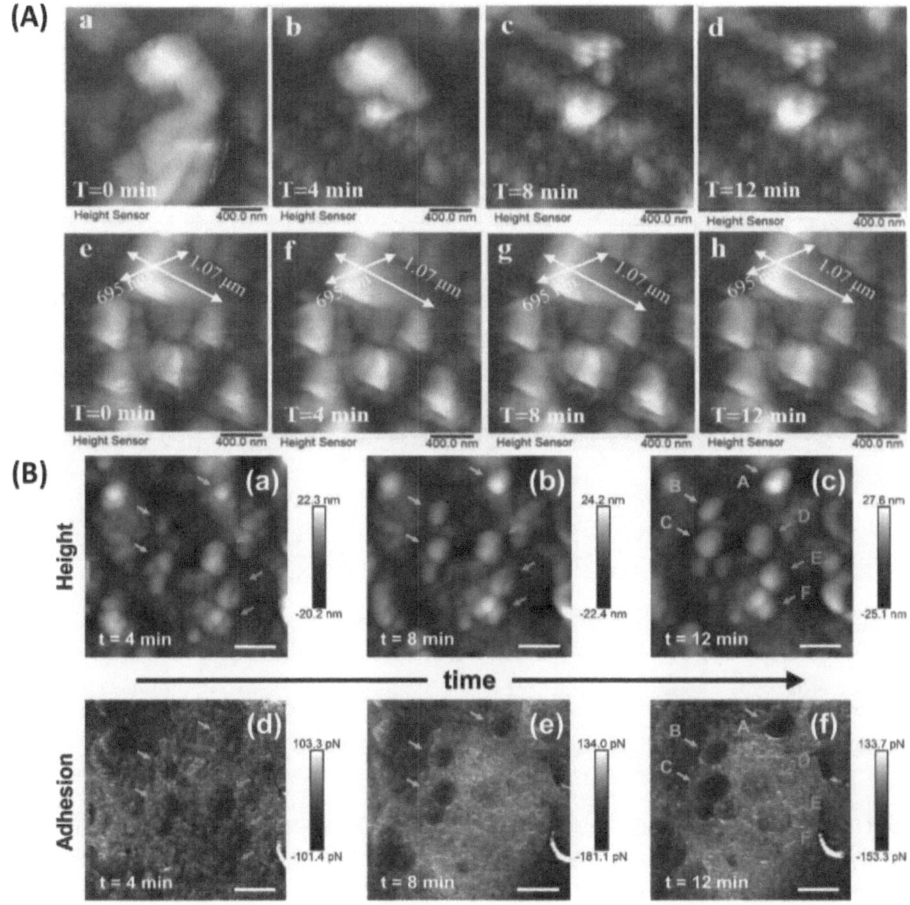


**FIGURE 7.19** (A) (a–d) AFM images of Li surface acquired at various times during Li deposition in EC/DMC electrolyte. (e–h) Topographic height images acquired at various times during Li deposition in FEC/DMC electrolyte.[26] (B) AFM images of the surface of the Li electrode during an electrochemical deposition experiment of Li. (a–c) Topographic height images acquired at various times during the Li deposition and (d–f) adhesion mapping images corresponding to images (a–c). The image acquisition was started at the beginning of the Li deposition process (t = 0min), which was confirmed by a voltage plateau in electrochemical charge profile. All the scale bars indicated a length of 100 nm.[25]

accumulation of the SEI shell layer at the anode/electrolyte interface increases the interface charge transfer resistance and ultimately leads to the degradation of battery performance (Figure 7.20). This work provides a direct visual basis for the morphology and *in situ* evolution of the SEI shell layer grown on the surface of the spherical metal lithium particles and reveals degradation mechanism of the lithium anode during the charge-discharge cycle. It provides the key insights of electrode design optimization and interface construction.

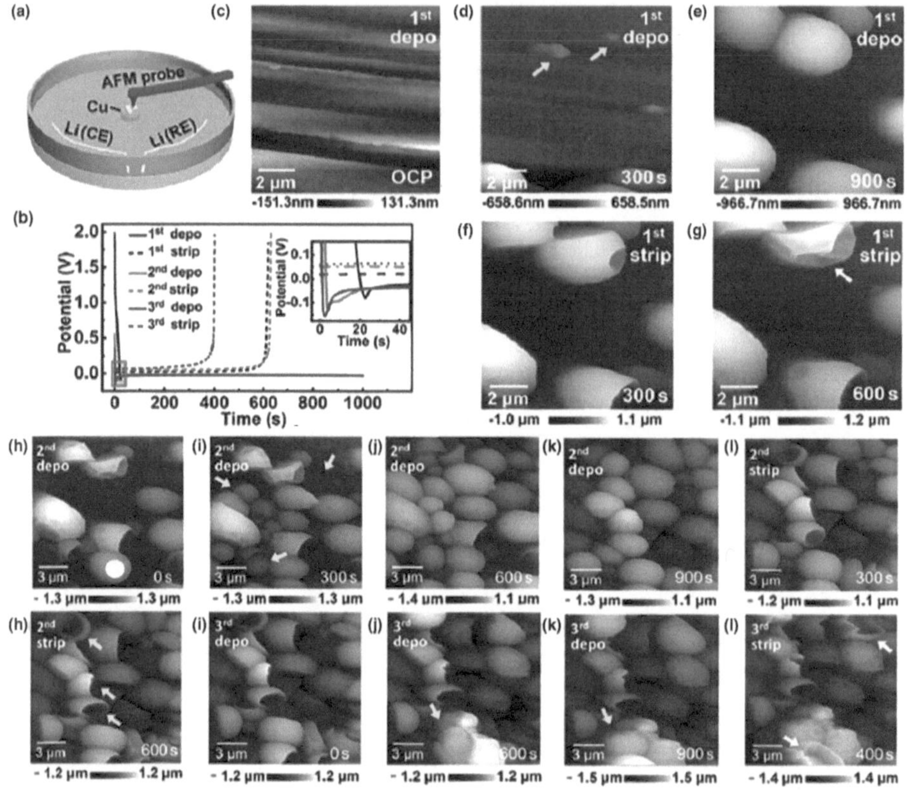


**FIGURE 7.20** (a) Schematic illustration of an *in situ* AFM cell. (b) The voltage profiles of a Cu electrode in PC containing 0.5 M LiTFSI at the current density of 0.5 mA/cm$^2$. The inset in (b) is an enlarged view of the red box area marked at initial Li deposition. *In situ* AFM images of the interfacial morphology variation at (c) OCP; during the first Li deposition/stripping process with Li deposition for 300 s (d), 900 s (e) and stripping for 300 s (f), 600 s (g); after Li deposition for 0 s (h), 300 s (i), 600 s (j), 900 s (k) and stripping for 300 s (l), 600 s (m) in the second cycle; and subsequently, Li deposition for 0 s (n), 600 s (o), 900 s (p) and stripping for 400 s (q) in the third cycle.[82]

After an in-depth understanding of the characteristics of SEI film and its relationship with the deposition/dissolution behavior of lithium metal and the nucleation of lithium dendrites, it is helpful to guide the construction of an effective artificial SEI film, so as to realize the optimization and protection of the lithium metal anode. Li et al.[83] prepared a layer of artificial lithium polyacrylic acid (LiPAA) SEI film on the surface of lithium metal. Through the adaptive interface control, it effectively realizes the dynamic control of lithium metal deposition/dissolution. The *in situ* AFM results of the lithium metal anode modified with the "smart" SEI film with/without this layer during the lithium deposition/dissolution process show that the modified metal lithium anode exhibits uniform and smooth lithium deposition/dissolution behavior during the charge/discharge cycle (Figure 7.21). Because the LiPAA polymer film has good adhesion and stability, it can effectively reduce side

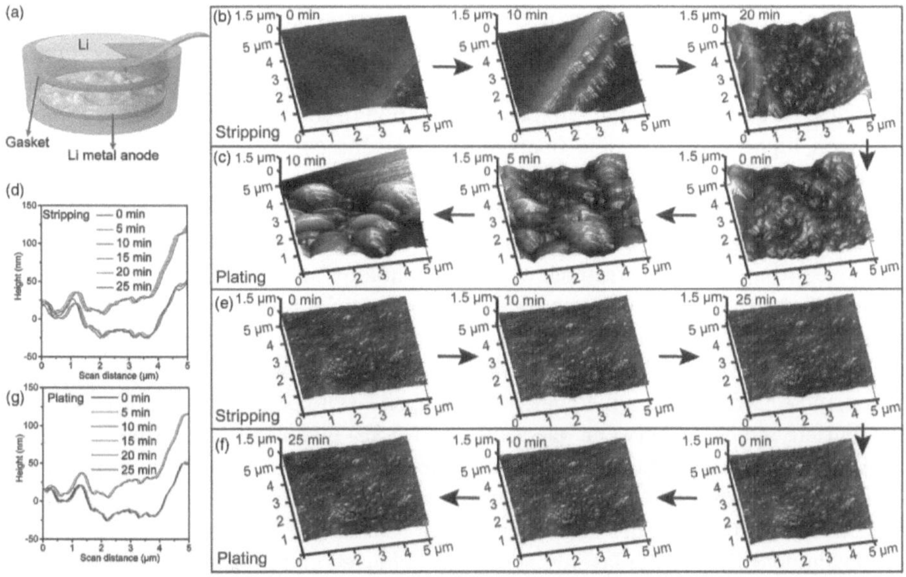


**FIGURE 7.21** (a) Schematic representation of the *in situ* AFM cell. The AFM images of pristine Li anode during (b) stripping and (c) plating processes. (d) The corresponding average height curves. (e) AFM images of the LiPAA-Li anode during the stripping process. The (f) AFM images and (g) corresponding heights of LiPAA-Li anode during plating process.[83]

reactions at interface and accommodate stress of Li-metal volume expansion, thereby significantly improving the electrochemical performance and cycle stability of the lithium metal electrode. Understanding and regulating the SEI film on the surface of the lithium metal electrode is expected to realize the practical application and development of lithium metal in the high specific energy secondary battery system.

## 7.6 DYNAMIC EVOLUTION OF THE ELECTRODE PROCESSES AND SOLID ELECTROLYTES IN SOLID-STATE LITHIUM BATTERIES

### 7.6.1 Introduction

Solid-state electrolytes (SSEs) with good mechanical strength and a wide operating temperature range are believed to be able to effectively inhibit the inherent uncontrollable lithium dendritic growth and puncture of SSEs, and prevent further safety hazards such as fire and explosion caused by the leakage of liquid electrolyte.[82,84–86] It is expected to address the many problems of lithium metal anode in the liquid electrolyte system, making it possible to build a solid-state lithium metal battery system with high energy density and high safety. However, related research results show that the growth of metallic lithium dendrites still exists in the SSEs,[87–89] and the unstable solid-solid electrode/electrolyte interface may aggravate the occurrence of interfacial side reactions and uneven Li deposition/dissolution. Therefore, an in-depth understanding of structure evolution of the electrode/electrolyte interface and reaction mechanism at the interface in solid electrolyte batteries will help to further guide the optimal design of battery materials and interfacial stability and improve the electrochemical performance and long-term stability of solid-state lithium metal batteries. At this stage, the *in situ* characterization of the solid-state batteries mainly relies on high-energy particle beam detection sources, and the main information obtained is the chemical composition of the sample.[90,91] Thus, there is a lack of real-time monitoring of the battery material structure on the micro-nanoscale upon charging and discharging process. The *in situ* AFM imaging characterization technology can reveal the evolution principle of the morphology at the interface between electrode and electrolyte on the nanoscale and further understand the profound reaction mechanism combined with the characterization of electromechanical properties. It is conducive to the realization of dynamic visualization research on the interfacial evolution and related reaction mechanism in the solid-state battery system. Considering the buried interface between electrode and SSE in solid-state lithium batteries (SSLBs), some electrochemical cells were designed for *in situ* AFM tracking of the dynamic evolution and CEI/SEI formation from the section of electrode/electrolyte interface. To further ensure the experimental accuracy, a high-quality cross section of the anode/SSE/cathode layered sheet was achieved and assembled into the *in situ* electrochemical AFM cell. AFM images were then acquired by using AFM probe scanning the cross-section surface of the electrode/SSE interfaces.

### 7.6.2 Cathode Electrolyte Interphase Evolution

For the study of cathode/SSE interface, *in situ* electrochemical AFM was employed to reveal the dynamic evolution of mechanical properties and the morphology of the interfacial layer on the $LiNi_{0.5}Co_{0.2}Mn_{0.3}O_2$ (NCM523) cathode particles upon charging/discharging.[92] During the charging process of the first cycle, filamentous products appear on local areas of the particle surface firstly, followed by the formation of flocculent substances. Quantitative nanomechanical measurements show that the average modulus increases from 0.5 GPa at OCP to 3 GPa and then decreases to 0.5 GPa (Figure 7.22). Combining *ex situ* XPS and time-of-flight secondary ion mass spectrometry (ToF-SIMS) characterization, it is believed that an "organic-inorganic structure" interface layer with a thickness of about 11 nm is formed on the surface. During the discharge process, the CEI film on-site formed on the surface of electrode gradually becomes dense and smooth, and the average modulus increases slightly. The *in situ* AFM morphology image is helpful for real-time tracking of the structure and morphology evolution of the CEI film during charging and discharging. The corresponding modulus image further quantitatively characterizes its mechanical properties, which helps to understand the properties of the CEI film and reveal the dynamic evolution of the structure, chemical and mechanical properties of the cathode materials in the SSLBs.

### 7.6.3 Structural Deformation and Ion Migration Mechanism of Solid Electrolyte

Regarding the evolution process of the intrinsic structure of solid electrolytes, *in situ* electrochemical AFM was employed to real-time monitor structural evolution

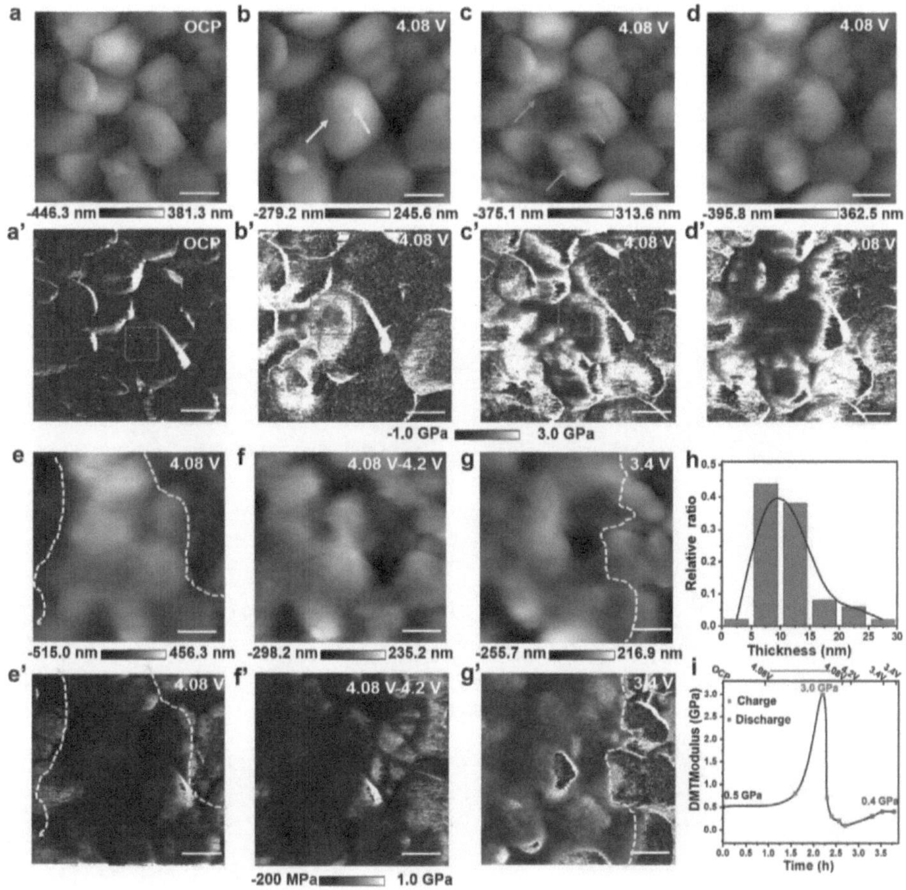


**FIGURE 7.22** *In situ* AFM images showing the topography (a–g) and mapping of the DMT modulus (a'–g') of NCM523 cathode material. The scale bars are 500 nm in (a, a', c–g, c'–g') and 400 nm in (b, b'). (h) The Gauss statistic distribution histograms of the film thickness from bearing analysis of Figure 2g. (i) Quantitatively measured average DMT modulus of the electrode surface at a certain position (red frame in a') during charge/discharge.[92]

and dynamic behavior of composite solid electrolyte (composed of organic polymer poly(ethylene oxide) (PEO) and inorganic ceramic $Li_{6.75}La_3Zr_{1.75}Ta_{0.25}O_{12}$ (LLZTO)) in the solid-state lithium-sulfur batteries (SSLSBs) at nanoscale.[93] It further reveals that the shuttle of polysulfides has a decisive effect on the mechanical stability of the electrolyte. AFM results indicate that during the dissolving process of polysulfides, the chain-like PEO evolves into a fuzzy amorphous structure; the granular LLZTO in the PEO matrix gradually becomes unstable and mobile, and the distance between the particles is increasing. It further results in the separation of LLZTO particles from the PEO network, leading to significant volume expansion of the electrolyte (Figure 7.23A). The structural deformation process of the above-mentioned polymer network and functional fillers severely damages the structural stability of the composite electrolyte, revealing the impact of the polysulfides shuttling in SSLSBs on the structure and performance of the SSEs at the nanoscale.

Moreover, *in situ* conductive atomic force microscopy (c-AFM) was further conducted to unravel the inhomogeneous migration of ions and electrons within the LLZO-PEO composite electrolyte with different weight ratios (0, 50, 75 wt.%) of LLZO under different temperatures (Figure 7.23B).[94] It is monitored that at low

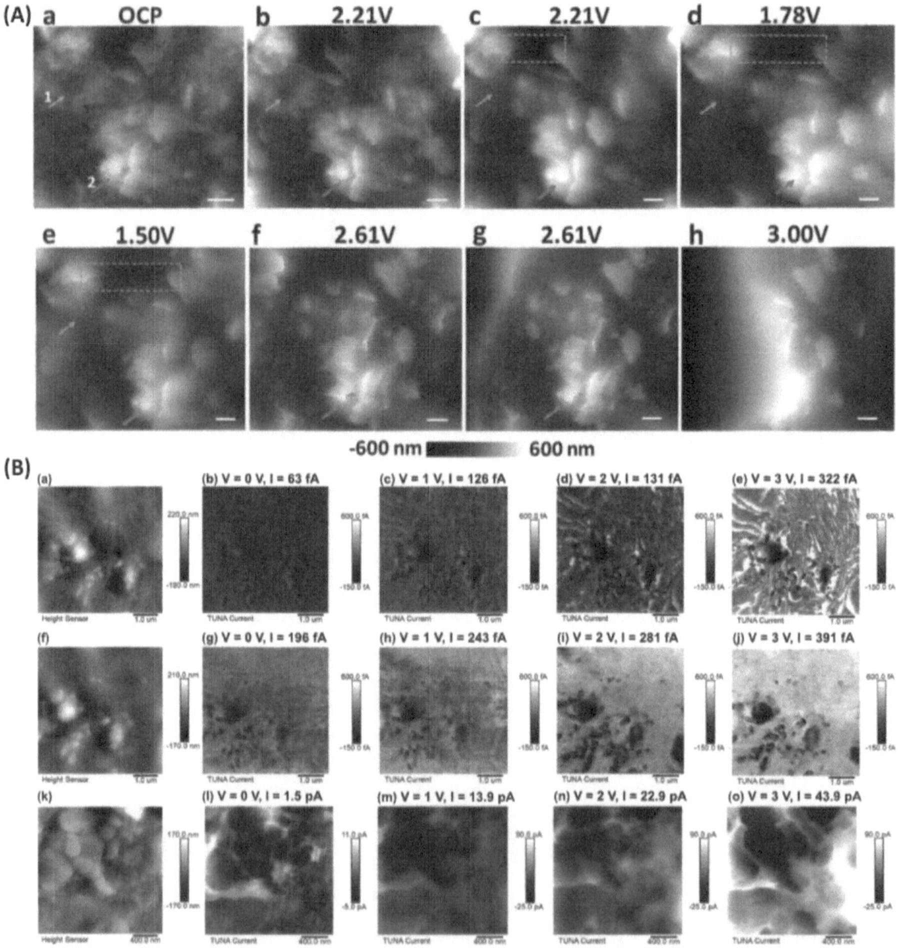


**FIGURE 7.23** (A) *In situ* AFM topography images of CE upon (a–e) discharging and (f–h) charging during the first cycles. The scale bars are 700 nm in (a–h).[93] (B) The *in situ* c-AFM characterization of Li ions migration in 50 wt. % LLZO-PEO ($LiClO_4$) composite electrolyte at 50°C: (a) topography; c-AFM current under (b) 0 V, (c) 1 V, (d) 2 V and (e) 3 V. The *in situ* c-AFM characterization of 50 wt. % LLZO-PEO ($LiClO_4$) at 55°C: (f) topography; c-AFM current under (g) 0 V, (h) 1 V, (i) 2 V and (j) 3 V. *In situ* c-AFM characterization of Li-ions migration in 75 wt. % LLZO-PEO($LiClO_4$): (k) topography at 30°C; c-AFM current under (l) 0 V, (m) 1 V, (n) 2 V and (o) 3 V at 55°C.[94]

temperature, Li ions can only migrate along the amorphous PEO phase, while the ion migration mechanism is mediated by the content of LLZO with temperature increased. At high temperature, Li ions migrate along the amorphous PEO in composite electrolyte with 0 and 50 wt.% LLZO, increased the LLZO content to 75 wt.%, LLZO particles form a continuous ionic network and Li-ions migrate through the LLZO particles. Direct tracking ion migration indicated by the current evolution of composite electrolyte provides more novel insights into the design and development of composite electrolytes for next-generation all-solid-state batteries.

### 7.6.4 Microscopic Mechanism of the Alloying-Regulated Lithium Precipitation

For the study of the anode/SSE interface, electrochemical AFM was employed to *in situ* study the electrode processes of the lithium metal, indium (In) metal and Li-In alloy upon the charging/discharging processes in the sulfide-based all-solid-state lithium batteries (ASSLBs).[95] It unravels the dynamic evolution of lithium deposition/dissolution and the micro-mechanism of the alloying effect. The kinetic behavior of uneven nucleation, growth, stacking and irreversible dissolution of the bulk deposited lithium on the lithium metal electrode was *in situ* observed, which demonstrates the dynamic deposition/dissolution processes of lithium in ASSLBs at nanoscale. It also elucidates the real-time two-dimensional uniform growth of the thin layered lithium indium alloy ($Li_xIn$) product on the surface of the In electrode. During the process of dealloying, the surface of $Li_xIn$ alloy on-site forms fold-like structures, revealing that the SEI layer covering on the alloy surface has favorable flexibility (Figure 7.24). The SEI is beneficial to realize the interfacial protection of the alloy electrode. It clarifies the interfacial micro-mechanism of In electrode for homogenizing the alloying process, further regulating the tunable growth of the flexible SEI layer.

### 7.6.5 Growth Behavior and Interphasial Property of Lithium Dendrites

In order to further understand the growth behavior and characteristics of lithium metal dendrites in SSEs, Shi et al.[96] used *in situ* optical microscopy (OM) to explore the electrochemical evolution behavior of lithium dendrites in a quasi-solid-state lithium metal battery, which realizes real-time monitoring of the dynamics evolution of the SEI layer at the Li dendrite surface. For deeply exploring the properties of the SEI layer on the surface of lithium dendrites observed in the *in situ* OM characterization, the AFM and electrochemical impedance spectroscopy (EIS) were further combined to investigate the separated SEI layer. Qualitative and quantitative research on physical and chemical properties such as morphology, mechanical modulus and local ionic conductivity were conducted (Figure 7.25A). It is revealed that the SEI layer on the surface of lithium dendrites presents a spherical shell-like morphology, a high Young's modulus and a local ionic conductivity of $6.02\times10^{-4}$ $Scm^{-1}$. The on-site formed SEI has high ionic conductivity and good interfacial wettability, making it possible to directly use it as an SSE in quasi-solid-state lithium batteries (QSSLBs). The above results are conducive to in-depth understanding of the influence of the SEI

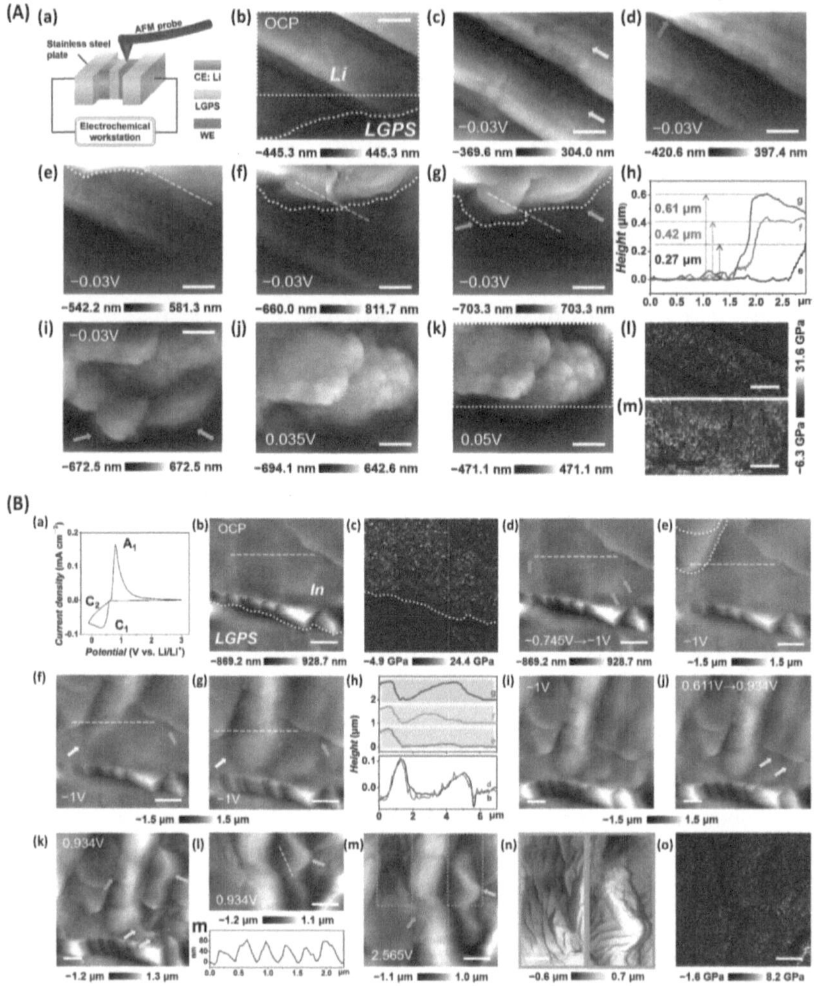

**FIGURE 7.24** (A) *In situ* AFM monitoring of the Li plating/stripping processes on the Li electrode. (a) Schematic illustration of the *in situ* electrochemical AFM cell. *In situ* AFM images on the Li electrode at different potentials of (b) OCP, cathodic (c–g, i) −0.03 V and anodic (j) 0.035 V and (k) 0.05 V. (h) Height section profiles of the deposited Li along the white dashed lines in (e–g). Corresponding DMT modulus mappings of the surfaces of the Li electrode (l) at OCP and (m) after Li plating/stripping processes. The scale bars are 1 μm.[95] (B) Morphological evolution of the uniform $Li_xIn$ lamellae and wrinkle-structure SEI shell on the In electrode upon lithiation/delithiation. (a) Cyclic voltammogram curve of the In electrode with LGPS as electrolyte and Li as CE. AFM images on the In electrode at different potentials of (b) OCP, cathodic (d) 0.745 V to −1 V, (e–g, i) 1 V and anodic (j) 0.611–0.934 V, (k, l) 0.934 V and (n) 2.565 V. Corresponding DMT modulus mappings of the (c) In/LGPS interface at OCP and (p) In electrode after Li–In alloying/dealloying reactions. (h) Height section profiles of the electrode surfaces along the white dashed lines in (b, d–g). (m) Height section profile of such a wrinkling morphology along the white dashed line in (l). (o) 3D AFM images of the wrinkle structures in blue and pink dotted boxes of (n). The scale bars are 2 μm in (b–g, i–l, n, and p) and 1 μm in (o).[95]

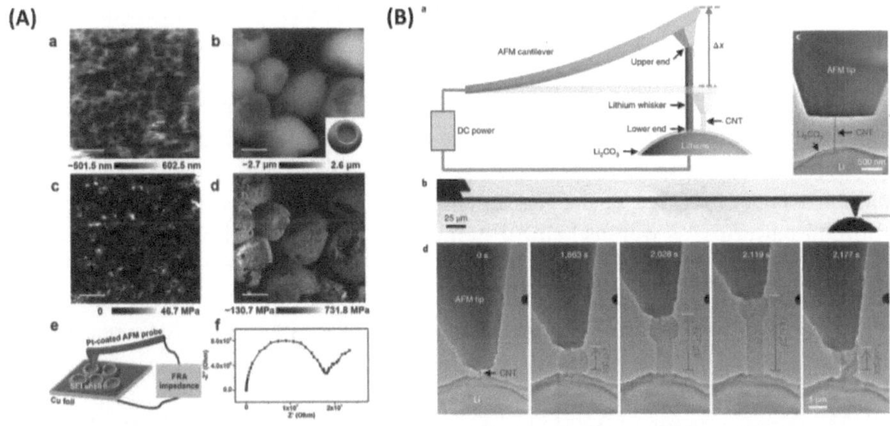


**FIGURE 7.25** (A) (a, b) AFM morphology and (c, d) modulus mappings of the (a, c) pristine GPE and (b, d) SEI shells. (e, f) Ion-conductive property of the SEI shell characterized by AFM combined with EIS. The scale bars are 2 µm in (a–d).[96] (B) (a) Schematic of the AFM–ETEM setup. TEM images showing (b) an AFM cantilever approaching the counter electrode of Li metal and (c) a CNT attached to a flattened AFM tip. (d) Time-lapse TEM images of Li whisker growth.[97]

layer on the evolution process of the lithium dendrites and the effective suppression of the lithium dendrites by chemically controlling the characteristics of the SEI layer.

In order to further deeply understand the growth process of lithium dendrites and the related electrical-chemical-mechanical properties, Zhang et al.[97] combined AFM and environmental transmission electron microscope (ETEM) for *in situ* characterization of the structure and stress evolution of the whisker-like metallic lithium dendrites. The research results show that the growth process of whiskers can be divided into three stages: (1) Firstly, spherical metallic lithium appears between the AFM probe and the $Li_2CO_3$/Li substance, and its size increases with the increasing applied voltage. (2) Then, lithium whiskers begin to increase in the axial direction. With the AFM probes that are constantly being lifted, the axial stress inside the whiskers continues to accumulate. (3) Finally, when the lithium whiskers are no longer longitudinally applied by the outside world, when the voltage increases and the growth continues, it usually collapses suddenly due to the release of axial stress (Figure 7.25B). The mechanical characterization of AFM is applied to quantitatively study the elastoplastic response of lithium whiskers; it is helpful to understand the internal stress generated during the growth of lithium dendrites, so as to further control the load of axial stress in the battery system, and provides new insights of design and strategy of suppressing lithium dendrites.

## 7.7 SUMMARY AND OUTLOOK

In summary, in this chapter, we introduced some typical examples of the application of *in situ* AFM in the fields of Li-S batteries, Li-$O_2$ batteries, SEI evolution and SSLBs. It is evident that AFM is very powerful for the morphological visualization of the electrode/electrolyte interfaces, providing significant insights into the interfacial evolution at both cathode and anode in both liquid- and solid-state electrolytes. Under various electrochemical environments, direct correlation can be established between interfacial morphology and battery function. Additional information, such as localized modulus, conductivity, surface potential and ion migration, can be further presented under specific modes of AFM. We believe that AFM has huge application prospects in terms of the fundamental understanding of mechanistic details in lithium batteries, as well as helpful supports of function optimization of state-of-the-art lithium battery community. More in-depth and interesting works in lithium battery field with the application of AFM can be expected in the near future. We listed some promising aspects on the applications of AFM in Li battery research as follows.

1. For the application of AFM in Li-$O_2$ batteries, we suggest that efforts should be made in the following three aspects. Firstly, combining AFM with characterization techniques (IR, Raman, etc.) with the capabilities for chemical species identification could achieve real-time monitoring of the conversion process of interfacial chemical components while obtaining the information of structural evolution, which is conducive to in-depth understanding of interfacial reaction pathways. Secondly, exploring the adsorption and conversion process of $O_2$ and intermediates at the active site, as well as the initial nucleation of solid products, is of great significance for understanding the Li-$O_2$ reaction mechanism. The use of high spatial and high temporal resolution AFM is expected to obtain this relevant information at the molecular scale. Finally, considering the complex chemical composition and structure of the $O_2$ electrode, it is important to construct a model catalytic electrode that can represent a real catalytic electrode and is suitable for AFM research. Reasonable use of sputtering, photolithography, atomic layer deposition, surface synthesis and other methods is expected to obtain model electrodes that can represent real catalytic electrodes.
2. On the basis of *in situ* visualization research of microscopic morphology and structure, expand more AFM test modes and develop technologies that are used in combination with characterization techniques with functions of simultaneous monitoring chemical compositions, and further develop the application of AFM-FTIR and AFM-Raman technology in the surface/interface analysis on electrochemical energy systems. For example, in the *in situ* exploration of the dynamic evolution of the microstructure and mechanical properties of the SEI film, while monitoring its corresponding change of chemical composition to obtain the corresponding correlation between its structure and composition, which is conducive to in-depth understanding of the characteristics of the SEI film and related thin film at interfaces in various rechargeable battery systems.
3. Cryo-transmission electron microscope (Cryo-TEM), as an advanced characterization technology for the lithium battery, can detect the microstructure of electron–beam sensitive materials such as metallic lithium dendrites and interface SEI films. Combined with atomic force microscopy, it is possible

to use Cryo-TEM to obtain sample micro-information while using AFM probes to monitor its corresponding force-electric properties. It reveals the microscopic mechanism of battery materials on atomic and nanoscale and realizes a more non-destructive characterization of the electrode surface under conditions close to the actual battery operation.

4. The long-term development and key applications of *in situ* electrochemical AFM in metallic lithium anode and SSLB have broad significance. It can help to understand the nucleation and growth evolution of the SEI film on the surface of the metal lithium anode and the kinetic process of lithium deposition/dissolution; it is conducive to monitoring the dynamic evolution process of the CEI film at the positive electrode/electrolyte interface; it is conducive to the realization of *in situ* visualization of the electrolyte structure during charging/discharging process. This further reveals the dynamic evolution law and internal failure mechanism at the solid-solid interface and provides a microscopic understanding and corresponding theoretical guidance, for wide application of solid lithium metal batteries with high energy density, high safety performance and long cycle life.

## REFERENCES


1. Cheng, X. B.; Zhang, R.; Zhao, C. Z.; Zhang, Q., Toward safe lithium metal anode in rechargeable batteries: A review. *Chem. Rev.* **2017**, *117* (15), 10403–10473.
2. Goodenough, J. B.; Kim, Y., Challenges for rechargeable li batteries. *Chem. Mater.* **2010**, *22* (3), 587–603.
3. Goodenough, J. B., Evolution of strategies for modern rechargeable batteries. *Acc. Chem. Res.* **2013**, *46*, 1053–1061.
4. Nagaura, T., Lithium ion rechargeable battery. *Prog. Batteries Sol. Cells* **1990**, *9*, 209.
5. Armand, M.; Tarascon, J. M., Building better batteries. *Nature* **2008**, *451* (7179), 652–657.
6. Ozawa, K., Lithium-ion rechargeable batteries with $LiCoO_2$ and carbon electrodes: The $LiCoO_2$/C system. *Solid State Ion.* **1994**, *69* (3), 212–221.
7. Goodenough, J. B.; Park, K. S., The Li-ion rechargeable battery: A perspective. *J. Am. Chem. Soc.* **2013**, *135* (4), 1167–1176.
8. Bruce, P. G.; Freunberger, S. A.; Hardwick, L. J.; Tarascon, J. M., Li-$O_2$ and Li-S batteries with high energy storage. *Nat. Mater.* **2011**, *11* (1), 19–29.
9. Yin, Y. X.; Xin, S.; Guo, Y. G.; Wan, L. J., Lithium-sulfur batteries: electrochemistry, materials, and prospects. *Angew. Chem. Int. Ed. Engl.* **2013**, *52* (50), 13186–13200.
10. Yang, X.; Luo, J.; Sun, X., Towards high-performance solid-state Li-S batteries: From fundamental understanding to engineering design. *Chem. Soc. Rev.* **2020**, *49* (7), 2140–2195.
11. Maleki Kheimeh Sari, H.; Li, X., Controllable cathode–electrolyte interface of $Li[Ni_{0.8}Co_{0.1}Mn_{0.1}]O_2$ for lithium ion batteries: A review. *Adv. Energy Mater.* **2019**, *9* (39), 1901597.
12. Zhang, Z.; Said, S.; Smith, K.; Jervis, R.; Howard, C. A.; Shearing, P. R.; Brett, D. J. L.; Miller, T. S., Characterizing batteries by in situ electrochemical atomic force microscopy: A critical review. *Adv. Energy Mater.* **2021**, *11*, 2101518.
13. Peled, E., The electrochemical behavior of alkali and alkaline earth metals in nonaqueous battery systems—The solid electrolyte interphase model. *J. Electrochem. Soc.* **1979**, *126* (12), 2047–2051.

14. Peled, E.; Golodnitsky, D.; Ardel, G., Advanced model for solid electrolyte interphase electrodes in liquid and polymer electrolytes. *J. Electrochem. Soc.* **1997**, *144* (8), L208–L210.
15. Aurbach, D.; Markovsky, B.; Levi, M. D.; Levi, E.; Schechter, A.; Moshkovich, M.; Cohen, Y., New insights into the interactions between electrode materials and electrolyte solutions for advanced nonaqueous batteries. *J. Power Sources* **1999**, *81–82*, 95–111.
16. Li, N. W.; Yin, Y. X.; Yang, C. P.; Guo, Y. G., An artificial solid electrolyte interphase layer for stable lithium metal anodes. *Adv. Mater.* **2016**, *28* (9), 1853–1858.
17. Cheng, X. B.; Zhang, R.; Zhao, C. Z.; Wei, F.; Zhang, J. G.; Zhang, Q., A review of solid electrolyte interphases on lithium metal anode. *Adv. Sci.* **2016**, *3* (3), 1500213.
18. Choi, N.-S.; Yew, K. H.; Lee, K. Y.; Sung, M.; Kim, H.; Kim, S.-S., Effect of fluoroethylene carbonate additive on interfacial properties of silicon thin-film electrode. *J. Power Sources* **2006**, *161* (2), 1254–1259.
19. He, Y.-B.; Liu, M.; Huang, Z.-D.; Zhang, B.; Yu, Y.; Li, B.; Kang, F.; Kim, J.-K., Effect of solid electrolyte interface (SEI) film on cyclic performance of $Li_4Ti_5O_{12}$ anodes for Li ion batteries. *J. Power Sources* **2013**, *239*, 269–276.
20. Peled, E.; Menachem, C.; Bar-Tow, D.; Melman, A., Improved graphite anode for lithium-ion batteries chemically: Bonded solid electrolyte interface and nanochannel formation. *J. Electrochem. Soc.* **1996**, *143* (1), L4–L7.
21. Zheng, J.; Zheng, H.; Wang, R.; Ben, L.; Lu, W.; Chen, L.; Chen, L.; Li, H., 3D visualization of inhomogeneous multi-layered structure and Young's modulus of the solid electrolyte interphase (SEI) on silicon anodes for lithium ion batteries. *Phys. Chem. Chem. Phys.* **2014**, *16* (26), 13229–13238.
22. Luchkin, S. Y.; Lipovskikh, S. A.; Katorova, N. S.; Savina, A. A.; Abakumov, A. M.; Stevenson, K. J., Solid-electrolyte interphase nucleation and growth on carbonaceous negative electrodes for Li-ion batteries visualized with in situ atomic force microscopy. *Sci. Rep.* **2020**, *10* (1), 8550.
23. Liu, X. R.; Deng, X.; Liu, R. R.; Yan, H. J.; Guo, Y. G.; Wang, D.; Wan, L. J., Single nanowire electrode electrochemistry of silicon anode by in situ atomic force microscopy: solid electrolyte interphase growth and mechanical properties. *ACS Appl. Mater. Interfaces* **2014**, *6* (22), 20317–20323.
24. Huang, S.; Cheong, L.-Z.; Wang, S.; Wang, D.; Shen, C., In-situ study of surface structure evolution of silicon anodes by electrochemical atomic force microscopy. *Appl. Surf. Sci.* **2018**, *452*, 67–74.
25. Kitta, M.; Sano, H., Real-time observation of Li deposition on a Li electrode with operand atomic force microscopy and surface mechanical imaging. *Langmuir* **2017**, *33* (8), 1861–1866.
26. Shen, C.; Hu, G.; Cheong, L. Z.; Huang, S.; Zhang, J. G.; Wang, D., Direct observation of the growth of lithium dendrites on graphite anodes by operando EC-AFM. *Small Methods* **2017**, *2* (2), 1700298.
27. Lu, W.; Zhang, J.; Xu, J.; Wu, X.; Chen, L., In situ visualized cathode electrolyte interphase on $LiCoO_2$ in high voltage cycling. *ACS Appl. Mater Interfaces* **2017**, *9* (22), 19313–19318.
28. Bi, Y.; Tao, J.; Wu, Y.; Li, L.; Xu, Y.; Hu, E.; Wu, B.; Hu, J.; Wang, C.; Zhang, J.-G.; Qi, Y.; Xiao, J., Reversible planar gliding and microcracking in a single-crystalline Ni-rich cathode. *Science* **2020**, *370* (6522), 1313–1317.
29. Xia, Y.; Zheng, J.; Wang, C.; Gu, M., Designing principle for Ni-rich cathode materials with high energy density for practical applications. *Nano Energy* **2018**, *49*, 434–452.
30. Liu, T.; Lin, L.; Bi, X.; Tian, L.; Yang, K.; Liu, J.; Li, M.; Chen, Z.; Lu, J.; Amine, K.; Xu, K.; Pan, F., In situ quantification of interphasial chemistry in Li-ion battery. *Nat. Nanotechnol.* **2019**, *14*, 50–56.

49. Wang, Y.; Lu, Y. C., Isotopic labeling reveals active reaction interfaces for electrochemical oxidation of lithium peroxide. *Angew. Chem. Int. Ed. Engl.* **2019**, *58* (21), 6962–6966.
50. Wang, J.; Zhang, Y.; Guo, L.; Wang, E.; Peng, Z., Identifying reactive sites and transport limitations of oxygen reactions in aprotic lithium-$O_2$ batteries at the stage of sudden death. *Angew. Chem. Int. Ed. Engl.* **2016**, *55* (17), 5201–5205.
51. Huang, J.; Tong, B.; Li, Z.; Zhou, T.; Zhang, J.; Peng, Z., Probing the reaction interface in Li-oxygen batteries using dynamic electrochemical impedance spectroscopy: Discharge-charge asymmetry in reaction sites and electronic conductivity. *J. Phys. Chem. Lett.* **2018**, *9* (12), 3403–3408.
52. Aetukuri, N. B.; McCloskey, B. D.; García, J. M.; Krupp, L. E.; Viswanathan, V.; Luntz, A. C., Solvating additives drive solution-mediated electrochemistry and enhance toroid growth in non-aqueous Li–$O_2$ batteries. *Nat. Chem.* **2014**, *7* (1), 50–56.
53. Tan, P.; Jiang, H. R.; Zhu, X. B.; An, L.; Jung, C. Y.; Wu, M. C.; Shi, L.; Shyy, W.; Zhao, T. S., Advances and challenges in lithium-air batteries. *Appl. Energy* **2017**, *204*, 780–806.
54. Wen, R.; Hong, M.; Byon, H. R., In Situ AFM Imaging of Li–$O_2$ electrochemical reaction on highly oriented pyrolytic graphite with ether-based electrolyte. *J. Am. Chem. Soc.* **2013**, *135* (29), 10870–10876.
55. Hong, M.; Yang, C.; Wong, R. A.; Nakao, A.; Choi, H. C.; Byon, H. R., Determining the facile routes for oxygen evolution reaction by in situ probing of Li–$O_2$ cells with conformal $Li_2O_2$ films. *J. Am. Chem. Soc.* **2018**, *140* (20), 6190–6193.
56. Shen, Z. Z.; Lang, S. Y.; Shi, Y.; Ma, J. M.; Wen, R.; Wan, L. J., Revealing the surface effect of the soluble catalyst on oxygen reduction/evolution in Li-$O_2$ batteries. *J. Am. Chem. Soc.* **2019**, *141* (17), 6900–6905.
57. Kwabi, D. G.; Batcho, T. P.; Feng, S.; Giordano, L.; Thompson, C. V.; Shao-Horn, Y., The effect of water on discharge product growth and chemistry in Li-$O_2$ batteries. *Phys. Chem. Chem. Phys.* **2016**, *18* (36), 24944–24953.
58. Shen, Z. Z.; Lang, S. Y.; Zhou, C.; Wen, R.; Wan, L. J., In situ realization of water-mediated interfacial processes at nanoscale in aprotic Li–$O_2$ batteries. *Adv. Energy Mater.* **2020**, *10*, 2002339.
59. Song, L.-N.; Zhang, W.; Wang, Y.; Ge, X.; Zou, L.-C.; Wang, H.-F.; Wang, X.-X.; Liu, Q.-C.; Li, F.; Xu, J.-J., Tuning lithium-peroxide formation and decomposition routes with single-atom catalysts for lithium–oxygen batteries. *Nat. Commun.* **2020**, *11* (1), 2191.
60. Gao, J.; Cai, X.; Wang, J.; Hou, M.; Lai, L.; Zhang, L., Recent progress in hierarchically structured $O_2$-cathodes for Li-$O_2$ batteries. *Chem. Eng. J.* **2018**, *352*, 972–995.
61. Xu, C.; Gallant, B. M.; Wunderlich, P. U.; Lohmann, T.; Greer, J. R., Three-dimensional Au microlattices as positive electrodes for Li-$O_2$ batteries. *ACS Nano* **2015**, *9* (6), 5876–5883.
62. Tu, F.; Hu, J.; Xie, J.; Cao, G.; Zhang, S.; Yang, S. A.; Zhao, X.; Yang, H. Y., Au-decorated cracked carbon tube arrays as binder-free catalytic cathode enabling guided $Li_2O_2$ inner growth for high-performance Li-$O_2$ batteries. *Adv. Funct. Mater.* **2016**, *26* (42), 7725–7732.
63. Wen, R.; Byon, H. R., In situ monitoring of the Li-$O_2$ electrochemical reaction on nanoporous gold using electrochemical AFM. *Chem. Commun.* **2014**, *50* (20), 2628–2631.
64. Shen, Z. Z.; Zhou, C.; Wen, R.; Wan, L. J., Surface mechanism of catalytic electrodes in lithium-oxygen batteries: How nanostructures mediate the interfacial reactions. *J. Am. Chem. Soc.* **2020**, *142* (37), 16007–16015.
65. Sun, D.; Shen, Y.; Zhang, W.; Yu, L.; Yi, Z.; Yin, W.; Wang, D.; Huang, Y.; Wang, J.; Wang, D.; Goodenough, J. B., A solution-phase bifunctional catalyst for lithium–oxygen batteries. *J. Am. Chem. Soc.* **2014**, *136* (25), 8941–8946.

31. v Cresce, A.; Russell, S. M.; Baker, D. R.; Gaskell, K. J.; Xu, K., In situ and quantitati characterization of solid electrolyte interphases. *Nano. Lett.* **2014**, *14* (3), 1405–1412.
32. Liu, X. R.; Wang, L.; Wan, L. J.; Wang, D., In situ observation of electrolyte-concentration-dependent solid electrolyte interphase on graphite in dimethyl sulfoxide. *ACS Appl. Mater. Interfaces* **2015**, *7* (18), 9573–9580.
33. Zhang, H.; Wang, D.; Shen, C., In-situ EC-AFM and ex-situ XPS characterization to investigate the mechanism of SEI formation in highly concentrated aqueous electrolyte for Li-ion batteries. *Appl. Surf. Sci.* **2020**, *507*, 145059.
34. Shi, Y.; Yan, H. J.; Wen, R.; Wan, L. J., Direct visualization of nucleation and growth processes of solid electrolyte interphase film using in situ atomic force microscopy. *ACS Appl. Mater. Interfaces* **2017**, *9* (26), 22063–22067.
35. Wan, J.; Hao, Y.; Shi, Y.; Song, Y. X.; Yan, H. J.; Zheng, J.; Wen, R.; Wan, L. J., Ultra-thin solid electrolyte interphase evolution and wrinkling processes in molybdenum disulfide-based lithium-ion batteries. *Nat. Commun.* **2019**, *10* (1), 3265.
36. Shi, Y.; Wan, J.; Li, J.-Y.; Hu, X.-C.; Lang, S.-Y.; Shen, Z.-Z.; Li, G.; Yan, H.-J.; Jiang, K.-C.; Guo, Y.-G.; Wen, R.; Wan, L.-J., Elucidating the interfacial evolution and anisotropic dynamics on silicon anodes in lithium-ion batteries. *Nano Energy* **2019**, *61*, 304–310.
37. Busche, M. R.; Adelhelm, P.; Sommer, H.; Schneider, H.; Leitner, K.; Janek, J., Systematical electrochemical study on the parasitic shuttle-effect in lithium-sulfur-cells at different temperatures and different rates. *J. Power Sources* **2014**, *259*, 289–299.
38. Lang, S.-Y.; Hu, X.-C.; Wen, R.; Wan L.-J., In situ/operando visualization of electrode processes in lithium-sulfur batteries: A review. *J. Electrochem.* **2019**, *25*(02), 141–159.
39. Lang, S.; Feng, X.; Seok, J.; Yang, Y.; Krumov, M. R.; Molina Villarino, A.; Lowe, M. A.; Yu, S.-H.; Abruña, H. D., Lithium–sulfur redox: Challenges and opportunities. *Curr. Opin. Electrochem.* **2021**, *25*, 100652.
40. Tian, J. H.; Jiang, T.; Wang, M.; Hu, Z.; Zhu, X.; Zhang, L.; Qian, T.; Yan, C., In situ/operando spectroscopic characterizations guide the compositional and structural design of lithium–sulfur batteries. *Small Methods* **2019**, *4*, 1900467.
41. Lang, S. Y.; Shi, Y.; Guo, Y. G.; Wang, D.; Wen, R.; Wan, L. J., Insight into the interfacial process and mechanism in lithium-sulfur batteries: An in situ AFM study. *Angew. Chem. Int. Ed. Engl.* **2016**, *55* (51), 15835–15839.
42. Lang, S. Y.; Shi, Y.; Guo, Y. G.; Wen, R.; Wan, L. J., High-temperature formation of a functional film at the cathode/electrolyte interface in lithium-sulfur batteries: An in situ AFM study. *Angew. Chem. Int. Ed. Engl.* **2017**, *56* (46), 14433–14437.
43. Lang, S. Y.; Xiao, R. J.; Gu, L.; Guo, Y. G.; Wen, R.; Wan, L. J., Interfacial mechanism in lithium-sulfur batteries: How salts mediate the structure evolution and dynamics. *J. Am. Chem. Soc.* **2018**, *140* (26), 8147–8155.
44. Zhang, P.; Zhao, Y.; Zhang, X., Functional and stability orientation synthesis of materials and structures in aprotic Li-$O_2$ batteries. *Chem. Soc. Rev.* **2018**, *47* (8), 2921–3004.
45. Kwak, W.-J.; Rosy, S.; Sharon, D.; Xia, C.; Kim, H.; Johnson, L. R.; Bruce, P. G.; Nazar, L. F.; Sun, Y.-K.; Frimer, A. A.; Noked, M.; Freunberger, S. A.; Aurbach, D., Lithium–oxygen batteries and related systems: Potential, status, and future. *Chem. Rev.* **2020**, *120*, 6626–6683.
46. Li, F.; Wu, S.; Li, D.; Zhang, T.; He, P.; Yamada, A.; Zhou, H., The water catalysis at oxygen cathodes of lithium-oxygen cells. *Nat. Commun.* **2015**, *6*, 7843.
47. Shen, Z. Z.; Zhou, C.; Wen, R.; Wan, L. J., Charge rate-dependent decomposition mechanism of toroidal $Li_2O_2$ in Li-$O_2$ batteries. *Chin. J. Chem.* **2021**, *39*, 2668–2672.
48. Shu, C.; Wang, J.; Long, J.; Liu, H. K.; Dou, S. X., Understanding the reaction chemistry during charging in aprotic lithium-oxygen batteries: Existing problems and solutions. *Adv. Mater.* **2019**, *31*, e1804587.

66. Bergner, B. J.; Schürmann, A.; Peppler, K.; Garsuch, A.; Janek, J., TEMPO: A mobile catalyst for rechargeable Li-$O_2$ batteries. *J. Am. Chem. Soc.* **2014**, *136* (42), 15054–15064.
67. Gao, X.; Chen, Y.; Johnson, L.; Bruce, Peter G., Promoting solution phase discharge in Li–$O_2$ batteries containing weakly solvating electrolyte solutions. *Nat. Mater.* **2016**, *15* (8), 882–888.
68. Liu, T.; Frith, J. T.; Kim, G.; Kerber, R. N.; Dubouis, N.; Shao, Y.; Liu, Z.; Magusin, P. C. M. M.; Casford, M. T. L.; Garcia-Araez, N.; Grey, C. P., The effect of water on quinone redox mediators in nonaqueous Li-$O_2$ batteries. *J. Am. Chem. Soc.* **2018**, *140* (4), 1428–1437.
69. Wang, S.; Yin, X.; Liu, D.; Liu, Y.; Qin, X.; Wang, W.; Zhao, R.; Zeng, X.; Li, B., Nanoscale observation of the solid electrolyte interface and lithium dendrite nucleation–growth process during the initial lithium electrodeposition. *J. Mater. Chem. A* **2020**, *8* (35), 18348–18357.
70. Gu, Y.; Wang, W. W.; Li, Y. J.; Wu, Q. H.; Tang, S.; Yan, J. W.; Zheng, M. S.; Wu, D. Y.; Fan, C. H.; Hu, W. Q.; Chen, Z. B.; Fang, Y.; Zhang, Q. H.; Dong, Q. F.; Mao, B. W., Designable ultra-smooth ultra-thin solid-electrolyte interphases of three alkali metal anodes. *Nat. Commun.* **2018**, *9* (1), 1339.
71. Wang, W.-W.; Gu, Y.; Yan, H.; Li, S.; He, J.-W.; Xu, H.-Y.; Wu, Q.-H.; Yan, J.-W.; Mao, B.-W., Evaluating solid-electrolyte interphases for lithium and lithium-free anodes from nanoindentation features. *Chem* **2020**, *6* (10), 2728–2745.
72. Meyerson, M. L.; Sheavly, J. K.; Dolocan, A.; Griffin, M. P.; Pandit, A. H.; Rodriguez, R.; Stephens, R. M.; Vanden Bout, D. A.; Heller, A.; Mullins, C. B., The effect of local lithium surface chemistry and topography on solid electrolyte interphase composition and dendrite nucleation. *J. Mater. Chem. A* **2019**, *7* (24), 14882–14894.
73. Busche, M. R.; Drossel, T.; Leichtweiss, T.; Weber, D. A.; Falk, M.; Schneider, M.; Reich, M. L.; Sommer, H.; Adelhelm, P.; Janek, J., Dynamic formation of a solid-liquid electrolyte interphase and its consequences for hybrid-battery concepts. *Nat. Chem.* **2016**, *8* (5), 426–434.
74. Cheng, X.-B.; Huang, J.-Q.; Zhang, Q., Review—Li metal anode in working lithium-sulfur batteries. *J. Electrochem. Soc.* **2017**, *165* (1), A6058–A6072.
75. Soto, F. A.; Ma, Y.; Martinez de la Hoz, J. M.; Seminario, J. M.; Balbuena, P. B., Formation and growth mechanisms of solid-electrolyte interphase layers in rechargeable batteries. *Chem. Mater.* **2015**, *27* (23), 7990–8000.
76. Xiong, S.; Xie, K.; Diao, Y.; Hong, X., Properties of surface film on lithium anode with $LiNO_3$ as lithium salt in electrolyte solution for lithium–sulfur batteries. *Electrochim. Acta* **2012**, *83*, 78–86.
77. Zhang, S. S., Role of $LiNO_3$ in rechargeable lithium/sulfur battery. *Electrochim. Acta* **2012**, *70*, 344–348.
78. Xiong, S.; Xie, K.; Diao, Y.; Hong, X., Characterization of the solid electrolyte interphase on lithium anode for preventing the shuttle mechanism in lithium–sulfur batteries. *J. Power Sources* **2014**, *246*, 840–845.
79. Li, W.; Yao, H.; Yan, K.; Zheng, G.; Liang, Z.; Chiang, Y. M.; Cui, Y., The synergetic effect of lithium polysulfide and lithium nitrate to prevent lithium dendrite growth. *Nat. Commun.* **2015**, *6*, 7436.
80. Lang, S.-Y.; Shen, Z.-Z.; Hu, X.-C.; Shi, Y.; Guo, Y.-G.; Jia, F.-F.; Wang, F.-Y.; Wen, R.; Wan, L.-J., Tunable structure and dynamics of solid electrolyte interphase at lithium metal anode. *Nano Energy* **2020**, *75*, 104967.
81. Dokukin, M. E.; Sokolov, I., On the measurements of rigidity modulus of soft materials in nanoindentation experiments at small depth. *Macromolecules* **2012**, *45* (10), 4277–4288.

82. Shi, Y.; Liu, G.-X.; Wan, J.; Wen, R.; Wan, L.-J., In-situ nanoscale insights into the evolution of solid electrolyte interphase shells: Revealing interfacial degradation in lithium metal batteries. *Sci. Chin. Chem.* **2021**, *64* (5), 734–738.
83. Li, N. W.; Shi, Y.; Yin, Y. X.; Zeng, X. X.; Li, J. Y.; Li, C. J.; Wan, L. J.; Wen, R.; Guo, Y. G., A flexible solid electrolyte interphase layer for long-life lithium metal anodes. *Angew. Chem. Int. Ed. Engl.* **2018**, *57* (6), 1505–1509.
84. Manthiram, A.; Yu, X.; Wang, S., Lithium battery chemistries enabled by solid-state electrolytes. *Nat. Rev. Mater.* **2017**, *2* (4), 16103.
85. Yang, C.; Fu, K.; Zhang, Y.; Hitz, E.; Hu, L., Protected lithium-metal anodes in batteries: From liquid to solid. *Adv. Mater.* **2017**, *29* (36), 1701169.
86. Quartarone, E.; Mustarelli, P., Electrolytes for solid-state lithium rechargeable batteries: recent advances and perspectives. *Chem. Soc. Rev.* **2011**, *40* (5), 2525–2540.
87. Han, F.; Westover, A. S.; Yue, J.; Fan, X.; Wang, F.; Chi, M.; Leonard, D. N.; Dudney, N. J.; Wang, H.; Wang, C., High electronic conductivity as the origin of lithium dendrite formation within solid electrolytes. *Nat. Energy* **2019**, *4* (3), 187–196.
88. Cheng, X.-B.; Zhao, C.-Z.; Yao, Y.-X.; Liu, H.; Zhang, Q., Recent advances in energy chemistry between solid-state electrolyte and safe lithium-metal anodes. *Chem* **2019**, *5* (1), 74–96.
89. Ren, Y.; Shen, Y.; Lin, Y.; Nan, C.-W., Direct observation of lithium dendrites inside garnet-type lithium-ion solid electrolyte. *Electrochem. Commun.* **2015**, *57*, 27–30.
90. Xiang, Y.; Li, X.; Cheng, Y.; Sun, X.; Yang, Y., Advanced characterization techniques for solid state lithium battery research. *Mater. Today* **2020**, *36*, 139–157.
91. Tan, D. H. S.; Banerjee, A.; Chen, Z.; Meng, Y. S., From nanoscale interface characterization to sustainable energy storage using all-solid-state batteries. *Nat. Nanotechnol.* **2020**, *15* (3), 170–180.
92. Guo, H. J.; Wang, H. X.; Guo, Y. J.; Liu, G. X.; Wan, J.; Song, Y. X.; Yang, X. A.; Jia, F. F.; Wang, F. Y.; Guo, Y. G.; Wen, R.; Wan, L. J., Dynamic evolution of a cathode interphase layer at the surface of $LiNi_{0.5}Co_{0.2}Mn_{0.3}O_2$ in quasi-solid-state lithium batteries. *J. Am. Chem. Soc.* **2020**, *142* (49), 20752–20762.
93. Song, Y. X.; Shi, Y.; Wan, J.; Liu, B.; Wan, L. J.; Wen, R., Dynamic visualization of cathode/electrolyte evolution in quasi-solid-state lithium batteries. *Adv. Energy Mater.* **2020**, *10* (25), 2000465.
94. Shen, C.; Huang, Y.; Yang, J.; Chen, M.; Liu, Z., Unraveling the mechanism of ion and electron migration in composite solid-state electrolyte using conductive atomic force microscopy. *Energy Stor. Mater.* **2021**, *39*, 271–277.
95. Wan, J.; Song, Y. X.; Chen, W. P.; Guo, H. J.; Shi, Y.; Guo, Y. J.; Shi, J. L.; Guo, Y. G.; Jia, F. F.; Wang, F. Y.; Wen, R.; Wan, L. J., Micromechanism in all-solid-state alloy-metal batteries: Regulating homogeneous lithium precipitation and flexible solid electrolyte interphase evolution. *J. Am. Chem. Soc.* **2021**, *143* (2), 839–848.
96. Shi, Y.; Wan, J.; Liu, G. X.; Zuo, T. T.; Song, Y. X.; Liu, B.; Guo, Y. G.; Wen, R.; Wan, L. J., Interfacial evolution of lithium dendrites and their solid electrolyte interphase shells of quasi-solid-state lithium-metal batteries. *Angew. Chem. Int. Ed. Engl.* **2020**, *59* (41), 18120–18125.
97. Zhang, L.; Yang, T.; Du, C.; Liu, Q.; Tang, Y.; Zhao, J.; Wang, B.; Chen, T.; Sun, Y.; Jia, P.; Li, H.; Geng, L.; Chen, J.; Ye, H.; Wang, Z.; Li, Y.; Sun, H.; Li, X.; Dai, Q.; Tang, Y.; Peng, Q.; Shen, T.; Zhang, S.; Zhu, T.; Huang, J., Lithium whisker growth and stress generation in an in situ atomic force microscope-environmental transmission electron microscope set-up. *Nat. Nanotechnol.* **2020**, *15* (2), 94–98.

# 8 Application of AFM in Solar Cell Research

*Ahmed Touhami*
University of Texas Rio Grande Valley

**CONTENTS**



## 8.1 INTRODUCTION

Solar energy is the most promising technology for affordable, clean, and reliable electricity for our modern lifestyle. A solar cell is essentially a p-n junction under illumination, where generation of charge-carriers by photons takes place.[1–3] The generated current due to light depends on how many photons are absorbed and how efficiently the resulting electron-hole pairs are collected.[1–3] Although this photovoltaic (PV) effect was first observed in 1839, the first modern PV solar cell was invented almost 100 years later.[4,5] Ever since, the fabrication of solar cells has passed through many improvement steps considering the technological and economic aspects. Several solar technologies including wafer, thin film, and organic have been researched to achieve reliability, cost-effectiveness, and high efficiency with huge success.[6–14] As illustrated in Figure 8.1, PV technology can be classified into three generations based on the absorber material and level of commercialization.[15] Currently, the first-generation PV cells based on crystalline silicon (c-Si) are the most common solar cells used in commercially available solar panels.[15,16] The second-generation PV cells, based on thin-film technologies, are more economical and include mainly amorphous silicon (a-Si), cadmium telluride (CdTe), copper indium gallium selenide (CIGS),[11,15] and kesterite photovoltaics such as $Cu_2ZnSn(S, Se)_4$ (CZTSSe).[17,18] Emerging solar cells, such as perovskite, organic, dye-sensitized, and quantum dot (QD) solar cells, are

DOI: 10.1201/9781003174042-8

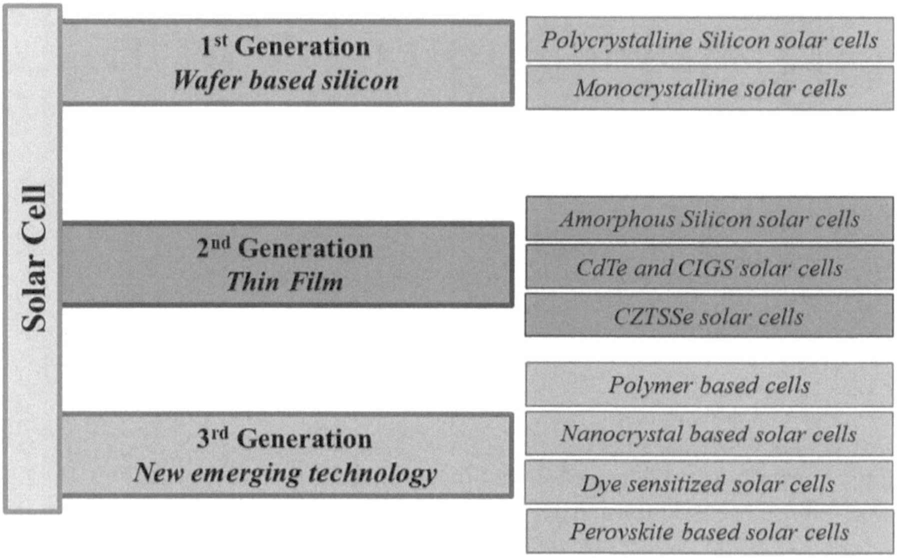


**FIGURE 8.1** Classification of solar cell technologies.

currently the most developed third-generation solar cell types.[11] The PV technology is progressing very fast, both in a new installed capacity, now reaching more than 400 GW worldwide, and in a big research effort to develop more efficient and sustainable technologies.[6] However, this progress dependent upon better nanoscale characterization methods to achieve enhanced spatial resolution given the growing use of nano-textured materials in solar cells. The nanoscale characteristics of the components and their interfaces often control critical processes of the device, such as charge-carrier generation, electron and ion transport, surface potentials, and electrocatalysis.[11] Understanding the spatial properties and structure–property relationships of these components can provide insight into designing scalable and efficient solar cell components and systems.

AFM with its various modalities offers a unique real-time visualization of the sample topography as well as whole device analytics down to the atomic level. Simultaneously, the sample morphological features can be correlated to its electronic, chemical, and optoelectronic properties, which better brings out the correlation between the structure and performance. A wide variety of AFM modalities have been used to explore morphological, electrical, and optical properties of solar cells, including conductive AFM (c-AFM), Kelvin probe force microscopy (KPFM), and electrostatic force microscopy (EFM).[19–23] For example, in thin-film inorganic solar cells, KPFM has revealed that charge separation does not occur at a heterojunction as expected, but instead occurs at a homojunction buried ~50 nm within the absorbing layer.[22] In organic photovoltaics (OPVs), submicron maps of photocurrent have contributed to the understanding of the interplay between processing conditions, blend morphology, and device performance.[22] Thermal and mechanical properties of solar cells have also been investigated at the nanoscale using PeakForce-QNM and scanning thermal microscopy modes.[24–26]

In this perspective, our goal in this chapter is to provide an overview of the recent progress in using AFM-based techniques to explore structural and functional properties of solar cell materials and devices. However, the challenges to review this research field is the speed of publications of new research data. With the risk of missing or leaving out interesting work, as well as studies of importance for the development of the field, we aim to cover the PV research in a broad sense. We will focus on the capabilities, considerations, and limitations of most important AFM modes and not explicitly review their technical details. The readers are encouraged to review the basis of AFM-based techniques in the first two chapters of this book. As mentioned above, PV devices are usually classified as first, second, and third generation to highlight the historical development of diverse PV technologies during the last three decades. However, this does not necessarily imply that the higher the generation the better the performance. Here, we reviewed the AFM impact in this research area following the same chronological classification of solar cells. After some brief notes on each type of solar cells, we report on the most important finding related to the use of AFM techniques in that specific solar cell type. Although cross-sectional AFM characterizations are becoming more popular in solar cell research, which contributes significantly in understanding the buried interface, most of the work reported in this chapter is focusing on the photoactive layer of the PV cell. Finally, we end up with a brief outlook on the future of this research field.

## 8.2 MONOCRYSTALLINE AND POLYCRYSTALLINE SILICON SOLAR CELLS

The oldest and the most popular solar cell technology is based on silicon wafers that account for 90% of the market. Particularly, the monocrystalline silicon solar cells are characterized by their high-power efficiencies reaching up to 26.7% recently (Figure 8.2a).[11,16,21] However, high cost and the sophisticated technological steps, required for manufacturing large single crystals, have led to use of polycrystalline silicon instead of the single crystal wafers (Figure 8.2b and c), of course, on the expense of the solar conversion efficiency. These cells are now marketed and produce solar conversion efficiencies between 12% and 14% according to the manufacturing procedures and wafer quality.[6,21] Since polycrystalline silicon solar cells are based on a mixture of different crystals, one of the causes of their low efficiency is attributable to the existence of grains and their boundaries. Electrical and optical properties may differ among the grains and the grain boundaries (GBs) may act as recombination sites or current leakage paths. In an earlier AFM study, Igarashi et al. used KPFM under light illumination to investigate the local distribution of photovoltages on polycrystalline silicon solar cells through potential measurements.[27] A photovoltage drop around the GB and a photovoltage difference between different grains were observed. GBs present in the material often exert a detrimental influence on the electrical properties because of the potential barriers associated with them.[27,28] However, these early KPFM measurements showed that not all GBs have similar properties, since they have their own character depending on the orientation and relationship between two adjoining grains.[29] It is quite common to classify GBs into

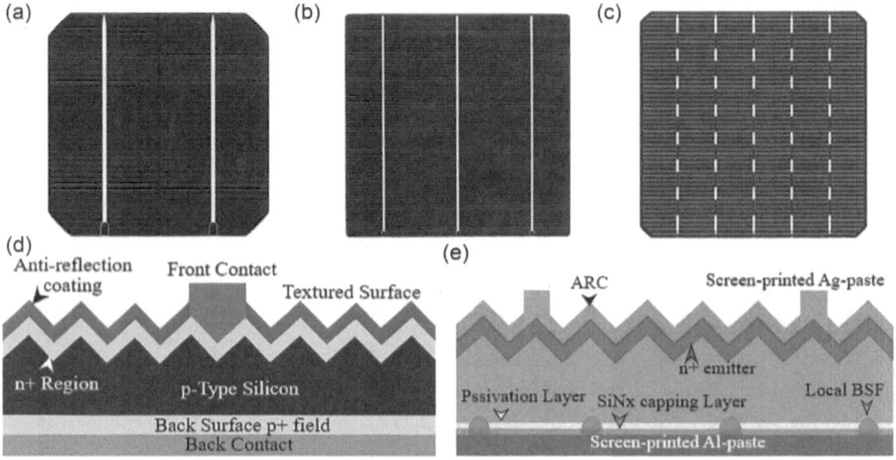


**FIGURE 8.2** (a) and (b) Typical monocrystalline and polycrystalline standard silicon solar cells. (c) An example of PERC solar cell (passivated emitter and rear contact). (d) Simplified cross-section of a commercial monocrystalline standard silicon solar cell that has been in use for decades and makes up around 80% of the world market. (Adapted from Saga[16]). (e) Simplified cross-section of PERC solar cell, which is a new technology aimed to achieve higher energy conversion efficiency by adding a dielectric passivation layer on the rear of the cell.

coincident lattice (tilt and twist boundaries) and general boundaries. As reported by Tsurekawa et al., random boundaries possess barrier heights almost twice as high as coincidence boundaries. The potential barrier height was found to depend on the GB inclination as well as its character.[29] In another study, Takahashi et al. used photo-assisted-KPFM (P-KPFM) to evaluate minority carrier lifetime (bulk carrier lifetime) in polycrystalline silicon solar cells.[30] The sample surface was illuminated by a modulated light and the minority carrier lifetime was extracted from a temporally averaged photovoltage at various modulation frequencies. The results from these measurements indicated that the lifetime significantly decreases in the vicinity of a GB of the polycrystalline silicon solar cell material. Thus, the GB degrades the solar cell performance by acting as a carrier recombination site and/or a leakage pass.[30]

Although AFM and its modalities were among the first techniques used to investigate solar cells properties, not many studies were reported on monocrystalline and polycrystalline silicon solar cells. In addition to morphological characterizations, the most important AFM measurements reported were in the use of KPFM to investigate GB properties as indicated above.

## 8.3 AMORPHOUS AND POLYCRYSTALLINE SILICON THIN-FILM SOLAR CELLS

Thin-film solar cells are based on an alternative technology, which uses less or no silicon in the manufacturing process. Several materials, such as a-Si, CdTe, CIGS, and $CuInSe_2$-based alloys, are commonly used in this type of PV technology.[15,31] The light-absorbing layers of first-generation solar cells and thin-film solar cells are of order 350 and 1 μm thickness, respectively.[31] Currently, enormous progress in device performance has been made in most of these technologies, and considerable effort is devoted to their commercialization.[31] However, these efforts are dependent upon better characterization methods and, in particular, the ability to achieve enhanced spatial resolution. Here we summarize the most important nanoscale characterization possibilities using AFM techniques mainly on polycrystalline CdTe and CIGS solar cells.

### 8.3.1 Amorphous Silicon Thin-Film (a-Si) Solar Cells

These cells can be prepared at low processing temperature, allowing low-cost polymers and other flexible substrates to be used. During the fabrication process, the backside of the substrate is coated by doped silicon. As shown in Figure 8.3a, these types of solar cells have a dark brown color on the reflecting side while a silverish

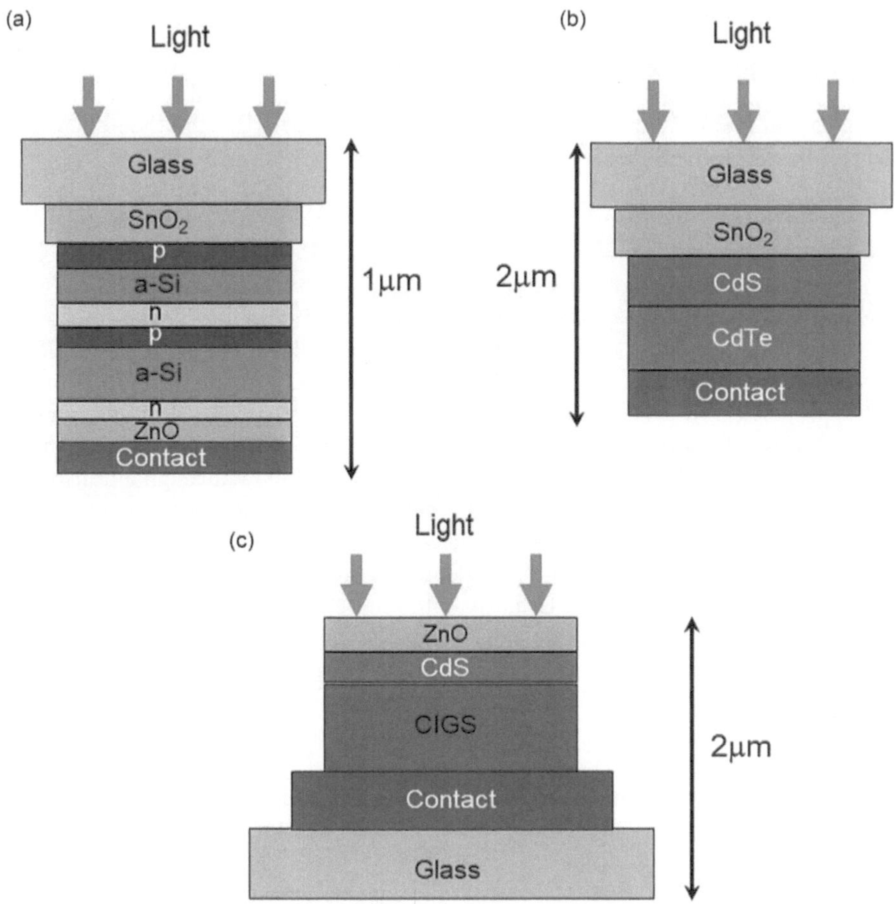


**FIGURE 8.3** Illustration of (a) amorphous silicon, (b) CdTe, and (c) CIGS solar cells.

color on the conducting side. The highest efficiency of these cells can reach 25%; however, their main limitation is their instability.[15] In recent study by Meyer et al., AFM and PeakForce-QNM were used to investigate the degradation of a-Si solar cells at the nanoscale.[32] Interface morphology, deformation, and internal delamination of the cells were analyzed. It was established that high roughness values resulted from mechanical degradation and other interface roughness parameters such as in phase, phase, and quadrature modules need to be employed to access the mechanical properties of (a-Si) solar cells.[32] Similarly to first-generation solar cells, there are almost no recent advanced AFM characterizations of this type of PV systems.

### 8.3.2 CdTe Thin-Film Solar Cells

CdTe/CdS solar cells are synthesized from polycrystalline materials and glass is chosen as a substrate, which make them particularly affordable for terrestrial thin-film PV applications.[11,15] As illustrated in Figure 8.3b, polycrystalline CdTe is the primary absorber in these single junction solar cells, with a direct band gap of 1.5 eV at room temperature.[15] CdS, with a direct band gap of 2.4 eV, functions as a window layer to the CdTe in a heterojunction and the p–n junction diode is formed between layers of cadmium sulfide.[15] Experimentally attained efficiencies for thin-film solar cells such as polycrystalline CdTe and silicon are consistently lower than the theoretical efficiencies based on the Shockley Quaeisser limit.[21,33] For both materials, the theoretical efficiency is 32%–33%, while the maximum experimentally obtained efficiency is 21% for CdTe and 25% for silicon.[34] This has been attributed to spatial non-uniformities such as composition, morphology, GB, and crystallographic orientations.[34] Even for single crystal films, high defect densities resulting from lattice-mismatch between CdS and CdTe layers can cause low minority-carrier lifetimes.[35] These defects can also trap carriers, greatly reducing the open-circuit voltage and efficiency of CdTe PV modules.[34] To investigate the influence of these microstructural defects on the device properties, Kutes et al. used photoconductive-AFM (pc-AFM) to measure the local response of micropatterned polycrystalline CdTe/CdS-based solar cells. In this pioneer work, CdTe/CdS μ-cells were illuminated from below through a transparent substrate and conducting electrode, while the patterned structures and photocurrent were interrogated with c-AFM from above.[36] As shown in Figure 8.4, these measurements confirmed that the surrounding dielectric is insulating while the islands exhibit a diode-like behavior that is particularly enhanced upon illumination due to photocarrier generation.[34,36] In addition, this technique can uniquely provide nanoscale maps of PV performance parameters such as the short-circuit current, open-circuit voltage, maximum power, and fill factor. The method was demonstrated with a stack of 21 images acquired during in situ illumination of the CdTe/CdS μ-cells, providing more than 42,000 I/V curves spatially separated by ~5 nm (Figure 8.5).[36] This approach is different from the most common approach that consists in recording "I/V" spectra by fixing a conducting probe at a desired location and then measuring the current while sweeping the tip or sample bias.[37] This higher spatial resolution was necessary in order to identify current nonlinearities for truly nanoscale features such as the grains, GBs, or possibly strain-relieved island edges which as presented in Figure 8.4. Each image is acquired with a distinct DC bias, incremented

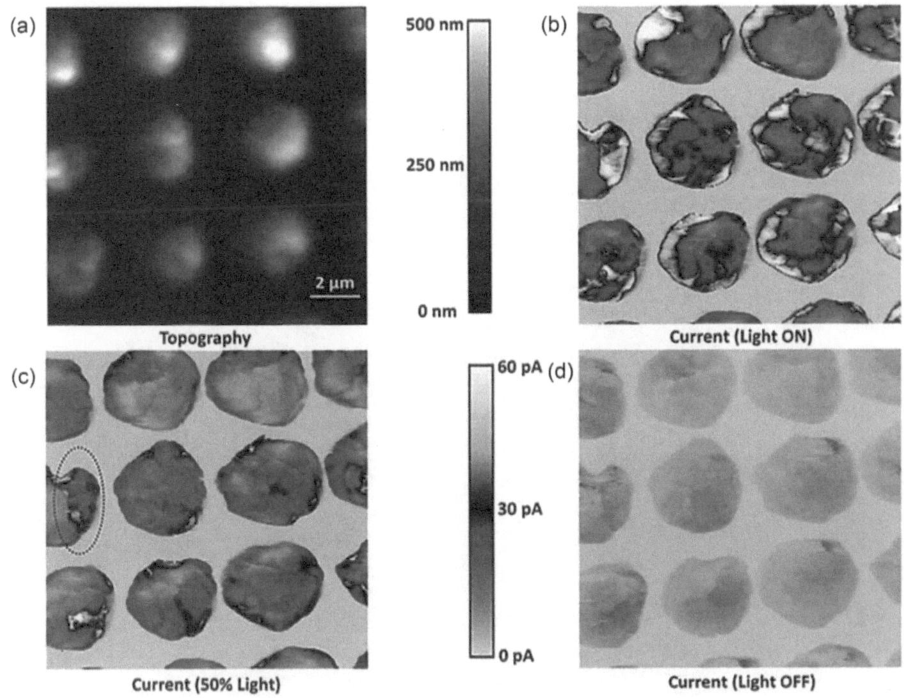


**FIGURE 8.4** (a) Topography of an array of µ-cells and their photocurrent images during (b) 160 mW/cm² (~1.6 suns) illumination, (c) 90 mW/cm² with overlain oval to identify a particular grain boundary, and (d) in dark conditions. Topography and current scales as noted.[36]

or decremented with every new frame as shown in Figure 8.5(top). Considering the current contrast for any given image pixel as a function of image frame (i.e., voltage) thereby yields spatially localized I/V spectra, sketched in Figure 8.5(bottom) to determine the conductance for two distinct grains.[36] Preferential photoconduction for certain grains, GBs, and percolation pathways was clearly resolved, confirming the importance of microstructural control for the optimization of ultimate solar cell properties down to the nanoscale. This coupled photoconductive AFM and I/V spectroscopy specifically demonstrated that the presence of extended inactive regions within any given µ-cell was found to be a regular feature for the polycrystalline CdTe/CdS solar cells.[36] Such effects could partially explain the differences between theoretical and maximum achieved efficiencies for these cells.[21] In another interesting study, the same research group used conductive tomographic-AFM (CT-AFM) to investigate charge transport pathways throughout three-dimensional grains and GBs with nanoscale resolution in the CdTe/CdS PV.[38] Tomography is achieved by applying forces on the order of micronewtons during scanning, such that the specimen is continuously sculpted by consecutive images, destructively removing slices of material within the scanned region only. The excavated material was swept out of the field of view by the rastering tip (Figure 8.6a). Images of electric current collected through the device thickness revealed spatially dependent short-circuit and open-circuit

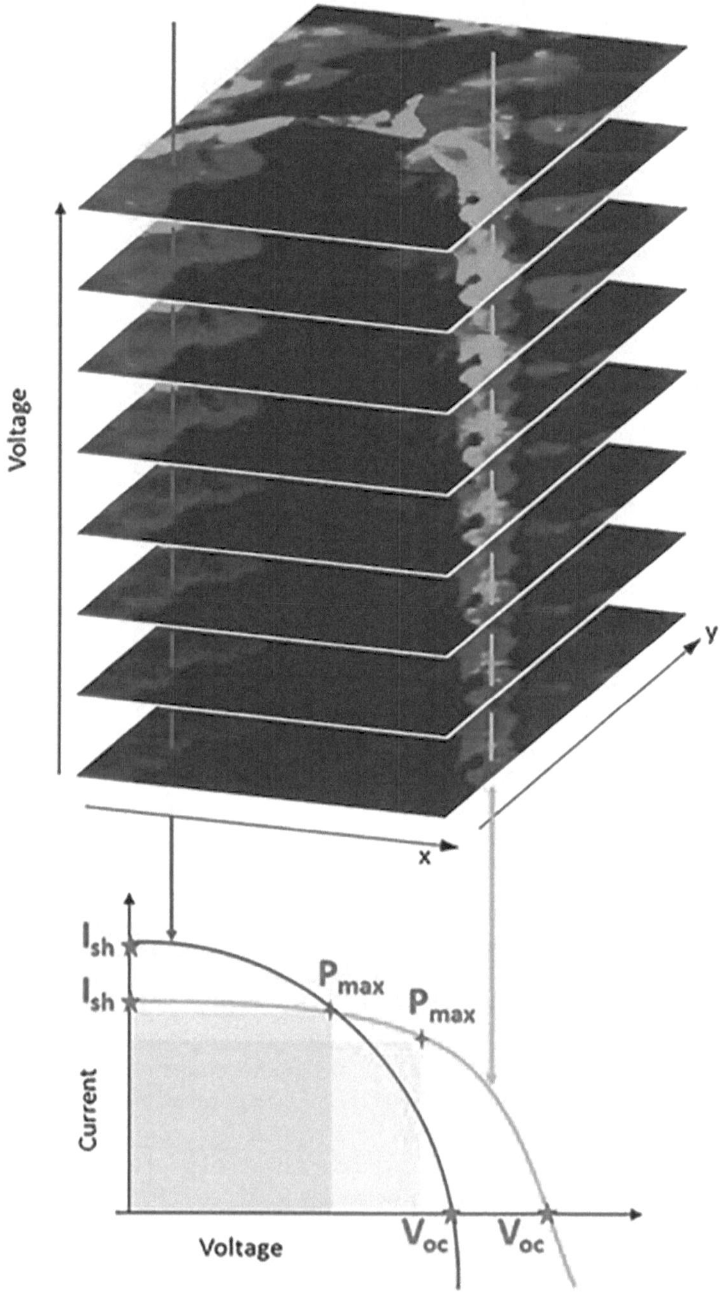


**FIGURE 8.5** Sketch of photocurrent AFM spectroscopy (pcAFMs) based on a series of consecutive pcAFM images. Each image was acquired with incrementally higher applied voltages (top). An array of photocurrent versus voltage spectra is easily extracted for each pixel in the image stack, allowing photoconduction and other photovoltaic performance measures to be quantified and mapped (bottom).[36]

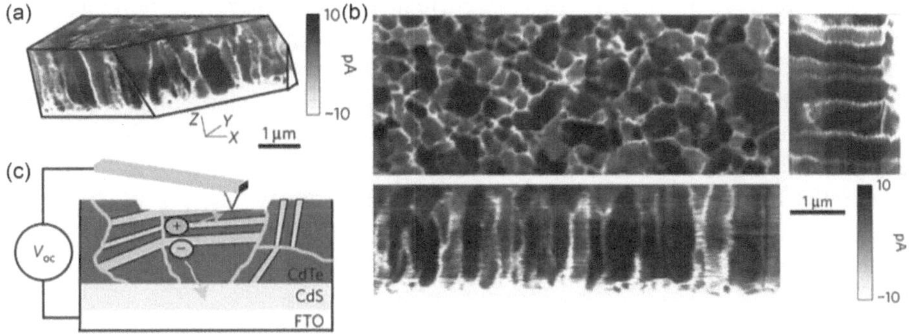


**FIGURE 8.6** (a) Tomographic view of sample with cadmium chloride treatment with tip biased at 0.7 V, near open-circuit condition. The inverted contrast scale highlights regions with more negative currents, where shunt resistance is lower (bright), versus areas still generating power (positive currents, dark). (b) XY (top-left), YZ (top-right), and XZ (bottom) views of tomography presented in (a). (c) Schematic view of CT-AFM and charge transport in CdTe at open-circuit voltage, when grain boundaries (orange) operate as electrical pathways for electrons (−), while planar defects (blue) are pathways for holes (+).[38]

performance, and confirmed that GBs are preferential pathways for electron transport.[38] Results on samples with and without cadmium chloride treatment revealed little difference in grain structure at the microscale, with samples without treatment showing almost no photocurrent either at planar defects or at GBs.[38] As shown in Figure 8.6, this result supported an energetically orthogonal transport system of GBs and interconnected planar defects as contributing to optimal solar cell performance, contrary to the conventional wisdom of the deleterious role of planar defects on polycrystalline thin-film solar cells.[38] Interconnected microstructural defects may therefore serve as p-type conduction pathways in high-quality CdTe solar cells, contrary to previous research that concludes they act only as photocarrier traps. Energetically and spatially orthogonal to these hole channels was a network of n-type corridors identified at GBs in CdTe.[38] The interplay between planar p-type conduction paths, n-type conduction at GBs, and their three-dimensional interconnections thus provides a possible pathway for engineering future enhancements in the performance of polycrystalline solar cells.[38]

### 8.3.3 CIGS Solar Cells

As illustrated in Figure 8.3c, Cu(In, Ga)$Se_2$ (CIGS) is a semiconductor which comprises four elements, copper, indium, gallium, and selenium.[39] In addition to their economical manufacturing cost and high yield process, CIGS solar cells have recently shown efficiency approaching 22%,[21] which make them promising materials for the next generation of high-efficiency thin-film solar.[15] In addition, the non-degrading nature and prolonged life are important benefits of CIGS solar cells technology.[6] In an early AFM study, Romero et al. reported on the anomalous behavior of GBs in p-CGS/n-CdS solar cells using tapping mode AFM based on tuning-fork sensors to investigate the electron transport and recombination upon local current injection.[40]

As show in Figure 8.7, GBs prevent the current spreading from the grain interior to adjacent grains upon local current injection from the tip.[40] This ability to perform conductive AFM in tapping mode is an attractive alternative to contact mode when the specimen is very sensitive to the force. To estimate the band profile around the GBs, Takihara et al. performed P-KPFM on CIGS thin-film solar cell.[41] As shown in Figure 8.8, abrupt potential drops around the GBs can be identified. Here "potential" means electron potential, and it is therefore to be expected that regions of low potential will easily attract electrons, which are the minority carriers in the CIGS layer. This was very consistent with the clear enhancement of the photovoltage around GBs

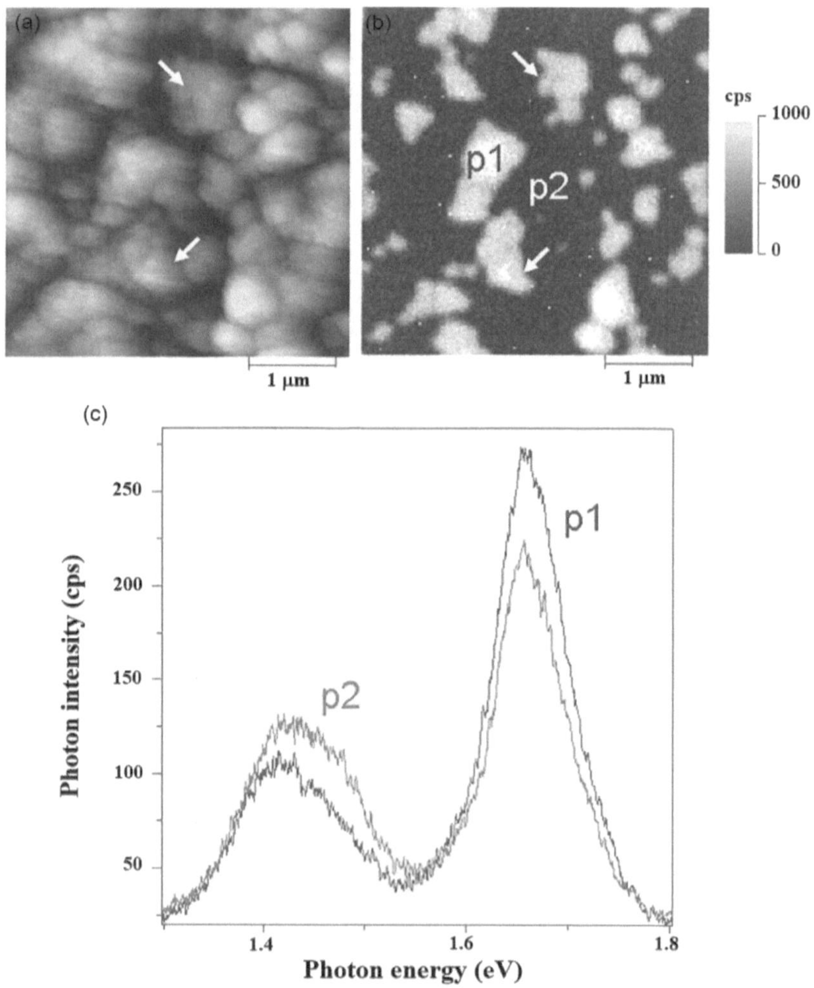


FIGURE 8.7 (a) a Tapping-mode AFM image of the p-CGS/n-CdS thin film with ultrathin ZnO front contacts. (b) Corresponding electroluminescence map when a forward bias of 3.5 V was applied to the tip. (c) Cathodoluminescence spectrum at locations $p_1$ and $p_2$ on the photon map $T=50$ K.[40]

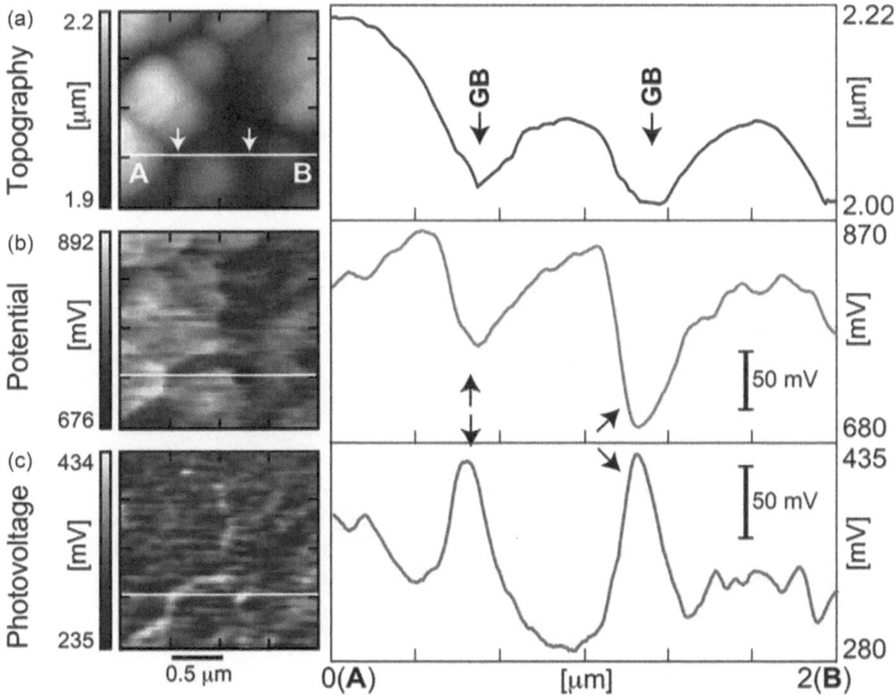


**FIGURE 8.8** Images and line profiles of (a) the topography, (b) the potential in darkness, and (c) the photovoltage under light illumination, obtained by photo-assisted KPFM on the $Cu(InGa)Se_2$ solar cell. The line profiles were taking along the white lines. Black arrows indicate the abrupt drops of the potential and the photovoltage increased around the GBs.[41]

that was observed by P-KPFM, as shown in Figure 8.8c. Assuming that the electron affinity was nearly uniform over the whole surface, the potential distribution shown in Figure 8.9b can be regarded as a profile of the conduction band edge of the CIGS material, except for their raw numerical values.[41] Thus, the distribution of the conduction band edge can be estimated from the surface potential measured by the P-KPFM method, which can yield highly localized PV properties of solar cells.[27–30] In addition to the P-KPFM measurements, Takihara et al. performed scanning tunneling spectroscopy (STS) measurements to confirm both downward bending of the conduction band edge and broadening of the band gap near GBs. Accordingly, the photo-carriers (electrons and holes) are easily separated by the built-in field near GBs, and their recombination rate at the GB is lowered, which is considered to be one of the big advantages of the CIGS materials.[41] Similarly, Hamamoto et al. performed Photothermal (PT)-AFM measurements on the CIGS solar cells with different band gaps to investigate their non-radiative recombination properties.[42] As shown in Figure 8.9, the PT signal images taken under above-gap excitation conditions indicated that the PT signals were enhanced near the GB on the CIGS solar cell with a band gap of 1.13 eV, while the areas where the PT signal was enhanced were broadened around the GB on CIGS with a band gap of 1.28 eV. These results

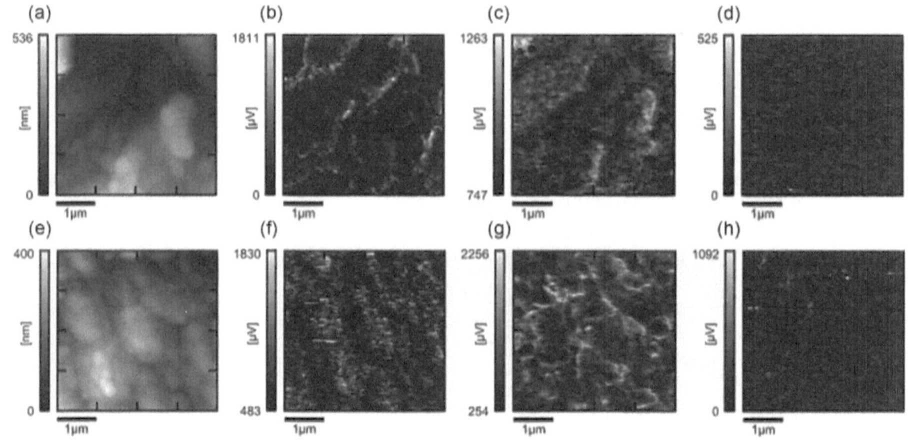


**FIGURE 8.9** (a) Topographic image of sample A, and (b)–(d) PT signal images taken under periodical light illumination at different photon energies that were above or below the CIGS bandgap (1.13 eV). (e) Topographic image of sample B, and (f)–(h) PT signal images taken under periodical light illumination at different photon energies that were above or below the CIGS bandgap (1.28 eV). Samples A and B are CIGS solar cells with 23% and 50% Ga contents respectively.[42]

were attributable to the accumulation of photo-generated free electrons around the GB, showing the existence of the built-in electric field around the GB, and were also attributable to the reduction of the built-in field in CIGS with a wider band gap.[42]

## 8.4 THIRD-GENERATION SOLAR CELL

Although silicon solar cells are well established, the main drawback of these inorganic PV cells is the energy and cost-intense manufacturing process and their weight. Third-generation cells are the new promising technologies but are not commercially investigated in detail. In recent years, this new concept solar cell becomes a global research hotspot. AFM techniques are taking a major part in investigating structure-function properties of these type of solar cells at the nanoscale. Here, we report examples of the use of AFM and its modalities to characterize the most developed third-generation solar cells.

### 8.4.1 Organic Solar Cells

Unlike most inorganic solar cells, OPV cells are composed of successively connected thin functional layers having coating of ribbon and polymer foil.[15] The OPV also works on the PV effect using molecular or polymeric absorbers, which results in localized excitons.[9] In these cells, the fundamental process of charge separation is the absorption of a photon in the active layer. However, photo-generated excitons in organic semiconductors are more difficult to separate into free charges compared to inorganic materials.[9,15] Recent efforts have achieved 13% power conversion efficiency, highlighting the potential of OPVs for large-scale commercial applications.[21]

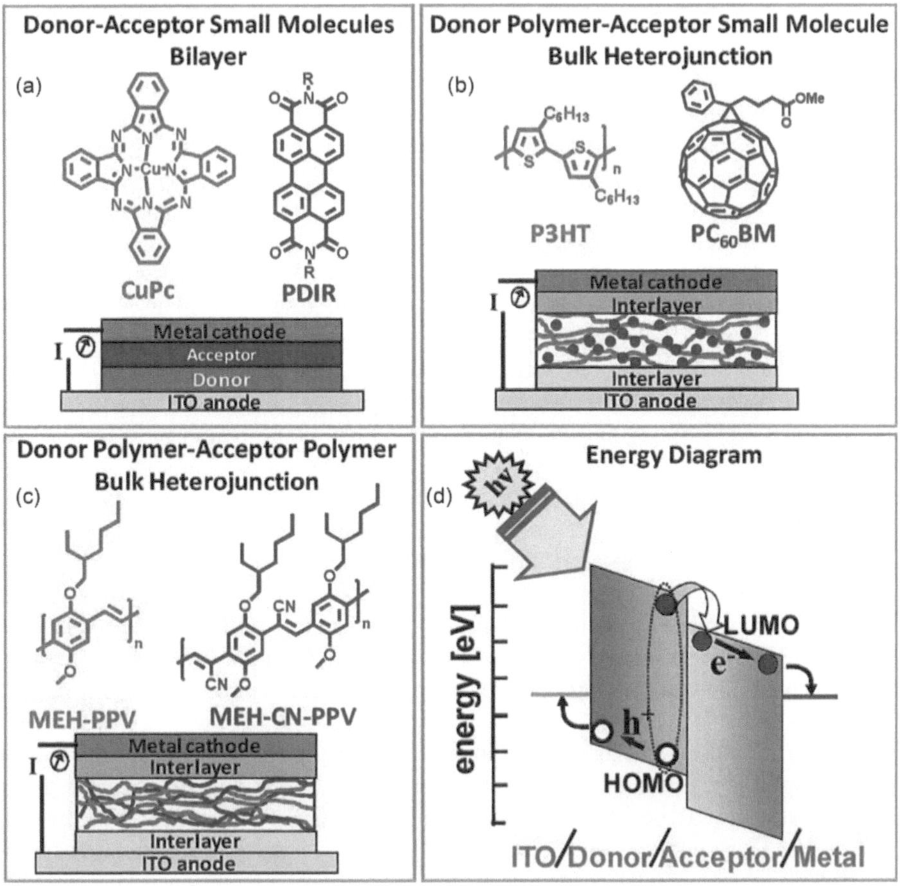


**FIGURE 8.10** (a) Structure and representative photoactive materials of a single heterojunction, (b) polymer-donor–molecular acceptor and (c) all-polymer bulk heterojunction photovoltaic cells. (d) Energy levels of the materials where light absorption/excitation dissociation/charge collection takes place.[43]

Over the last two decades, the efficiency of these devices has improved significantly, in particular through the development of solution-processed bulk heterojunction (BHJ) OPV cells. Typical BHJ OPV cells are shown schematically in Figure 8.10.[43] While fullerenes have been the most intensively studied acceptor materials in BHJ OPVs, research is currently underway in several groups investigating non-fullerene molecular acceptors. In an OPV device, semiconducting polymers or small organic molecules are used to accomplish the functions of collecting solar photons, converting the photons to electrical charges, and transporting the charges to an external circuit as a useable current.[42] The BHJ morphology upon which OPVs rely is extremely complex. The mixing of an electron donor and acceptor in a common solution, followed by spin coating, yields a morphology that has features on a variety of length scales. These features in turn affect the ability of the device to split excitons and the ability of the resulting charges to navigate a route through the film to emerge as useful photocurrent. As a result, the performance of OPVs is inherently a local property. Measurements on bulk devices that are several $mm^2$ in area will implicitly involve a great deal of averaging of the device properties and much of the local microscopic detail will subsequently be lost.[44] Several groups have analyzed OPV systems using AFM,[19,23,45,46] conducting AFM,[47] EFM,[47] and KPFM.[48–53] Uniquely, AFM and its modalities allow probing electrical properties of organic solar cells such as the contact potential difference (CPD), variations in electrical conductivity, and local photocurrents.[44] Thus, structural information can be directly correlated with electrical information on a nanometer scale. For example, Giridharagopal et al. used tow extension techniques to the conducting-AFM, time-resolved EFM (trEFM), a non-contact technique that utilizes time-resolved measurements, and pc-AFM on OPV layers to analyze the local variations in photoinduced charge generation, collection, discharge, and the local morphology and its relation to the local photo-response in BHJs.[44] As shown in Figure 8.11, the combined techniques allowed for morphology and electrical and optical properties of BHJs to be investigated all on the nanoscale and on the same area of the device. This methodology was crucial in making significant steps forward in our understanding of how OPV systems operate in terms of local morphology. Thus, with only a single calibration factor, a trEFM image of a polymer blend can be used to accurately predict the efficiency of the polymer solar cell that will be fabricated from a particular film. One can imagine using such a method both to screen new materials in the lab, or as a rapid quality control diagnostic in a production facility.[44]

It is worth pointing out that, assuming domain identity from AFM phase contrast or topographical features can often lead to inaccurate morphological conclusions. In this context, Gu et al. developed a technique known as photo-induced force microscopy (PiFM) for imaging organic solar cell BHJs with nanoscale chemical specificity. PiFM is a relatively recent scanning probe microscopy technique that combines an AFM tip with a tunable infrared laser to induce a dipole for chemical imaging (Figure 8.12). By coupling the nanometer resolution of AFM with the chemical specificity of a tuned IR laser, they were able to map the donor and acceptor domains in a model all-polymer BHJ with resolution approaching 10 nm. Domain size from PiFM images were compared to bulk-averaged results from resonant soft X-ray scattering, indicating excellent quantitative agreement. In addition, this study showed poor correlation between AFM topography, AFM phase, and PiFM, highlighting the need to move beyond standard AFM for morphology characterization of BHJs.[54]

The choice of the solvent has a major influence on the OPV device performance, next to the polymer blend ratio, the solution's concentration, the thermal treatment, and the molecular structure.[9] These factors are expected to influence the domain size in the donor and acceptor blend and with that to affect the efficiency of charge extraction in the percolating network in a BHJ OPV.[55] Several studies demonstrated that the device performance could be considerably enhanced by treatment with polar solvents that are commonly used in interface engineering through solution processing in OPVs before deposition of metal electrodes.[56–58] For example, Schopp[55] demonstrated that pc-AFM equipped with a tunable light source is a powerful technique and can be used to reveal the solvent effects in a complex morphology of the BHJ solar cells

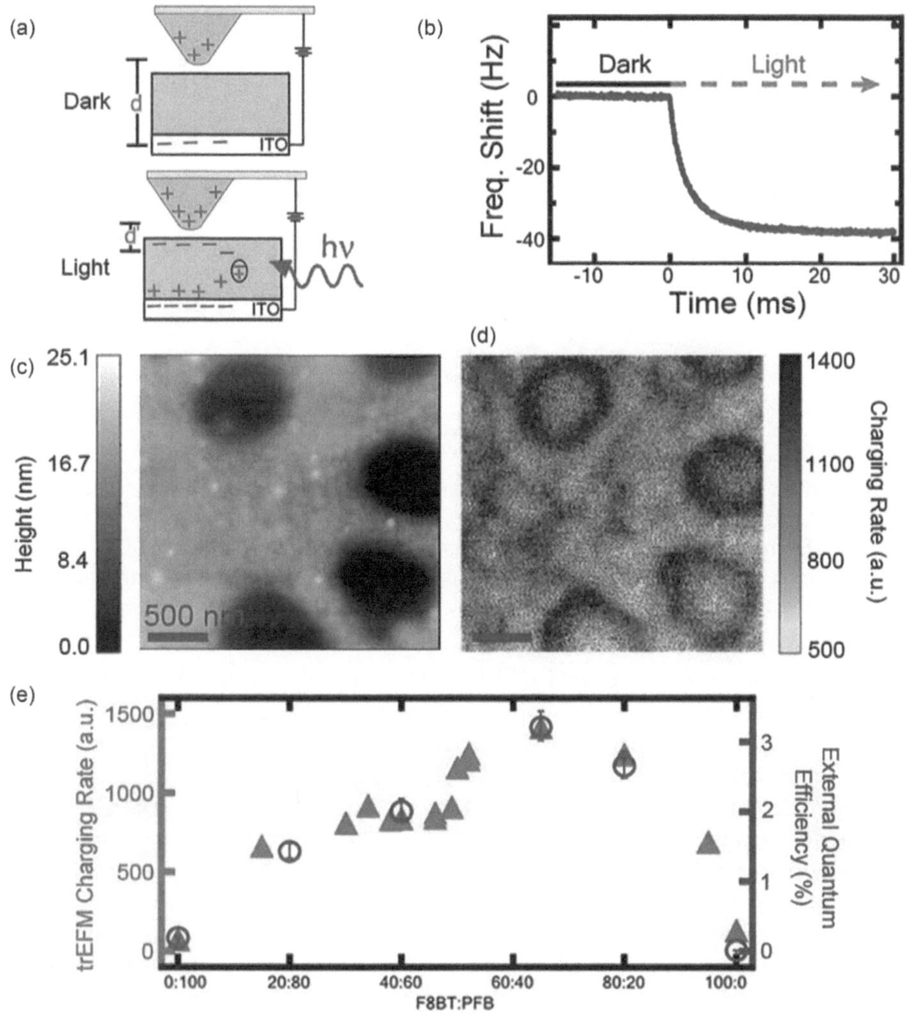


**FIGURE 8.11** (a) Schematic depiction of how photogenerated charge carriers cause an increase in the capacitive gradient and a change in the surface potential and thus a shift in the resonance frequency. The time rate of change in this shift is what is measured by trEFM. (b) Representative plot of the resonance frequency shift versus time following photoexcitation. At time $t=0$ ms, the LED is turned on, causing an exponential decay in the frequency shift. By finding the time constant of this decay, we can extract a relative charging rate. (c) Topography and (d) charging rate image for the same area of a PFB:F8BT sample, dissolved in xylene with 1:1 composition. (e) Spatially-averaged charging rates in films with different PFB:F8BT ratios are quantitatively consistent with the trend exhibited by EQE measurements.[44]

on the nanoscale. High photocurrent of Chlorobenzene-cast devices compared with Toluene-cast devices was explained by homogeneous donor/acceptor distribution, leading to small local variation in the photocurrent generation. More interestingly, Zhou et al.[53] used KPFM to understand the effect of Methanol treatment on BHJ

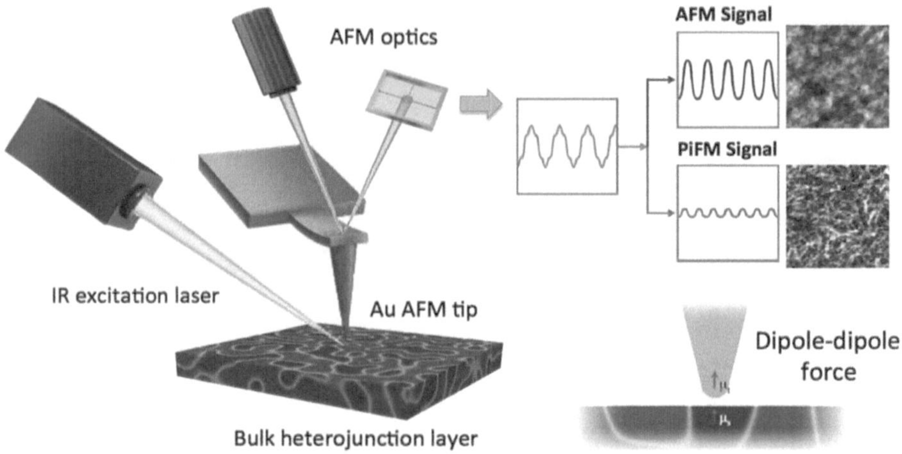


**FIGURE 8.12** Simplified schematic of the PiFM setup. PiFM measures the dipole force between the absorbing sample and the AFM tip. The sample is excited at a specified IR wavelength, inducing a dipole resulting from the vibrational motion of chemical bonds. A mirror dipole is induced in the tip leading to an attractive force between the sample molecules and the tip.[54]

solar cells based on [6,6]-phenyl C71-butyric acid methyl ester (PC70BM) and a low-bandgap donor material thieno[3,4-b]-thiophene/benzodithiophene (PTB7). As shown in Figure 8.13a and b, KPFM measurements on the active layer revealed a $101 \pm 19\,mV$ shift in average surface potential occurring after methanol treatment. However, Methanol treatment had no effect on the surface morphology of PTB7:PC70 BM as indicated by the AFM images in Figure 8.13c and d. The effects of methanol treatment on the enhancement of device performance were shown to originate from an increase in built-in voltage across the device due to passivation of surface traps and a corresponding increase of surface charge density. All these effects induced a simultaneous enhancement in the open-circuit voltage, short-circuit current density, and fill factor of device performance after methanol treatment.[53] These findings shed light on when evaluating new organic solar cell materials; it is therefore of utmost importance to consider polarity and volatility of the solvents and to test additives to eventually obtain well-mixed films.

Another strategy to improve the efficiency of OPVs has emerged by incorporating a ferroelectric polymer layer into the device, which eliminates the need for an external bias. Generally, an external bias voltage is required to efficiently separate the electrons and holes and thus prevent their recombination, which is a main cause of energy loss in OPVs devices. Yuan et al. used c-AFM to show that a large, permanent internal electric field (50 V μ/m) can be ensured by incorporating a ferroelectric polymer layer into the device, which eliminates the need for an external bias.[46] In addition, the induced electric field significantly increased the device efficiency of several types of OPV. These enhanced efficiencies were 10%–20% higher than those achieved by other methods, such as morphology and electrode work-function optimization.[46]

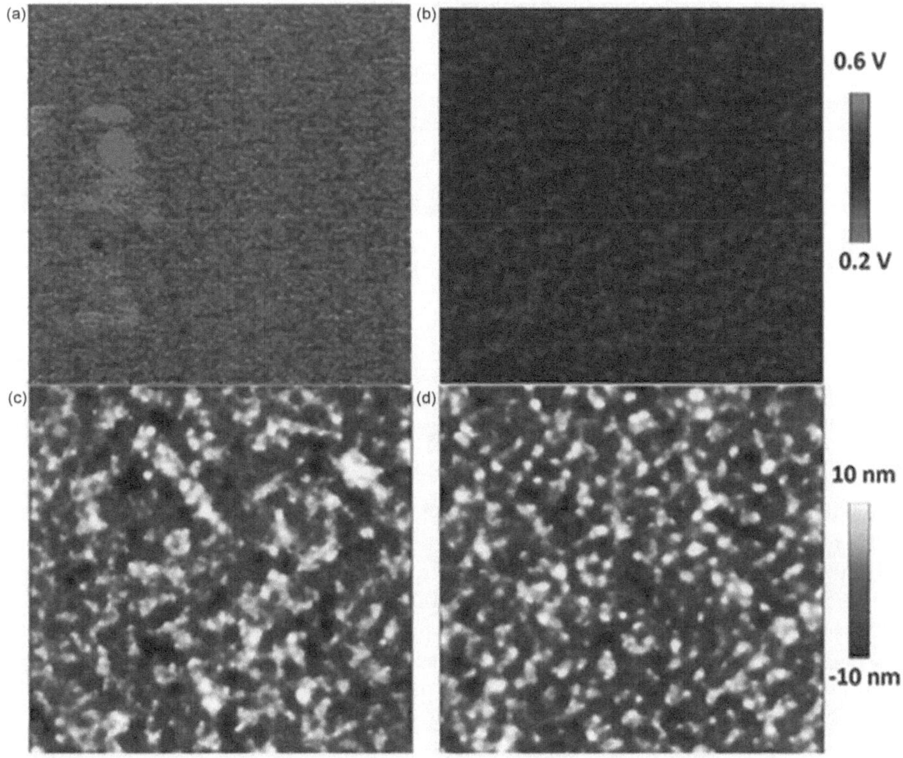


**FIGURE 8.13** The effect of methanol treatment on surface potential for PTB7:PC70BM solar cell. (a) and (b) Contact potential difference (CPD) images of the active layer without (a) and with methanol treatment (b) obtained from KPFM. (c) and (d) Surface topographic AFM images (size: 5 μm ×5 μm) without (c) and with methanol treatment (d).[53]

Recently, inverted structure OPVs with reverse polarity from conventional ones are becoming increasingly important due to their high-power conversion efficiency, good long-term device stability, and the adaptability to tandem device structures.[49,59] The energy band depth profile of inverted structure devices has its unique features compared to that of conventional structure devices. The thin electrode interlayers in the inverted devices are expected to introduce abrupt energy level offsets in the depth profile.[49,59] In a pioneer work, Qi Chen et al. used cross-sectional KPFM technique to determine the true energy band alignment and built-in field distribution on inverted structure poly(3-hexylthiophene) ($P_3HT$): [6,6]-phenyl-C61-butyric acid methyl ester (PCBM) OPVs (Figure 8.14). They revealed that the apparent inconsistency in cross-sectional KPFM investigations of inverted OPV devices is due to tip-to-tip variations on the probe geometry. The occasionally observed conflict between the KPFM measured built-in field, and the J-V characteristics is the result of the low spatial resolution in combination with the tip/cantilever induced convolution effect, which can mask abrupt energy level offsets caused by thin interlayers (Figure 8.14). The effects of these artifacts can be minimized by calibrating the transfer function of

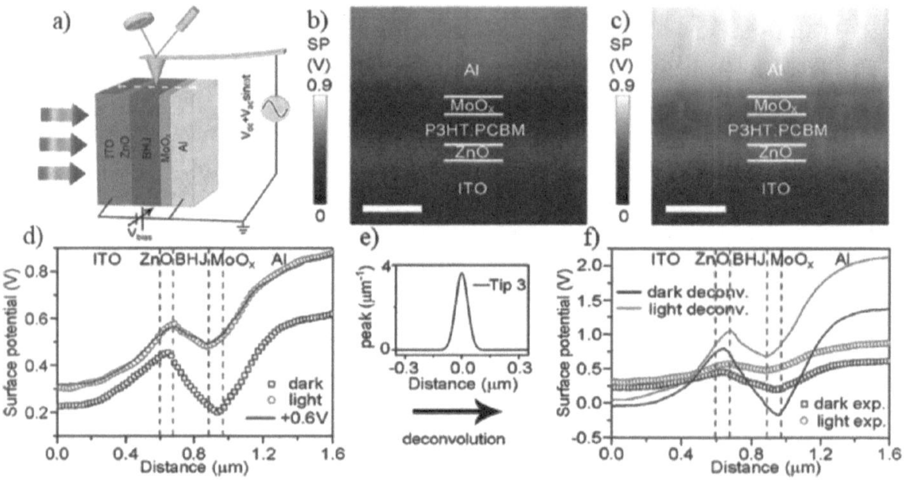


**FIGURE 8.14** (a) Schematic illustration of KPFM measurements of the vacuum level depth profile of operando cross-sectional devices. (b) and (c) AFM images of the ITO/ZnO(80 nm)/P3HT:PCBM/MoOx(80 nm)/Al device cross-section in dark state (b) and under light illumination (c) in open-circuit condition (scale bar: 400 nm). (d) AFM profiles extracted from (b) (black square) and (c) (red circle), and the AFM profile at +0.6 V bias in dark (blue line). (e) The transfer function of Tip 3 derived from numerical convolution fitting. (f) Deconvoluted AFM profiles from (d): black line for the dark and red line for the illuminated state of the device. (Based on Q. Chen[49])

sharp tips and performing numerical deconvolution.[49,59] The authors clarify that the built-in field direction in inverted devices is consistent with the device polarity; more importantly, the work function of the interlayers and the interfacial states between the active layer and the interlayers are critically important in determining the band bending in the active layer. However, the challenges of the cross-sectional KPFM method are to keep the device under investigation in operando states and to improve the imaging resolution.[49,59]

## 8.4.2 Dye-Sensitized Solar Cells

As illustrated in Figure 8.15, the heart of Dye-Sensitized Solar Cell (DSSC) is composed of the dye Self-Assembled Monolayer (SAM) adsorbed on a wide-bandgap semiconductor, usually a high surface area mesoporous $TiO_2$ photoanode, infiltrated with an electrolyte containing the redox shuttle molecule.[60] The primary role of the dye SAM is to sensitize the $TiO_2$ semiconductor, similar to sensitization of silver halides in paper photography. Upon illumination, the dye goes into a photo-excited state and can inject an electron into the conduction band of the semiconductor.[60] The oxidized dye is regenerated by a hole conductor, traditionally a liquid electrolyte, covering the dye and containing a redox mediator.[60] The molecular dye film has also a secondary function: it must act as an electronic barrier that prevents the

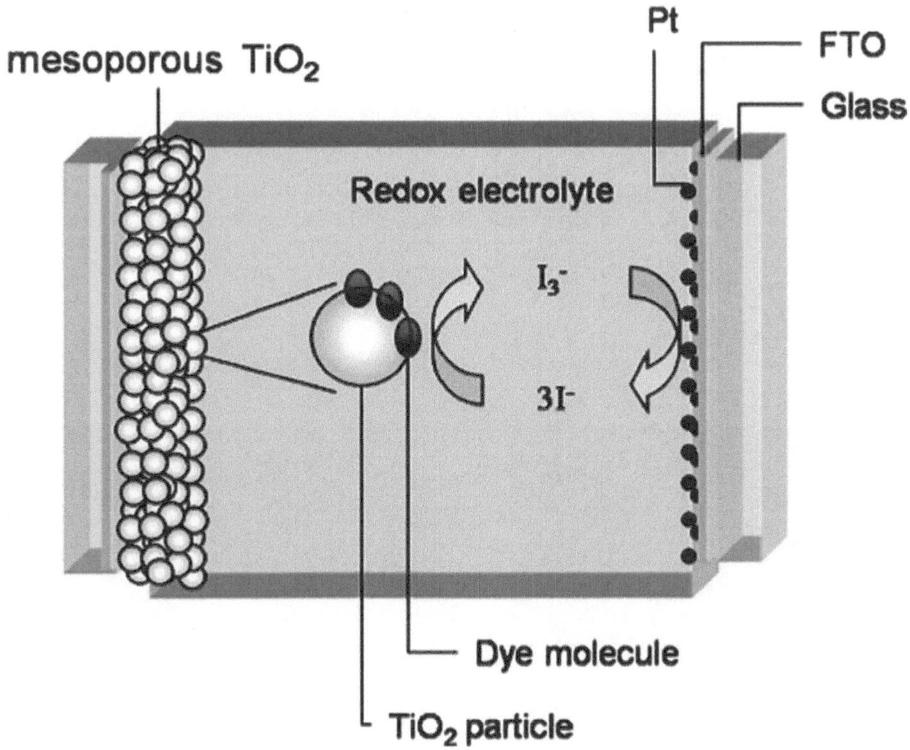


**FIGURE 8.15** Schematic overview of a dye-sensitized solar cell.[60]

photo-injected electrons to recombine with the oxidized form of the redox mediator present in the electrolyte. For typical dye molecules, this is ensured by hydrophobic alkyl chains that hinder the redox-mediator from accessing the semiconductor surface, and prevent lateral aggregation of dye molecules.[11] The anchoring groups of the dye, usually carboxylic acids, are hydrophilic, which gives the dye an amphiphilic character and behavior often similar to anionic surfactants.[60] The simple conventional processing methods such as printing techniques, flexibility, transparency and multicolor options, and low cost make DSSC attractive.[39] However, deterioration of DSSC and stability issues are certain challenges suffered by these kinds of cells.[39] The DSSC macroscopic efficiency is known to depend on both the molecular arrangement of the adsorbed dye layer and the contacting electrolyte.[60] This is generally true for most SAM-functionalized surfaces, which requires linking in situ molecular-level details with macroscopic observations in order to derive a full understanding.[61] Charge separation at the dye/$TiO_2$ interface in DSSCs is strongly influenced by the thickness and homogeneity of the sensitizing dye layer, as this controls the potential drop across the interface, and the probability of an excited electron being transferred from the dye to the $TiO_2$.[60] As shown in Figure 8.16, an ex-situ AFM measurements indicated the existence of large inhomogeneous dye aggregates on the flat $TiO_2$ substrate for standard device preparation procedure.[62] In addition, the pores between

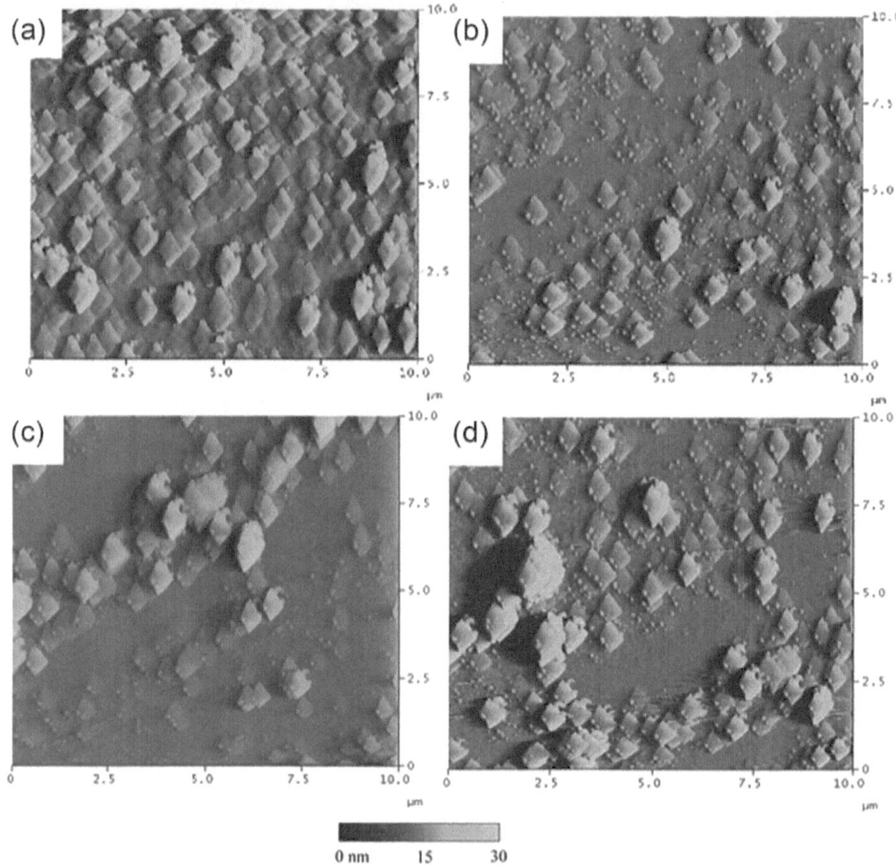


**FIGURE 8.16** AFM topography images (10 μm × 10 μm) showing the progress of dye adsorption on sputtered $TiO_2$ surfaces over 24 hours, showing key surface changes: (a) $t = 0$ hours, (b) $t = 3$ hours, (c) $t = 24$ hours, (d) after rinsing with pure acetonitrile.[62]

particles on the $TiO_2$ substrate were filled during dye adsorption.[62] To elucidate the molecular arrangement and the density of the dye in DSSCs, Voïtchovsky et al. performed in situ molecular-level AFM measurements of adsorbed dye molecules at mesoporous and flat $TiO_2$ surfaces in functionally relevant solvents.[61] They demonstrated that high-resolution AFM imaging could be used to explore the formation of the dye monolayer, showing several molecular conformations on the surface at different dye concentrations (Figure 8.17). The molecular conformation of the dye depends on the density coverage with domains of different molecular conformation able to coexist within the adsorbed submonolayer.[61] Recently, Freund et al. presented high-resolution topographic measurements using bimodal non-contact-AFM at room temperature of the anchoring part of a larger dye molecule 4,4′-di(4-carboxyphenyl)-6,6′-dimethyl-2,2′-bipyridine (DCPDMbpy) adsorbed on a NiO(001) crystal surface.[63] The surface structure of NiO(001) was resolved with atomic resolution using the first resonance and the torsional resonance. As shown in Figure 8.18, depending

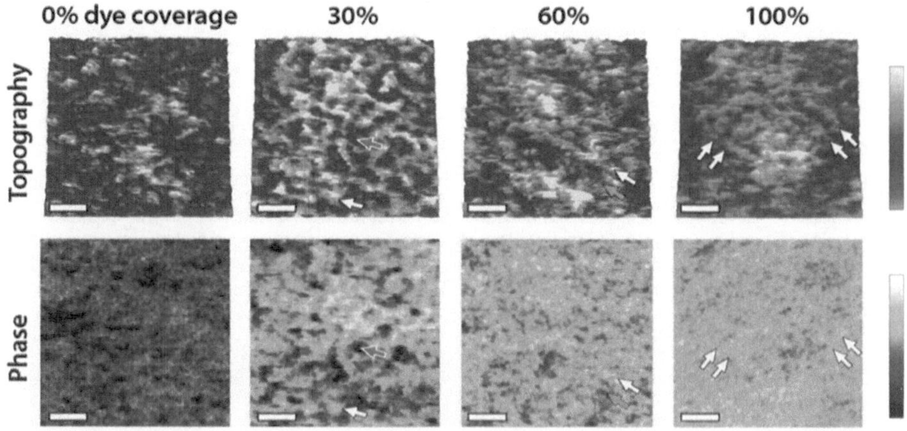


**FIGURE 8.17** High-resolution AFM images of the dye layer at 0%, 30%, 60%, and 100% mass coverage obtained in similar imaging conditions. At 0%, the substrate appears rough and noisy due to short-range tip–sample attractive interactions. At 30%, the dye molecules (white arrow) assemble in a soft sponge-like disordered structure that can easily be disrupted by the AFM tip. Multiples holes are visible in the layer (red arrow) and appear darker in the phase. At 60%, the layer is partially ordered with the apparition of rows (white arrow) and less ordered lower regions (green arrow). At 100%, the surface is fully covered; it appears smooth and only dye rows are visible (white arrows). The scale bar is 2 nm and the color scales are 1 nm (topography) and 15° in all images.[61]

on the deposition rate, single molecules, molecular clusters, and molecular islands have been imaged. Through the so-called multipass technique, submolecular resolution could be achieved and direct evidence of flat-lying molecules on the substrate with trans-conformation could be demonstrated.[63] Furthermore, DCPDMbpy exhibits a chiral character upon confinement on a surface leading to the appearance of two different surface enantiomers. Upon increasing the coverage, molecular islands with two symmetric orientations (*V* and *H*) appeared based on flat-lying molecules with alternating enantiomeric form (Figure 8.18).[63]

### 8.4.3 Perovskite Solar Cells

Perovskite solar cells (PSCs) are a newly developed solar cell research group, which has several advantages compared with thin-film solar cells and conventional silicon.[13] As illustrated in Figure 8.19, perovskites are class of compounds defined by the formula $ABX_3$ where X denotes a halogen such as $I^-$, $Br^-$, $Cl^-$ and A and B are cations of different sizes.[15] The efficiency of these cells is approaching that of single-crystalline silicon solar cells reaching up to 31% despite the presence of large GB area in the polycrystalline thin films.[21] However, due to degradation of the material used in this cell, the efficiency can be significantly reduced.[15]

As in other polycrystalline solar cells, GBs are present in perovskite as well and play an important role that could be benign or detrimental to solar-cell performance. To understand the electrical structure and behavior of GBs in PSCs, Li

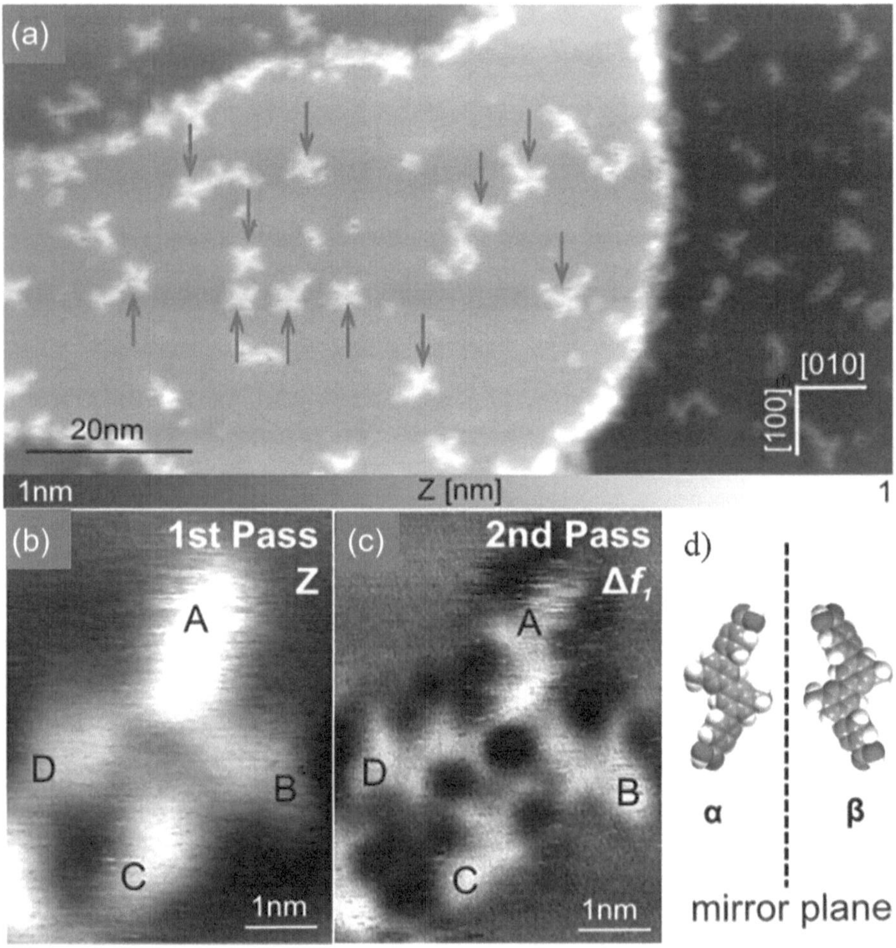


**FIGURE 8.18** Windmill-shaped cluster on NiO(001). (a) Topographic image of DCPDMbpy forming molecular clusters on NiO(001). Violet arrows show that DCPDMbpy often form windmill-shaped clusters. (b) Topographic image of a molecular cluster formed by four DCPDMbpy molecules acquired in the first scanning pass. (c) Frequency shift image recorded in the second scanning pass with an offset of –350 pm showing that DCPDMbpy can exhibit different conformations. (d) Sketch of the two different surface enantiomers of trans-DCPDMbpy.[63]

et al. performed comparative nanoscale electrical characterizations of both typical and inverted organolead halide PSCs, using KPFM and c-AFM. As shown in Figure 8.20, by comparing the CPD of these two devices in the dark and under illumination, KPFM measurements confirmed that the downward band bending at the GBs thus forms a low barrier, which predominantly attracts electron under illumination. In addition, the c-AFM measurements showed that the photocurrent flows through the GBs are negligible at 0 V bias, while the major photocurrents form on the grains. However, the photocurrents at the GBs become much higher than those

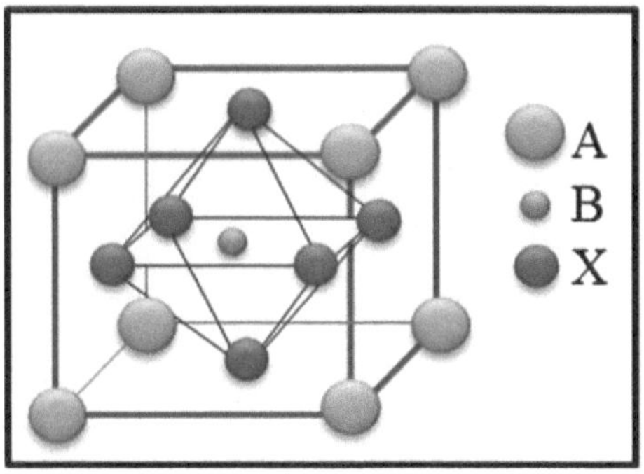


**FIGURE 8.19** Crystal structure of organometal perovskite.[15]

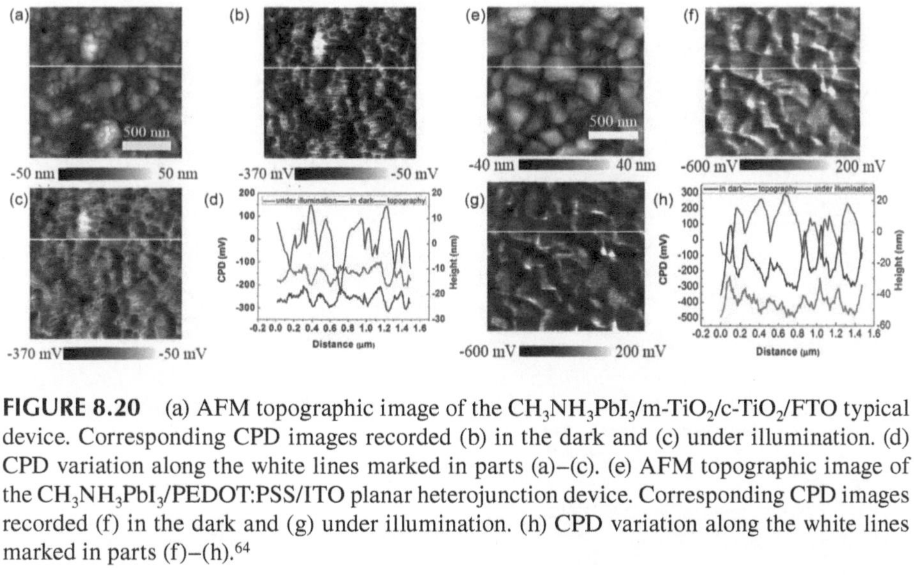

**FIGURE 8.20** (a) AFM topographic image of the $CH_3NH_3PbI_3$/m-$TiO_2$/c-$TiO_2$/FTO typical device. Corresponding CPD images recorded (b) in the dark and (c) under illumination. (d) CPD variation along the white lines marked in parts (a)–(c). (e) AFM topographic image of the $CH_3NH_3PbI_3$/PEDOT:PSS/ITO planar heterojunction device. Corresponding CPD images recorded (f) in the dark and (g) under illumination. (h) CPD variation along the white lines marked in parts (f)–(h).[64]

of the grains when the bias overcomes the barrier of the GBs. These two measurements demonstrate the enhanced photoinduced electron collections taking place at the GBs that act as effective charge dissociation interfaces and photocurrent transduction pathways.[64] Similarly, Yun et al. demonstrated that GBs in $CH_3NH_3PbI_3$ film in a $CH_3NH_3PbI_3$/$TiO_2$/FTO/glass heterojunction structure play a beneficial role. As shown in Figure 8.21, KPFM measurement showed that a potential barrier is formed along the GBs and higher surface photovoltage is found along the GBs. In addition, the c-AFM measurement indicated higher short-circuit current collection near GBs compared with that within grain interior. Thus, photogenerated carriers are effectively separated and collected at GBs confirming the beneficial roles GBs play in collecting carriers efficiently in these types of solar cells.[65]

Among all the intriguing properties of perovskites, ion migration has drawn most attention due to its correlation with the unique photocurrent hysteresis behavior.[66] As

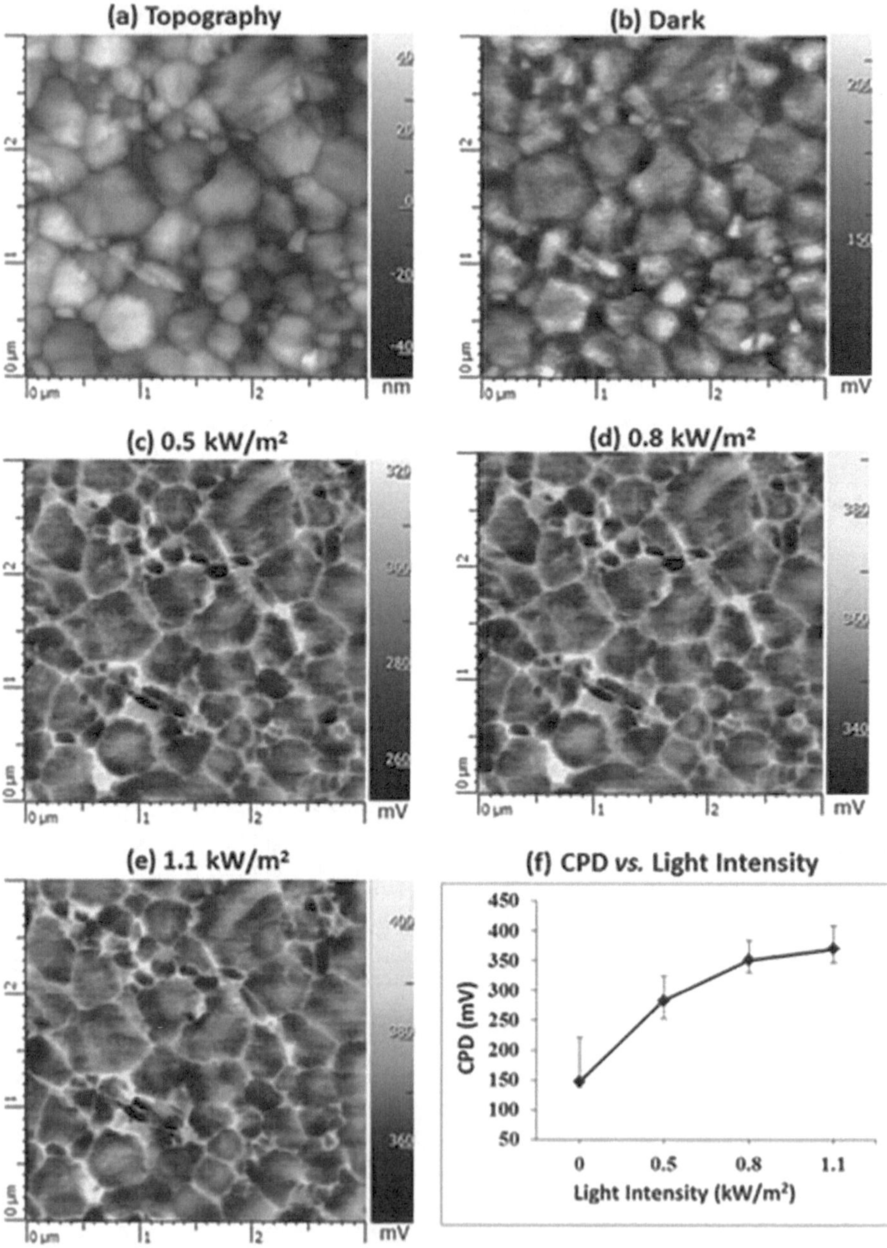


**FIGURE 8.21** KPFM measurements performed on a $CH_3NH_3PbI_3/TiO_2$/FTO/glass structure over an area of 3 μm². (a) Topography map and (b) CPD images taken in the dark. (c)–(e) CPD maps under various laser illumination intensities at a wavelength of 500 nm. (f) Intensity dependence of CPD of the sample at a wavelength of 500 nm as measured by KPFM.[65]

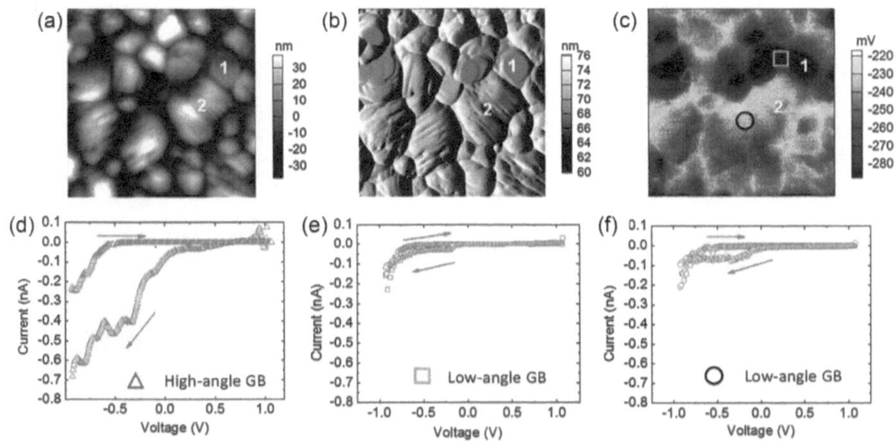


**FIGURE 8.22** Grain boundary morphology-dependent current measured by c-AFM. (a) Height and (b) amplitude images of the same area of a perovskite thin film; (c) KPFM image showing the locations where c-AFM tip measured the photocurrents. (d)–(f) Photocurrents measured at the location labeled in (c) with forward and reverse scanning at a scan rate of 0.14 V/s.[66]

shown in Figure 8.22, c-AFM measurements performed by Shao et al. revealed much stronger hysteresis both for photocurrent and dark-current at the GBs than on the grains interiors, which was explained by faster ion migration at the GBs.[67] The dramatically enhanced ion migration results in a redistribution of ions along the GBs after electric poling, in contrast to the intact grain area. In contrast, perovskite single-crystal devices without GBs showed negligible current hysteresis and no ion-migration signal.[67] These findings were crucial for the understanding of the role of GBs in PV device performance. In another interesting study on the role of GBs in perovskite systems, Adhikari et al. reported on the effects of an interface between $TiO_2$–perovskite and grain–GBs of perovskite films prepared by single step and sequential deposited technique using different annealing times at optimum temperature. They performed quantitative measurement of GB potential and $TiO_2$–perovskite interface surface potential in PSCs. KPFM measurement showed that charge transport in PSC depends upon annealing conditions. The KPFM results of single step and sequentially deposited films indicated that the barrier increases due to the formation of $PbI_2$ that suppresses the back-recombination between electrons in $TiO_2$ and holes in perovskite. Surface potential spatial maps of perovskite absorber layer show a higher surface potential at GBs than within grains. GB potential was also found to decrease from no annealing to 15 minutes annealing, then increases with longer annealing time at 30 and 60 minutes. Transient photovoltage measurement results show charge-carrier lifetime is the longest at 40 μs for 15 minutes annealing at 100°C and then decreases with longer annealing time.[68] Using pc-AFM spectroscopy, Kutes et al. were able to map variations of PV performance at the nanoscale for planar PSCs based on hole-transport-layer-free methylammonium lead triiodide ($MAPbI_3$), which is the most well studied organolead trihalide perovskite (OTP) thin films and is relatively easy to

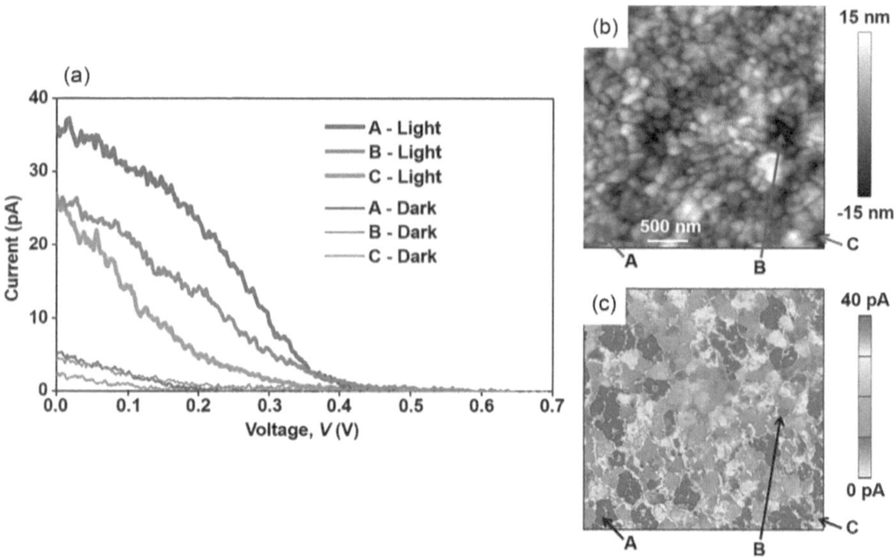


**FIGURE 8.23** $MAPbI_3$ thin-film (a) current measurements as a function of applied bias from +1 to 0 V in light and dark conditions for grains marked A, B, and C in (b) AFM topographic image and (c) ISC map of the same area.[69]

solution-process.[69] Simultaneous specimen illumination and photoconduction-AFM measurement approach allowed the acquisition of solar-cell I–V data efficiently with true nanoscale spatial resolution.[36] As shown in Figure 8.23, the recorded photoresponse parameters, short-circuit current (ISC), open-circuit voltage (VOC), and maximum power point (PMAX), at the nanoscale, revealed substantial variations in the photo-response that correlate with thin-film microstructural features such as intragrain planar defects, grains, GBs, and notably grain aggregates.[69] Notably, in the case of ISC, individual grains as well as aggregates of several adjacent grains exhibit similar values. Regions with the strongest ISC and VOC, of unknown origins but clearly coupled to microstructural features, also occupy <31% and 13% of the investigated area, respectively (Figure 8.23). The abrupt changes in the photo-response parameters from one location to another strongly indicate that transport of photocarriers is promoted by high conductivity pathways (grains, interfaces), which are often disconnected from neighboring regions by other low conductivity grains and interfaces acting as barriers. When transport is effectively unimpeded, photocarriers in $MAPbI_3$ thin films are demonstrated to exhibit sufficiently high mobilities and diffusion lengths to cause uniform ISC signals across as much as ~400 nm grains and/or grain-aggregates. Finally, linear features in the ISC and VOC contrast were often observed within individual grains likely due to the presence of planar defects (facets, twin boundaries, stacking faults, ferroelectric, or ferroelastic domains).[70]

To understand the roles of illumination-induced polarization and ionic migration in PV hysteresis, Xia et al. used dynamic-strain–based scanning probe microscopy to map nanoscale photocurrents under a series of biases. As shown in Figure 8.24, strong linear piezoelectricity arising from photo-enhanced polarization

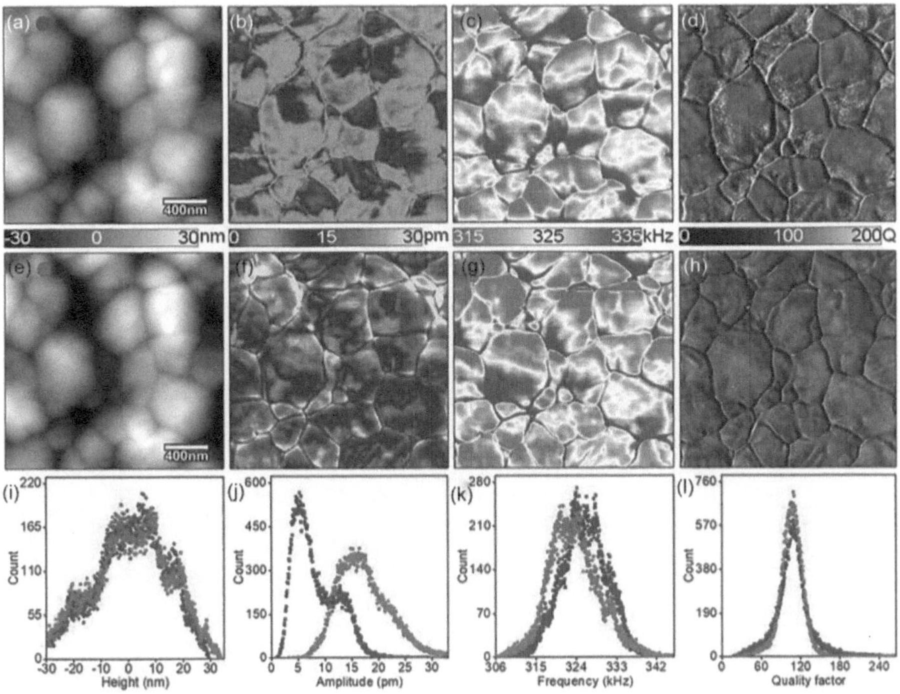


**FIGURE 8.24** Comparison of piezoresponse mappings in the dark and under illumination: (a)–(d) mappings of topography, amplitude, contact resonant frequency, and quality factor in the dark; (e)–(h) mappings of topography, amplitude, contact resonant frequency, and quality factor under light illumination; (i)–(l) comparison of histogram distributions in the dark and under illumination. The AC voltage applied is 2 V.[71]

was observed, while ionic migration was found to be not significantly increased by lightening.[71] Negligible difference between forward and backward scans, and local IV curves reconstructed from principal component analysis, showed minimal hysteresis of just 1%. These observations at the nanoscale were confirmed in a macroscopic similar PSC, exhibiting a high efficiency of 20.11% and with hysteresis index as small as 3%. Ionic migration, polarization, and photocurrent hysteresis were thus directly correlated at the nanoscale, and photo-enhanced polarization in triple-cation mixed-halide perovskites was established, which does not contribute to the PV hysteresis.[71] As a result, only small PV hysteresis was observed at both nano- and macroscale, demonstrating that even with the presence of strong polarization, the PV hysteresis is minimal. The study thus established polarization order in triple-cation mixed-halide perovskite under light illumination, and supported the concept that the primary mechanism responsible for PV hysteresis is ionic migration[72] instead of polarization.[71] In another pioneer study, Jiang et al. performed KPFM measurement on micrometer scale to explore ion migration in 2D lead halide perovskites of varying dimensionality as functions of both light illumination and temperature. As shown in Figure 8.25, the investigated perovskite films, prototypical 2D $BA_2PbI_4$ ($n = 1$), and

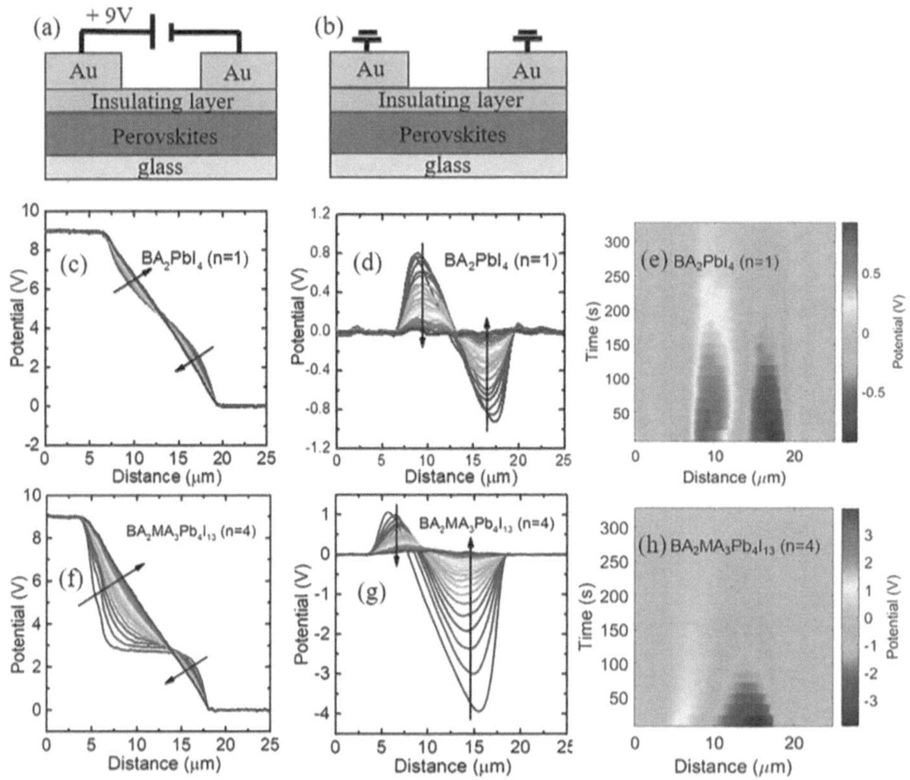


**FIGURE 8.25** Device architectures for SKPM measurement during the (a) charging and (b) discharging process. The measured potential profiles, evolving from dark purple to dark red lines with a 16–17 s time scale for each scanning trace, during the charging process for (c) BA2PbI$_4$ ($n=1$) and (f) $BA_2MA_3Pb_4I_{13}$ ($\sim\langle n\rangle=4$) perovskite films, and during the discharging process for (d) $BA_2PbI_4$ ($n=1$) and (g) $BA_2MA_3Pb_4I_{13}$ ($\sim\langle n\rangle=4$) perovskite films. The arrows in these panels indicate the evolution with respect to time. The corresponding time-resolved potential decay for (e) $BA_2PbI_4$ ($n=1$) and (h) $BA_2MA_3Pb_4I_{13}$ ($\sim\langle n\rangle=4$) perovskite films during the discharging process. All of the potential profiles have been shifted so that the grounded electrode is fixed at zero (essentially removing the small CPD between the tip and the gold electrode).[73]

methylammonium-incorporated quasi-2D $BA_2MA_3Pb_4I_{13}$ ($\sim\langle n\rangle=4$) showed differently shaped potential profiles during the charging process, which was attributed to a difference in their background electronic carrier population, as well as the carrier redistribution due to photogeneration.[73] Furthermore, a difference in the relaxation dynamics between $BA_2PbI_4$ ($n=1$) and $BA_2MA_3Pb_4I_{13}$ ($\sim\langle n\rangle=4$) films was observed. The authors explained these phenomena by proposing that the ion motion in pure 2D $BA_2PbI_4$ ($n=1$) perovskites could be dominated by paired halide and halide vacancy motion, whereas for methylammonium-incorporated quasi-2D $BA_2MA_3Pb_4I_{13}$ ($\sim\langle n\rangle=4$) perovskite films, the ion motion results from the interplay between paired halide/vacancy and methylammonium/vacancy. In contrast to earlier reports, these results indicate that, just as in 3D perovskites, ion transport can and does take place,

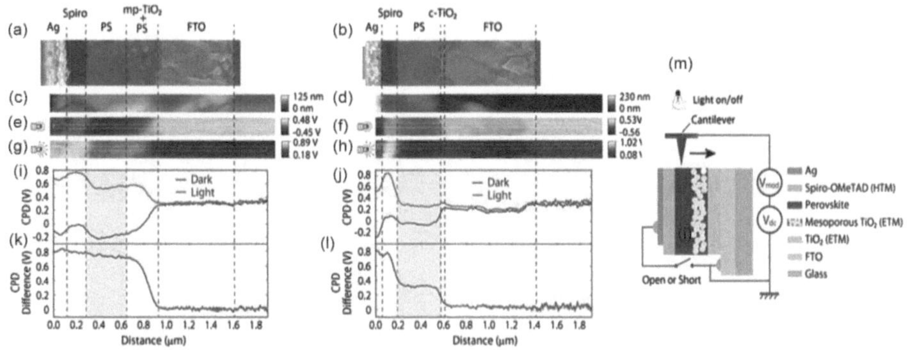


**FIGURE 8.26** KPFM characterizations of perovskites for the mesoporous and planar structures. (a) and (b) Scanning electron microscopy images of the cross-section of the mp-$TiO_2$ (a) and c-$TiO_2$ (b) devices. (c) and (d) Topographic image took on the cross-sectional surface of the mp-$TiO_2$ (c) and c-$TiO_2$ (d) devices. (e)–(h) CPD maps (KPFM images) taken in the same position as the topographic image of mp-$TiO_2$-based (e) and c-$TiO_2$-based (f) devices in short-circuit configuration in dark conditions and of mp-$TiO_2$-based (g) and c-$TiO_2$-based (h) devices in open-circuit configuration under light irradiation conditions. (i) and (j) Line profiles taken at the position marked by white rectangles in the KPFM images for mp-$TiO_2$ (i) and c-$TiO_2$ (j). (k) and (l) Difference CPD profile between dark and illuminated conditions in (i) and (j) for mp-$TiO_2$ (k) and c-$TiO_2$ (l), respectively. (m) Schematic of the setup used for KPFM measurements under dark and light conditions. c, compact; CPD, contact potential difference; FTO, fluorine-doped tin oxide; mp, mesoporous; PS, perovskite; Spiro, 2,2′,7,7′-tetrakis[N,N-di(4-methoxyphenyl)amino]-9,9′-spirobifluorene; ETM, electron transport material; FTO, fluorine-doped tin oxide; HTM, hole transport material; $V_{DC}$, DC voltage; $V_{mod}$, modulation voltage.[74]

although dimensionality clearly affects the character of both electronic and ionic transport in 2D perovskites.[73]

The electrical potential distribution within PV devices plays critical roles in governing charge-carrier dynamics and the device performance. Cai et al. measured the electrical potential distribution within highly efficient PSCs in working status.[74] As shown in Figure 8.26, they found a major potential drop at the perovskite/mesoporous $TiO_2$ interface, indicating one diode junction in mesoporous perovskite cells. In contrast, the major potential drop was found on both sides of the perovskite layer in planar cells, exhibiting two diode junctions where the charge carriers could be recombined more easily than single-junction cells. They further controlled the position of diode junctions in planar perovskites by changing the dopant concentration.[74] The undesirable charge recombination was largely suppressed, which enabled a better ideality factor and a higher efficiency 20.12% for planar perovskite cells with the increase of donor concentration. The findings from this interesting study will be helpful for understanding the fundamental mechanism within PSCs and achieving high device efficiency.[48,49,74–76]

In an effort to improve moisture-related stability issues that can affect 3D perovskites, several groups have reported the use of reduced dimensional phases, including 2D-layered Ruddlesden–Popper systems.[77] For example, Giridharagopal

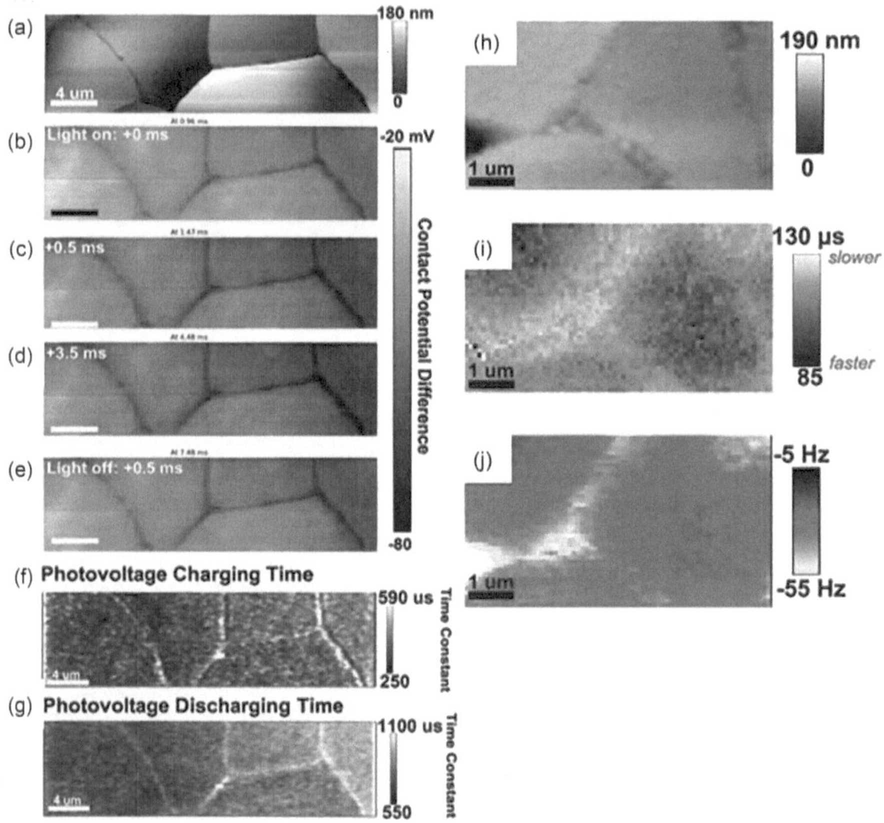


**FIGURE 8.27** Time-resolved G-KPFM on BAPI films. (a) Topography and (b)–(e) time slices of contact potential images on the same location showing the evolution of contact potential over the image. (f) Reconstructed CPD charging time constant and the (g) discharging time constant over the image. Illumination source is 455 nm and ~60 mW/cm$^2$. The right side is fast free time-resolved electrostatic force microscopy (FF-trEFM) on BAPI. (h) Topography, (i) FF-trEFM time constant, and (j) FF-trEFM frequency shift images under 455 nm illumination (~100 mW/cm$^2$) and $V_{tip} = +7$ V. Brighter regions exhibit slower charging in (i) and greater total charge (j). Regions adjacent to grain boundaries exhibit slower charging and are therefore consistent with the G-KPFM image data.[78]

et al. studied light-induced dynamics in thin films comprising Ruddlesden–Popper phases of the layered 2D perovskite $(C_4H_9NH_3)_2PbI_4$.[78] Two complementary scanning probe methods, time-resolved G-mode KPFM, and fast free time-resolved electrostatic force microscopy (FF-trEFM) were employed to probe ionic and electronic carrier dynamics as function of position, time, and illumination.[78] As shown in Figure 8.27, layered perovskites of $(C_4H_9NH_3)_2PbI_4$ (BAPI) exhibit heterogeneous optoelectronic properties. Surprisingly, these methods appear to show that BAPI films exhibit faster surface charge buildup in grain centers rather than at GBs, possibly due to a mixture of ionic and trap-mediated electronic transport. In G-KPFM, this effect happens with time scales on the order of <~500 μs. Qualitatively similar behavior, with response time estimate at ~70–100 μs, was observed in FF-trEFM, which in principle can achieve sub-microsecond time resolution. The authors proposed that this effect is dominated by a change in acceptor-type vacancies and associated screening under illumination due to the generation of photo-induced field that occurs primarily at GBs, along with a higher density of traps at the GBs resulting in slower electronic charging rates.[78]

Among the many anomalous properties of perovskite materials, of special interest are the ferroelectric properties including both classical and relaxor-like components, as a potential origin of slow dynamics, field enhancement, and anomalous mobilities. Kutes et al. performed the first investigation of ferroelectric domains in solution-processed, high-quality β-$MAPbI_3$ thin films using piezo-force microscopy (PFM), which is an ideal tool for probing local ferroelectric response at the domain scale.[70] PFM is based on AFM, where an AC electric field is applied to a scanning conducting probe in contact with a ferroelectric thin-film surface.[79] Since all ferroelectric materials are also piezoelectric,[80] the region in contact with the tip mechanically vibrates, which is simultaneously detected by the AFM during scanning.[70] The phase difference between the piezoactuation and the applied AC field is then used to detect ferroelectric domain orientations, while the amplitude reveals domain wall positions. Superimposing a DC voltage on the domain-mapping AC signal can be used to switch the domains in situ during scanning.[81] Similarly, as employed here, poling can be achieved during scans applying continuous DC-biases (or pulse patterns), which are then alternated with zero DC-bias PFM domain mapping scans.[79,81] As shown in Figure 8.28, the authors were able to directly image domains and domain walls in β-$MAPbI_3$ thin films directly using PFM, and an attempt to switch the ferroelectric domains reversibly using scanning DC-biases has been successful.[70] The domains are approximately equal in size to the grains (~100 nm). Recently, Ahmadi et al. used band excitation PFM and contact mode KPFM to explore ferroelectric properties of three representative formamidinium-based perovskites. As shown in Figure 8.29, these measurements provided strong evidence for the presence of ferroelectric domains in these systems; however, the domain dynamics were suppressed by fast ion dynamics. These materials hence present the limit of ferroelectric materials with spontaneous polarization dynamically screened by ionic and electronic carriers.[82] Such investigations are crucial to shed light on the fundamental PV mechanisms for perovskite-based solar cells, potentially suggesting routes to enhance the performance of this promising class of novel PVs.

Doping in organic–inorganic perovskite semiconductors is an effective method to tailor their optoelectronic properties. Recently Wu et al. used multi-functional AFM techniques, such as KPFM, EFM, and c-AFM, to correlate the photo-generated current and surface potential variation before and after manganese ($Mn^{2+}$) doping of $MAPbI_3$ perovskite films. They demonstrated that the device performance could be improved for a low $Mn^{2+}$ dopant concentration in $MAPbI_3$ perovskite via enhancing the VOC and FF, while the JSC always continued to decrease.[83] Similarly, Faraji et al. used KPFM to investigate the nanoscale effects of photochemically active additives of benzoquinone (BQ), hydroquinone (HQ), and tetracyanoquinodimethane (TCNQ) on GBs in $CH_3NH_3PbI_3$ solar cells under laser light illumination. The recently found improvement in the efficiency of BQ-added solar cells were clearly

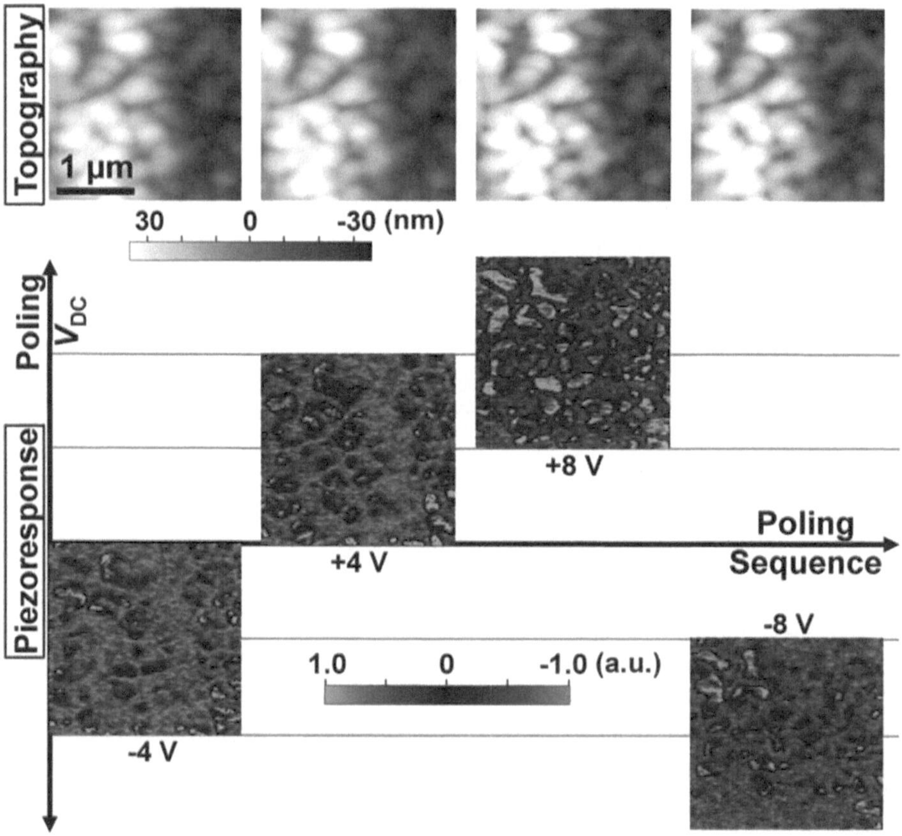


**FIGURE 8.28** Four AFM topography images (top row) of a single 2.5×2.5 μm$^2$ area, with simultaneously acquired corresponding PFM images beneath mapping the A·sin($\varphi$) piezoresponse, each after scanned DC poling at the biases indicated (VDC), evincing partial, reversible ferroelectric domain switching in as-processed β-$MAPbI_3$ thin film.[70]

seen in vanishing CPDs at GBs under illumination, rendering the material more uniform under solar cell operating conditions. As shown in Figure 8.30, these effects were observed for BQ, but not for HQ and CNQ.[84] This study demonstrated that the GBs in $CH_3NH_3PbI_3$ films with additives of BQ, HQ, and TCNQ play a crucial role in halide PSC performance. KPFM measurements showed that a potential barrier is formed along the GBs and higher CPD changes at GBs are found for HQ and TCNQ, whereas the opposite is true for BQ. These measurements confirmed that photogenerated carriers interact differently with GBs in these samples, leading to differences in overall device efficiencies.[84] In another example, Panigrahi et al. had successfully incorporated gold nanoparticles (Au NPs) into $MAPbI_3$ perovskite layer, which significantly increased the device photocurrent and changed internal surface potential. Pc-AFM was used for mapping the photocurrent at the nanoscale across the perovskite layer. The results showed the improvement of the photo-response across the plasmon-induced perovskite layers due to enhancement of carrier generation

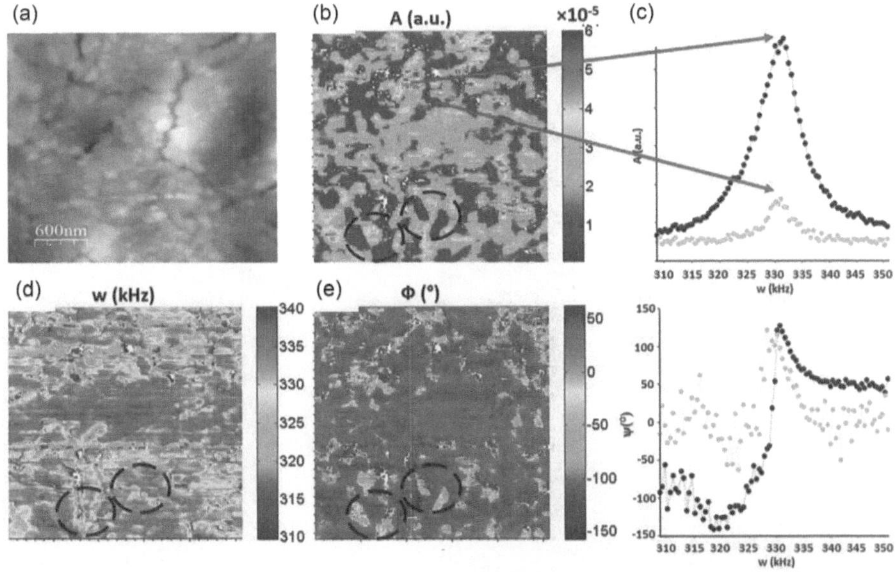


**FIGURE 8.29** Band excitation PFM images of $FA_{0.85}MA_{0.15}PbI_3$ (a) AFM topography, (b) PFM amplitude, (c) top: amplitude versus frequency and bottom: phase versus frequency, (d) frequency, and (e) phase. Note that circles illustrating perfect domain in the bottom left corner of (d) and (e) guide to eye that show regions with difference in amplitude and phase and no frequency changes, consistent with ferroelectric domains.[82]

through plasmonic scattering of Au NPs. Furthermore, the KPFM measurements of the potential distribution in two types of cells with different perovskite layers supported the argument that plasmonic effect has an influence on the charge-carrier dynamics inside the solar cells.[85]

Finally, it is worth mentioning that highly luminescent $CsPbBr_3$ perovskite QDs have gained huge attention in research due to their various applications in optoelectronics, including as a light absorber in PV solar cells. To improve the performances of such devices, it requires a deeper knowledge on the charge transport dynamics inside the solar cell, which are related to its power-conversion efficiency. In this context, Panigrahi et al. developed fully inorganic $CsPbBr_3$ perovskite QD-sensitized solar cells and investigated their electrical potential distribution across the layers under different wavelengths of light in the solar spectrum. As shown in Figure 8.31, using KPFM, they were able to observe the anomalous surface potential depth profile and correlated with the local structure of the layers of the solar cell.[86] Carrier generation, separation, and transport capacity inside the cells were dependent on the light illumination. Large differences in surface potential between electron and hole transport layers with unbalanced carrier separation at the junction have been observed under white light (full solar spectrum) illumination. However, under monochromatic light (single wavelength of solar spectrum) illumination, poor charge transport occurred across the junction as a consequence of less difference in surface potential between the active layers.[86] These findings provide fundamental understanding for the basic

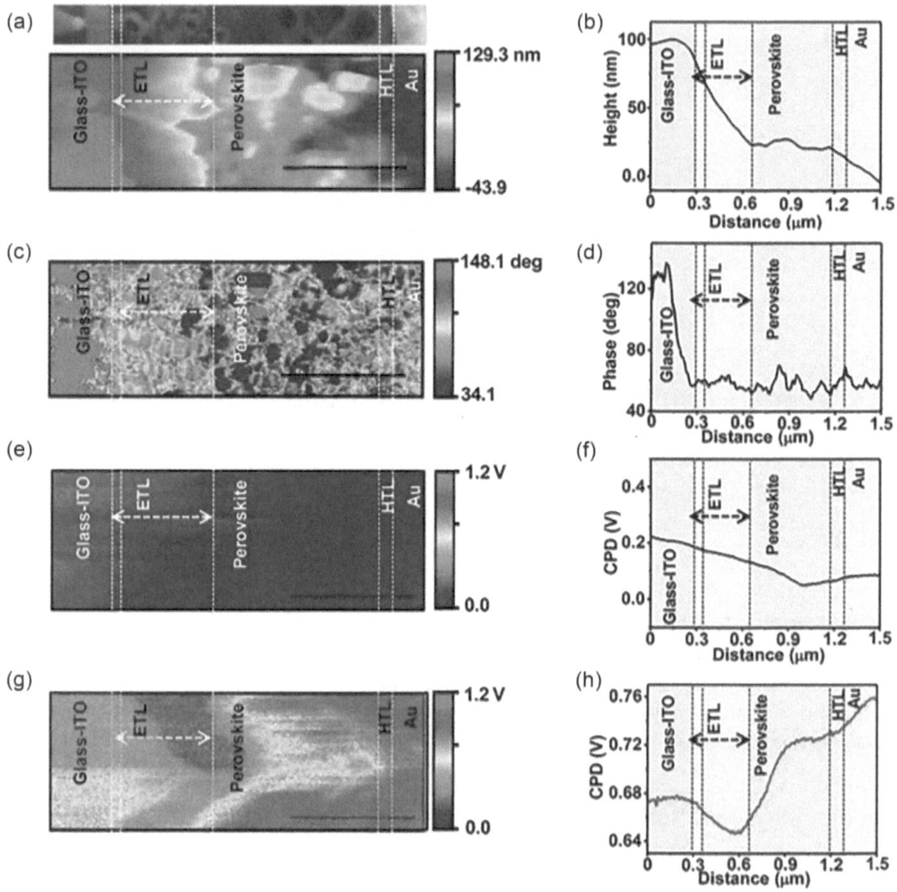


**FIGURE 8.30** (a) and (b) AFM topography and the corresponding line profile across the cross section of the solar cell. (c) and (d) Phase contrast image and the corresponding line profile of the solar cell. (e) and (f) Surface potential image and potential depth profile across the layers of the solar cell under dark condition. (g) and (h) Surface potential image and potential depth profile across the layers of the solar cell under solar illumination.[84]

charge transport mechanism across the interfaces and show the possibility to design high-performance fully inorganic perovskite QD-sensitized solar cells in the future.

## 8.5 OUTLOOK

It is obvious that most of the recent AFM-PV research is focusing more on thin-film technologies based on CdTe and CIGS and the emerging halide perovskites, all of which are higher quality optoelectronic materials than a-Si. Further research and development in a-Si technology will likely be discouraged owing to the rapid progress of these alternatives. Thus, there is a pressing need for a better developed understanding of the sub-processes at play in emerging PV materials and technologies. Currently

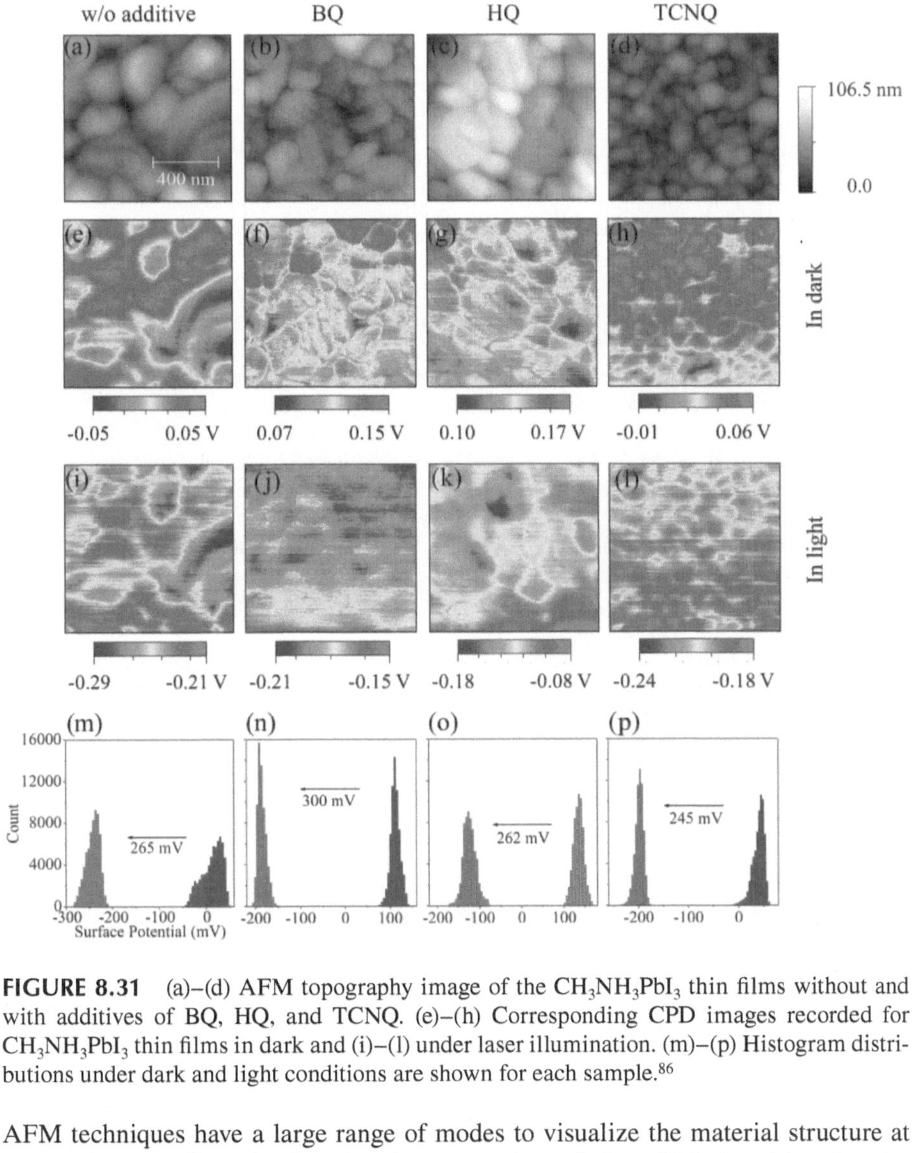


**FIGURE 8.31** (a)–(d) AFM topography image of the $CH_3NH_3PbI_3$ thin films without and with additives of BQ, HQ, and TCNQ. (e)–(h) Corresponding CPD images recorded for $CH_3NH_3PbI_3$ thin films in dark and (i)–(l) under laser illumination. (m)–(p) Histogram distributions under dark and light conditions are shown for each sample.[86]

AFM techniques have a large range of modes to visualize the material structure at nanoscale, as well as the functional response in real time. This is achieved under both dark and variable illumination conditions. When this is considered in conjunction with the better spatial resolution, more rapid imaging, and better environmental control possible, AFMs are seen to be a crucial part of PV research and development. While some of the measurements described here require specialized hardware and expertise, an increasing number of sophisticated scanning probe microscopy measurements can be carried out on commercially available instruments. Since no single technique can fully analyze an operating solar cell, AFMs need to be combined to other techniques such as fluorescence, IR, and Raman spectroscopies to provide crucial and unique data and should become more widely adopted in PV research.

## REFERENCES


1. Pulfrey, D. L., *Photovoltaic Power Generation.* New York: Van Nostrand Reinhold: 1978.
2. Fahrenbruch, A. L.; Bube, R. H.; D'Aiello, R. V., Fundamentals of solar cells (photovoltaic solar energy conversion). *Journal of Solar Energy Engineering*, 1984; Vol. 106,pp. 497–498.
3. Merrigan, J. A., *Sun Light to Electricity.* Cambridge, MA: MIT Press: 1975.
4. Robert, C., *Solar Panel Processing*, First Edit ed. Philadelphia: Old City Publishing Inc: 2010; p. 92.
5. Yadav, A.; Pawan Kumar, A.; Tech, M., Enhancement in efficiency of Pv cell through P&O algorithm. *International Journal for Technological Research in Engineering*, 2015; Vol. 2, pp. 2347–4718.
6. Badawy, W. A., A review on solar cells from Si-single crystals to porous materials and Quantum dots. *Journal of Advanced Research*, Cairo University, 2015; Vol. 6, pp. 123–132.
7. Chander, A. H.; Krishna, M.; Srikanth, Y., Comparison of different types of solar cells: A review. *IOSR Journal of Electrical and Electronics Engineering Ver. I*, 2015; Vol. 10, pp. 151–154.
8. Gul, M.; Kotak, Y.; Muneer, T., Review on recent trend of solar photovoltaic technology. *Energy Exploration and Exploitation*, 2016; Vol. 34, pp. 485–526.
9. Hoppe, H.; Sariciftci, N. S., Organic solar cells: An overview. *Journal of Materials Research*, 2004; Vol. 19, pp. 1924–1945.
10. Lee, T. D.; Ebong, A. U., A review of thin film solar cell technologies and challenges. *Renewable and Sustainable Energy Reviews*, Elsevier Ltd: 2017; Vol. 70, pp. 1286–1297.
11. Nayak, P. K.; Mahesh, S.; Snaith, H. J.; Cahen, D., Photovoltaic solar cell technologies: Analysing the state of the art. *Nature Reviews Materials*, Springer US: 2019; Vol. 4, pp. 269–285.
12. Sharma, S.; Jain, K. K.; Sharma, A., Solar cells: In research and applications—a review. *Materials Sciences and Applications*, 2017; Vol. 29, pp. 762–770.
13. Tang, H.; He, S.; Peng, C., A short progress report on high-efficiency perovskite solar cells. *Nanoscale Research Letters*, 2017; Vol. 12, pp. 410–417
14. Urbina, A., The balance between efficiency, stability and environmental impacts in perovskite solar cells: a review. *Journal of Physics: Energy*, IOP Publishing: 2020; Vol. 2, 022001.
15. Rathore, N.; Panwar, N. L.; Yettou, F.; Gama, A., A comprehensive review of different types of solar photovoltaic cells and their applications. *International Journal of Ambient Energy*, Taylor & Francis: 2019; Vol. 0, pp. 1–18.
16. Saga, T., Advances in crystalline silicon solar cell technology for industrial mass production. *NPG Asia Materials*, 2010; Vol. 2, pp. 96–102.
17. Li, J. B.; Chawla, V.; Clemens, B. M., Investigating the role of grain boundaries in CZTS and CZTSSe thin film solar cells with scanning probe microscopy. *Advanced Materials*, 2012, Vol. 24(6), pp. 720–723.
18. Zhou, H.; Hsu, W.-C.; Duan, H.-S.; Bob, B.; Yang, W.; Song, T.-B.; Hsu, C.-J.; Yang, Y., CZTS nanocrystals: A promising approach for next generation thin film photovoltaics. *Energy & Environmental Science*, 2013, Vol. 6, pp. 2822–2838.
19. Chen, X.; Lai, J.; Shen, Y.; Chen, Q.; Chen, L., Functional scanning force microscopy for energy nanodevices. *Advanced Materials*, 2018; Vol. 30, pp. 1–22.
20. Giridharagopal, R.; Cox, P. A.; Ginger, D. S., Functional scanning probe imaging of nanostructured solar energy materials. *Accounts of Chemical Research*, 2016; Vol. 49, pp. 1769–1776.

21. Green, M. A.; Hishikawa, Y.; Warta, W.; Dunlop, E. D.; Levi, D. H.; Hohl-Ebinger, J.; Ho-Baillie, A. W. H., Solar cell efficiency tables (version 50). *Progress in Photovoltaics: Research and Applications*, 2017; Vol. 25, pp. 668–676.
22. Hieulle, J.; Stecker, C.; Ohmann, R.; Ono, L. K.; Qi, Y., Scanning probe microscopy applied to organic–inorganic halide perovskite materials and solar cells. *Small Methods*, 2018; Vol. 2, pp. 1–17.
23. Pingree, L. S. C.; Reid, O. G.; Ginger, D. S., Electrical scanning probe microscopy on active organic electronic devices. *Advanced Materials*, 2009; Vol. 21, pp. 19–28.
24. Bonnell, D. A., *Scanning Probe Microscopy for Energy Research*. World Scientific Series in Nanoscience and Nanotechnology. Singapore: World Scientific Publishing: 2013.
25. Li, C.; Minne, S. C.; Pittenger, B.; Mednick, A., Simultaneous electrical and mechanical property mapping at the nanoscale with PeakForce TUNA. Bruker Application Note AN132, 2011; Vol. AN132.
26. O'Dea, J.; Brown, L.; Hoepker, N.; Marohn, J.; Sadewasser, S., Scanning probe microscopy of solar cells: From inorganic thin films to organic photovoltaics. *MRS Bulletin*, 2012; Vol. 37, pp. 642–650.
27. Igarashi, T.; Ujihara, T.; Takahashi, T., Photovoltage mapping on polycrystalline silicon solar cells through potential measurements by atomic force microscopy with piezoresistive cantilever. *Japanese Journal of Applied Physics, Part 1: Regular Papers and Short Notes and Review Papers,* 2006; Vol. 45, pp. 2128–2131.
28. Takihara, M.; Igarashi, T.; Ujihara, T.; Takahashi, T., Photovoltage mapping on polycrystalline silicon solar cells by Kelvin probe force microscopy with piezoresistive cantilever. *Japanese Journal of Applied Physics, Part 1: Regular Papers and Short Notes and Review Papers*, 2007; Vol. 46, pp. 5548–5551.
29. Tsurekawa, S.; Kido, K.; Watanabe, T., Measurements of potential barrier height of grain boundaries in polycrystalline silicon by Kelvin probe force microscopy. *Philosophical Magazine Letters*, 2005; Vol. 85, pp. 41–49.
30. Takihara, M.; Takahashi, T.; Ujihara, T., Minority carrier lifetime in polycrystalline silicon solar cells studied by photoassisted Kelvin probe force microscopy. *Applied Physics Letters*, 2008; Vol. 93, pp. 1–4.
31. Dutta, K. C. A. P. P. A. V., Thin-film solar cells: An overview. *Progress in Photovoltaics*, 2004; Vol. 12, pp. 69–92.
32. Meyer, E. L.; Osayemwenre, G. O., Interfacial assessment of degraded amorphous silicon module using scanning probe microscopy. *International Journal of Mechanical and Materials Engineering*, 2020; Vol. 15, 9.
33. Shockley, W.; Queisser, H. J., Detailed balance limit of efficiency of p-n junction solar cells. *Journal of Applied Physics*, 1961; Vol. 32, pp. 510–519.
34. Kutes, Y., Nanoscale photovoltaic performance of thin film solar cells by atomic force microscopy. Docutotal Dissertatons, University of Connecticut Graduate School, 2015.
35. Wolden, C. A.; Kurtin, J.; Baxter, J. B.; Repins, I.; Shaheen, S. E.; Torvik, J. T.; Rockett, A. A.; Fthenakis, V. M.; Aydil, E. S., Photovoltaic manufacturing: Present status, future prospects, and research needs. *Journal of Vacuum Science & Technology A: Vacuum, Surfaces, and Films*, 2011; Vol. 29, p. 030801.
36. Kutes, Y., Aguirre, B. A., Bosse, J. L., Cruz-Campa, J. L., Zubia, D., and Huey, B. D., Mapping photovoltaic performance with nanoscale resolution. *Progress in Photovoltaics: Research and Applications*, 2016; Vol. 24, pp. 315–325.
37. Kelley, T. W.; Frisbie, C. D., Point contact current–voltage measurements on individual organic semiconductor grains by conducting probe atomic force microscopy. *Journal of Vacuum Science & Technology B: Microelectronics and Nanometer Structures*, 2000; Vol. 18, p. 632.

38. Luria, J.; Kutes, Y.; Moore, A.; Zhang, L.; Stach, E. A.; Huey, B. D., Charge transport in CdTe solar cells revealed by conductive tomographic atomic force microscopy. *Nature Energy*, 2016; Vol. 1, pp. 1–6.
39. Mohammad Bagher, A., Types of solar cells and application. *American Journal of Optics and Photonics*, 2015; Vol. 3, p. 94.
40. Romero, M. J.; Jiang, C. S.; Abushama, J.; Moutinho, H. R.; Al-Jassim, M. M.; Noufi, R., Electroluminescence mapping of CuGa Se 2 solar cells by atomic force microscopy. *Applied Physics Letters*, 2006; Vol. 89, pp. 2004–2007.
41. Takihara, M., Minemoto, T., Wakisaka, Y. and Takahashi, T., An investigation of band profile around the grain boundary of $Cu(InGa)Se_2$ solar cell material by scanning probe microscopy. *Prog. Photovolt: Res. Appl.*, 2013; Vol. 21, pp. 595–599.
42. Hamamoto, Y.; Hara, K.; Minemoto, T.; Takahashi, T., Photothermal spectroscopy by atomic force microscopy on $Cu(In, Ga)Se_2$ solar cell materials. *Solar Energy Materials and Solar Cells*, Elsevier: 2015; Vol. 141, pp. 32–38.
43. Facchetti, A., Polymer donor: Polymer acceptor (all-polymer) solar cells. *Materials Today*, Elsevier Ltd.: 2013; Vol. 16, pp. 123–132.
44. Giridharagopal, R.; Shao, G.; Groves, C.; Ginger, D. S., New SPM techniques for analyzing OPV materials. *Materials Today*, Elsevier Ltd: 2010; Vol. 13, pp. 50–56.
45. Moulé, A. J.; Meerholz, K., Controlling morphology in polymer-fullerene mixtures. *Advanced Materials*, 2008; Vol. 20, pp. 240–245.
46. Yuan, Y.; Reece, T. J.; Sharma, P.; Poddar, S.; Ducharme, S.; Gruverman, A.; Yang, Y.; Huang, J., Efficiency enhancement in organic solar cells with ferroelectric polymers. *Nature Materials*, Nature Publishing Group: 2011; Vol. 10, pp. 296–302.
47. Douhéret, O., Swinnen, A., Bertho, S., Haeldermans, I., D'Haen, J., D'Olieslaeger, M., Vanderzande, D. and Manca, J.V., High-resolution morphological and electrical characterisation of organic bulk heterojunction solar cells by scanning probe microscopy. *Progress in Photovoltaics: Research and Applications*, 2007; Vol. 15, pp. 713–726.
48. Bergmann, V. W.; Weber, S. A. L.; Javier Ramos, F.; Nazeeruddin, M. K.; Grätzel, M.; Li, D.; Domanski, A. L.; Lieberwirth, I.; Ahmad, S.; Berger, R., Real-space observation of unbalanced charge distribution inside a perovskite-sensitized solar cell. *Nature Communications*, 2014; Vol. 5, pp. 5001–5009.
49. Chen, Q.; Ye, F.; Lai, J.; Dai, P.; Lu, S.; Ma, C.; Zhao, Y.; Xie, Y.; Chen, L., Energy band alignment in operando inverted structure $P_3HT$:PCBM organic solar cells. *Nano Energy*, Elsevier Ltd: 2017; Vol. 40, pp. 454–461.
50. He, Z.; Zhong, C.; Huang, X.; Wong, W. Y.; Wu, H.; Chen, L.; Su, S.; Cao, Y., Simultaneous enhancement of open-circuit voltage, short-circuit current density, and fill factor in polymer solar cells. *Advanced Materials*, 2011; Vol. 23, pp. 4636–4643.
51. Liscio, A.; Palermo, V.; Samorì, P., Nanoscale quantitative measurement of the potential of charged nanostructures by electrostatic and Kelvin probe force microscopy: Unraveling electronic processes in complex materials. *Accounts of Chemical Research*, 2010; Vol. 43, pp. 541–550.
52. Maturová, K.; Van Bavel, S. S.; Wienk, M. M.; Janssen, R. A. J.; Kemerink, M., Morphological device model for organic bulk heterojunction solar cells. *Nano Letters*, 2009; Vol. 9, pp. 3032–3037.
53. Zhou, H.; Zhang, Y.; Seifter, J.; Collins, S. D.; Luo, C.; Bazan, G. C.; Nguyen, T. Q.; Heeger, A. J., High-efficiency polymer solar cells enhanced by solvent treatment. *Advanced Materials*, 2013; Vol. 25, pp. 1646–1652.
54. Gu, K. L.; Zhou, Y.; Morrison, W. A.; Park, K.; Park, S.; Bao, Z., Nanoscale domain imaging of all-polymer organic solar cells by photo-induced force microscopy. *ACS Nano*, 2018; Vol. 12, pp. 1473–1481.
55. Schopp, N., Characterization of organic semiconductor materials and devices for organic solar cell applications with conductive and photoconductive atomic force microscopy. Master Thesis, RWTH Aachen University, 2018.

56. Li, H.; Tang, H.; Li, L.; Xu, W.; Zhao, X.; Yang, X., Solvent-soaking treatment induced morphology evolution in $P_3HT$/PCBM composite films. *Journal of Materials Chemistry*, 2011, Vol. 21, pp. 6563–6568.
57. Liu, X.; Wen, W.; Bazan, G. C., Post-deposition treatment of an arylated-carbazole conjugated polymer for solar cell fabrication. *Advanced Materials*, 2012, Vol. 24(33),pp. 4505–4510.
58. Nam, S.; Jang, J.; Cha, H.; Hwang, J.; An, T. K.; Park, S.; Park, C. E., Effects of direct solvent exposure on the nanoscale morphologies and electrical characteristics of PCBM-based transistors and photovoltaics. *Journal of Materials Chemistry*, 2012, Vol. 22(12).
59. Chen, Q.; Mao, L.; Li, Y.; Kong, T.; Wu, N.; Ma, C.; Bai, S.; Jin, Y.; Wu, D.; Lu, W.; Wang, B.; Chen, L., Quantitative operando visualization of the energy band depth profile in solar cells. *Nature Communications*, 2015; Vol. 6, pp. 1–9.
60. Hagfeldt, A.; Boschloo, G.; Sun, L.; Kloo, L.; Pettersson, H., Dye-sensitized solar cells. *Chemical Reviews*, 2010; pp. 6595–6663.
61. Voïtchovsky, K.; Ashari-Astani, N.; Tavernelli, I.; Tétreault, N.; Rothlisberger, U.; Stellacci, F.; Grätzel, M.; Harms, H. A., In situ mapping of the molecular arrangement of amphiphilic dye molecules at the $TiO_2$ surface of dye-sensitized solar cells. *ACS Applied Materials and Interfaces*, 2015; Vol. 7, pp. 10834–10842.
62. Marquet, P.; Andersson, G.; Snedden, A.; Kloo, L.; Atkin, R., Molecular scale characterization of the titania-dye-solvent interface in dye-sensitized solar cells. *Langmuir*, 2010; Vol. 26, pp. 9612–9616.
63. Freund, S.; Hinaut, A.; Marinakis, N.; Constable, E. C.; Meyer, E.; Housecroft, C. E.; Glatzel, T., Anchoring of a dye precursor on NiO(001) studied by non-contact atomic force microscopy. *Beilstein Journal of Nanotechnology*, 2018; Vol. 9, pp. 242–249.
64. Li, J. J.; Ma, J. Y.; Ge, Q. Q.; Hu, J. S.; Wang, D.; Wan, L. J., Microscopic investigation of grain boundaries in organolead halide perovskite solar cells. *ACS Applied Materials and Interfaces*, 2015; Vol. 7, pp. 28518–28523.
65. Yun, J. S.; Ho-Baillie, A.; Huang, S.; Woo, S. H.; Heo, Y.; Seidel, J.; Huang, F.; Cheng, Y. B.; Green, M. A., Benefit of grain boundaries in organic-inorganic halide planar perovskite solar cells. *Journal of Physical Chemistry Letters*, 2015; Vol. 6, pp. 875–880.
66. Xiao, Z.; Yuan, Y.; Shao, Y.; Wang, Q.; Dong, Q.; Bi, C.; Sharma, P.; Gruverman, A.; Huang, J., Giant switchable photovoltaic effect in organometal trihalide perovskite devices. *Nature Materials*, 2015, Vol. 14 (2), pp. 193–198.
67. Shao, Y.; Fang, Y.; Li, T.; Wang, Q.; Dong, Q.; Deng, Y.; Yuan, Y.; Wei, H.; Wang, M.; Gruverman, A.; Shield, J.; Huang, J., Grain boundary dominated ion migration in polycrystalline organic-inorganic halide perovskite films. *Energy and Environmental Science*, 2016; Vol. 9, pp. 1752–1759.
68. Adhikari, N.; Dubey, A.; Khatiwada, D.; Mitul, A. F.; Wang, Q.; Venkatesan, S.; Iefanova, A.; Zai, J.; Qian, X.; Kumar, M.; Qiao, Q., Interfacial study to suppress charge carrier recombination for high efficiency perovskite solar cells. *ACS Applied Materials and Interfaces*, 2015; Vol. 7, pp. 26445–26454.
69. Kutes, Y.; Zhou, Y.; Bosse, J. L.; Steffes, J.; Padture, N. P.; Huey, B. D., Mapping the photoresponse of $CH_3NH_3PbI_3$ hybrid perovskite thin films at the nanoscale. *Nano Letters*, 2016; Vol. 16, pp. 3434–3441.
70. Kutes, Y.; Ye, L.; Zhou, Y.; Pang, S.; Huey, B. D.; Padture, N. P., Direct observation of ferroelectric domains in solution-processed $CH_3NH_3PbI_3$ perovskite thin films. *Journal of Physical Chemistry Letters*, 2014; Vol. 5, pp. 3335–3339.
71. Xia, G.; Huang, B.; Zhang, Y.; Zhao, X.; Wang, C.; Jia, C.; Zhao, J.; Chen, W.; Li, J., Nanoscale insights into photovoltaic hysteresis in triple-cation mixed-halide perovskite: Resolving the role of polarization and ionic migration. *Advanced Materials*, 2019; Vol. 31, pp. 1–9.

72. Yuan, Y.; Huang, J., Ion migration in organometal trihalide perovskite and its impact on photovoltaic efficiency and stability. *Accounts of Chemical Research*, 2016; Vol. 49, pp. 286–293.
73. Jiang, F.; Pothoof, J.; Muckel, F.; Giridharagopal, R.; Wang, J.; Ginger, D. S., Scanning Kelvin probe microscopy reveals that ion motion varies with dimensionality in 2D halide perovskites. *ACS Energy Letters*, 2021; Vol. 6, pp. 100–108.
74. Cai, M.; Ishida, N.; Li, X.; Yang, X.; Noda, T.; Wu, Y.; Xie, F.; Naito, H.; Fujita, D.; Han, L., Control of electrical potential distribution for high-performance perovskite solar cells. *Joule*, Elsevier Inc.: 2018; Vol. 2, pp. 296–306.
75. Bergmann, V. W.; Guo, Y.; Tanaka, H.; Hermes, I. M.; Li, D.; Klasen, A.; Bretschneider, S. A.; Nakamura, E.; Berger, R.; Weber, S. A. L., Local time-dependent charging in a perovskite solar cell. *ACS Applied Materials and Interfaces*, 2016; Vol. 8, pp. 19402–19409.
76. Zhang, M.; Chen, Q.; Xue, R.; Zhan, Y.; Wang, C.; Lai, J.; Yang, J.; Lin, H.; Yao, J.; Li, Y.; Chen, L.; Li, Y., Reconfiguration of interfacial energy band structure for high-performance inverted structure perovskite solar cells. *Nature Communications*, Springer US: 2019; Vol. 10, pp. 1–9.
77. Chen, Y.; Sun, Y.; Peng, J.; Tang, J.; Zheng, K.; Liang, Z., 2D Ruddlesden–Popper perovskites for optoelectronics. *Advanced Materials*, 2018; Vol. 30, pp. 1–15.
78. Giridharagopal, R.; Precht, J. T.; Jariwala, S.; Collins, L.; Jesse, S.; Kalinin, S. V.; Ginger, D. S., Time-resolved electrical scanning probe microscopy of layered perovskites reveals spatial variations in photoinduced ionic and electronic carrier motion. *ACS Nano*, 2019; Vol. 13, pp. 2812–2821.
79. Gruverman, A.; Auciello, O.; Tokumoto, H., Imaging and control of domain structures in ferroelectric thin films via scanning force microscopy. *Annual Review of Materials Science*, 1998; Vol. 28, pp. 101–123.
80. Lines, M. E., and A. M. Glass, *Principles and Applications of Ferroelectrics and Related Materials*. Oxford: Oxford University Press: 2001.
81. Huey, B. D.; Nath Premnath, R.; Lee, S.; Polomoff, N. A., High speed SPM applied for direct nanoscale mapping of the influence of defects on ferroelectric switching dynamics. *Journal of the American Ceramic Society*, 2012; Vol. 95, pp. 1147–1162.
82. Ahmadi, M.; Collins, L.; Puretzky, A.; Zhang, J.; Keum, J. K.; Lu, W.; Ivanov, I.; Kalinin, S. V.; Hu, B., Exploring anomalous polarization dynamics in organometallic halide perovskites. *Advanced Materials*, 2018; Vol. 30, pp. 1–10.
83. Wu, Y.; Chen, W.; Wan, Z.; Djurišić, A. B.; Feng, X.; Liu, L.; Chen, G.; Liu, R.; He, Z., Multifunctional atomic force probes for Mn2+ doped perovskite solar cells. *Journal of Power Sources*, 2019; Vol. 425, pp. 130–137.
84. Faraji, N.; Qin, C.; Matsushima, T.; Adachi, C.; Seidel, J., Grain boundary engineering of halide perovskite $CH_3NH_3PbI_3$ solar cells with photochemically active additives. *Journal of Physical Chemistry C*, 2018; Vol. 122, pp. 4817–4821.
85. Panigrahi, S.; Jana, S.; Calmeiro, T.; Nunes, D.; Deuermeier, J.; Martins, R.; Fortunato, E., Mapping the space charge carrier dynamics in plasmon-based perovskite solar cells. *Journal of Materials Chemistry A*, 2019; Vol. 7, pp. 19811–19819.
86. Panigrahi, S.; Jana, S.; Calmeiro, T.; Nunes, D.; Martins, R.; Fortunato, E., Imaging the anomalous charge distribution inside $CsPbBr_3$ perovskite quantum dots sensitized solar cells. *ACS Nano*, 2017; Vol. 11, pp. 10214–10221.

# 9 Application of AFM for Analyzing the Microstructure of Ferroelectric Polymer as an Energy Material

*Dong Guo*
University of Shanghai for Science and Technology

*Kai Cai*
Beijing Center for Physical and Chemical Analysis

*Jingshu Xu*
Beihang University

**CONTENTS**



DOI: 10.1201/9781003174042-9

## 9.1 INTRODUCTION

Ferroelectric materials are special types of functional materials with a spontaneous electric polarization that can be reversed by the application of an external electric field in opposite direction. This unique switchable polarization endows the materials a broad range of applications including capacitors, non-volatile memories, oscillators, and filters. While the most important function of ferroelectrics may be their piezoelectricity, namely, the ability for electro-mechanical energy conversion owing to the lack of inversion symmetry, which makes the materials and their applications closely linked to energy, an issue of increasing priority for human. In addition to the traditional piezoelectric transducers, actuators, and force sensors, various emerging devices like energy harvesters, nanogenerators, electro-optic systems, and catalysts are getting global attention.[1]

The mainstream of ferroelectric materials are inorganic oxides, in contrast, only very limited organic materials show ferroelectricity. Among these, the fluoropolymer PVDF and the copolymer of VDF-TrFE may be seen as the prototype and the most widely used organic ferroelectric/piezoelectric materials. Compared to the inorganic materials, the properties of ferroelectric polymers like ease fabrication, mechanical flexibility and large area processing are highly desirable for the many emerging portable and wearable sensors and other devices. As a consequence, the fluoropolymer has become the focus of an overwhelming amount of work, particularly as energy conversion materials for new flexible devices and self-powered systems.[2–5]

The ferroelectricity and piezoelectricity of the fluoropolymer originates from the regularly aligned H-C-F dipoles in the molecular chain. Polymorph is a common for polymer. The ferroelectric fluoropolymer mainly shows four phases, while only the polar phases show piezoelectricity, particularly the β-phase with all-trans conformation. However, the phase change and microstructural characteristics of the copolymer are very complicated and still far from being fully understood. Analyzing the complicated phase change and microstructural characteristics of the fluoropolymer is a key issue of paramount importance for their application. Therefore, one of the main topics for the ferroelectric polymer research is the phase change. Appropriate microscope is an indispensable tool for microstructure analysis and for material performance modification. For polymer, the widely used high-resolution facility like electron microscopes have serious disadvantages. The radiation damage by electron beam is an inevitable and serious factor that limits the amount of microstructural information that can be collected. In comparison, atomic force microscopy (AFM)-based facilities are much more desirable because of the damage-free nanometer resolution imaging ability.[6–8]

Since the invention of AFM, the scientific world has envisioned the quick development of the technique and its revolutionary role for nanoscale surface characterization and analysis on the basis of various AFM variants. Over the past decades, advanced new modes were developed, and two particularly interesting modes are PFM and nano IR or AFM-IR. Piezoresponse force microscopy is a particularly useful AFM variant first appeared in 1992. It is indispensable nowadays for imaging piezoelectric/ferroelectric materials and manipulation of the domains. PFM basically works by detecting the surface deformation by converse piezoelectric effect stimulated via an AC bias through a sharp conductive probe in contact with the surface of a ferroelectric or piezoelectric material. The resulting deflection of the probe cantilever is detected through standard split photodiode detector and then demodulated by a lock-in amplifier. The technology can not only provide simultaneous high-resolution topography and ferroelectric domain imaging, but it also capable of chemical imaging via thermally induced surface change. These unique features open the opportunities for analyzing the target samples for researchers in many fields from ferroelectrics, semiconductors, and even biology by incomparable ways. More recently, a new AFM-based technique named nano-IR or AFM-IR appeared. Basically, this technique can provide unique chemical mapping at nanometer or micron resolution and hence the site-specific microstructural details based on detection of thermal expansion caused by IR absorption. This is expected to have a great impact on the investigation of the materials.[9,10]

In this chapter, main issues concerning the current study of ferroelectric polymer microstructure and their application in energy conversion and recent advances by using AFM-based techniques are discussed. The discussion is focused mainly on most relevant techniques, e.g., PFM and AFM-IR, and their application for new energy conversion devices or self-powered systems so as to provide some new hints and clues for understanding the materials and their potential application.

## 9.2 CURRENT CHALLENGE

Polymer has much more complicated structural characteristics than inorganic materials mainly because of the partial crystallinity and the many possible conformation of the long molecular chains. Polymer-based materials have some special characteristics; two properties relevant to their phase content are confinement effect and filler-induced transition between different polymorphs.[11]

### 9.2.1 Crystallographic Structure of PVDF and P(VDF-TrFE)

To understand the application of the fluoropolymer as an energy material, it is necessary to understand their microstructural characteristics. PVDF and P(VDF-TrFE) have similar crystallographic structure, and both have four polymorphs related to different conformations of the molecules as listed in Table 9.1. The lattice structure

**TABLE 9.1**
**Structural Information of the Polymorphs of PVDF**

| | | | | | Lattice Parameters | | | |
|---|---|---|---|---|---|---|---|---|
| Names | Conformation | Character | Space Group | Lattice | *a* [Å] | *b* [Å] | *c* [Å] | $\beta$[Å] |
| Form I $\beta$ | TT, all *trans* | Ferroelectric | Cm2m, $C_{2v}$ | Orthorhombic | 8.58 | 4.91 | 2.56 | |
| Form II $\alpha$ | TGTG' antiparallel | Nonpolar | $P2_1/c$, $C_{2h}$ | Monoclinic | 4.96 | 9.64 | 4.62 | 90 |
| Form $II_p$, $\delta=\alpha_p$ | TGTG' parallel | Ferroelectric | $P2_1cn$, C | Monoclinic | Similar as form II | | | |
| Form III $\gamma$ | $T_3GT_3G$' parallel | Ferroelectric | Cc, $C_s$ | Monoclinic | 4.96 | 9.58 | 9.23 | 92.9 |

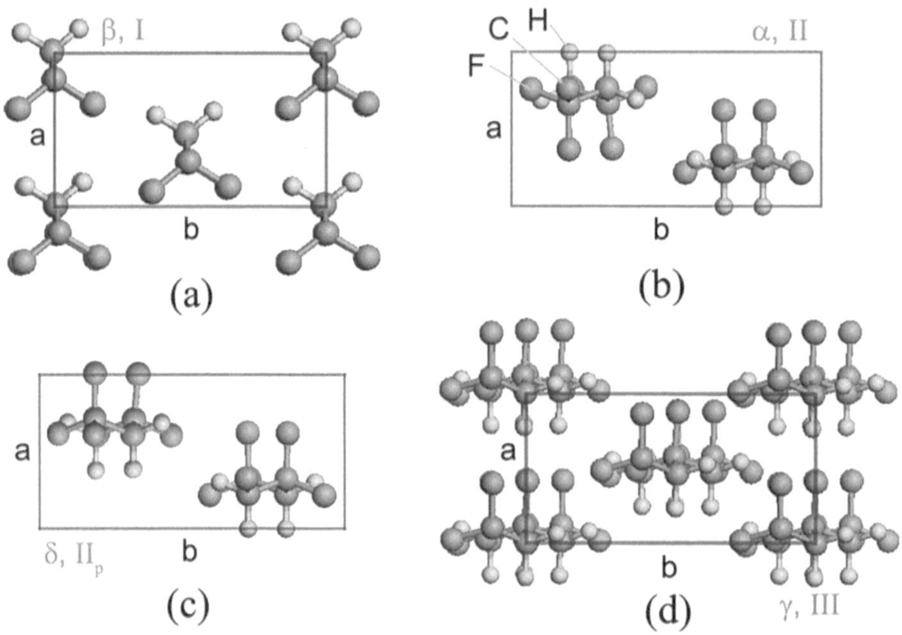


**FIGURE 9.1** Schematic illustration of the structure of four polymorphs of PVDF and P(VDF-TrFE).

of different polymorphs is schematically illustrated in Figure 9.1. Note that the lattice parameters are different for the copolymer with different VDF/TrFE ratios. The phase transition between the polymorphs can be induced by stress or thermal treatment. In order to display ferroelectric behaviors, the chains of the ferroelectric polymers must be able to crystallize in a manner in which the F-C-H molecular dipoles do not cancel out. Consequently, only the polar phases are ferroelectrics.

The usual melt-crystallized monopolymer PVDF forms the nonpolar α-phase (also designated form II) with chains in the TGTG' conformation as illustrated in Figure 9.1b. This phase is the most stable one because its gauche conformation carbon backbone can provide more spaces for the fluorine atoms (atomic radius 0.27 nm) than that offered by other (0.256 nm) conformation. Transformation of the α-phase into the ferroelectric β-phase with the all-trans (TT) conformation requires stretching or ultrahigh pressure, which imposes serious limitations for its practical application. The γ-phase is also non-ferroelectric and has not received much interest. The δ-phase is a ferroelectric but rarely studied due to the difficulty in fabrication, while new methods concerning its fabrication have been reported recently. In contrast, the copolymer P(VDF-TrFE) (partial substitution of F for H) spontaneously forms the all-trans ferroelectric β-phase because the steric hindrance between F atoms in the gauche-conformation is stronger than that between neighboring atoms in the all-trans conformation due to large size of F than H. The more stable TT conformation in the copolymer also results in increased crystallinity. The maximum crystallinity of pure PVDF is 50% and may reach values up to 90% in the copolymer. The direct crystallization from the melt or solution in the ferroelectric β-phase and the higher crystallinity are important advantages of the copolymers over pure PVDF. Although the α-phase is non-ferroelectric and cannot be used for piezoelectric devices, they are more desirable for applications as capacitive energy storage devices because of their relatively lower dielectric loss and energy dissipation as well as a higher energy storage density due to the slim polarization electric field loop.[12–16]

### 9.2.2 Current Challenge in Polar Phase PVDF Fabrication and Phase Content Adjustment

Although the copolymer P(VDF-TrFE) only forms in β phase that is inherently ferroelectric, it is much more expensive than the homopolymer PVDF. As a common thermal plastic, PVDF is available from various manufacturers in the world, and it has been widely used in as pipes, sheets, and coatings for packing and filtration. Therefore, the phase change issue in PVDF or energy saving stretching and ultra-high voltage-free fabrication of β phase PVDF has been a long-lasting challenge for this material. In addition, for capacitive energy storage applications, a high ratio of α phase is more desirable because the lack of ferroelectric loop may enhance the energy storage density and reduce the dielectric dissipation.[17]

New energy saving ways for direct fabrication of piezoelectric PVDF and control of PVDF phase content have been the focus of many studies. By combing uniaxial stretching and high electrical field, electrospinning can produce β-phase PVDF nanowires with a high yield. In an interesting recent study, kilometer-long, parallel, electroactive stable PVDF micro- and nanoribbons were produced via environment-friendly thermal fiber drawing technique that can iteratively reduce the size of the product. This method has a high efficiency, and no high electrical poling is required although high temperature and high stress are needed during thermal drawing. These electro-spun electroactive PVDF nanowires have been used in energy harvesters and nano-generators; however, the method seems not feasible for large area devices. The rarely reported ferroelectric δ-phase PVDF with a high $d_{33}$ of −36 pm/V measured under a high field of 250 MV/m was also obtained recently by a solid state route, while the required high pressure (~20 kN/cm$^2$) is also likely prohibitive for large-area fabrication. Therefore, a stress and high-voltage free technique that enables highly efficient fabrication of large area flexible piezoelectric PVDF film is still challenging.[18,19]

A widely observed phenomenon in polymer is filler-induced phase chance. Fortunately, PVDF also showed similar phenomenon, and a wide range of fillers can induce phase change in PVDF. Possible mechanisms are electrostatic interactions and filler-induced crystallization. A recent work reported an energy saving solution route for fabricating electroactive PVDF-based nanocomposite film by using spontaneously aligned molybdenum disulfide ($MoS_2$) nanosheets as the filler via super-2D-confinement. Only 3.4 vol % of the aligned $MoS_2$ nanosheets induced a very high ratio of β-phase PVDF up to ~86% in the matrix. A proposed possible mechanism for that is the disturbed crystallization and a synergistic effect between the enhanced electrostatic interaction and 2D geometry confinement by the flat nanosheets, as schematically illustrated in Figure 9.2. The best nanocomposite with the optimum

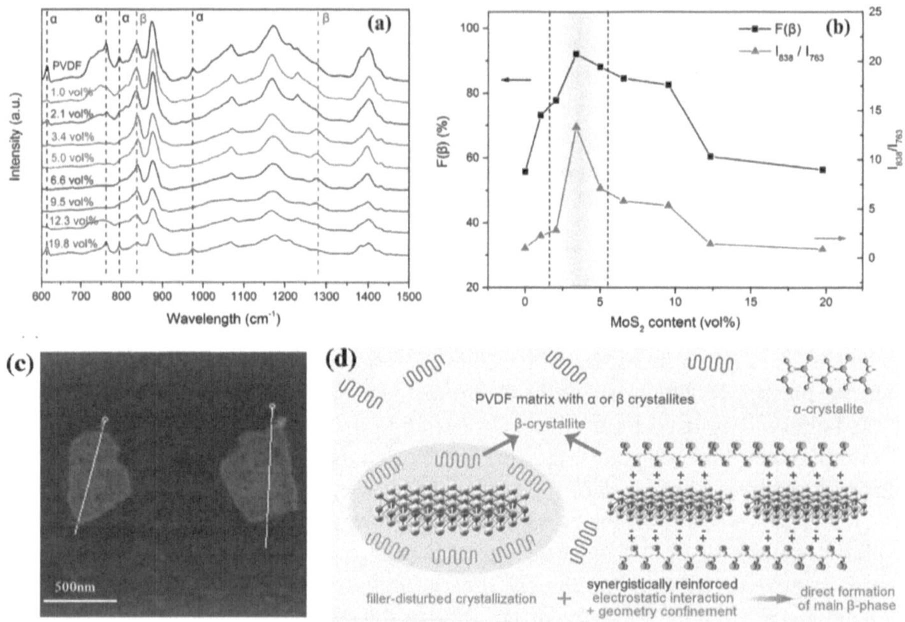


**FIGURE 9.2** (a) ATR-IR spectra of the $MoS_2$/PVDF composites. (b) $MoS_2$ content dependence of β phase fraction $F(\beta)$ and relative intensity ratio I838/I763 for the bands at 838 and 763 cm$^{-1}$. (c) AFM of monolayer $MoS_2$ (d) Schematic illustration of the mechanisms for the filler-induced formation of polar β-phase in the PVDF matrix. (Reprinted with permission from Cai, et al., *ACS Sustainable Chemistry & Engineering*, 2018, 6, 5043–5052.)

composition showed improved mechanical properties and increased capacitive energy storage density. Such composites are envisioned to be promising for applications requiring different functionalities. However, the increased conductivity by the semiconducting filler caused problems in poling, which is necessary to get piezoelectricity in PVDF. Further work is expected to overcome this challenge.

Despite the many advances so far, further understanding of the materials, structural characteristics requires to analyze the structural change under different conditions, especially by external factors. Therefore, new characterization techniques other than electron microscopes are highly desired; this will be discussed in the following section.

### 9.2.3 AFM-Based Characterization Techniques for Fluoropolymer

The microstructure of polymer can be analyzed by various techniques, including X-ray scattering like small-angle X-ray scattering (SAXS) and wide-angle X-ray scattering (WAXS), X-ray diffraction (XRD) with out-of-plane and in-plane geometry (GIXD) and vibrational spectroscopy like IR and Raman. Microstructural information may also be derived by analyzing the electrical properties. Compared to these, high-resolution microscopy is a straightforward method and has the advantage of direct morphology and lattice structure imaging. However, the problem for conventional electron microscopy (SEM, TEM) is that the polymer sample may be destroyed during measurement. In contrast, SPM-based microscopy has the advantages of high-resolution imaging without destroying the samples. In this category, AFM and the variant are particularly useful. Based on the selection of the specific polymeric systems, PFM can be used to get a broad range of information about structural characteristics and electrical properties, which may lead to a better understanding of the growth process. More recently appeared AFM-IR can give chemical imaging such as dipole and phase content distribution mapping. Thus, these techniques are highly desirable for further analyzing the detailed mechanisms of phase change in ferroelectric polymer. As aforementioned, because the microstructure and phase content are key factors for the functionality and performance of the relevant devices, these AFM-based techniques have been demonstrated to be rather ideal tools for the fluoropolymer for energy applications.[20]

## 9.3 APPLICATION OF PFM FOR PIEZOELECTRIC POLYMER

### 9.3.1 Combined PFM and KFM Study of the Surface Charge Dynamics in Tribological Nanogenerators

One of the main functionalities of the ferroelectric polymer is piezoelectricity, that is the ability to transfer mechanical energy into electricity or vice versa. Therefore, PVDF and P(VDF-TrFE) have been the main functional components used in many new self-powered and energy storage or conversion devices such as nanogenerators, and many studies have been reported. What deserves noting is that PVDF can also be used for its ferroelectricity or self-polarization in nanogenerators. An interesting work is by Lee et al. By using ferroelectric polarization, it is found that both the amount and direction of charge transfer in triboelectric materials can be controlled. The authors studied the ferroelectric-dependent triboelectricity in P(VDF-TrFE) thin films and the tribological nanogenerator performance by combining Piezoresponse Force Microscope (PFM) and Kelvin Force Microscope (KFM). They rubbed the ferroelectric surfaces with the AFM tip after different bias poling and found that the surface potential of the positively and negatively poled area, respectively, become smaller and larger. It is also found that the power output from the triboelectric nanogenerator is dependent on the polarization state. The results indicate that the amount and direction of the charge transfer in triboelectricity can be controlled by the ferroelectric polarization state. The work provides a way to control the charge transfer behavior of the ferroelectric polymer through triboelectric effect for systems like next-generation memory technologies that requires a higher output performance for charging.[21]

As shown in Figure 9.3, two different poling states have been achieved by applying positive and negative bias voltages via the conductive probe on the P(VDF-TrFE) film. In Figure 9.3a, the positively and negatively poled areas induced by bias voltages on the P(VDF-TrFE) film exhibit a clear contrast in PFM phase image that corresponds to downward and upward polarizations, which indicates switched polarizations. The surface potentials caused by injected charges were visualized by KFM images as shown in Figure 9.3b. The surface after 12 hours relaxation was imaged again and potential of different signs were obtained as shown in Figure 9.3c. Further

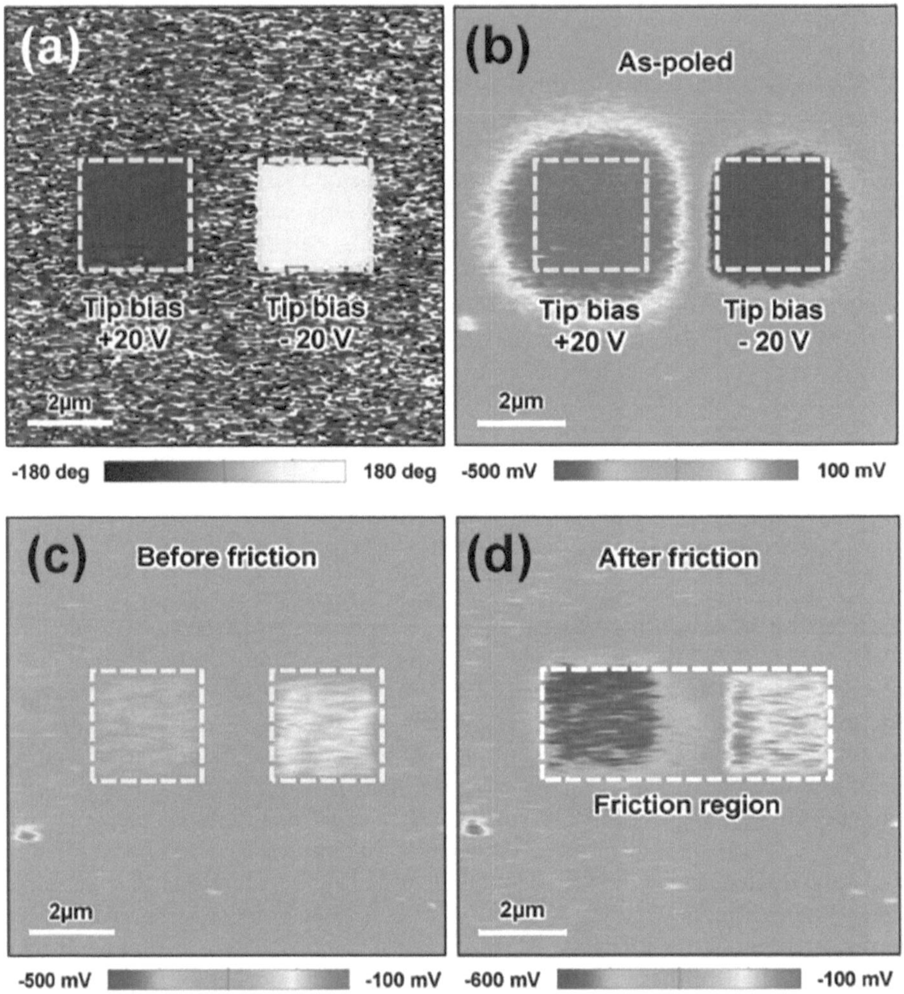


**FIGURE 9.3** Polarization switching and surface potential distribution of P(VDF-TrFE) thin film: (a) and (b) PFM phase and KFM images after poling, respectively; (c) KFM image after 12 hours relaxation; (d) KFM image after friction over an area indicated by dashed line.

imaging after friction of the poled area with Pt-coated AFM tip for 20 times indicate that efficient triboelectric charge transfer has occurred, giving rise to high potential difference. Similar experiments on non-ferroelectric PMMA confirm that the sign of charge transfer on the ferroelectric polymer films can be controlled because of the polarization. Device characterization further indicates that the positively and negatively poled P(VDF-TrFE)-based nanogenerator shows output voltages of opposite signs, which confirmed that the polarization direction indeed determines the sign of the charge transfer by the triboelectric process. Moreover, as shown in Figure 9.4, the positively and negatively poled P(VDF-TrFE)-based nanogenerator under a compressive force of 2 kg shows output voltages up to a 400 and 100 V, respectively, much

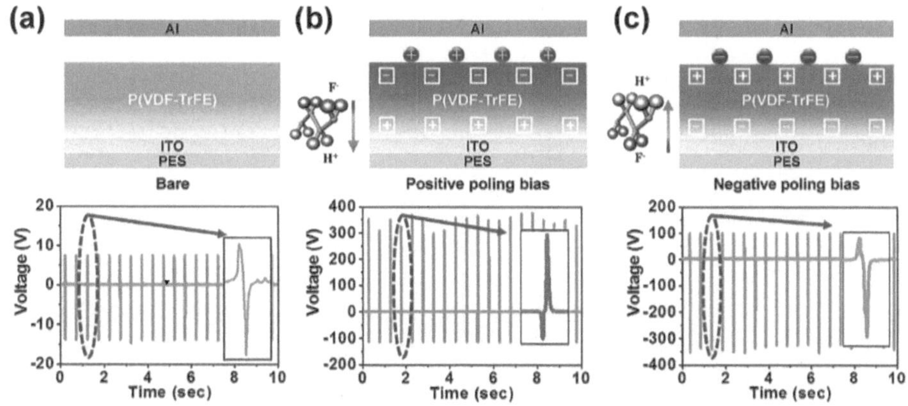


**FIGURE 9.4** Structure and performance of the nanogenerators based on P(VDF-TrFE) films after different poling treatment: (a) bare P(VDF-TrFE). (b) and (c) Positively and negatively poled P(VDF-TrFE) films, respectively.

large than that of an un-poled device (15 V). The results demonstrated an interesting way to analyze surface charge dynamics during the triboelectric process and to fabricate new nanogenerators by combining PFM and KFM.

### 9.3.2 Vector PFM Characterization of P (VDF-TrFE) Nanowires

Vector PFM is a method that can be used to measure the in-plane component of polarization by detecting the angular torsion of the cantilever in addition to the usual vertical displacement. Based on combined vertical and lateral data sets obtained at different scanning directions, this method can be used to reconstruct three-dimensional polarization However, a problem is that the lateral and vertical PFM data were generally obtained with different sensitivities and therefore could not be directly compared to get a quantitative estimation of the electric mechanical response. Therefore, a comprehensive comparison of the parameters under the same conditions should be used. An example is a recent study that compares the P(VDF-TrFE) films obtained after different thermal treatments. The vector-PFM amplitude and phase images of an SVA (Solvent Vapor Annealed) film and a TA (Thermal Annealed) film with a same thickness of 160 nm are shown in Figure 9.5. Then the distribution profiles derived from the PFM images can be used for a qualitative judgment of the piezoelectricity. Namely, the larger average value of the vertical (V-AM) and lateral amplitude (L-AM) distribution profiles of the SVA film than that of the TA one indicates that the former sample has a stronger bidirectional piezoelectricity.[23] Because the ferroelectricity in the polymer originates from the F-C-H dipoles parallel to the b-axis of its lattice and the piezoelectric response is approximately proportional to the dipole component along external strain, the researchers concluded from the higher average V-AM in SVA film that the F-C-H dipoles have been on average oriented along the film surface normal. From the much higher lateral to vertical amplitude ratio of SVA sample (8.07) than that of TA one (2.97), where the dipoles have

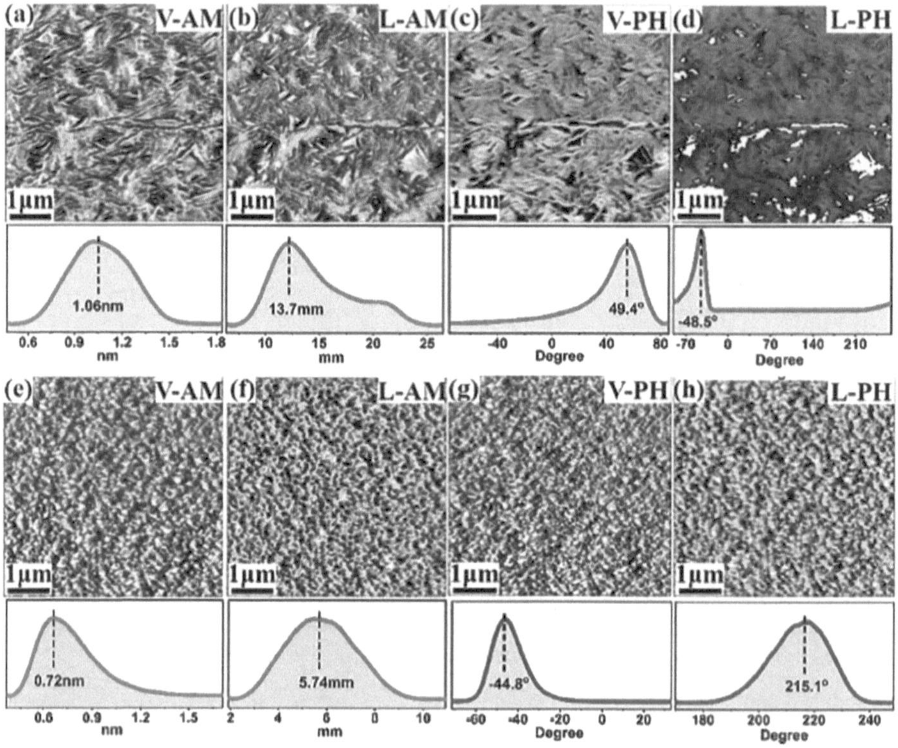


**FIGURE 9.5** PFM images for the SVA film annealed at 140°C for 3 hours: vertical (a) and lateral (b) PFM amplitude images and their distribution profiles. Vertical (c) and lateral (d) PFM phase images and their distribution profiles; PFM images for the TA film annealed at 140°C for 3 hours: vertical (e) and lateral (f) PFM amplitude images and their distribution profiles. Vertical (g) and lateral (h) PFM phase images and their distribution profiles.

no vertical orientation, the study further derived a lateral ratio of 2.72 (lateral vs. vertical) in the nanowire films formed by SVA. The work demonstrated an "internal comparison" method by using sample measured under same conditions to analyze dipole orientation in the ferroelectric polymer by using vector-PFM. The practical use of this anisotropy and the high lateral response is for energy harvesters based on bending ($d_{31}$) strains. Therefore, the work demonstrated an "internal comparison" method based on samples measured under the same experimental conditions to analyze dipole orientation in the ferroelectric polymer and evaluate the potentials of a new processing technique or ferroelectric material for energy harvesting application by using vector-PFM.

A further example is the use of vector-PFM for dipole orientation mapping in ferroelectric polymer in a recent study. P(VDF-TrFE) film with a nominal thickness of 8 nm with a nanorod morphology was imaged as shown in Figure 9.6a–e, where the distribution profiles are also shown. The lateral PFM amplitude image of the nanorods in Figure 9.6c shows higher contrast than that of the vertical one

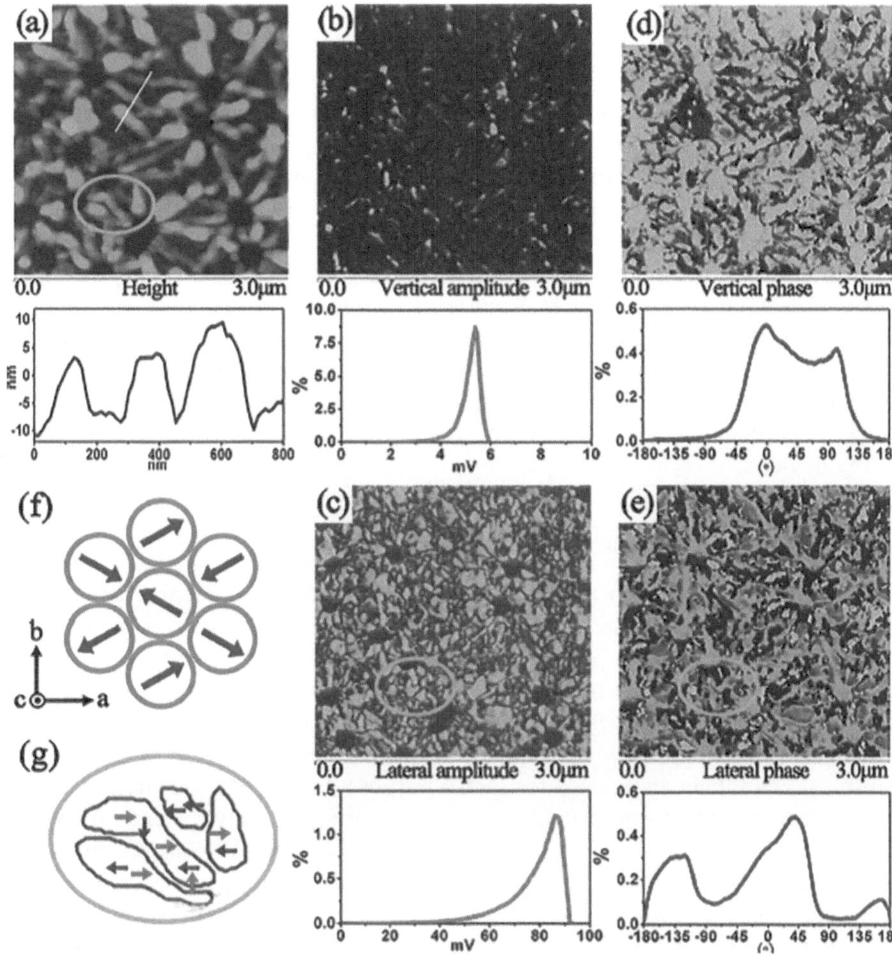


**FIGURE 9.6** (a)–(e) are the AFM topography and PFM images of the P(VDFTrFE) nanorod assemblies: (a) AFM topography and the height profile of the line shown in the image, (b) vertical amplitude image and the amplitude distribution profile, (c) lateral amplitude image and the amplitude distribution profile, (d) vertical phase image and the phase distribution profile, (e) lateral phase image and the phase distribution profile, (f) schematic illustration of the dipole orientation distribution in the plain perpendicular to the c-axis of the lattice, and (g) schematic illustration of the dipole distribution map of a region marked by the ellipse shown in (a).

in Figure 9.6b, and the former also has a better correlation with the topography in Figure 9.6a. This difference is clear from their amplitude signal profiles. The dark areas with exposed substrate with no piezoresponse may affect the PFM signals. From the double peak distribution of the phase signal profile below (Figure 9.6d), the work found two possible average preferential dipole orientations (up or down) as schematically illustrated in Figure 9.6f. The profile below (Figure 9.6e) indicates a similar orientation, but the additional central peak was attributed to complicated reasons like the adhesive or viscoelastic tip-sample interactions.

Because the lateral amplitude image coming from the torsion of the cantilever mainly detect the domains with in-plane polarization (dipole) orthogonal to the cantilever axis, by combining the lateral PFM amplitude and phase images in Figure 9.6c and e, particularly the phase contrast, the study got an in-plane dipole orientation distribution, as schematically illustrated in Figure 9.6g for an area marked by the ellipse. It seems that a further example is the use of vector-PFM for dipole orientation mapping in ferroelectric polymer in a recent study. P(VDF-TrFE) film with a nominal thickness of 8 nm with a nanorod morphology was imaged. The work further confirmed the molecular orientation by using a reflection IR technique. Although the mapping is only a rough estimation of orientation, the work demonstrated an indirect method for deriving site-specific orientation states in ferroelectric polymer, which may be useful for modifying the performance of relevant energy devices.

### 9.3.3 In Situ Hot Stage PFM

The crystallinity and grain morphology are critical factors governing device performance. To understand the crystallization behaviors and the correlation between the microstructure and the performance of the ferroelectric polymer thin films or nanostructures, a real-time observation of the grain growth process and their corresponding change in piezoelectric response during heating is much more advantageous than other post-annealing characterization techniques. In situ hot stage PFM is a special tool that can be used for such a purpose. It has been accepted as a standard measurement that can visualize the free, unperturbed, polymer crystal growth. A work of this type of characterization about spin-coated P(VDF-TrFE) films was carried out by using an NT-MDT NTEGRA Prima microscope with a sample stage with heating chamber. As shown in Figure 9.7, different leaf-like grains of the sample show sharp contrast in the phase image (Figure 9.7c) and the uniform grain color, the study concluded that the grains were highly ordered but with different orientations. Interestingly, when the measuring temperature is increased to 150°C, melt phase appears around the leaf grains (Figure 9.7d), and the center parts of the grains still show a strong piezoelectric response even the temperature is higher than the Curie temperature, that is smaller than 140°C for the 70/30 copolymer.[24,25] The authors ascribed the contrast of more than 180°C in the phase image (Figure 9.7f) to an enhanced mechanical adhesion as a consequence of partial melting.[26] The work also revealed that the ferroelectric phase transition in P(VDF-TrFE) has a diffusive nature due partly to the partial crystalline and partial amorphous microstructural characteristics of the copolymer. What need to be noted is the elliptical region in Figure 9.7b,c,e, and f. From the observation that the amplitude may decrease, increase, or keep relatively constant for different grains when the temperature is increased from 135°C to 150°C, the work further concluded that molecular dipole reorientation in some grains had occurred above 135°C. The work demonstrated that the structural change of individual grains at elevated temperature can be directly visualized by using hot stage PFM.

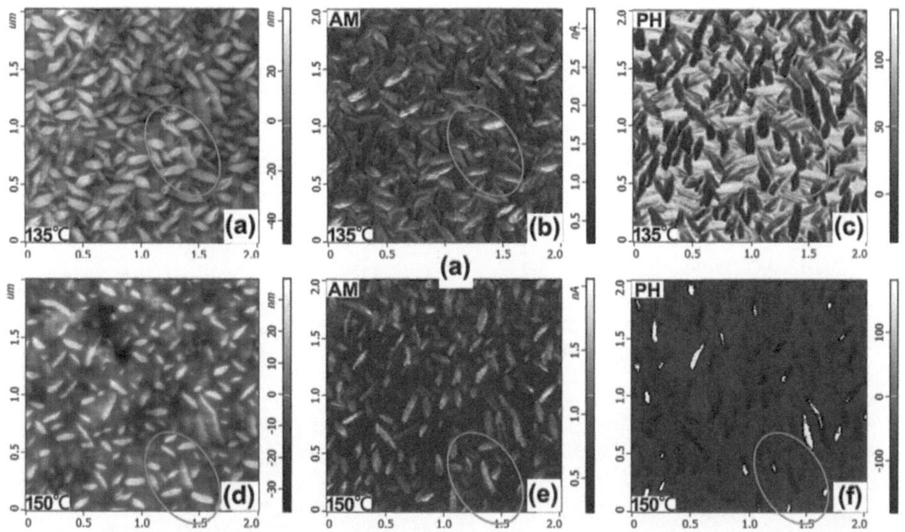


**FIGURE 9.7** AFM morphology and PFM amplitude (AM) and phase (PH) images obtained at different temperatures. The elliptical region compares more details about the piezoelectric response of individual grains.

## 9.4 AFM-IR TECHNIQUES FOR FERROELECTRIC MATERIALS STUDY

### 9.4.1 Application for Copolymer Films in Low Energy Consumption Ferroelectric Memory

One of the most important applications of ferroelectric materials is in random-access memory (FeRAM), which has advantages of fast switching and low power consumption. Compared to FeRAMs based on inorganic oxide ferroelectrics, FeRAMs with organic ferroelectric layers have advantages of ease processing, large area, and easy integration with flexible electronics. An interesting representative work is the study of single domain switching kinetics in 2D-confined ferroelectric PVDF-TrFE nanodots for low-energy consumption FeRAMs by Shen et al. The researchers got high-resolution structural mapping by using AFM-IR.

A representative work is the study of ferroelectric polymer films for low power FeRAM.

The nanoscale infrared(IR) absorption at wavelength of 1,119 $cm^{-1}$ due to paraelectric phase are shown in Figure 9.8a. Low scattering amplitude and high absorption intensity are related to the high paraelectric phase. The results confirm that samples with different VDF-TrFE ratios (Figure 9.8b and c) have different ferroelectric phase and content of defects. A higher HTLT ferroelectric phase content with relatively less ordered conformations is present in Figure 9.8c.

The work further studied the domain switching mechanism of the P(VDF-TrFE) nanodots and demonstrated the ability of PFM in studying the Precise programmability and data stability of the materials for application as FeRAm. As shown in

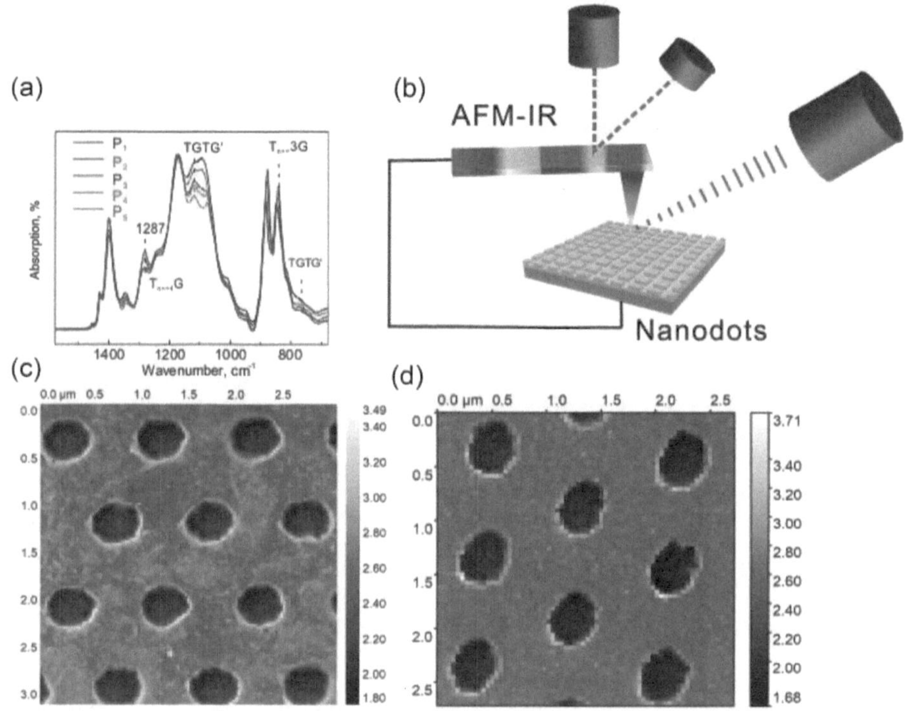


**FIGURE 9.8** (a) FTIR spectra of P(VDF-TrFE) bulk with various molar ratios of the VDF unit. (b) Schematic illustration of the AFM-IR configuration. (c) and (d) Nanoscale IR mapping amplitude at a wavelength of 1,119 $cm^{-1}$ of the P3 and P1 nanodots, respectively.

Figure 9.9, a simple write–erase–rewrite model was constructed in situ. The reading out of "1" and "0" states are shown by the red and purple zones, respectively.

### 9.4.2 Application in Investigating Ferroelectric Polymer Films for Energy Harvester

PVDF-based Translucent porous composites containing ZnO and Ag particles with high β phase PVDF has been reported. Enhanced β phase formation is attributed to the electrostatic bonding between fillers and $-CH_2$ and $-CF_2$ chains, yet, local conformational disorder at the ZnO-matrix interface region caused by strong interface effect results in local stabilization of β phase. Coupling of high β phase content and microporous structure is believed to be essential for achieving considerable piezoelectric outputs (7.1 $\mu W/cm^2$) and excellent force sensitivity (1.155 V/kPa).

Atomic force microscopy-infrared spectroscopy (AFM-IR) is used to provide the direct structure mapping to clarify the inhomogeneous influence of interface effects on β phase formation and stabilization of the polymers. The β phase distribution and interfacial conformation were directly measured by the AFM-IR (nanoIR2-fs, Anasys Instruments, America) at the tapping mode (spatial resolution is 10 nm). The

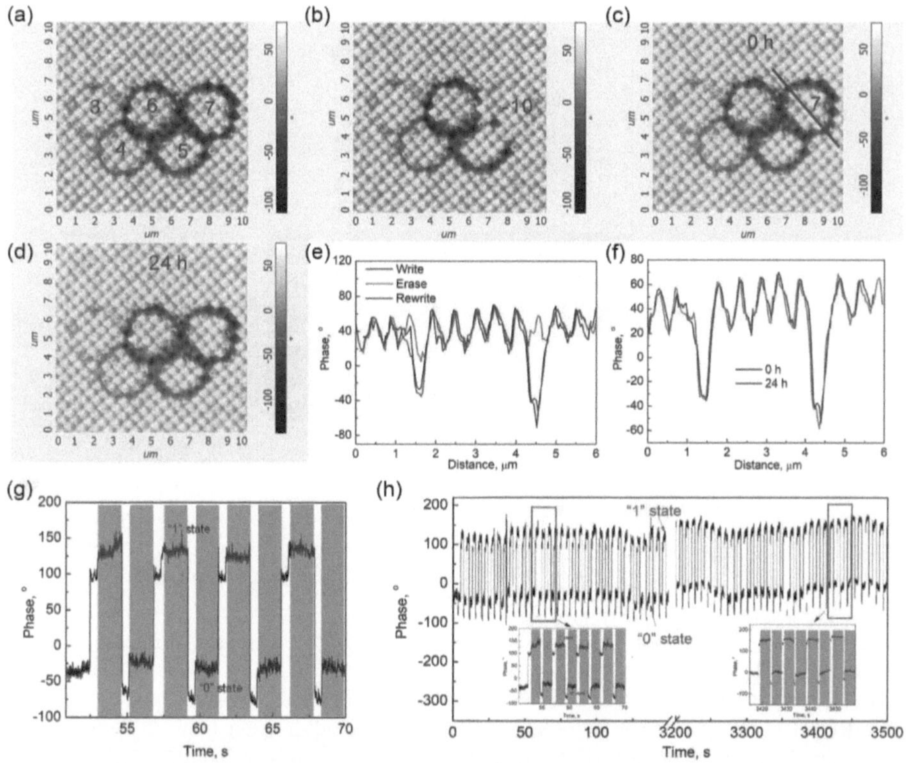

**FIGURE 9.9** Precise programmability and data stability. (a) PFM phase images showing "the Olympic rings" written on the P1 nanodots at positive voltages of 3, 4, 5, 6, and 7 V at a writing speed of 0.5 μm/s. (b) PFM phase images showing the "7 V" ring precisely erased by a negative voltage of −10 V. (c) Erased ring rewritten by a voltage of 7 V at the same location. (d) PFM phase images were read again after 24 hours. PFM-phase value of the blue lines is abstracted in panels (a–d) in the write-erase-rewrite mode (e), and at the storage time of 0 and 24 hours (f). (g) Phase value of a certain nanodot-time graph was determined in multiple write–erase modes. (h) The phase angle dependence on time under multiple write–erase cycles; the insets show the comparison of the original state with the final state of phase angle variation.

spectra in the range of 900–1900 $cm^{-1}$ were recorded by placing the AFM probe on top of the composites.[27]

As illustrated in Figure 9.10, The AFM-IR is based on the photothermal effect induced by a pulsed IR quantum cascade laser source with a wave length of 1,288 $cm^{-1}$ that corresponds to β phase) of PVDF, which is then detected by the AFM tip. The chemical map in Figure 9.10c shows a uniformly dispersed β phase (light region).[28] The sites 0 and 0′ were chosen to confirm the difference of β phase (light region) and conformational disorder (dark region) by the local IR spectra (Figure 9.10d). The map image after addition of ZnO NPs (1.5 wt% Ag-30 wt% ZnO/PVFE) clearly indicates that the ZnO-PVFE interface strongly effects the formation and stabilization of β phase.

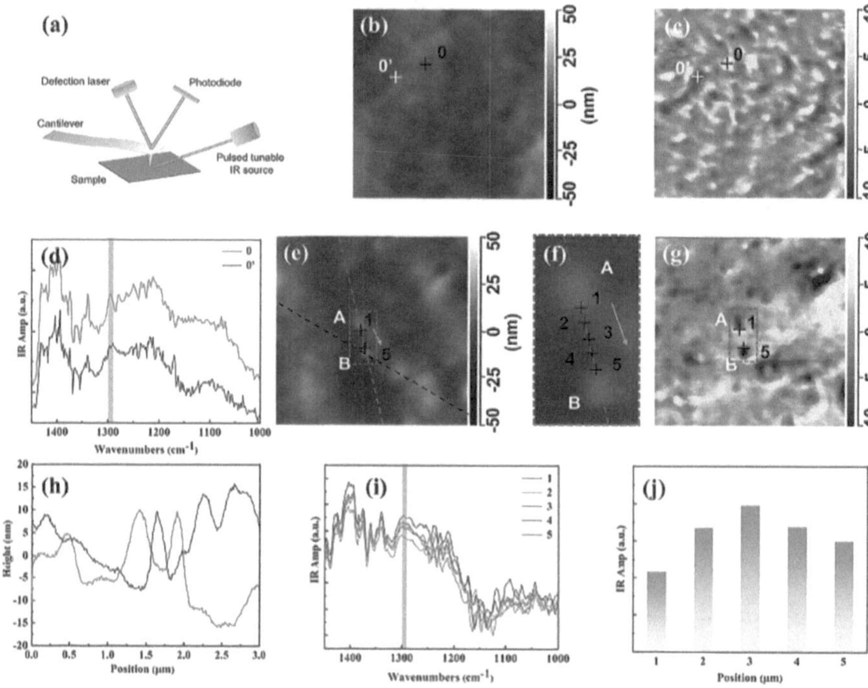


**FIGURE 9.10** AFM-IR characterization of conformational disorder induced by ZnO NPs. (a) Schematic of AFM-IR test. (b) Simultaneously measured topography ($3 \times 3\,\mu m^2$) and (c) AFM-IR chemical map irradiated by a $1{,}288\,cm^{-1}$ laser of pure P(VDF-TrFE). (d) Local IR spectra of the sites marked in (b). (e) Simultaneously measured topography ($3 \times 3\,\mu m^2$). (f) Zoomed-in view of the blue rectangle in (e) and (g) AFM-IR chemical map irradiated by a $1{,}288\,cm^{-1}$ laser of 1.5Ag-30ZnO/PVFE. (h) AFM height profiles of the dark line in (e). (i) Local IR spectra of the sites marked in (e) and (f). (j) Comparison of the local IR response in different sites at $1{,}288\,cm^{-1}$.

The different trends of local IR spectra at different sites also demonstrate the inhomogeneous nature of interface effect of the filler in inducing phase change. The direct structural analysis at the molecular level demonstrates the positive effects of ZnO-PVFE interfaces on β phase nucleation and stabilization.

The work also demonstrated the efficiency of the composite-based Nanogenerators in harvesting mechanical energy from human motions and fabricated a self-powered flexible coded lock with $3 \times 2$ pixel tactile sensor array (Figure 9.11).

## 9.5 OTHER NEW APPLICATIONS

Charge Gradient Microscopy (CGM) is a new application of conductive AFM (CAFM), which enables the imaging of ferroelectric domains at a much higher speed than PFM by scraping, collecting, and quantifying the screen charges on the surfaces. An advantage of high speed scanning is the ability to measure the dynamics of the screening process under non-equilibrium conditions.[29,30] The scraped charges

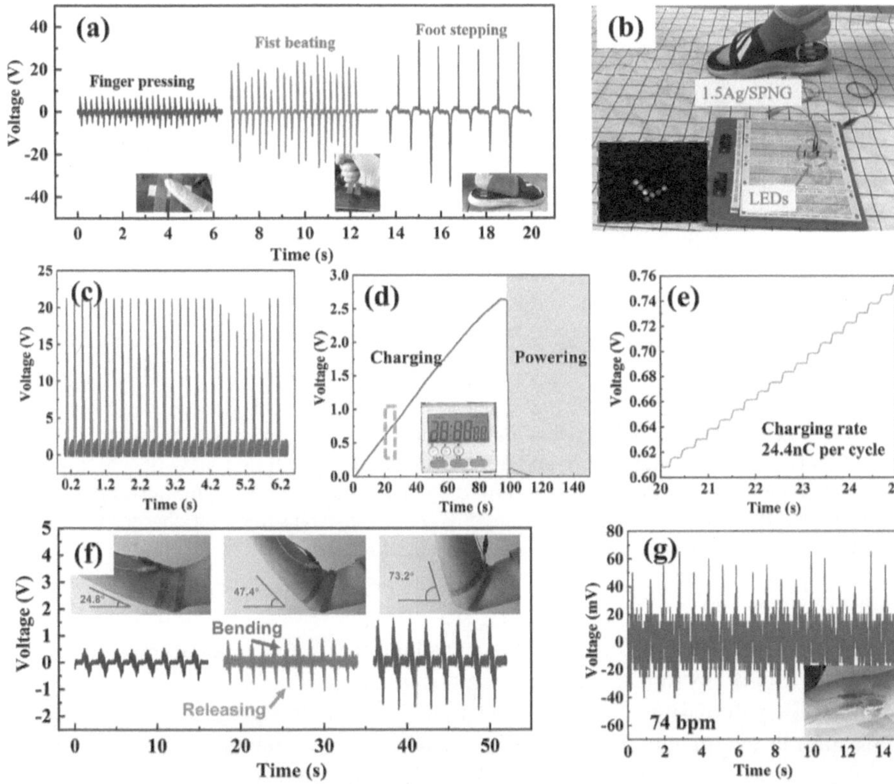


**FIGURE 9.11** Practical applications of the 1.5 Ag/SPNG for energy harvesting and sensing. (a) The output voltages of 1.5 Ag/SPEH under finger pressing, fist beating, and foot stepping. (b) The photograph of LED lighting test by foot stepping, inset: 7 LEDs lighting by 1.5 Ag/SPNG via foot stepping. (c) Rectified voltage signals generated by fist beating. (d) Charging curve of 3.3 μF capacitor, inset: photograph for powering a digital watch. (e) The amplifying view of charging curve from 20 to 25 seconds (f) Output voltages of the 1.5 Ag/SPNG attached to elbow joint to monitor different bending angles. (g) Pulse voltages obtained from 1.5 Ag/SPNG in response to the arterial pulsation of volunteer.

can be measured as current that scales with scraping rate. It induces a charge gradient that leads to the immediate relocation or refilling of the screen charges in the vicinity of the CGM probe. Tong et al. further investigated the kinetics behind the mechanical removal of externally bonded screening charges and the kinetics of rescreening in ambient conditions by using CGM and electrostatic force microscopy (EFM).[31] They found that a minimum pressure is required to mechanically remove the screening charges and increasing the pressure leads to further removal of charges until a critical pressure where all screening charges are removed. The authors also found an exponential recovery rescreening with a single time constant after charge scraping with different pressures. This implies that the screening charge degree on a ferroelectric surface can be mechanically controlled without affecting the polarization underneath.

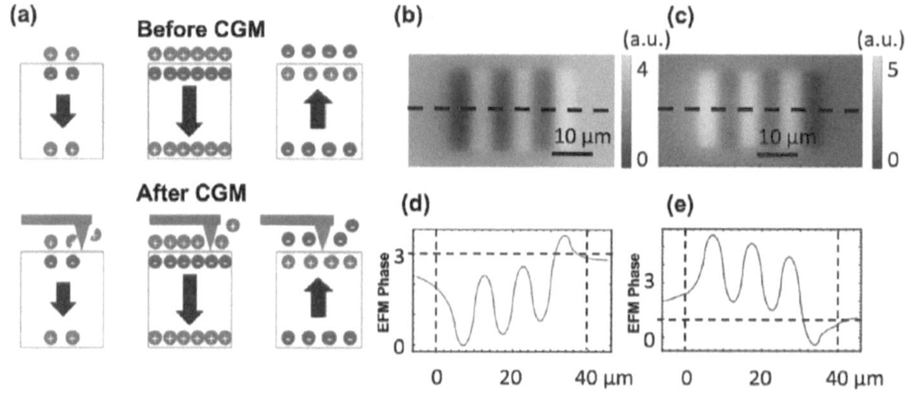


**FIGURE 9.12** Schematic of hypothetical mechanism for screening charge distribution on up and down poled domains before and after CGM: (a) Cross-section view of screening and polarization charges in pristine state, followed by down and up poled domains (upper) before and (lower) after CGM scan. (b) Variation of expected surface screening charges before and after CGM scans on periodically poled ferroelectric domains.

Hong et al. also investigated the transduction of mechanical motion of the nanoscale CGM tip into electric current from the poled ferroelectric domains in P(VDF-TrFE) thin films. It was found that the current signals of the copolymer films mainly originated from the ferroelectric domains when the surface screening charges were scraped with the CGM tip and the direction and area of injected charges modulated both the direction and the amount of the generated current. The researchers attributed the symmetric current generation in periodically poled regions to the different binding between external and polarization charges. The mechanical charge scraping on the polymer film almost totally disappeared after 20 scan cycles, and then such a degradation was attributed to the increase of the chemical bonding strength between the external screening charges and the polarization charges as a consequence of change in chemical properties of the film surface after the CGM scans. The findings present a new challenge for the reliability of mechanical charge scraping and which in turn may provide hints on energy harvesting by charge scraping from the ferroelectric polymer materials.

The authors also gave a hypothetical mechanism as schematically shown in Figure 9.12a, which can explain the nearly symmetric variation of the surface potential in periodically poled regions and the current generation over periodically poled domains. Figure 9.12b and c shows the expected EFM phase images, based on the hypothetical mechanism and electrostatic potential calculations, before and after CGM scans on the up and down domains. The expected EFM images and the line profiles shown in Figure 9.12d and e are in good qualitative agreement with EFM images before and after CGM scans and are consistent with the asymmetric collected charges from the up and down domains. These works demonstrate a new AFM variation similar but different to PFM, which can be used for high speed characterization of surface charging process.

## 9.6 FURTHER DEVELOPMENT AND OUTLOOK

The rapid development of AFM-based techniques has enabled the investigation of a wide range of samples by getting various types of images or data that are impossible by any other characterization techniques. In this chapter, we focused on those new AFM variants related to ferroelectric polymer as an energy material, including vector-PFM, hot stage PFM, and AFM-IR, and discussed how the techniques can give comprehensive information about molecular organization and surface polarization and charging properties, which are critical for improving material performance for devices like nanogenerator, energy harvester, or other energy conversion of storage devices. The technique has the advantages of easy sample preparation, nondestructive and high speed imaging, high resolution and the ability to determine a variety of properties at the nanoscale level. For a better tool for multi-functional material investigation, multimode or the ability to get different type of data other than topography is highly desired. Challenges may still exist concerning high speed imaging, high stability with less cross-talk in different type of signals. Nevertheless, the technique has been well recognized as an indispensable tool, and it will have a stronger impact for investigating new energy materials and devices.

## REFERENCES

1. Feng, T., Xie, D., Zang, Y.,... & Pan, W. (2013). Temperature control of P(VDF-TrFE) copolymer thin films. *Integrated Ferroelectrics*, 1. doi: 10.1080/10584587.2012.694748.
2. Sekine, T., Sugano, R., Tashiro, T., Fukuda, K., Kumaki, D., Dos Santos, F. D., ... & Tokito, S. (2016). Fully printed and flexible ferroelectric capacitors based on a ferroelectric polymer for pressure detection. *Japanese Journal of Applied Physics*, 10S. doi: 10.7567/JJAP.55.10TA18.
3. Chen, X., Yamada, H., Horiuchi, T., & Matsushige, K. (1998). Structures and local polarized domains of ferroelectric organic films studied by atomic force microscopy. *Japanese Journal of Applied Physics*, 37(Part 1, No. 6B), 3834.
4. Yi, Q. & McAlpine, M. C. (2010). Nanotechnology-enabled flexible and biocompatible energy harvesting. *Energy & Environmental Science*, 9, 1275–1285.
5. Soin, N., Shah, T. H., Anand, S. C., Geng, J., Pornwannachai, W., Mandat, P., et al. (2014). Novel "3-d spacer" all fibre piezoelectric textiles for energy harvesting applications. *Energy & Environmental Science*, 1670–1679.
6. Kim, D., Hong, S., Hong, J.,... & No, K. (2013). Fabrication of vertically aligned ferroelectric polyvinylidene fluoride mesoscale rod arrays. *Journal of Applied Polymer Science*, 6. doi: 10.1002/app.39415.
7. Kumar, C., Gaur, A., Rai, S. K., & Maiti, P. (2017). Piezo devices using poly(vinylidene fluoride)/reduced graphene oxide hybrid for energy harvesting. *Nano-Structures & Nano-Objects*. doi: 10.1016/j.nanoso.2017.10.006.
8. Lakbita, I. & El-Hami, K. (2014). Effects of annealing process on the crystalline morphologies of P(VDF/TrFE) studied by atomic force microscopy. *Chemical Science Review and Letters*, 3(12), 990–994.
9. Calahorra, Y., Whiter, R. A., Jing, Q., Narayan, V., & Kar-Narayan, S. (2016). Localized electromechanical interactions in ferroelectric P(VDF-TrFE) nanowires investigated by scanning probe microscopy. *APL Materials*, 4(11), 116106.
10. Wang, Z., Sun, B., Lu, X., Wang, C., & Su, Z. (2019). Molecular orientation in individual electrospun nanofibers studied by polarized AFM–IR. *Macromolecules*, 52(24) 9639–9645.

11. Jiyoung, C., Michael, D., et al. (2012). Piezoelectric nanofibers for energy scavenging applications. *Nano Energy*, 1(3), 356–371
12. Jung, H. B., Kim, J. W., Lim, J. H., Kwon, D. K., & Jeong, D. Y.. (2019). Energy storage properties of blended polymer films with normal ferroelectric P(VDF-HFP) and relaxor ferroelectric P(VDF-TrFE-CFE). *Electronic Materials Letters*, 16(4), 57–64.
13. Veeralingam, S., & Badhulika, S. (2020). $Bi_2S_3$/PVDF/PPY-based freestanding, wearable, transient nanomembrane for ultrasensitive pressure, strain, and temperature sensing. *ACS Applied Bio Materials*, 4(1), 14–23.
14. Gebrekrstos, A., Sharma, M., Madras, G., & Bose, S. (2017). Critical insights into the effect of shear, shear history, and the concentration of a diluent on the polymorphism in poly(vinylidene fluoride). *Crystal Growth & Design*, 17(4), 1957–1965.
15. Ting, Y., Suprapto, B. N., Sivasankar, K., & Aldori, Y. R.. (2020). Using annealing treatment on fabrication ionic liquid-based PVDF films. *Coatings,* 10(1), 44.
16. Sukumaran, S., Rouxel, D., Zineb, T. B., Chatbouri, S., & Tisserand, E. (2020). Recent advances in flexible PVDF based piezoelectric polymer devices for energy harvesting applications. *Journal of Intelligent Material Systems and Structures*.
17. Safarnia, M., Pakizeh, M., & Namvar-Mahboub, M. (2020). Assessment of structural and separation properties of a PVDF/PD composite membrane incorporated with $TiO_2$ nanotubes and $SiO_2$ particles. *Industrial & Engineering Chemistry Research*.
18. Prasad, P. D., & Hemalatha, J. (2021). Energy harvesting performance of magnetoelectric poly(vinylidene fluoride)/$NiFe_2O_4$ nanofiber films. *Journal of Magnetism and Magnetic Materials*. doi: 10.1016/j.jmmm.2021.167986.
19. Cui, Y., Feng, Y., Zhang, T., Zhang, C., & Lei, Q. (2020). Excellent energy storage performance of ferroconcrete-like all-organic linear/ferroelectric polymer films utilizing interface engineering. *ACS Applied Materials & Interfaces*, 12(50), 56424–56434.
20. Wen, Q., Sun, S., Song, J., Charles, N., Stephen, D., & Turner, J. A. (2018). Focused electron-beam-induced deposition for fabrication of highly durable and sensitive metallic afm-ir probes. *Nanotechnology*, 29(33), 335702.
21. Kim, J., Lee, J. H., Ryu, H., Lee, J. H., Khan, U., & Han, K., et al. (2017). High-performance piezoelectric, pyroelectric, and triboelectric nanogenerators based on P(VDF-TrFE) with controlled crystallinity and dipole alignment. *Advanced Functional Materials*.
22. Kalinin, S. V., Rodriguez, B. J., Jesse, S., Shin, J., Baddorf, A. P., & Gupta, P., et al. (2006). Vector piezoresponse force microscopy. *Microscopy and Microanalysis*, 12(3), 206–220.
23. Ong, W. L., Ke, C. M., Lim, P., Kumar, A., Zeng, K. Y. and Ho, G. W. (2013). Direct stamping and capillary flow patterning of solution processable piezoelectric polyvinylidene fluoride films. *Polymer*, 54, 5330–5337.
24. Tashiro, K., Takano, K., Kobayashi, M., Chatani, Y., & Tadokoro, H. (1981). Structural study on ferroelectric phase transition of vinylidene fluoride-trifluoroethylene random copolymers. *Polymer*, 22(10), 1312–1314.
25. Barique, M. A., & Ohigashi, H. (2001). Annealing effects on the curie transition temperature and melting temperature of poly(vinylidene fluoride/trifluoroethylene) single crystalline films. *Polymer*, 42(11), 4981–4987.
26. Jackson, C. L., & Mckenna, G. B. (1990). The melting behavior of organic materials confined in porous solids. *Journal of Chemical Physics*, 93(12), 9002–9011.
27. Karsten, H., & Shaykhutdinov, T. (2018). Polarization-dependent atomic force microscopy–infrared spectroscopy (AFM-IR): Infrared nanopolarimetric analysis of structure and anisotropy of thin films and surfaces. *Applied Spectroscopy*, 72(6), 817–832.
28. Liu, Y., Zhang, B., Xu, W., Haibibu, A., & Wang, Q. (2020). Chirality-induced relaxor properties in ferroelectric polymers. *Nature Materials*, 19(11), 1169–1174.

29. Qingfeng, Z., Ehsan, N., Shuhong, E., et al. (2020). Minimizing electrostatic interactions from piezoresponse force microscopy via capacitive excitation. *Theoretical & Applied Mechanics Letters*, 10(01), 30–33.
30. Gushchina, E., Zaitseva, N. V., Delimova, L. A., Orlov, G. A., & Vorotilov, K. A. (2020). Atomic force microscopy of porous ferroelectric PZT films. *Journal of Physics Conference Series*, 1697, 012090.
31. Tong, S., Park, W. I., Choi, Y. Y., Stan, L., Hong, S., & Roelofs, A. (2015). Mechanical removal and rescreening of local screening charges at ferroelectric surfaces. *Physical Review Applied*, 3(1), 014003.

# 10 Application of AFM in Microbial Energy Systems

*Xiaochun Tian*
Institute of Urban Environment,
Chinese Academy of Sciences

## CONTENTS



## 10.1 INTRODUCTION

Microbial energy systems, such as microbial electrochemical systems [1], biocapacitors [2], and microbial solar cells [3], have attracted much interest in the past two decades because they can achieve the conversion of chemicals or solar energy to electrical energy. More importantly, these systems are types of green electrochemical devices that can work under mild conditions with high activity and selectivity. Therefore, theories and technologies on bioenergy have become a common focus in the fields of energy sources, materials, biology, chemistry, and environmental science.

Compared with abiotic energy conversion systems (such as fuel cells and solar cells), the key differences of microbial energy systems are microbial activity and interface reactions, while the performance of these systems also depends on material properties, e.g., electrodes or catalysts. The schematic diagram in Figure 10.1a shows the structure and principle of microbial energy systems. Taking a microbial fuel cell as an example, functional microbes metabolize organic matter and produce

DOI: 10.1201/9781003174042-10

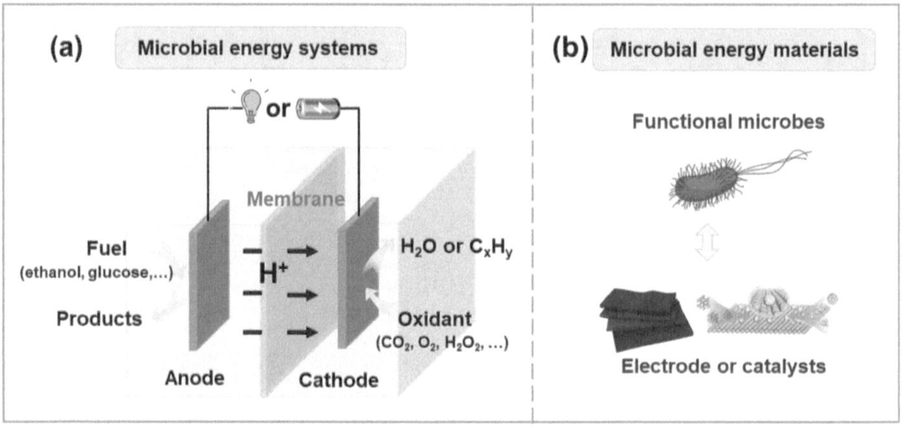


**FIGURE 10.1** Schematic diagram for (a) the structure and principle in microbial energy systems, and (b) biomaterials in microbial energy systems.

electrons at an anode, where the electrons transfer along an external circuit to perform work and then transfer to the cathode of the cells, where some chemical species are reduced. Protons flow from the bioanode to the cathodic chamber across a proton-exchange membrane to balance the charges in both chambers. As a result, microbial fuel cells can be used to generate power and harness energy from wastewater. Otherwise, microbial electrolysis cells enable the generation of different chemical products, e.g., hydrogen, methane, and acetate, at the biocathode either using an external power source or applying an electrode potential by a potentiostat. Therefore, anodic or cathodic interfaces are essential for improving the output and efficiency of microbial energy systems.

Microbial energy systems usually include functional microbes and electrodes or various materials on the interface, as shown in Figure 10.1b. For example, nonphotosynthetic bacteria in bioelectrosynthesis systems can convert $CO_2$ to value-added products; various materials, such as nanomaterials, conductive polymers, and functional molecules, are modified on electrode surfaces, or they are coupled with cells, both of which are available methods to accelerate electron transfer for electrode reactions. Otherwise, microbial synthetic materials, or microbial cells themselves, can be used as catalysts in bioenergy systems, and their good properties and performance are given serious attention.

With the development of microbial electrochemical technologies, studying the microcosmic interface has become increasingly important, especially for understanding the roles of microorganisms in electron transfer, which has been promoted by advanced electrochemical and nanoprobe technologies. Atomic force microscopy (AFM) is a scanning probe microscope that can be used to image interfacial structures in situ, visually and microcosmically. More importantly, the advantages of AFM can accelerate experimentation on live biological samples in solution.

This chapter will introduce the roles of AFM in microbial energy systems, including the characterization of the microbial morphology, conductivity, mechanical properties, and electrochemistry. This introduction will reveal the characterization of microbes and materials from a microscopic perspective and impel the application of AFM in the fields of bioenergy conversion, microbial synthesis, and environmental remediation.

## 10.2 MORPHOLOGY CHARACTERIZATION

Advanced nanoscience and nanotechnology are promoting AFM to become an imaging tool with multiple functionalities and high spatial resolution. Compared with transmission or scanning electron microscopy, AFM has an evident advantage for measuring biological samples on a solid-solution interface and maintaining an equivalent spatial resolution. More importantly, AFM enables imaging at three-dimensional scales based on the interaction force between the probe and the sample. Therefore, multiple modes of AFM (contact mode, dynamic mode, and PeakForce tapping) have been used to characterize the morphology of microbes and materials.

### 10.2.1 Microbial Morphologies

#### 10.2.1.1 Size and Morphology of Single Cell

In microbial energy systems, microorganisms usually take the form of biofilms to convert chemical energy from organic matter into electricity. Figure 10.2 shows a schematic of multistage biofilm formation [4]. Revealing microscopic bacterial cells is valuable for understanding the deep mechanisms and improving the performance of microbial energy systems.

The size of the AFM probe is nanoscale, and it is suitable for distinguishing individual cells. As a result, AFM is mostly used to image and measure the three-dimensional shape and length of a single cell. Many functional microorganisms have been reported to have the ability to transfer electrons, and many of them have lengths of several micrometers and diameters of hundreds of nanometers. Among them, *Shewanella oneidensis* MR-1 and *Geobacter sulfurreducens*, as model strains, have been well-studied, and they were both rod-like, as shown by AFM. In addition, the lengths of *S. oneidensis* MR-1 and *G. sulfurreducens* were both ~2 μm.

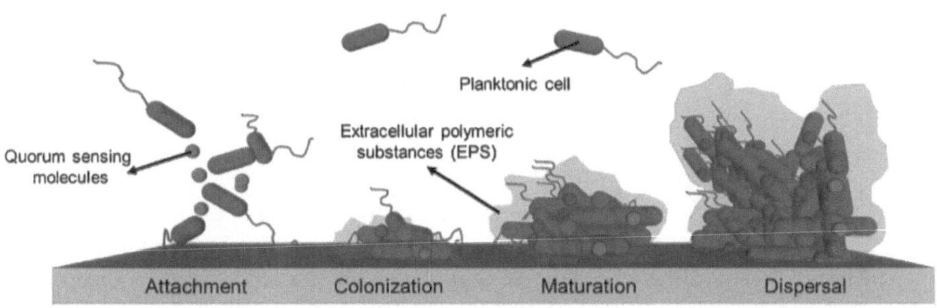


**FIGURE 10.2** Schematic of multistage biofilm formation. (The figure is reproduced from Ref. [4] with permission.)

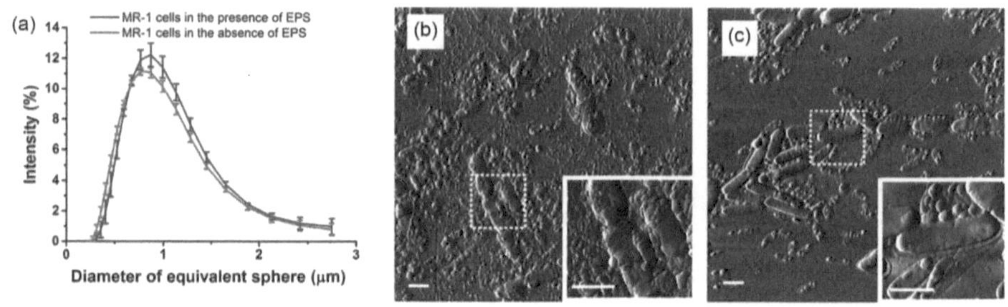


**FIGURE 10.3** EPS efficiently removed from *Shewanella oneidensis* MR-1 cells. (a) Sphere determination shows a size decrease in *S. oneidensis* MR-1 cells after EPS extraction. AFM images in contact mode (b) before and (c) after EPS extraction. (The figure is reproduced from Ref. [5] with permission.)

Biofilms and almost every microbial cell are enveloped by extracellular polymeric substances (EPS), which assist in biofilm formation and protection from unfavorable environments. When a microbial energy system is working, how electron transfer across the EPS occurs should not be ignored. A $35 \pm 15$ nm EPS layer was efficiently removed from *S. oneidensis* MR-1 by heating at 38°C for 30 minutes [5]. Before and after EPS extraction, the AFM images in the contact mode are shown in Figure 10.3b and c, respectively. After removing EPS, the results showed that the cell size of *S. oneidensis* MR-1 decreased (Figure 10.3a), and the cell surface became smooth. In addition, EPS can store electrochemically active substances (such as flavins and c-type cytochromes), which act as electron transit media. More importantly, the roles of EPS in microbial electron transfer are not only for gram-negative *S. oneidensis* MR-1 but also for gram-positive *Bacillus* sp. WS-XY1 and yeast *Pichia stipites.*

In recent years, gram-positive bacteria have also been reported to have the ability to transfer extracellular electrons in microbial energy systems. Due to the difference in the outer membrane structure and composition (shown in Figure 10.4), there were obviously different properties between gram-positive and gram-negative bacteria. For example, surface charge values of −80 to −140 mC/m$^2$ were found for gram-negative *Escherichia coli*, while gram-positive *B. subtilis* showed a much higher conductivity around the cell wall and surface charge values between −350 and −450 mC/m$^2$ [6]. Otherwise, the PeakForce tapping mode of AFM was used to characterize the mature surface of live cells (*Staphylococcus aureus* and *Bacillus subtilis* species), and the results showed a landscape of large (up to 60 nm in diameter) and deep (up to 23 nm) pores constituting a disordered gel of peptidoglycan; the same sacculus fragment in air and liquid by AFM showed that there is a higher spatial resolution in liquid [7].

### 10.2.1.2 Extracellular Appendages and Vesicles

Both *S. oneidensis* MR-1 and *G. sulfurreducens* have been reported to have wire-like extracellular appendages, and the morphologies of these appendages can be imaged by AFM. For the wirelike appendages, there were obvious differences between the two model species. The typical wire height of *S. oneidensis* MR-1 was measured to be ~10 nm in the contact mode [8]. However, the height profiles of AFM results for pili and flagella of *G. sulfurreducens* were ~3 and ~12 nm, respectively [9]. Although

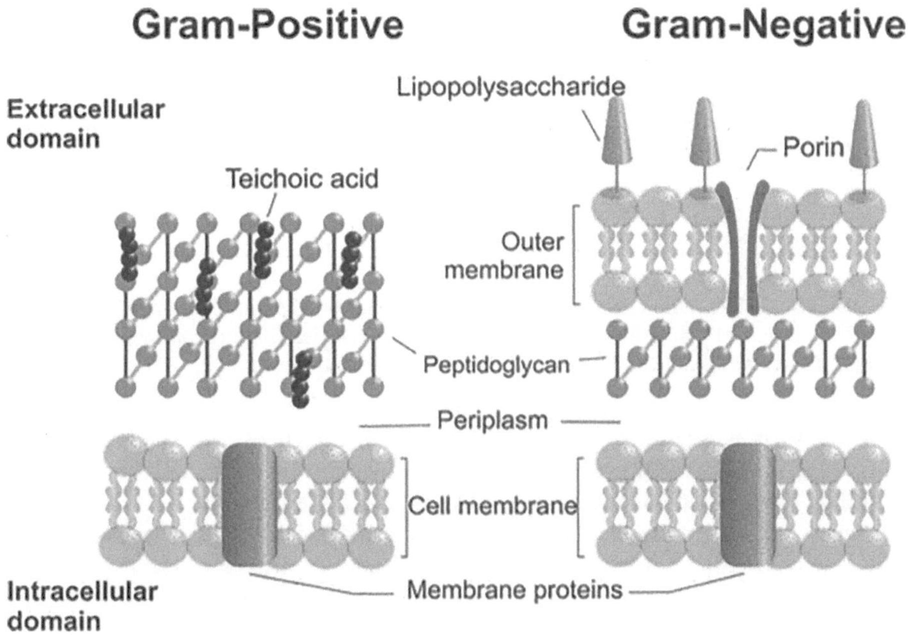


**FIGURE 10.4** Cartoon illustration for the difference in the outer membrane of gram-positive and gram-negative bacteria. (The figure is reproduced from Ref. [6] with a Creative Commons Attribution (CC-BY) License.)

the roles of these wirelike appendages in electron transfer mechanisms remain controversial, increasingly similar shapes of appendages have been identified in other microbial species. For example, a unicellular filamentous gram-positive bacterium, *Lysinibacillus varians* GY32, can use acetate or formate to produce electrons to an electrode in microbial fuel cells or sediment microbial fuel cells [10]. In addition, the diameter of individual appendages attached to the GY32 strain was observed to be 4–6 nm by the PeakForce QNM mode of AFM with ScanAsyst Air probes. According to the shape, size, and conductivity, these pili-like or nanowire-like appendages were considered to be similar to both type IV pilin and multiheme c-type cytochrome nanowires of *G. sulfurreducens*.

Extracellular vesicles are present in many types of biological systems, and their hydrated or desiccated topography can be visualized by AFM [11]. Bacterial vesicles are composed primarily of the outer membrane and periplasm. A visual result from correlated AFM and live-cell membrane fluorescence showed that *S. oneidensis* MR-1 nanowires formed from vesicles to continuous filaments, and the filaments were outer membrane and periplasmic extensions [12]. In addition, AFM images of nanowire formation are shown in Figure 10.5a–c.

### 10.2.2 Catalysts and Electrode Materials

Materials in microbial energy systems usually include all types of electrode and catalyst materials. When these materials are just inorganic nanomaterials, the roles

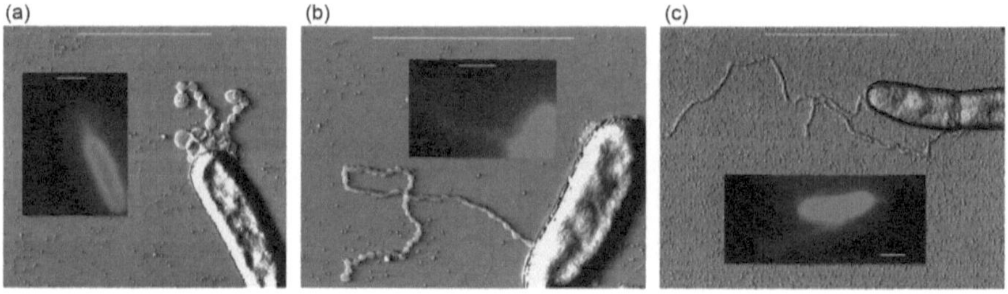

FIGURE 10.5 The nanowire morphologies of *S. oneidensis* MR-1 cells range from vesicle chains (a) to partially smooth filaments incorporating vesicles (b) and continuous filaments (c). (The figure is reproduced from Ref. [12] with permission.)

of AFM, such as measuring the shape, height, and roughness, can be brought into full play as AFM in other energy systems. The different situations of biomaterials in microbial energy systems are hybrid inorganic-microbial materials. To improve the efficiency and selectivity of microbial energy systems, inorganic materials coupled with microbial cells have achieved an improvement in microbial electrosynthesis, microbial fuel cells, and semiartificial photosynthetic systems. There are two common routes to hybridize microbial cells and inorganic materials: one is that microorganisms enable the synthesis of inorganic materials on the cell surface through the extracellular electron transfer of metabolic reduction; the other is that microbial cells take adsorption or endocytosis by mixing with synthesized inorganic materials.

AFM has been used to characterize the morphologies of hybrid inorganic-microbial materials, including nanoparticles, electrode materials, and various materials on microbial surfaces, to assist in revealing the mechanism of electron transfer in microbial energy systems. For example, carbon dot-fed *S. oneidensis* MR-1 exhibited accelerated extracellular electron transfer, and the maximum and power output of bioelectrical systems obviously increased [13]. In this work, AFM images of carbon dots showed an average height of ~2.2 nm, and the increase in surface roughness and excretion of these carbon dot-fed *S. oneidensis* MR-1 were also measured and shown by AFM. It can be concluded that the microscopic morphology of biological materials is beneficial for understanding the changes in reactions in microbial energy systems.

## 10.3 MECHANICAL PROPERTIES

AFM imaging is based on the force interaction between the probe and samples, allowing the determination of both the mechanical and adhesive properties of living cells [14]. The surface characteristics of a substrate, such as hardness, porosity, roughness, and hydrophobicity, play crucial roles in the formation of biofilms. The gram-positive bacterium *Lactococcus lactis* is a strain able to display pili, and AFM force spectroscopy was used to investigate the mechanical properties of biofilms. The results showed that strains devoid of pili displayed smoother and stiffer biofilms (Young's Modulus of 4–100 kPa) than piliated strains (Young's modulus of ~0.04–0.1 kPa) [15]. When inorganic materials couple with microorganisms, the mechanical

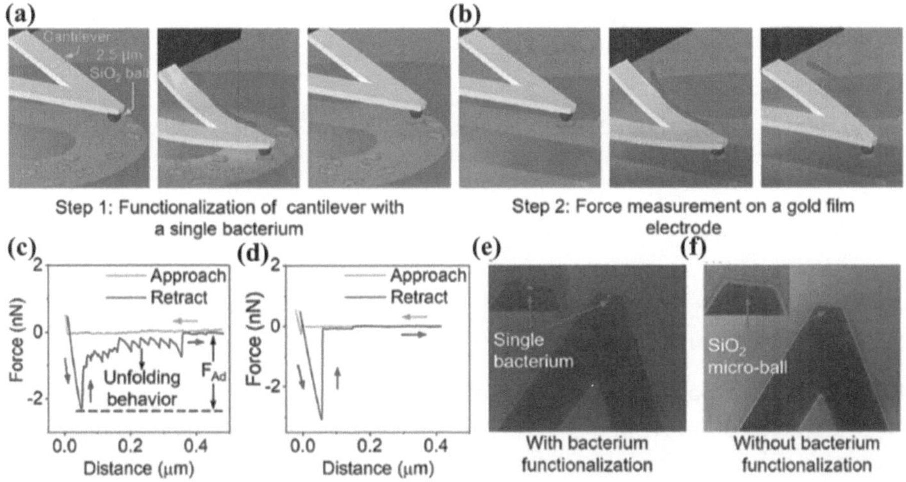


**FIGURE 10.6** (a) and (b) Two steps for single-cell force spectroscopy measurement. (c) and (d) Representative force–distance curves obtained on the working electrode using probes with and without single-bacterium functionalization. (e) and (f) Fluorescence microscopy images of the triangular cantilever with and without single-bacterium functionalization. (Figure reproduced from Ref. [17] with permission.)

properties may be an important factor. For example, when photocatalytic graphitic carbon nitride is mixed with *Staphylococcus epidermidis*, the stiffness of biofilms increases after photocatalysis [16].

To detect the adhesion force of a single-celled electricigen, Zhang et al. realized the immobilization of a single bacterium on a $SiO_2$ microball using polydopamine as an adhesive by contact mode AFM, and the process and results are shown in Figure 10.6. Single-cell force spectroscopy precisely detected the nanonewton scale adhesion force between the gold electrode surface and single-celled *S. oneidensis* MR-1 [17] and representative force-distance curves using probes with and without single-bacterium functionalization. As a result, the anodic potential, redox mediator, and $O_2$ are all factors that influence the process of extracellular electron transfer.

## 10.4 ELECTRON TRANSFER MECHANISMS

Electron transfer is essential for the reactions of energy conversion systems. The mechanisms of microorganisms in microbial fuel cells were specified as direct electron transfer and mediated electron transfer, as shown in Figure 10.7. Microbial direct electron transfer [18] refers to microbial cells or biofilms in close association with the electrode surface through either redox-active proteins and extracellular matrix or conductive "pili". Mediated electron transfer occurs via soluble electron mediators released by the cell, oxidized mediators are reduced at the outer cell surface, and then the reduced mediators donate electrons to the electrode. Therefore, the AFM methods to measure the electron transfer in microbial energy systems are mainly to measure electrical conductivity and electrochemical redox activity.

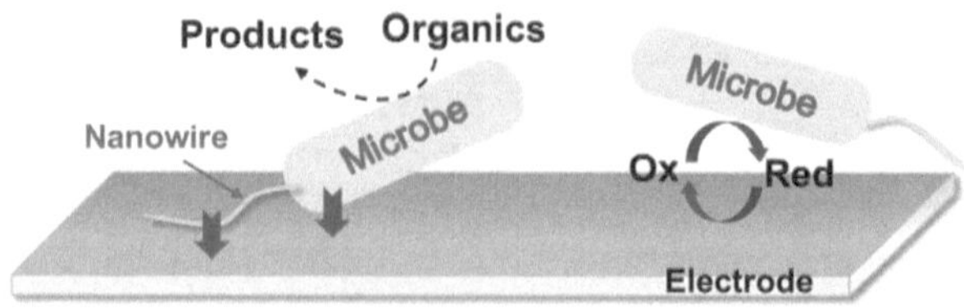


**FIGURE 10.7** The mechanism of direct electron transfer and mediated electron transfer in microbial fuel cells.

## 10.4.1 Electrical Conductivity

### 10.4.1.1 Microbial Cells and Nanowires

Electric techniques of AFM include conducting probe microscopy, electrostatic force microscopy, and Kelvin probe force microscopy, and all of them have had experience in being applied to the investigation of microbial cells or nanowires.

First, the conducting probe AFM played a critical role in the extracellular electron transfer mechanism of microorganisms. A breakthrough in the mechanism of microbial direct electron transfer was reported by Reguera et al. in 2005; the pili of *G. sulfurreducens* were imaged in the contact mode of AFM, and its correspondence between the current and applied voltage was measured by conducting-probe AFM, indicating that pili of *G. sulfurreducens* might serve as nanowires to transfer electrons extracellularly [19]. *Geobacter* nanowires are presumed to be conductive as a result of the amino acid sequence of the type IV pilin subunit PilA and, possibly, the tertiary structure of the assembled pilus. Since then, the conductive mechanisms of nanowires have attracted much interest and debate. However, electrical measurements of conducting probe AFM gave current-voltage (I-V) curves as important evidence. For example, Figure 10.8 shows the electrical transport along a bacterial nanowire of *S. oneidensis* MR-1, electron transport rates up to $10^9$/s at 100 mV of applied bias, and a measured resistivity on the order of 1 Ω·cm [20]. In addition, the ΔmtrC/omcA mutants produced appendages morphologically consistent with wild-type nanowires but that were found to be nonconductive. Otherwise, conducting probe AFM was also used to investigate the electronic transport characteristics of *S. oneidensis* MR-1 nanowires, and the nanowires exhibit p-type, tunable electronic behavior with a field-effect mobility on the order of $10^{-1}\,cm^2/(V\ s)$, comparable to devices based on synthetic organic semiconductors [21].

Electrostatic force microscopy has also been used to directly visualize charge propagation along pili of *G. sulfurreducens* by injecting charges at a single point and gently touching the pili [9]. It is worth noting that the imaging principle of electrostatic force is two-pass lift mode measurement, which is similar to Kelvin probe force microscopy but different from conductive probe AFM. Although evidence of microbial nanowires has been revealed by electric techniques of AFM, the mechanisms in molecular biology are still controversial. Otherwise, electron transfer on the surface of microbial cells is also an important and unclear research interest. Kelvin probe force microscopy represents a powerful tool that can be utilized to examine the changes in microbial membrane surface potential upon adhesion to various substrate

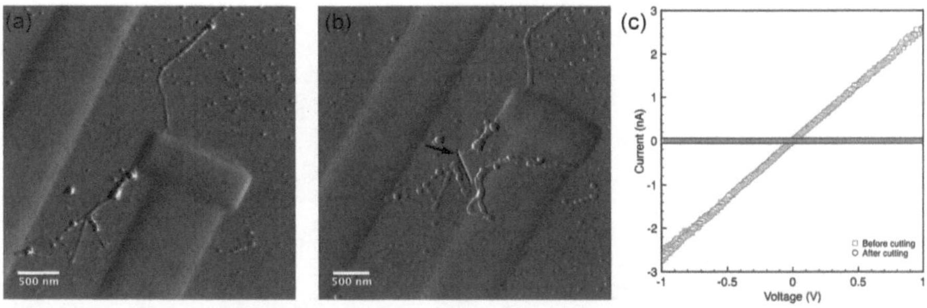


**FIGURE 10.8** Measuring electrical transport along a bacterial nanowire. (The figure is reproduced from Ref. [20] with permission.)

surfaces [17]; therefore, it will probably assist in measuring microbial cell conductivity in the future.

#### 10.4.1.2 Inorganic-Microbial Materials

Biosynthesis of metal or metallic compounds emerged with the advantage of mild conditions and exhibited superior activity for the electrochemical oxidation of fuel molecules. For example, the biogenic synthesis of Pd nanoparticles was possible using *S. oneidensis* MR-1 for the electrochemical oxidation of formate with high selectivity, but it is interesting that there is no catalytic activity for the other fuel molecules methanol, ethanol, and acetate [22]. The contact mode on Agilent AFM with a current-sensing module, known as current sensing AFM, was utilized to simultaneously obtain topography, deflection, and current images. Images of typical topography and current are simultaneously shown for natural (A) *S. oneidensis* MR-1 cells and (B) *S. oneidensis* MR-1 cells with Pd NPs in Figure 10.9. Pure bacterial cells appeared dark in the current image, indicating poor conductivity; in the presence of Pd NPs, the conductivity increased obviously. Thus, it can be concluded that inorganic microbials have the advantages of microbial selectivity and the catalytic performance of inorganic materials.

Another effective pathway to improve the output of microbial energy systems is to advance the surface properties of electrodes and to decrease electron transfer loss. Conductive AFM was used to provide a better understanding of electron transfer between bacteria and a modified electrode. For example, a self-doped conjugated polyelectrolyte was designed as an electrode surface, and it is helpful for the good distribution of bacterial cells. Then, conductive AFM was used to quantitatively study electron transfer from single cells of *G. sulfurreducens* to the polyelectrolyte electrode, and these results showed that dead cells, poor distribution, and performance of membrane cytochromes are associated with overpotential losses of a microbial fuel cell [23].

### 10.4.2 Electrochemical Redox Activity

Enzymes in microbial cells, mediators, mostly have redox properties. Thus, electrochemical technologies as direct and effective analytical methods enable us to reveal the process of matter and energy conversion from the perspective of electron transfer.

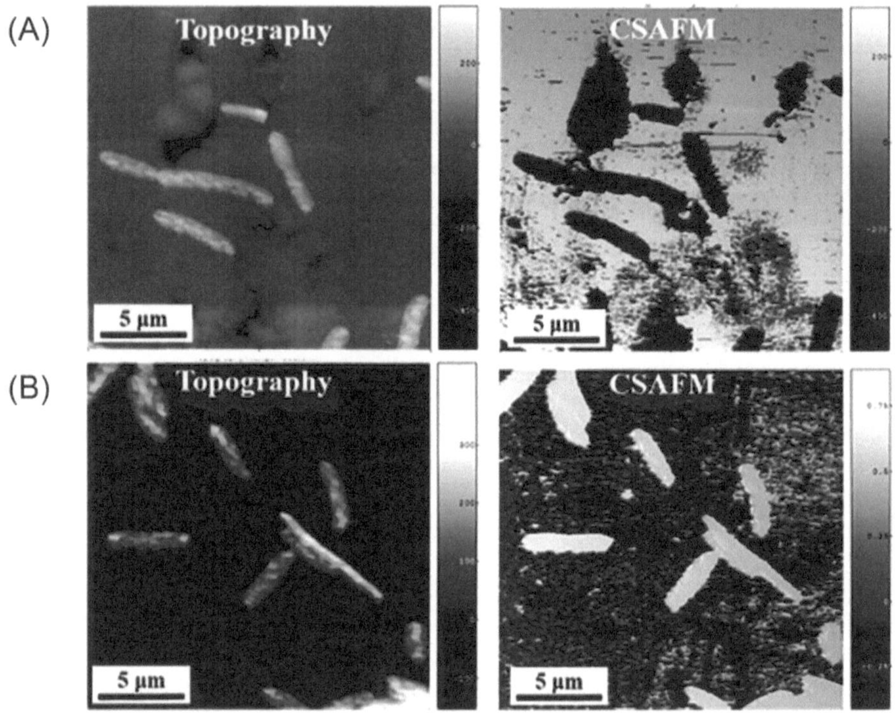


**FIGURE 10.9** Images of typical topography and conductivity current for (a) *S. oneidensis* MR-1 cells and (b) *S. oneidensis* MR-1 cells with Pd NPs. (Figure reproduced from Ref. [22] with permission.)

Moreover, electrochemical scanning probe technologies, such as Scanning Electrochemical Microscopy (SECM), electrochemical AFM, and SECM-AFM, are rapidly developing and preliminarily play important roles in imaging biological redox reactions at the microscale and even nanoscale.

#### 10.4.2.1 Microenvironment of the Biofilm

SECM has been applied to monitor microbial metabolic activity, biofilm formation, and the microenvironment of biofilms (pH, $O_2$, and $H_2O_2$), revealing the electron transfer pathways of microorganisms in bioenergy systems [24]. Studies of microbial electrochemistry using SECM are summarized in Table 10.1, containing the working principles, tips, microbial strains, and mediators.

At present, the commonly used microelectrode is a platinum disk with a diameter of 3~25 μm. Pt is stable and has electrocatalytic activity for oxygen reduction and $H_2O_2$ oxidation, and thus, a Pt microelectrode has been used to measure the concentrations of dissolved oxygen and reactive oxygen species. Au or carbon fiber microelectrodes can be used when the redox potentials of mediators are in accordance with the potential range of adsorption or desorption for H on Pt. In addition, three simple principles for the electron transfer of microbial energy systems are shown in Figure 10.10, and they are similar to the common principles of SECM. For the

**TABLE 10.1**
**Studies of Microbial Electrochemistry Based on SECM**

| No. | Principle | Microorganisms | Tip (Diameter) | Mediators | References |
|---|---|---|---|---|---|
| 1 | Positive feedback mode | *Escherichia coli* | 3 μm Pt | $K_3Fe\ (CN)_6$ | [25] |
| 2 | | *Shewanella oneidensis* MR-1 | 15 μm Pt | $Ru(NH_3)_6Cl_3$ | [26] |
| 3 | | *Shewanella oneidensis* MR-1 | 20 μm Pt | $H^+/H_2$ | [27] |
| 4 | | *Methanosarcina barkeri* – CdS | 25 μm Pt | FcMeOH | [28] |
| 5 | | *Escherichia coli* | Ion-selective microelectrode | $Ag^+$ | [29] |
| 6 | Negative feedback mode | *Sporosarcina pasteurii* | Ion-selective microelectrode | $H^+$ and $Ca^{2+}$ | [30] |
| 7 | | *Streptococcus gordonii* | Ion-selective microelectrode dual Pt | $H^+$ $H_2O_2$ | [31] |
| 8 | | *Rahnella aquatilis* RA1 | Ion-selective microelectrode | $H^+$ | [32] |
| 9 | Penetration mode | *Escherichia coli* | 25 μm Pt | FcMeOH | [33] |
| 10 | | *Shewanella oneidensis* MR-1 | 20 μm P | FcMeOH | [34] |
| 11 | Competition mode | *Escherichia coli* JM109 | 4 μm Pt | $O_2$ | [35] |
| 12 | | *Escherichia coli* | 25 μm Pt | $O_2$ | [36] |
| 13 | | *Vibrio fischeri* | 10 μm Pt | $H_2O_2$ | [37] |
| 14 | Generation-collection mode | *Pseudomonas aeruginosa* PAO1 | 10 μm IrO2/Pt disk | pH | [38] |
| 15 | | | 10 μm Pt wire | $O_2$ ROS[a] | |
| | | *Pseudomonas aeruginosa* | 5 μm Pt | Pyocyanin | [39] |
| 16 | Feedback cascade mode | *Saccharomyces cerevisiae* | 12.5 μm Pt | pBQ, PD, DCPIP or PQ[b] $K_3[Fe(CN)_6]$ | [40] |
| 17 | | *Aspergillus niger* *Rhizoctania* sp. | 10 μm Pt | PQ or PD $K_3[Fe(CN)_6]$ | [41] |
| 18 | | *Pseudomonas aeruginosa* | 10 μm Pt | Pyocyanin $K_3[Fe(CN)_6]$ | [42] |
| 19 | Generation-collection and penetration mode | *Streptococcus gordonii* *Aggregatibacter actinomycetemcomitans* | 25 μm Au | $H_2O_2$ | [43] |
| 20 | FFT-SEIM[c] | *Saccharomyces cerevisiae* | 23 μm Pt | $O_2$ Vit-$K_1$ or Vit-$K_3$[d] | [44] |

[a] ROS: Reactive oxygen species.

[b] pBQ: p-benzoquinone; PQ: 9,10-phenanthrenequinone; PD: 1,10-phenanthroline-5,6-dione; DCPIP: dichlorophenolindophenolsodium salt hydrate.

[c] FFT-SEIM: Fourier transform scanning electrochemical impedance microscopy.

[d] Vit-$K_1$: Sodium 1,4-Naphthoquinone-2-Sulfonate; Vit-$K_3$: 2-Methyl-1,4-Naphthoquinone.

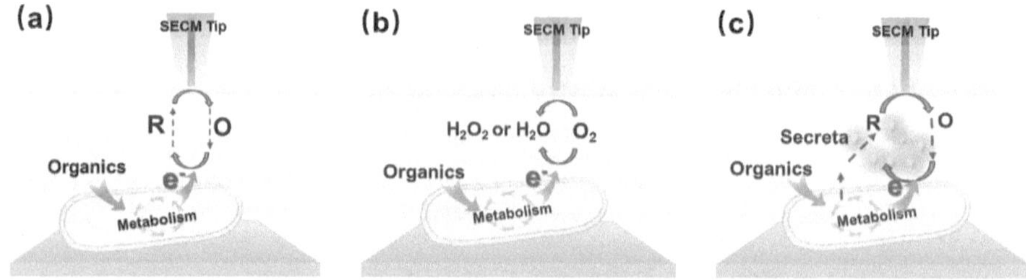


**FIGURE 10.10** Three simple principles of Scanning Electrochemical Microscopy for microbial electrochemistry. (a) Feedback mode, (b) competitive mode, (c) generation-collection mode. (The figure is reproduced from Ref. [24] with permission.)

feedback mode in Figure 10.10a, redox mediators include $K_3Fe(CN)_6$, FcMeOH, and $Ru(NH_3)_6Cl_3$, and they were mostly used to measure the reductive ability and electricity production of the species *E. coli* and *S. oneidensis*. The competitive mode in Figure 10.10b is often used to study substances that inhibit metabolic activity, for example, the toxicity of low concentrations of $Ag^+$ on *E. coli* and the dependence of the catalase activity of *Vibrio fischeri* on the $H_2O_2$ concentration. The generation-collection mode has an advantage in detecting the redox active electron mediators secreted by microorganisms. Taking Figure 10.10c as an example, pyocyanin is secreted by *Pseudomonas aeruginosa*, diffuses to the tip surface, and is then oxidized. This process is substrate generation-probe collection. In addition, oxidized pyocyanin can diffuse to the substrate and be reduced again by microorganisms. With the accumulation of pyocyanin in the bulk solution, a positive feedback process occurs between the SECM tip and microbes. In practical applications, it is usually difficult to reveal the reaction process of electron transfer by a certain mode, but two steps or types of principles may be involved. Therefore, more complex electron transfer pathways are also summarized in Table 10.1, which will guide SECM to analyze the electrode interface reaction and kinetic process in bioenergy systems.

#### 10.4.2.2 Bioelectrochemistry at the Nanoscale

Electrochemical AFM refers to incorporating an electrochemical cell into AFM. Since electrochemical reactions occur on an electrode interface in liquid, various functions of AFM are conducted in liquid, and extra potential or current can be applied or recorded by an electrochemical working station. When the electrochemical cell is a three-electrode system, the working electrode is the usual substrate with biological samples, and the roles of the AFM probe are the same as those of imaging samples in solution. When the AFM probe is used to record electrical signals, the probe is a nanoelectrode of SECM-AFM.

Although electrochemical AFM has been used to image some nanomaterials at different potentials, there are limited studies characterizing the redox properties of microbial cells in bioenergy systems. Although imaging living cells in solution is tough, single-cell force microscopy of *S. oneidensis* MR-1 at different potentials has been studied by an integrated electrochemical single-cell force microscopy system. In addition, Figure 10.11a shows the in situ electrochemical single-cell force

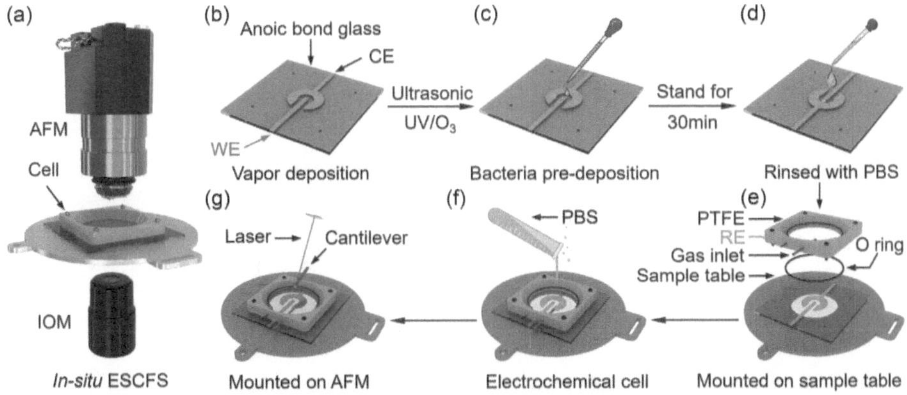


**FIGURE 10.11** (a) In situ electrochemical single-cell force spectroscopy system with an inverted optical microscope, and (b)–(g) the fabrication process of the in situ electrochemical cell. (The figure is reproduced from Ref. [17] with permission.)

spectroscopy system with an inverted optical microscope, and Figure 10.11b–g shows the fabrication process [17]. When the applied potentials were −0.1, 0, 0.1, and 0.2 V at the working electrode, the maximum adhesion forces between a single cell of *S. oneidensis* MR-1 and the electrode were 1.227, 2.242, 2.646, and 3.472 nN, respectively, with sodium lactate under anaerobic conditions. However, the electricity generation ability and conduction of single cells, whether associated with adhesion, can still be detected by electrochemical AFM or SECM-AFM.

In regard to SECM-AFM, the fabrication of conductive nanoprobes early in development is difficult, and thus, having a homemade probe is the largest advantage to carry out studies on redox reactions at the nanoscale using electrochemical AFM. An exciting situation of SECM-AFM in recent years is that commercial conductive nanoprobes are produced and reliable in imaging current electrochemical redox reactions. More importantly, measuring the electrical properties of materials in liquid is also feasible using this probe for the commercialization of SECM modules for AFM instruments. Although SECM plays a role in developing a mediated electron transfer mechanism of biofilms, it is still unreported at a single-cell scale using SECM-AFM. Nevertheless, early applications of SECM-AFM in biological fields were carried out by mapping the enzyme (oxidase activity) glucose while simultaneously recording the interfacial topography of the enzyme on the substrate electrode [45]. Therefore, it is an effective tool to research the electron transfer mechanism of enzyme fuel cells.

Based on SECM-AFM, Demaille and coworkers developed an approach of molecule touching (Mt)/AFM-SECM or mediator tethering (Mt)/AFM-SECM by the modification of redox-active molecules on either the sample surface [46] or the AFM-SECM tip [47], respectively. These related methods even allowed the detection of single-protein molecules and viruses. Figure 10.12 shows the enzymatic activity of individual biocatalytic *fd*-viral particles by (Mt)/AFM-SECM [48]. Moreover, a current detection sensitivity of ~10 fA and a spatial resolution of ~10 nm could be achieved, which would be enough in theory for mapping the redox activity of single-celled *S. oneidensis* and *G. sulfurreducens*.

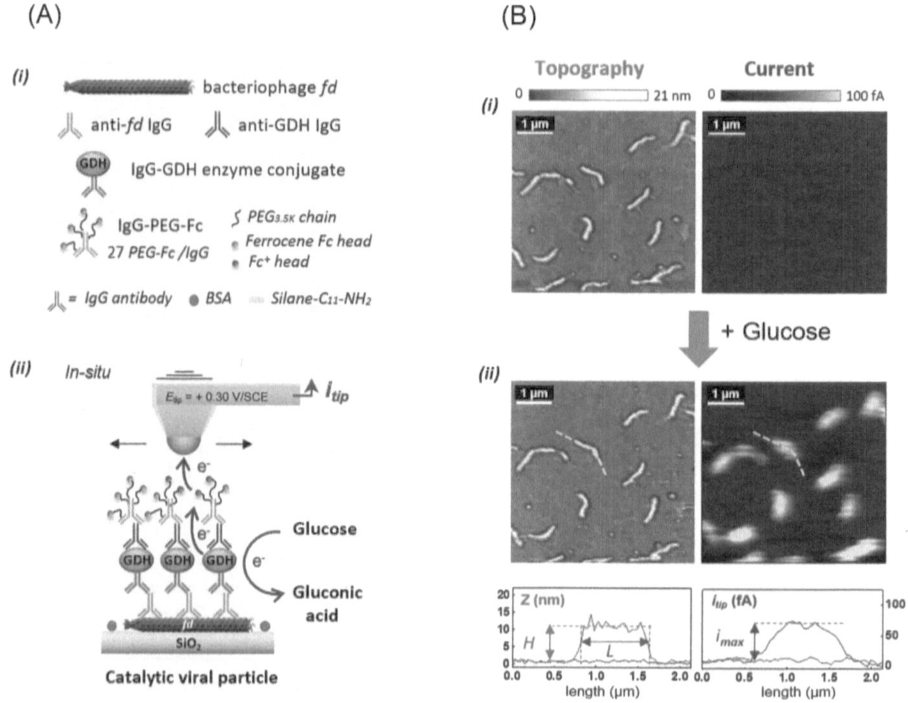


**FIGURE 10.12** The enzymatic activity of individual biocatalytic *fd*-Viral particles by (Mt)/AFM-SECM. (The figure is reproduced from Ref. [48] with permission.)

## 10.5 SUMMARY AND FUTURE PROSPECTS

AFM is developing as a comprehensive imaging tool with a high spatial revolution, and its multifunctional modes, such as force curve, conductivity, magnetic force, electrochemistry, etc., have effectively promoted AFM application in microbial energy systems. At present, there are still many key scientific and technical problems in research on electron transfer mechanisms and biological sample imaging requiring extensive cooperation among scientific researchers around the world. Moreover, the study of microbial energy systems using AFM is facing great chances and challenges in multidisciplinary fields.

1. AFM has been widely used to obtain information of materials and microbial cells with regard to morphology, mechanical properties, and electrical conductivity at the single-cell scale in atmospheric environments. However, it is much more important for AFM to measure the microenvironment of bioenergy materials in situ because biological cells are usually alive in solution and have an active change with the time scale. In particular, it is difficult for moving microbes to stably adhere to a solid surface in solution. Therefore, monitoring the properties of living biological samples in solution in real time becomes a challenge for AFM

in biological applications. Although the modification of the solid-liquid interface has been used to capture a cell or enzyme, inorganic-microbial hybrid systems are still very limited. Otherwise, AFM tips have achieved great improvement in resolution and function, and many types of functional tips have been achieved by nanofabrication, such as heat, light, and liquid flow, which make the characterization of biomaterials much more feasible and easier in solution.

2. Providing both image and property data in a single platform is a development tendency for studying living biological samples in a concise, visual, effective manner. AFM integrated techniques have advantages in studying electron transfer mechanisms at the single-cell or molecular structural level. For example, the coupling of electrochemical AFM and fluorescence microscopy can quantitate the electron transfer reactions and electricity production of biomaterials and simultaneously monitor cell activity in situ at the electrode/solution interface. Meanwhile, by designing a tip, AFM-SECM is a powerful electrochemical technique to monitor the microenvironmental pH and reactive oxygen species of a single cell. In addition, it is beneficial for harvesting the kinetic processes of electron transfer reactions and microbial metabolism. Due to the limitations of AFM at the molecular structure level, the combination of spectroscopy and AFM, AFM-IR or AFM-Raman, has emerged and been applied in accurately obtaining the morphology and bonding changes of a specific protein or enzyme. Furthermore, to combine spectroscopy with electrochemical AFM and even SECM-AFM, in situ real-time monitoring of bioenergy conversion can be improved with regards to the resolution of time, space and energy simultaneously. Otherwise, with the development of technology for single-cell separation and sequencing, combination platforms of electrochemical AFM and single-cell sequencing technology will be constructed and anticipated in the future.

AFM has significantly promoted the development of bioenergy systems and environmental microbiology. It is also helpful to reveal the roles of biomaterials from the physical and chemical points of view to promote the application of biological electron transfer in the fields of bioenergy utilization, nanomaterial synthesis, and environmental remediation.

## REFERENCES

1. B.E. Logan, K. Rabaey, Conversion of Wastes into bioelectricity and chemicals by using microbial electrochemical technologies, *Science*, 337 (2012) 686–690.
2. K. Sode, T. Yamazaki, I. Lee, T. Hanashi, W. Tsugawa, BioCapacitor: A novel principle for biosensors, *Biosensors and Bioelectronics*, 76 (2016) 20–28.
3. L. Liu, S. Choi, Miniature microbial solar cells to power wireless sensor networks, *Biosensors and Bioelectronics*, 177 (2021) 112970.
4. G. Caniglia, C. Kranz, Scanning electrochemical microscopy and its potential for studying biofilms and antimicrobial coatings, *Analytical and Bioanalytical Chemistry*, 412 (2020) 6133–6148.

5. Y. Xiao, E. Zhang, J. Zhang, Y. Dai, Z. Yang, H.E.M. Christensen, J. Ulstrup, F. Zhao, Extracellular polymeric substances are transient media for microbial extracellular electron transfer, *Science Advances*, 3 (2017) e1700623.
6. K. Cremin, B.A. Jones, J. Teahan, G.N. Meloni, D. Perry, C. Zerfass, M. Asally, O.S. Soyer, P.R. Unwin, Scanning ion conductance microscopy reveals differences in the ionic environments of gram-positive and negative bacteria, *Analytical Chemistry*, 92 (2020) 16024–16032.
7. L. Pasquina-Lemonche, J. Burns, R.D. Turner, S. Kumar, R. Tank, N. Mullin, J.S. Wilson, B. Chakrabarti, P.A. Bullough, S.J. Foster, J.K. Hobbs, The architecture of the gram-positive bacterial cell wall, *Nature*, 582 (2020) 294–297.
8. M.Y. El-Naggar, Y.A. Gorby, W. Xia, K.H. Nealson, The molecular density of states in bacterial nanowires, *Biophysical Journal*, 95 (2008) L10–L12.
9. N.S. Malvankar, S.E. Yalcin, M.T. Tuominen, D.R. Lovley, Visualization of charge propagation along individual pili proteins using ambient electrostatic force microscopy, *Nature Nanotechnology* 9 (2014) 1012–1017.
10. Y. Yang, Z. Wang, C. Gan, L.H. Klausen, R. Bonné, G. Kong, D. Luo, M. Meert, C. Zhu, G. Sun, J. Guo, Y. Ma, J.T. Bjerg, J. Manca, M. Xu, L.P. Nielsen, M. Dong, Long-distance electron transfer in a filamentous gram-positive bacterium, *Nature Communications*, 12 (2021) 1709.
11. M. Skliar, V.S. Chernyshev, Imaging of extracellular vesicles by atomic force microscopy, *JoVE*, 151 (2019) e59254.
12. S. Pirbadian, S.E. Barchinger, K.M. Leung, H.S. Byun, Y. Jangir, R.A. Bouhenni, S.B. Reed, M.F. Romine, D.A. Saffarini, L. Shi, Y.A. Gorby, J.H. Golbeck, M.Y. El-Naggar, *Shewanella oneidensis* MR-1 nanowires are outer membrane and periplasmic extensions of the extracellular electron transport components, *Proceedings of the National Academy of Sciences*, 111 (2014) 12883–12888.
13. C. Yang, H. Aslan, P. Zhang, S. Zhu, Y. Xiao, L. Chen, N. Khan, T. Boesen, Y. Wang, Y. Liu, L. Wang, Y. Sun, Y. Feng, F. Besenbacher, F. Zhao, M. Yu, Carbon dots-fed *Shewanella oneidensis* MR-1 for bioelectricity enhancement, *Nature Communications*, 11 (2020) 1379.
14. T. Hohmann, F. Dehghani, Measuring mechanical and adhesive properties of single cells using an atomic force microscope, *Methods in Molecular Biology (Clifton, N.J.)*, 2294 (2021) 81–92.
15. I. Drame, C. Lafforgue, C. Formosa-Dague, M.-P. Chapot-Chartier, J.-C. Piard, M. Castelain, E. Dague, Pili and other surface proteins influence the structure and the nanomechanical properties of *Lactococcus lactis* biofilms, *Scientific Reports*, 11 (2021) 4846.
16. H. Shen, E.A. Lopez-Guerra, R. Zhu, T. Diba, Q. Zheng, S.D. Solares, J.M. Zara, D. Shuai, Y. Shen, Visible-light-responsive photocatalyst of graphitic carbon nitride for pathogenic biofilm control, *ACS Applied Materials and Interfaces*, 11 (2019) 373–384.
17. S. Zhang, L. Wang, L. Wu, Z. Li, B. Yang, Y. Hou, L. Lei, S. Cheng, Q. He, Deciphering single-bacterium adhesion behavior modulated by extracellular electron transfer, *Nano Letters*, 21 (2021) 5105–5115.
18. D.R. Lovley, Electromicrobiology, *Annual Review of Microbiology*, 66 (2012) 391–409.
19. G. Reguera, K.D. McCarthy, T. Mehta, J.S. Nicoll, M.T. Tuominen, D.R. Lovley, Extracellular electron transfer via microbial nanowires, *Nature*, 435 (2005) 1098–1101.
20. M.Y. El-Naggar, G. Wanger, K.M. Leung, T.D. Yuzvinsky, G. Southam, J. Yang, W.M. Lau, K.H. Nealson, Y.A. Gorby, Electrical transport along bacterial nanowires from *Shewanella oneidensis* MR-1, *Proceedings of the National Academy of Sciences of the United States of America*, 107 (2010) 18127–18131.
21. K.M. Leung, G. Wanger, M.Y. El-Naggar, Y. Gorby, G. Southam, W.M. Lau, J. Yang, *Shewanella oneidensis* MR-1 bacterial nanowires exhibit p-type, tunable electronic behavior, *Nano Letters*, 13 (2013) 2407–2411.

22. R. Wu, X. Tian, Y. Xiao, J. Ulstrup, H.E. Mølager Christensen, F. Zhao, J. Zhang, Selective electrocatalysis of biofuel molecular oxidation using palladium nanoparticles generated on *Shewanella oneidensis* MR-1, *Journal of Materials Chemistry A*, 6 (2018) 10655–10662.
23. D.X. Cao, H. Yan, V.V. Brus, M.S. Wong, G.C. Bazan, T.-Q. Nguyen, Visualization of charge transfer from bacteria to a self-doped conjugated polymer electrode surface using conductive atomic force microscopy, *ACS Applied Materials & Interfaces*, 12 (2020) 40778–40785.
24. X. Tian, Y. Li, Q. Pan, F. Zhao, Research progress and prospects of microbial electrochemistry based on scanning electrochemical microscopy, *Chinese Journal of Analytical Chemistry,* 49 (2021) 858–866.
25. T. Kaya, K. Nagamine, D. Oyamatsu, H. Shiku, M. Nishizawa, T. Matsue, Fabrication of microbial chip using collagen gel microstructure, *Lab on a Chip*, 3 (2003) 313–317.
26. W.J. Zhang, H.K. Wu, I.M. Hsing, Real-time label-free monitoring of *Shewanella oneidensis* MR-1 biofilm formation on electrode during bacterial electrogenesis using scanning electrochemical microscopy, *Electroanalysis*, 27 (2015) 648–655.
27. R. Moreira, M.K. Schütz, M. Libert, B. Tribollet, V. Vivier, Influence of hydrogen-oxidizing bacteria on the corrosion of low carbon steel: Local electrochemical investigations, *Bioelectrochemistry*, 97 (2014) 69–75.
28. J. Ye, J. Yu, Y. Zhang, M. Chen, X. Liu, S. Zhou, Z. He, Light-driven carbon dioxide reduction to methane by *Methanosarcina barkeri*-CdS biohybrid, *Applied Catalysis B: Environmental*, 257 (2019) 117916.
29. D. Zhan, F.-R.F. Fan, A.J. Bard, The $K_v$ channel blocker 4-aminopyridine enhances $Ag^+$ uptake: A scanning electrochemical microscopy study of single living cells, *Proceedings of the National Academy of Sciences of the United States of America*, 105 (2008) 12118–12122.
30. D. Harris, J.G. Ummadi, A.R. Thurber, Y. Allau, C. Verba, F. Colwell, M.E. Torres, D. Koley, Real-time monitoring of calcification process by *Sporosarcina pasteurii* biofilm, *Analyst,* 141 (2016) 2887–2895.
31. V.S. Joshi, P.S. Sheet, N. Cullin, J. Kreth, D. Koley, Real-time metabolic interactions between two bacterial species using a carbon-based pH microsensor as a scanning electrochemical microscopy probe, *Analytical Chemistry*, 89 (2017) 11044–11052.
32. D. Oulkadi, S. Banon, C. Mustin, M. Etienne, Local pH measurement at wet mineral-bacteria/air interface, *Electrochemistry Communications*, 44 (2014) 1–3.
33. Z. Hu, J. Jin, H.D. Abruña, P.L. Houston, A.G. Hay, W.C. Ghiorse, M.L. Shuler, G. Hidalgo, L.W. Lion, Spatial distributions of copper in microbial biofilms by scanning electrochemical microscopy, *Environmental Science & Technology*, 41 (2006) 936–941.
34. X. Tian, X. Wu, D. Zhan, F. Zhao, Y. Jiang, S. Sun, Research on electron transfer in the microenvironment of the biofilm by scanning electrochemical microscopy, *Acta Physico-Chimica Sinica at Science*, 35 (2019) 22–27.
35. T. Kaya, M. Nishizawa, T. Yasukawa, M. Nishiguchi, T. Onouchi, T. Matsue, A microbial chip combined with scanning electrochemical microscopy, *Biotechnology and Bioengineering*, 76 (2001) 391–394.
36. K.B. Holt, A.J. Bard, Interaction of silver(I) ions with the respiratory chain of Escherichia coli: An electrochemical and scanning electrochemical microscopy study of the antimicrobial mechanism of micromolar Ag+, *Biochemistry*, 44 (2005) 13214–13223.
37. E. Abucayon, N. Ke, R. Cornut, A. Patelunas, D. Miller, M.K. Nishiguchi, C.G. Zoski, Investigating catalase activity through hydrogen peroxide decomposition by bacteria biofilms in real time using scanning electrochemical microscopy, *Analytical Chemistry*, 86 (2014) 498–505.
38. M.H. Lin, S. Mehraeen, G. Cheng, C. Rusinek, B.P. Chaplin, Role of near-electrode solution chemistry on bacteria attachment and poration at low applied potentials, *Environmental Science & Technology*, 54 (2020) 446–455.

39. J.L. Connell, J. Kim, J.B. Shear, A.J. Bard, M. Whiteley, Real-time monitoring of quorum sensing in 3D-printed bacterial aggregates using scanning electrochemical microscopy, *Proceedings of the National Academy of Sciences of the United States of America*, 111 (2014) 18255–18260.
40. A. Ramanavicius, I. Morkvenaite-Vilkonciene, A. Kisieliute, J. Petroniene, A. Ramanaviciene, Scanning electrochemical microscopy based evaluation of influence of pH on bioelectrochemical activity of yeast cells: *Saccharomyces cerevisiae*, *Colloids and Surfaces B: Biointerfaces*, 149 (2017) 1–6.
41. A. Kisieliute, A. Popov, R.M. Apetrei, G. Carac, I. Morkvenaite-Vilkonciene, A. Ramanaviciene, A. Ramanavicius, Towards microbial biofuel cells: Improvement of charge transfer by self-modification of microorganisms with conducting polymer - polypyrrole, *Chemical Engineering Journal*, 356 (2019) 1014–1021.
42. D. Koley, M.M. Ramsey, A.J. Bard, M. Whiteley, Discovery of a biofilm electrocline using real-time 3D metabolite analysis, *Proceedings of the National Academy of Sciences of the United States of America*, 108 (2011) 19996–20001.
43. X. Liu, M.M. Ramsey, X. Chen, D. Koley, M. Whiteley, A.J. Bard, Real-time mapping of a hydrogen peroxide concentration profile across a polymicrobial bacterial biofilm using scanning electrochemical microscopy, *Proceedings of the National Academy of Sciences of the United States of America*, 108 (2011) 2668–2673.
44. A. Valiūnienė, J. Petronienė, M. Dulkys, A. Ramanavičius, Investigation of active and inactivated yeast cells by scanning electrochemical impedance microscopy, *Electroanalysis*, 32 (2020) 367–374.
45. A. Kueng, C. Kranz, A. Lugstein, E. Bertagnolli, B. Mizaikoff, Integrated AFM–SECM in tapping mode: Simultaneous topographical and electrochemical imaging of enzyme activity, *Angewandte Chemie International Edition*, 42 (2003) 3238–3240.
46. L. Nault, C. Taofifenua, A. Anne, A. Chovin, C. Demaille, J. Besong-Ndika, D. Cardinale, N. Carette, T. Michon, J. Walter, Electrochemical atomic force microscopy imaging of redox-immunomarked proteins on native potyviruses: From subparticle to single-protein resolution, *ACS Nano*, 9 (2015) 4911–4924.
47. A. Anne, E. Cambril, A. Chovin, C. Demaille, C. Goyer, Electrochemical atomic force microscopy using a tip-attached redox mediator for topographic and functional imaging of nanosystems, *ACS Nano*, 3 (2009) 2927–2940.
48. T.O. Paiva, K. Torbensen, A.N. Patel, A. Anne, A. Chovin, C. Demaille, L. Bataille, T. Michon, Probing the enzymatic activity of individual biocatalytic FD-viral particles by electrochemical-atomic force microscopy, *ACS Catalysis*, 10 (2020) 7843–7856.

# 11 Practical Guidance of AFM Operations for Energy Research

*Yang Liu, Xin Guo, Yaolun Liu, Xin Wang, Chen Liu, Wenhui Pang, Fei Peng, Shurui Wang, Youjie Fan, and Hao Sun*
Bruker (Beijing) Scientific Technology Co., Ltd

## CONTENTS



DOI: 10.1201/9781003174042-11

## 11.1 INTRODUCTION

With increase of the energy demand and decrease of the energy resources, energy problem becomes more and more prominent. In response to the problem, numerous new energy materials were developed to overcome the shortcoming of traditional materials, for instance, energy density, conversion efficiency, cycle life, weight, etc. These properties are highly related to structure and physical/chemical processes at sub-micron or nanoscale, such as interface structure, defects, local crystallinity, adsorption, corrosion, catalysis, and so on.

To reveal relationship between energy materials structure and their properties at nanoscale level, atomic force microscopy (AFM) is an ideal tool due to its unique capabilities of acquiring high-resolution multiple dimensional surface information in various environments in real time and at nanoscale.

AFM is a system which utilizes interactions between AFM probe and sample surface to generate images. A commercial AFM usually takes a feedback system to ensure that an AFM probe accurately tracks surface topography. According to different feedback signals, there are three commonly used primary working modes of AFM in atmosphere and liquid environments, i.e., contact mode, tapping mode, and PeakForce tapping (PFT) mode. The principles, scopes of applications, and advantages/disadvantages have been thoroughly discussed in previous paper.[1] In different imaging modes and different working conditions, the force control mechanisms are different from each other, which means probes with different parameters are required.[2] Furthermore, for different applications, there are also some different requirements to the probe and sample conditions.

Various factors can affect AFM imaging quality, including sample, probe, system, software settings, data processing, and so on. In this chapter, we will first discuss the best practice of AFM sample preparation and probe selection for different application in energy research, followed by artifact recognition and data processing. Then a series of examples will be used as case studies to illustrate how to carry out high-quality AFM experiments on energy materials with the rules discussed in this chapter.

## 11.2 AFM SAMPLE PREPARATION

### 11.2.1 Common Rules for AFM Sample Preparation

The common rules for AFM sample preparation are very simple. The key is to FIX the sample to a RIGID support and make the sample surface CLEAN.

The AFM probe scans over the sample surface, so it is necessary to immobilize the sample on a substrate and keep the sample surface as stable as possible. Macroscopic samples (such as silicon wafers or polymer membranes) can be attached directly to a stainless-steel sample disk or other rigid substrates using an adhesive.[3–6] The sample is usually immobilized by some kinds of "glues". The commonly used glue for AFM imaging in atmosphere is double-sided adhesive tape. The tape with main ingredient cyanoacrylate is not recommended because it has large distortion after applying force on it and needs longer time to reach stable state. Epoxy is recommended to fix the sample onto substrate instead. If the sample is not fixed, the AFM

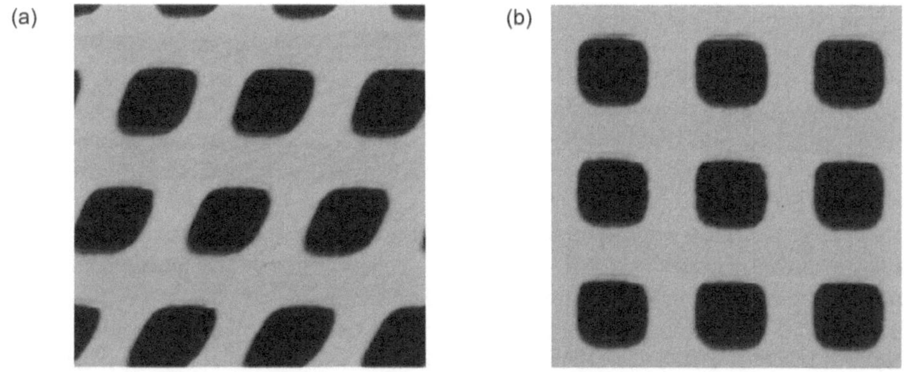


**FIGURE 11.1** AFM topography image of (a) a grating sample fixed by double-sided adhesive tape and (b) a grating sample fixed by epoxy. Large distortion can be observed in (a) while no distortion in (b).

image will be distorted due to the drift. Figure 11.1a shows distorted AFM image of a grating sample fixed by double-sided adhesive tape. Large distortion can be observed because the tape is deforming. Figure 11.1b shows a normal AFM image of a grating sample fixed by epoxy. No distortion is observed in this image.

Bad glue also leads to the drift on vertical direction. If the glue continues deforming during AFM imaging (usually expanding on vertical direction after pressing the sample onto the substrate), the height of the sample will also change (becomes higher and higher if glue is expanding), consequently, the z piezo of the AFM scanner tracks an additional height data change which is introduced by the glue deformation. In worse case, the deformation of the glue is so large that the z piezo often retracts to its limit, leading to a failed imaging.

The glue is only suitable for fixing bulk samples. If we would like to image samples like nano particles[7–9] (NPs) (such as Au NPs, Pt NPs, perovskite NPs) and nano fibers[10] (such as carbon nano fibers, ZnO nano fibers), in atmosphere condition, "drop and dry" is a good way to immobilize the particles or fibers onto substrates using van der Waals interaction. The best practice is to dilute the sample with volatile organic solvent, which is also inert to the sample.[11] Commonly used solvents are ethanol, isopropanol, acetone, etc. Pure water is not the best choice because it is more difficult to evaporate and needs longer time for sample preparation. The solvents must be clean, free of insoluble and nonvolatile impurities, so gas chromatography (GC)/high-performance liquid chromatography (HPLC) purity or metal-oxide-semiconductor (MOS) grade reagents are recommended. To avoid generation of aggregates, the sample should be diluted to a very low concentration; a rule of thumb is to dilute the sample till just perceptible color can be seen, or about 1mmol magnitude concentration. Then use ultrasonic cleaner to further disperse the sample in the solvent, drop the sample solution to a clean substrate, wait for 3–5 minutes for sample deposition, and finally dry with pure $N_2$ or air.

The commonly used substrates for AFM imaging include mica,[12] highly oriented pyrolytic graphite (HOPG),[13] gold,[14] silicon,[15] and glass.[16] The surfaces of mica and

HOPG are ultra-flat at atomic level after a layer is freshly cleaved. Freshly prepared mica surface is negatively charged and hydrophilic, while fresh HOPG surface is conductive and hydrophobic. The nature of the surface can be easily changed by modification. This characteristic makes them ideal substrates for high-resolution AFM imaging. Gold and many other metals or alloys could be sputter-coated on silicon surface to make a flat and conductive surface. It is a good substrate for thiol-molecule AFM experiments since hydrosulfuryl can form strong covalent bond with gold surface. Silicon wafer with or without oxide layer can be cut into different shape and size by glass cutter, and then is cleaned by several steps to be a good substrate for AFM imaging, especially for some electrical measurement of MOS structures. Glass is a much cheaper choice, which has many commercially available sizes, such as glass slide, cover glass, glass bottom petri dish, and so on. It has rougher surface than mica, HOPG, and silicon wafers and is only used for samples with large surface fluctuation. The range of z piezo on most of commercial AFM systems is <20 μm. The surface fluctuation must be lower than the z range for successful imaging. Similar to silicon wafers, glass should also be cleaned before using.

Cleaning process of substrate is vital for high-quality AFM imaging. The best practice of the cleaning process of silicon wafer is as follows: (1) Clean the silicon wafer in ultrasonic cleaner with ultrapure water for 5 minutes; (2) Clean the silicon wafer in ultrasonic cleaner with ethanol (recommended: Sigma Aldrich product #459844) for 5 minutes; (3) Immerse silicon wafer into piranha solution, heat to ~80°C and maintain for 30 minutes; (4) Repeat step (1) and (2); (5) Dry the wafer with highly pure $N_2$. Now a very clean and hydrophilic silicon surface is obtained. To get a hydrophobic surface, after step (4), immerse the wafer into 5% hydrofluoric acid solution (recommended: Sigma Aldrich product #1.00335) for 5 seconds, then repeat step (4) and (5). Heating the wafer in tube furnace at ~1,000°C for ~ 1 hour can get ~300 nm oxide on the silicon wafer surface.

Before imaging, check if there is any soft or loose residue remain on the sample after sample preparation, which can easily contaminate the tip. If yes, clean it as needed. If the sample is hydrophilic and exposed in ambient air for long time, some dust or debris will be absorbed to the surface, leading to surface degrades and difficult to be imaged. If the sample is not in good condition and hard to recover, get a freshly cleaned sample.

To reduce imaging noise, there are some tips. If sample is hold on chuck by magnetic mount, make sure the sample is not oversized compared to mounting surface. Otherwise too much overhang will cause sample to vibrate and introduce noise. Figure 11.2 shows good and bad sample mounting. If sample loosely sits on the chuck or mount the puck with double-sided adhesive tape, make sure there no air pocket underneath the sample, otherwise will cause sample to resonant.

When doing AFM imaging in fluid, the basic rules are the same but much more complicated. Sample preparation in fluid requires more attentions beyond appropriate substrates and immobilization. First, solutions and any other objects that could contact substrates, such as AFM probe, holder, tweezer, pipette tip, and so on, all these elements should be cleaned. Debris in solution could contaminate probe tip or sample surface. A typical cleaning procedure has been discussed in previous paper.[2] Cyanoacrylate glue (e.g. Super Glue) cannot be used for mounting samples in liquid,

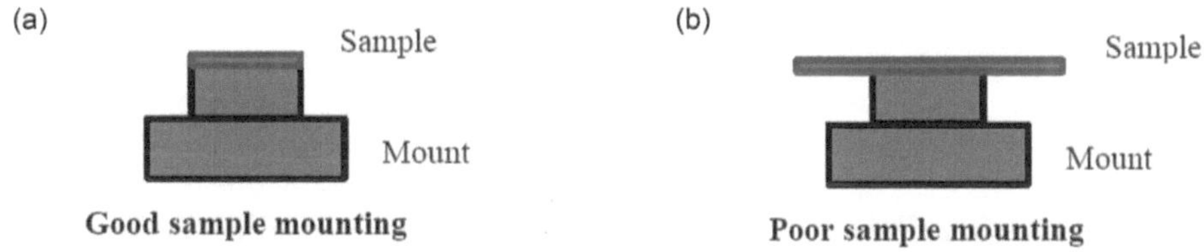


**FIGURE 11.2** (a) Good sample mounting, the sample is smaller compared to the mounting surface. (b) Bad sample mounting, the sample is oversized compared to the mounting surface and it will introduce noise during imaging.

because soluble glue may contaminate samples and cause sample position drifts. Epoxy glue is a better option to adhere substrates to steel disks. For non-critical applications, Devcon 2-Ton Epoxy works well. For applications where contamination control is more critical, use a more inert, solvent-free epoxy like Master Bond EP21LV, EP21AR, or a hot melt adhesive. Second, the whole imaging environment should be stable. Evaporation and temperature variation will also cause drift. The vertical and lateral deflection may change continuously when the tip and the holder were immersed into solution at first. It may take about half an hour to make the system stable.

In fluid, electrical interaction is the most common method for sample immobilization. For example, the mica surface is negatively charged, so the positive charged samples like proteins and some lipids with isoelectric point larger than seven can be absorbed on it. Mica surface can be positively charged when divalent cation like $Ni^{2+}$ or $Mg^{2+}$ exists in solution or it is modified by silane such as 3-Aminopropyl trimethoxy silane (APTES). If electrical interaction is not strong enough, chemical bond or physical frame can also be used to fix the sample. Mica, glass, and gold surface can be easily modified by N-hydroxy succinimide (NHS) esters, click chemistry, thiol modification, and so on. For living cells or bacteria samples, they are usually immobilized on poly-*L*-lysine medicated glass or petri dish. Porous membrane or porous hydrogel can also be used to immobilize cells or bacteria at appropriate size.

### 11.2.2 Sample Preparation for Different Applications

Based on following common rules of sample preparation we discussed above, for different applications, there are additional requirements for high-quality imaging. In this section we will discuss the particularities of sample preparation for force measurement, electrical measurement, AFM-IR, and electrochemistry experiment.

#### 11.2.2.1 Force Measurement

AFM force modes can provide mechanical properties of the sample and play important roles in energy research. Take battery research for example, mechanical properties are very useful to help researchers understand the physical or chemical properties changing during battery working. Processed anodes can help to inhabit growth of Li dendrite, by analysis of the force curve of processed or bare Li anodes, which can characterize the contents of Li crystal with higher stiffness.[17–19] Force curve is also useful to reveal the mechanical properties of vertical-layer structures in a solid-electrolyte interphase (SEI).[20–22] The distribution of mechanical properties, especially Young's modulus, is important to illustrate the heterogeneity in one battery sample[23–26] or the heterogeneity before and after charging cycles.[27–29] Other environmental factors, like temperature, can also modulate the physical or chemical structure in batteries, which can be illustrated in modulus or adhesion mapping.[30]

The commonly used force characterization modes include force-distance curve, force mapping techniques (ForceVolume, quantitative imaging (QI mode)) which are based on "trigger" to control the maximum applied force, as well as PeakForce Quantitative Nanomechanical Mapping (PFQNM) mode and its derivative modes, which are based on "feedback" to control the maximum applied force. In recent years, contact resonance mode has been commercialized to characterize the modulus of hard samples. Nano dynamic mechanical analysis mode (nano-DMA) was also realized to reveal viscoelastic properties by storage modulus, loss modulus, and loss tangent with various drive frequencies at nanoscale.[31,32] The measurement is based on contact mechanics. Researchers should consider parameters of contact mechanics when preparing the sample for accurate mechanical characterization.

Take Hertz model as an example, which is used to deal with the contact of two elastic spheres without adhesion force.[33] In practice, we usually use a probe to indent a flat surface; the key parameters of Hertz model in this situation are defined in Figure 11.3.

In Hertz model, the indentation depth $\delta$ and contact radius $a$ can be written as

$$\delta = \frac{a^2}{R^*} \tag{11.1}$$

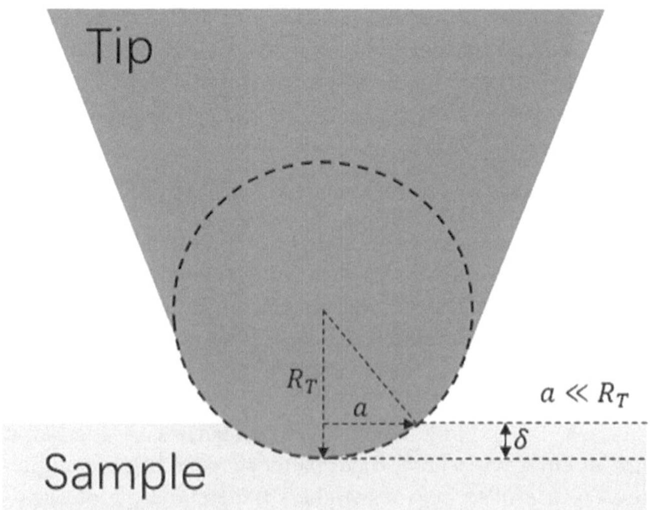


**FIGURE 11.3** Key parameters in Hertz model. $\delta$ is indentation depth of the tip on the sample, $a$ is the contact radius of contact area, $R_T$ is the tip radius. The model is valid when $a \ll R_T$.

$$a = \sqrt[3]{\frac{3FR^*}{4E^*}} \tag{11.2}$$

Effective radius $R^*$ can be written as

$$\frac{1}{R^*} = \frac{1}{R_T} + \frac{1}{R_S} \tag{11.3}$$

where $R_T$ and $R_S$ are radius of the tip and the sample. For flat sample surface, $R_S$ is infinite and $R^* = R_T$.

Reduced modulus $E^*$ can be expressed as

$$\frac{1}{E^*} = \frac{1-\upsilon_T^2}{E_T} + \frac{1-\upsilon_S^2}{E_S} \tag{11.4}$$

where $E_T$ and $E_S$ are Young's modulus of the tip and the sample, $\upsilon_T$ and $\upsilon_S$ are Poisson's ratio of the tip and the sample.

From equations (11.1) and (11.2), the relationship between force and indentation depth can be written as

$$F = \frac{4}{3} E^* \sqrt{R^*} \delta^{3/2} \tag{11.5}$$

where $F$ is the force applied on the sample surface.

When using a relative hard tip (e.g. modulus of diamond is ~1,220 GPa and silicon ~165 GPa,) to measure a relative soft flat sample, so $E^* \approx \dfrac{E_S}{1-\upsilon_S^2}$.

Thus, equation (11.5) has been transformed into, $E_T \gg E_S$

$$F = \frac{4}{3} \frac{E_S}{1-\upsilon_S^2} \sqrt{R_T} \delta^{3/2} \tag{11.6}$$

For accurate modulus measurement, measuring accurate indentation depth $\delta$, applied force $F$, and tip radius $R_T$ is very critical when using Hertz model. Except contact resonance mode, in other force modes, AFM performs force curve to measure the cantilever deflection $D$ and uses deflection sensitivity (DS) and spring constant to calculate the applied force $F$. The related parameters of probe will be discussed in detail in Section 11.3.1 of this chapter. And sample indentation depth $\delta$ is calculated by the cantilever deflection $D$ and scanner movement $Z$, as shown in Figure 11.4. If we introduce a relative calibration error $e$ to DS, the measured deflection will be $(1+e)D$, and calculated indention depth is $z-(1+e)D$, so the indentation depth error will be $\dfrac{eD}{z-D} = eD/\delta$. It means, the indentation depth error induced by defection sensitivity error could be amplified by a factor of $D/\delta$. To decrease the indentation depth error, $\delta$ cannot be too small. Usually, at least 2 nm indentation is necessary for accurate measurement. However, it does not imply that the smaller $D$ the better result. Further analysis will be done in Section 11.3.1 of this chapter.

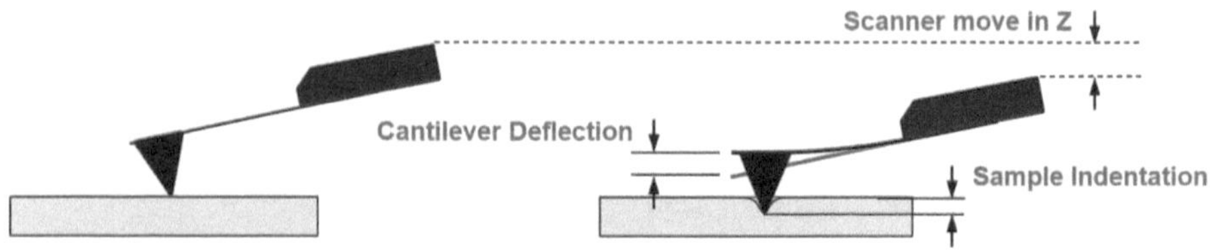


**FIGURE 11.4** After the AFM tip contacts the sample surface, the scanner movement in direction $Z$ causes two effects: cantilever defection $D$ and sample indentation $\delta$. $Z = D + \delta$.

Based on above discussion, for accurate modulus measurement, the thickness of sample is critical. Generally, the indentation depth should be controlled to be less than the tenth of the total thickness of sample to avoid "substrate effect". It implies that at least 20 nm thickness is required for an accurate modulus measurement of a homogeneous sample. It does not mean the thicker sampler we have, the better result we get. For multi-phase composite samples, modulus is a good indicator to identify the components in different domains, but sometimes too thick sample may lead to the inaccurate modulus of core domain because of the matrix below the core areas, especially when the matrix is softer than the core domain. By reducing the thickness of sample (cannot reduce to <20 nm), the mechanical properties of core domain will be more accurate after excluding the matrix below the core area, as shown in Figure 11.5.

For other contact mechanics models, they are more complex and suitable for different situations, but the concepts are similar. Derjaguin-Muler-Toporov (DMT) model considers an elastic sphere with rigid surface and includes van der Waals forces outside the contact region, so it is valid for stiff samples ($a \ll R_T$) with low adhesion. Johnson-Kendall-Roberts (JKR) model neglects long-range interaction outside the contact area but includes short range forces in the contact area and is valid to soft samples with large adhesion. Sneddon model considers a rigid indenter (sphere or other rigid shapes) on a linearly elastic half-space and is valid when $\delta \gg R_T$.

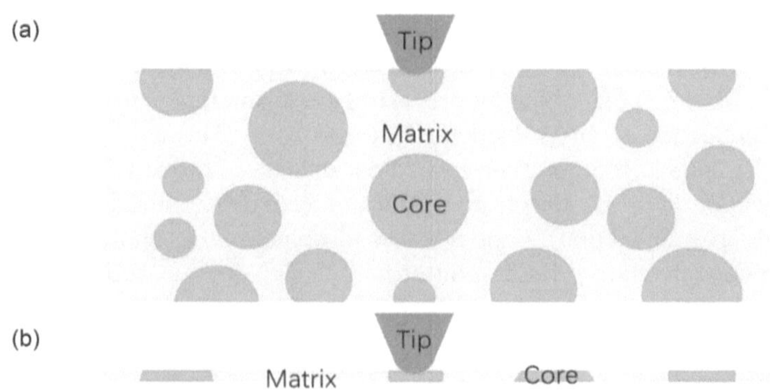


**FIGURE 11.5** (a) Thick sample may lead to inaccurate modulus of core domain because there is matrix below the core domain. (b) By reducing the thickness of sample, the modulus measurement can be more accurate.

Researchers must pay attention to the conditions of suitable contact mechanics models and make the sample conditions match the requirements of the model used.

Relative flat and homogenous surface is also required to match the postulation that $R_S$ is infinite. If the surface is not flat, when the probe scans over the surface, $R_S$ is changing and effective radius $R^*$ is changing accordingly. It is difficult to make an absolute flat surface but control the roughness to be small is a good practice for force measurement. A common method to create a perfectly flat surface is polishing and grinding the hard samples like carbon fiber,[34,35] shale,[36] silica,[37] and alloy.[38,39] Argon ion polishing[40] and focused ion beam[41] are also recommended methods to create flat surface of hard samples. However, softer samples, especially polymers, cannot be polished. Low-temperature microtome, which can cut sample into pieces with a thickness from micrometers to nanometers at controlled temperature, is a good solution instead.[42–44] Temperature below the glass transition point guarantees the strength of polymer for microtome. Even the softest biomaterials can also be cut into well-defined surface for AFM imaging.[45,46] Moreover, both polishing-grinding and microtome are necessary processes to produce cross-section surface of any sample.[47]

Again, clean sample surface is critical in mechanical characterization, especially for adhesion force and modulus measurement. The contamination on surface can be absorbed onto the tip and changes the effective tip radius and modulus. For adhesion measurement, if the tip radius changes, the contact area will be also changed at the same applied force. The deformation of contamination can also be introduced into the modulus calculation. Freshly polished, grinded, or microtome surface is required for high-quality force measurement. For hard surfaces like metal and silica, First Contact Polymer (Photonic Cleaning Technologies) is recommended to remove the contaminants from the surface.

#### 11.2.2.2 Electrical Measurement

Electrical modes of AFM enrich its application in solar cells,[48] energy storage,[49] energy conversion,[50] and electrocatalytic activity.[51] Electric force microscopy (EFM) and Kelvin probe force microscope (KPFM) can provide information on charge accumulation, charge transfer, work function, and dielectricity.[52] Conductive atomic force microscopy (cAFM) and Tunneling AFM (TUNA) make it possible to probe photocurrent and resistance.[53] Piezoresponse force microscopy (PFM) visualizes the ferroelectric domain structure and hysteresis loop.[54] The electrical modes of AFM are very inclusive for samples. Insulators such as bismuth ferrite ($BiFeO_3$, BFO) and $TiO_2$;[55] semiconductors such as InN and GaAs;[56] and conductors such as graphene[57] are all deeply investigated by electrical modes of AFM.

For electrical modes of AFM, sample should be electrically connected to the sample stage to make sure the sample can be held at ground potential or applied bias through the stage. It is recommended to electrically connect the sample by mounting it to a standard conductive sample puck or stage using conducting epoxy or silver paint, as shown in Figure 11.6a. If only the sample surface is conductive, but the base of the sample is not, the conductive epoxy or paint should contact both the conductive surface layer and the conductive sample mount. In this situation, the conducting epoxy or silver paint may be higher than the sample surface because it needs to contact with the conductive layer, and just make sure it is not located directly underneath

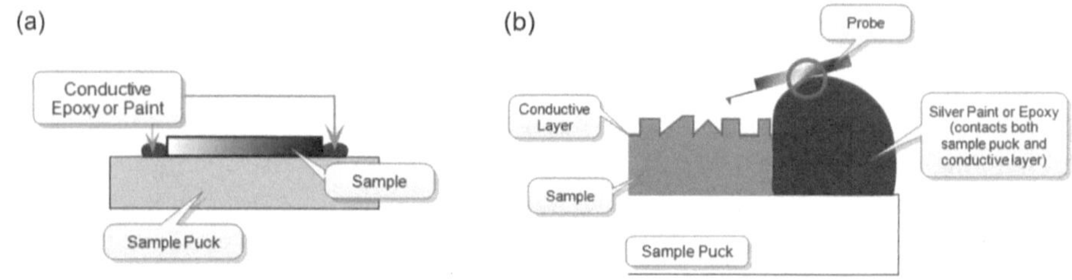


**FIGURE 11.6** (a) Schematic of sample preparation for electrical measurement. (b) If only sample surface is conducive, silver paint or epoxy should contact both sample puck and conductive layer but need to make sure the large glob is not directly under the cantilever.

the cantilever to avoid undesired contact, as shown in Figure 11.6b. Then verify the connection with a multimeter. If conducting epoxy or silver paint is not available, copper tape or carbon tape also works. For nanodevices, bonding machine can be used to attach metal wire on the sample surface. When performing electrical measurement in fluid, sample preparation rules for both electrical measurement and fluid imaging should be taken account of at the same time.

Similar to conductive AFM, photo-induced conductive AFM (pcAFM) requires conductivity between sample and sample mount. In pcAFM, optical transparence should be considered, because in typical pcAFM setup, the light is introduced by optical fiber under the sample. As such, indium-tin oxide (ITO) glass is a widely used substrate for this application. Sample could be fixed using silver paint or sputter coated using gold on the ITO substrate to enable electrical contact. Sample mounting on a commercial pcAFM solution on Bruker Dimension Icon system is shown in Figure 11.7.

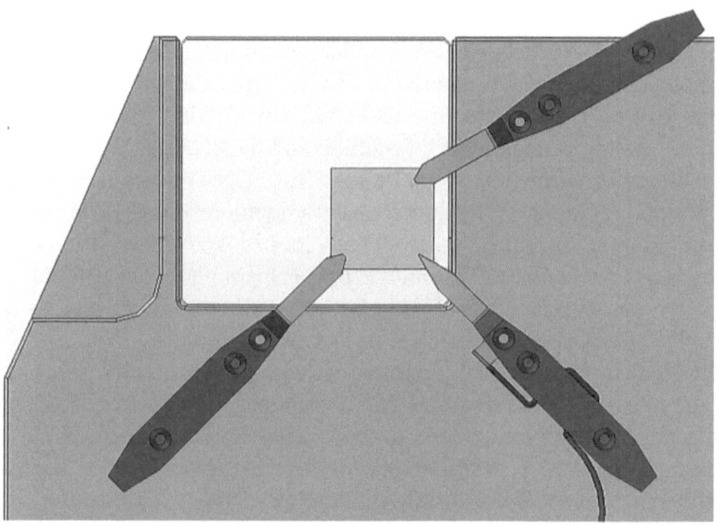

**FIGURE 11.7** Sample mounting for pcAFM measurement on Bruker Dimension Icon system. The transparent ITO glass is the substrate to make sure the light from optical fiber in bottom can get in on the sample.

### 11.2.2.3 AFM-IR

AFM combined with infrared (IR) spectroscopy, known as AFM-IR, is a novel technique to acquire the information of the nanoscale spectrum. Wide range of cross related research activities in photovoltaics, photo electrocatalysis novel LED materials, can be categorized in energy research fields.[58–61] AFM-IR with its advantages in revealing nanoscale chemical information can help to investigate nanoscale heterogeneity and degradation in organic hybrid perovskite material. Frontier research topics such as ferro-elasticity in perovskite and site varying activity in novel catalytic materials can now be studied with high resolution.[62,63] Future applications might include nanoscale chemical identification of interfacial defects and corresponding inhibition in materials or device.

The recommended core principle of sample preparation for AFM-IR is to use of IR transparent substrates such as ZnS/ZnSe/gold[64,65] and use of cryo-microtomy[66,67] to transfer samples to substrates. The purpose of using IR transparent substrates is to avoid interference of the sample signal caused by the IR absorption on substrate. Cryo-microtomy preparation can ensure that AFM testing is simple and easy as well as IR light is not blocked due to the high fluctuation of the sample surface. A typical cryo-microtomy setup is shown in Figure 11.8. The sample is fixed on specimen disc and sliced into thin slices by the blade under low temperature environment. The smoother a sample is, the easier it is to get good results. Generally, it is best to keep the roughness below 100 nm (root mean square, RMS).

According to the different IR optical path setups, including top-down method and bottom-up method, the sample preparation methods are also slightly different.

Top-down method means IR light is incident onto the sample from above. A schematic diagram of top-down method is shown in Figure 11.9. A pulsed, tunable IR laser is incident to the region where the tip interacts with the sample. With the IR illumination, rapid photo thermal expansion of the sample drives the AFM cantilever

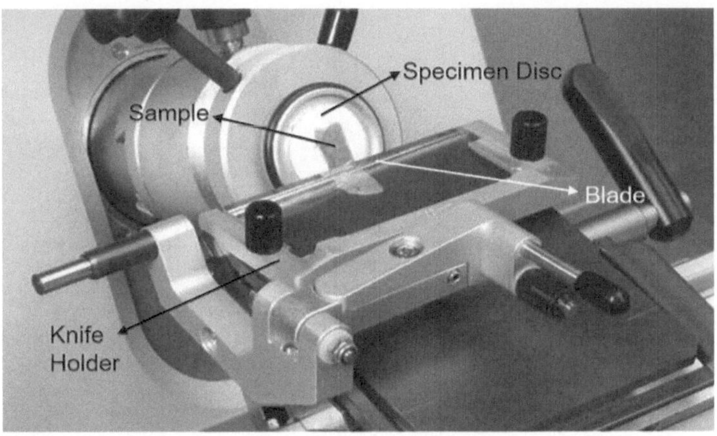


**FIGURE 11.8** Schematic diagram of cryo-microtomy. The sample is fixed on specimen disc and sliced into thin slices by the blade in low temperature environment. (The picture is from https://www.leicabiosystems.com.)

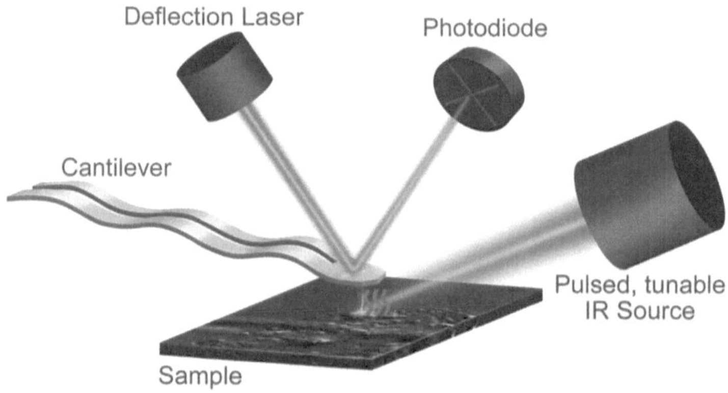


**FIGURE 11.9** Schematic diagram of top-down method. The intensity of IR absorption can be obtained by detecting the oscillation of the deflection laser on the photodiode. (The picture is from https://www.bruker.com.)

oscillation. By detecting the deflection laser oscillation on the photodiode, the intensity of IR absorption can be obtained.

In general, IR technique with top-down method combined with AFM is carried out on ZnS/ZnSe/gold substrate which has no absorption of IR. It is very popular to use the cryo-microtomy for sample preparation, and samples are also usually drop-casted or spin-casted on the substrates from solution.[65,67,68] For instance, Figure 11.10 illustrates the deposition procedure for perovskite thin films fabricated by anti-solvents. Perovskite precursor solution is first dropped on the $TiO_2$ coated fluorine-doped tin oxide (FTO) glass substrate and then spinned at a fixed rotation speed. Chlorobenzene (CB) or methoxybenzene (PhOMe) is then dropped onto the wet film surface before precipitation of perovskite solid during the spin coating process.

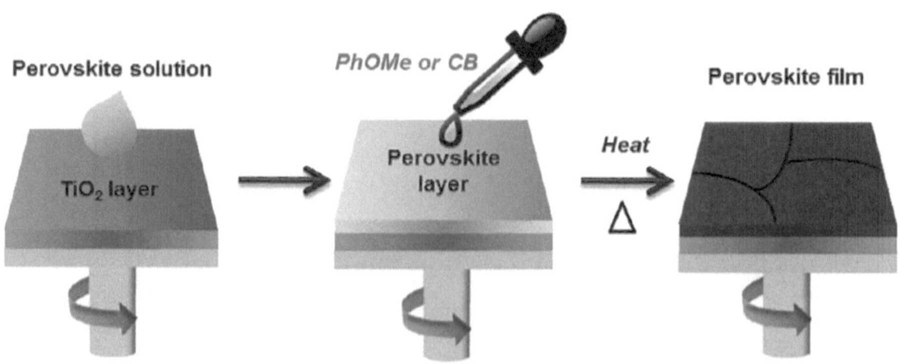


**FIGURE 11.10** Schematic diagram of perovskite films deposition with drop and spin method. Perovskite precursor solution is firstly dropped on the $TiO_2$ layer. CB or PhOMe is then dropped onto the wet film surface during the spin coating process. The rapid and homogenous crystallization with heating creates a uniform, dense, and pinhole-free crystalline thin film. (Reprinted with permission from Zhang et al.[69] Copyright 2018 John Wiley and Sons.)

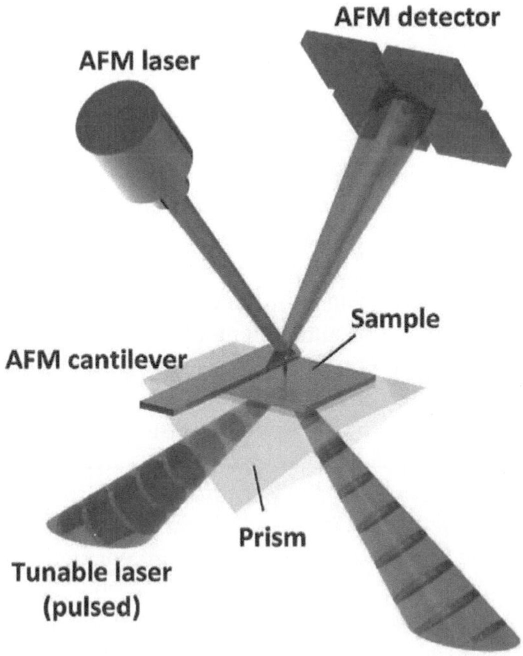


**FIGURE 11.11** Schematic diagram of bottom-up method. ZnS/ZnSe prism substrate allows IR light to pass through with no IR absorption. The intensity of IR absorption can be obtained by detecting the deflection laser oscillation on the photodiode. (Reprinted with permission from Kurouski et al.[70] Copyright 2020 Royal Society of Chemistry.)

The rapid and homogenous crystallization with heating creates a uniform, dense, and pinhole-free crystalline thin film. It should be noted that anti-solvents can affect supersaturation as well as the nucleation and crystal growth rates, thereby strongly affecting the crystallization kinetics and final morphology. Therefore, the solvent used for sample preparation should be chosen carefully.[69]

Bottom-up method means IR light is incident onto the sample from below, so optical transparence should be taken into consideration. A schematic diagram of bottom-up method is shown in Figure 11.11.[70] The detection principle is the same as the top-down method. In general, IR technique with bottom-up method combined with AFM is carried out on ZnS/ZnSe prism substrate which allows IR light to pass through with no IR absorption. Cryo-microtomy is widely used in this application. Samples should be frozen sectioned and transferred to the top surface of prism. The sample thickness is usually <1 μm, to make sure IR light penetrates the whole sample.[66,71,72]

#### 11.2.2.4 Electrochemistry

Electrochemical atomic force microscope (EC-AFM) combines AFM and electrochemical control to study the structure and electrochemical properties at solid–electrolyte interface in situ under different electrochemical conditions.[73] EC-AFM is mainly used to study in situ electrodeposition,[74] Li ion batteries,[75–78] biological science,[79] supercapacitors,[80] corrosion and protection,[81] and so on. There are several groups that are pioneering in EC-AFM for batteries study. Shen, Wen, and Mao's groups used EC-AFM to study the dynamic evolution of the solid-electrolyte interfaces in working environments.[75,82,83]

The sample for EC-AFM usually acts as working electrode (WE) during EC experiment. Flat, clean, and conductive surface is required for a good WE. Flat metal surface (such as Au, Pt, Cu) can be achieved by annealing,[84] polishing,[85,86] and physical vapor deposition on flat substrates.[87,88] HOPG is another widely used substrate for EC-AFM because it is conductive and ultra-flat after being freshly cleaved.[89,90] To fix the sample and keep the WE at ground potential, electrical contact with the sample mount is required. Conductive epoxy can be used to fix the sample, but the epoxy must be inert to the EC reaction and not contaminate the solution.

For EC sample preparation, the rules are quite simple; however, to create an EC environment, sample must be placed in the electrochemical cell (EC Cell). Figure 11.12 shows a commercial EC cell on sample-scanning AFM, Bruker MultiMode system. There are four channels on the cell, three in the side and one on the top. Two channels in the side are for fluid flow in and out, while the third one in the side is for reference electrode (RE). The top one is used for inserting the counter electrode (CE). The sample-scanning AFM system only allows small sample, but the sample size must be large enough to allow O-ring to seal the cell. As a WE, the sample must have electrical contact with the sample mount, which is connected to EC workstation, so the WE potential can be maintained at ground in EC experiments. Similar setup was used in previous research.[91,92]

Figure 11.13 shows another commercial design of EC cell on tip-scanning AFM, Bruker Dimension Icon system. It has a nickel-plated tungsten alloy base for improving thermal conductivity, with four nickel-plated NdFeB magnets in the base bottom to hold the base firmly onto the sample chuck, a Teflon clamp to secure and seal the sample to the base, a glass cover to control electrolyte evaporation, and two O-rings to seal the sample and glass cover. Tip-scanning AFM does not limit sample size; if this EC cell is used, the maximum sample size can be up to 41 mm. If only the sample surface is conductive, it must contact the gold-plated pogo pins, which are electrically connected to the EC cell base.

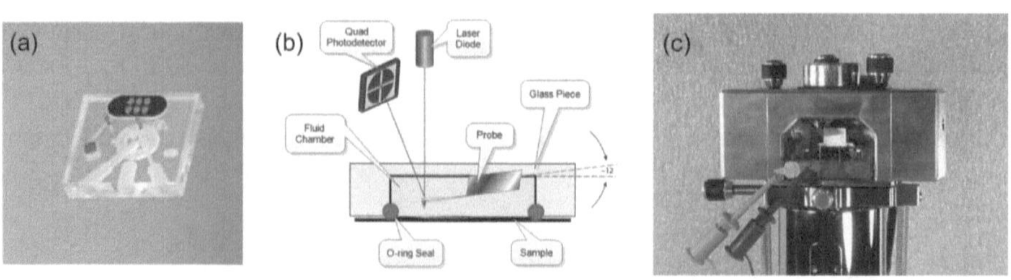


**FIGURE 11.12** Commercial EC cell on sample-scanning AFM. (a) Photo of Bruker MultiMode EC cell. (b) Schematic diagram of a closed EC environment on MultiMode system. (c) Photo of fully assembled MultiMode EC cell.

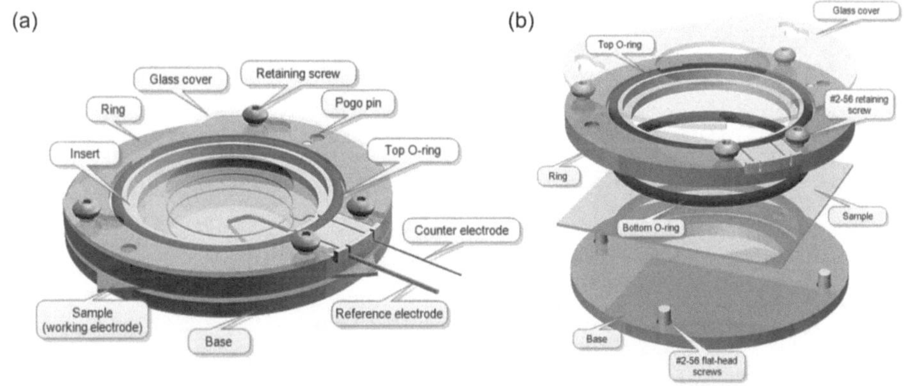


**FIGURE 11.13** Commercial EC cell on tip-scanning AFM. (a) The general-purpose EC cell of Bruker Dimension Icon system. Shown with a gold sample. (b) The exploded view of Dimension Icon EC cell.[86]

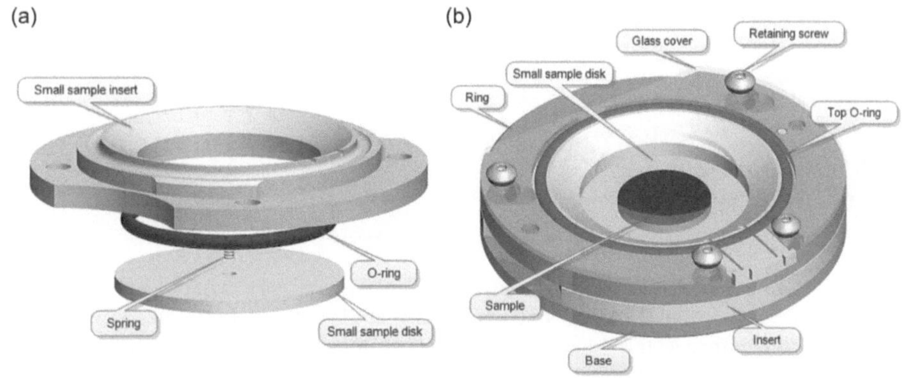


**FIGURE 11.14** (a) The small sample adapter for Dimension Icon EC cell. (b) Fully assembled dimension icon EC cell with the small sample adapter and a black sample.

Small sample adapter is designed for mounting small EC-AFM sample, as shown in Figure 11.14. The small sample is glued with Torr Seal to a Teflon sample disk. A conductive spring is used to make electrical contact with the back of the sample through the hole in the disk. Then the sample can be controlled by EC workstation as WE.

For different purposes, researchers may need to design or modify EC cell. For example, Luchkin et al. designed a new EC cell to measure SEI formation on cross-sections of composite battery electrodes.[77] Figure 11.15 schematically illustrates a standard AFM EC cell and the new EC cell design. In the new cell shown in Figure 11.15b, the sample could be bulky and does not require sealing by an O-ring, which allows cross-sections of composite battery electrodes embedded in epoxy resin. The epoxy resin fixes the composite electrode sample, and the polished surface

**FIGURE 11.15** Schematic illustration of a standard (a) and the new (b) EC cells. In the standard cell the flat sample (A-1) is fixed and sealed at the bottom of the cell (A-2). The cell body is filled with the electrolyte and the cantilever is immersed in the electrolyte bath (A-3) for scanning. In the new cell the sample (B-1) is embedded into epoxy resin (B-2), polished/cross-sectioned (B-3), connected to the substrate by a conductive silver or carbon paint (B-4), and installed into an AFM. The cantilever is positioned above the sample (B-5), and the electrolyte is injected through the tube to form the meniscus between the sample and the cantilever holder (B-6).

additionally serves as a support for the electrolyte meniscus, RE and WE. The meniscus is formed by injecting the electrolyte through the tubing fixed from the top in the window of the cantilever holder.

Here is another example for EC modification. Nellist et al. designed a contact-based potential-sensing electrochemical atomic force microscopy (PS-EC-AFM) technique to directly measure the surface electrochemical potential in heterogeneous electrochemical systems in operando.[93] Figure 11.16 is the schematic diagram of the modified EC cell and LED light source in the bottom. The LED light source on AFM stage illuminated the sample from the hole on Teflon baseplate. A standard Ag/AgCl RE was also integrated in the Teflon baseplate for easy usage and replacement. This design expanded the EC cell from EC to photo-electrochemistry study.

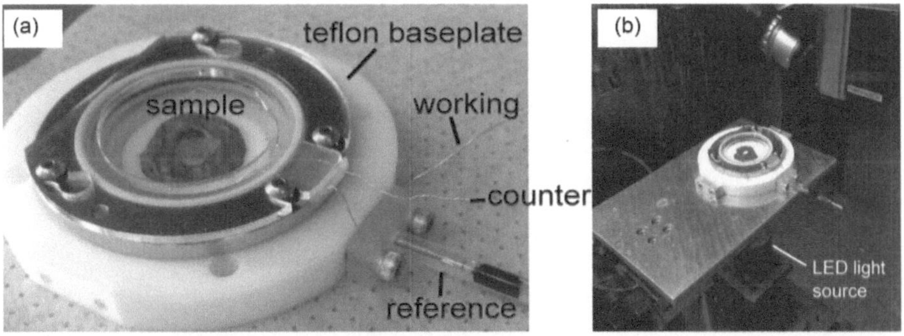


**FIGURE 11.16** (a) Modified EC cell for AFM. (b) Illumination from the bottom of the EC cell on the AFM stage. (Reprinted with permission from Nellist et al.[93] Copyright 2017 Springer Nature.)

## 11.3 AFM PROBE SELECTION

AFM probe is one of critical parts in an AFM system, which plays an important role in high quality imaging. As shown in Figure 11.17, AFM probe consists of three components: the substrate which is used to hold the probe in probe holder, the cantilever which is a force sensor to "feel" the interaction between the tip and the sample, and the tip which directly interacts with sample surface to generate AFM images. V-shape shown in Figure 11.17b and rectangle-shape shown in Figure 11.17c are two common cantilever geometries. Both the cantilever and the tip affect AFM imaging quality significantly. In this section, we will review the key parameters of them and the impact on the imaging first and then we will discuss the rules of probe selection for different applications.

### 11.3.1 Key Parameters of AFM Probes

The key parameters for AFM probe cantilever include geometrical and kinetic parameters.

The geometrical parameters are related to cantilevers' dimension, namely width $w$, length $L$, and thickness $t$. The dimension of the cantilever varies different probe designs and applications, and the definition of these three parameters of different shape cantilevers is shown in Figure 11.18.

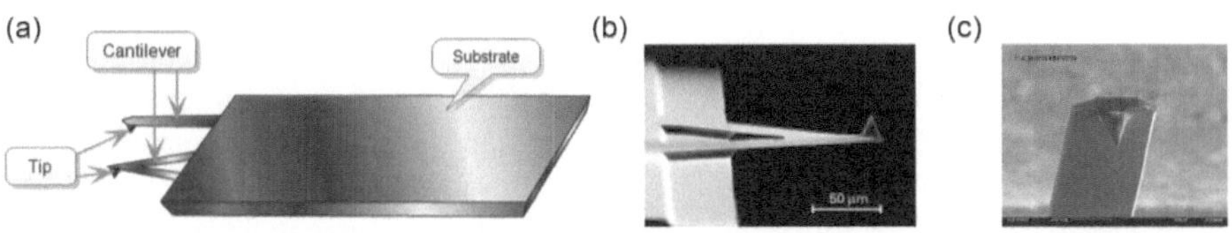


**FIGURE 11.17** (a) Schematic diagram of AFM probe. Three major components: substrate, cantilever, and tip. (b) SEM image of a V-shape cantilever. (c) SEM image of rectangle-shape cantilever.

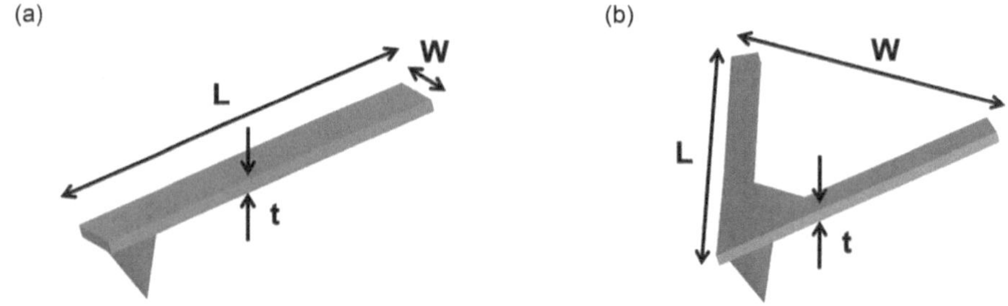


**FIGURE 11.18** Definition of cantilever width $w$, length $L$, and thickness $t$ in (a) rectangle-shape cantilever and (b) V-shape cantilever.

The geometrical parameters of the cantilever affect kinetic parameters a lot. There are two very important kinetic parameters: spring constant $k$ and resonance frequency $f_0$.

The spring constant $k$ is determined by the dimension and Young's modulus of cantilever. For rectangular cantilever, the spring constant can be expressed as

$$k = \frac{E}{4}\frac{wt^3}{L^3} \tag{11.7}$$

where $E$ is the Young's modulus of the cantilever, and $w$, $t$, and $L$ are width, thickness, and length of the cantilever, respectively. For V-shape cantilever, the expression of spring constant is much more complicated. Using parallel beam approximation (PBA), finite element analysis, (FEA) or other analytical models, we can get different expressions of V-shape cantilever spring constant with varying degrees of approximation. The simplest one is Butt's approximation, shown as equation (11.8); other methods introduce some other dimensional factors to make the result more accurate.

$$k = \frac{E}{2}\frac{wt^3}{L^3} \tag{11.8}$$

The spring constant $k$ directly affects the imaging force. In contact and PFT mode, since the operating frequency is far from cantilever resonance frequency, according to Hooke's Law shown in equation (11.9), the imaging force is proportional to $k$:

$$F = kD \tag{11.9}$$

where $F$ is the applied force on the sample surface, and $D$ is cantilever deflection. While in tapping mode, since the operating frequency is near the resonance frequency of the cantilever, the peak force on a vibrating cantilever is approximately proportional to $k/Q$, which can be obtained from vibration equation of AFM probe,[94,95] where $Q$ is the quality factor of the cantilever. Therefore, the conclusion is that large $k$ indicates stiffer probe and stronger interaction between the tip and the sample, and vice versa. Strong force between the tip and the sample results in worse lateral resolution. The reason was explained in another paper.[2] However, it does not mean the smaller the $k$ is, the better resolution we get. In tapping mode, to avoid the attractive force pull the cantilever to jump to contact the surface, the following condition must be met for stable operations[96]:

$$\left|\frac{d^2V_{ts}}{dz^2}\right| < k, \quad \left|-\frac{dV_{ts}}{dz}\right| < kA_0 \tag{11.10}$$

where $V_{ts}$ is the tip-sample potential and $A_0$ is the amplitude of the cantilever. That's the reason why for tapping mode in atmosphere, the commonly used probe has larger $k$ than contact and PFT mode. In fluid operation, because the attractive force is minimized, probe with smaller $k$ can be used for tapping mode.

Resonance frequency is another key kinetic parameter of cantilever. A cantilever has several resonance frequencies, but only the fundamental frequency is used in tapping mode; in some dual-frequency operations, the second order resonance frequency also can be used to drive the cantilever vibrating at that frequency. The higher order resonance frequencies usually exceed AFM operation frequency, so we rarely discuss about them. The fundamental resonance frequency of the cantilever $f_0$ (in Hz) or $\omega_0$ (in rad) is determined by the following equation:

$$f_0 = \frac{1}{2\pi}\sqrt{\frac{k}{m}} \text{ or } \omega_0 = \sqrt{\frac{k}{m}} \tag{11.11}$$

where $m$ is the effective mass of the cantilever.

The resonance frequency affects cantilever response time. In tapping mode, the cantilever response time constant $\tau$ can be written as

$$\tau = 2Q/f_0 \tag{11.12}$$

which means the change of amplitude with a change in tip-sample interaction is on a time scale $2Q/f_0$. The conclusion it that higher resonance frequency and lower quality factor lead to faster response time of the cantilever and allow faster scan. If one operates the AFM system in vacuum environment where $Q$ is too large (usually $10^4$ ~$10^5$), tapping mode cannot work properly due to the very slow response of the cantilever. In this case, frequency modulation AFM works well because the response time is on a time scale $1/f_0$ and not dependent on $Q$.[94] Operating tapping mode in liquid allows faster scan speed due to the low $Q$ factor.

Another important parameter is DS, which is used to convert the deflection signal in volts to the signal in nanometer, with the unit nm/V. This is not an inherent property of the cantilever, but it is related to the cantilever property and AFM design.

In most of the commercial AFM, optical level technique is used to detect the deflection. A laser beam from a laser diode is focused onto the cantilever and then reflected to the position sensitive photodetector (PSPD). The force applied to the cantilever $F$ leads to the bend of it with the change of the end slope $dZ_c/dx$. According to equation (11.13) and Hooke's law, the deflection of the cantilever can be expressed as

$$Z_c = \frac{F}{k} = \frac{4FL^3}{Ewt^3} \tag{11.13}$$

And the end slope of the cantilever can be expressed as[97]

$$\frac{dZ_c}{dx} = \frac{6FL^2}{Ewt^3} \tag{11.14}$$

This result implies that the vertical deflection is proportional to the end slope of the cantilever if the bending is small. The reflected laser beam on PSPD moves through an angle which equals to twice of the change of end slope $\alpha = \Delta(dZ_c/dx)$. Thus, the laser spot moves on the detector through a distance which can be written as[97]

$$\Delta\text{PSD} \approx 2d\tan\alpha = \frac{12FL^2d}{Ewt^3} \tag{11.15}$$

where $d$ is the distance between PSPD and laser reflecting point on the cantilever, shown in Figure 11.19b. Combined with equation (11.13), we can get $DS$

$$\text{DS} = \frac{Z_c}{\Delta PSD} = \frac{L}{3d} \tag{11.16}$$

It means the DS is proportional to the cantilever length and inversely proportional to the distance of the PSPD away from the laser reflecting point on the cantilever. In fact, the reflected beam is diverging as a light cone, not the ideal focused beam as discussed above, so if the distance $d$ becomes longer, the power of the laser per area at PSPD decreases. This effect cancels the influence of $d$ on DS.[98] The other thing that needs to be paid attention is that the cantilever length here is the effective length, which can be changed if the laser spot position changes.

Therefore, the conclusion is that the shorter the cantilever, the more sensitive it is and the smaller changes of the topography can be detected.

The cantilever is usually made of silicon or silicon nitride, and some of the cantilever will have backside coating, which is usually used to increase the reflectivity of the AFM laser so that images with good signal noise ratio can be obtained. The

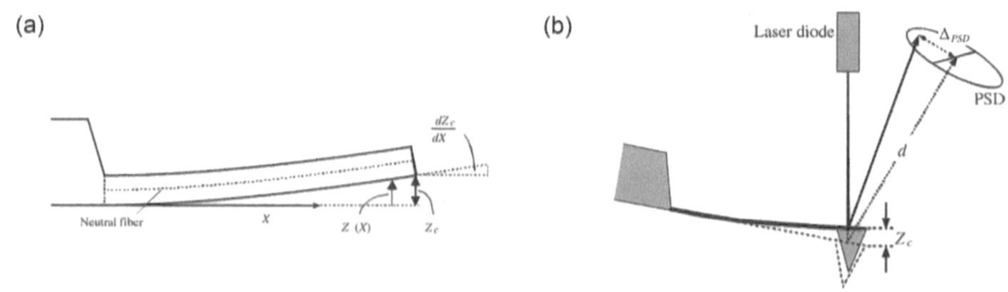


**FIGURE 11.19** (a) Schematic side view of a bending cantilever when a force applied on its end. $X$ is the horizontal coordinate originating at the basis of the cantilever, $Z(X)$ is the cantilever deflection at the position $X$, $Z_c$ is the cantilever deflection at its end. (b) Schematic of optical level to detect cantilever deflection. ΔPSD is the detector signal and d is the distance of PSPD away from the reflecting point on the cantilever. (Reprinted with permission from Butt et al.[97] Copyright 2005 Elsevier.)

backside coating could be Al, Au, Pt/Ir, Co/Cr, etc., all of them can increase the reflectivity, but the usage scenarios are different. For AFM imaging in liquid, one cannot use the cantilever which coating can react with the solution. For example, Al-coated cantilever cannot be used in aqueous solution. Poor backside coating will lead to some imaging quality problems. If the backside coating of the cantilever is not good, usually intensity of the laser signal detected by AFM photodetector will be lower than normal, and the AFM image could have more noise and interference pattern because more lasers go to the sample.

The key parameters for the AFM probe tip are geometrical parameters. In Figure 11.20, ten major tip geometries are shown. Different geometries are preferred for different applications. For example, a rotated AFM tip is used for symmetric imaging; a tipless tip is used for tip modification for force measurement purpose; a visible apex tip is used for precise positioning; a high aspect ratio tip is used for deep trench measurement; a super sharp tip is used for high resolution imaging.

The geometry of the tip is described as follows: tip height $h$, front angle (FA), back angle (BA), side angle (SA), tip setback (TSB), and tip radius $R$. The first five parameters are shown in Figure 11.21b, which should be considered to avoid artifacts from tip-sample convolution and in some contact mechanics models.

The end radius of the tip intensively affects the lateral resolution and needs to be carefully considered when doing high-resolution imaging. Usually, a sharper tip leads to better lateral resolution. For example, Figure 11.22 shows the same gold grains sample imaged by two probes with different end tip radii. However, in different applications the same probe may have different tip radii. We will discuss this situation in next Section 11.3.2.3.

The last thing that needs to be discussed about the AFM tip is tip coating. The front side tip coating is mainly used for specific applications. For example, a tip with conductive coating, like Pt/Ir, Pt, Co/Cr, etc., can be used for electrical measurement; a tip with magnetic coating, like Co/Cr, can be used for magnetic force microscopy (MFM).

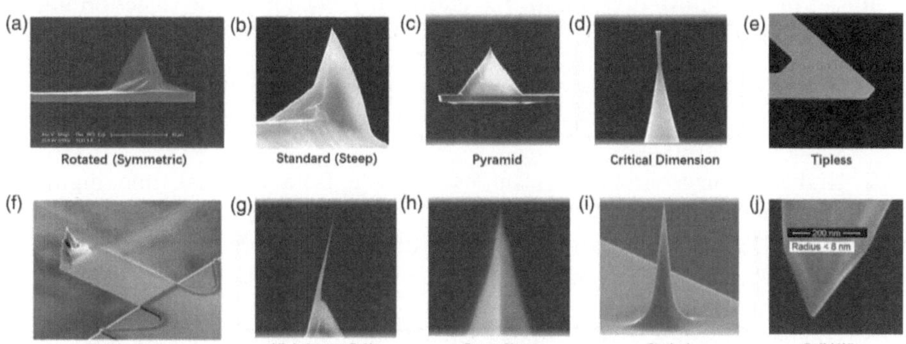


**FIGURE 11.20** AFM tip geometry. (a) Rotated (symmetric); (b) standard (steep); (c) pyramid; (d) critical dimension; (e) tipless; (f) visible apex; (g) high aspect ratio; (h) super sharp; (i) conical; (j) solid wire. (The pictures are from www.brukerafmprobes.com.)

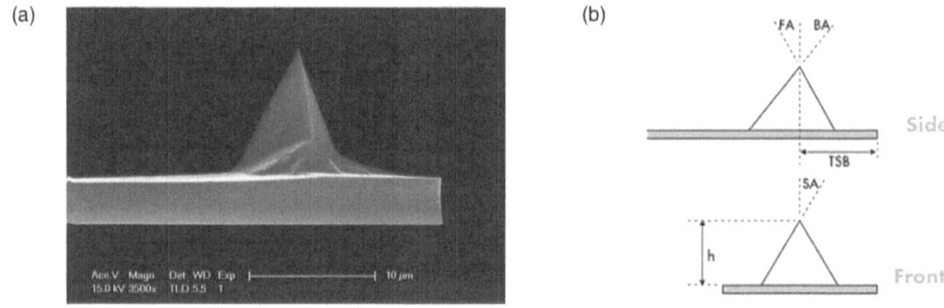


**FIGURE 11.21** AFM tip geometrical parameters. (a) SEM probe of a Bruker RTESPA-300 probe. (b) The tip schematic to illustrate tip height $h$, front angle (FA), back angle (BA), side angle (SA), and tip setback (TSB). (The pictures are from www.brukerafmprobes.com.)

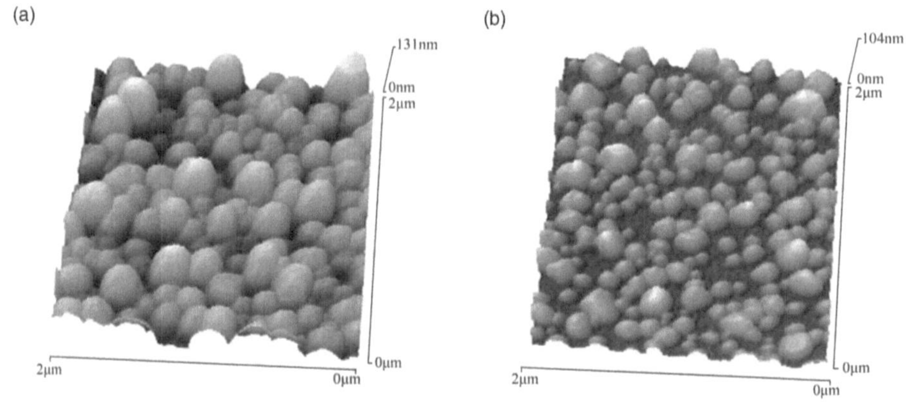


**FIGURE 11.22** The same gold grains sample imaged by (a) AFM probe with a normal tip and (b) AFM probe with a sharpened tip. The sharper tip can resolve more details of the sample surface.

**Based on the above discussion, the common rules for probe selection are listed below:**

1. Higher spring constant cantilever leads to larger interaction force between the tip and the sample when the deflection is constant: Softer samples prefer softer cantilevers.
2. Higher resonance frequency cantilever leads to faster response time (higher bandwidth): Fast scan applications prefer higher resonance cantilevers.
3. Lower DS leads to higher sensitivity to detect small deflection.
4. Backside coating improves cantilever reflection: High SNR applications prefer cantilever with backside coating.
5. Smaller tip radius leads to higher later resolution: High resolution imaging applications prefer small tip radius.

In the next section we will use these rules in different applications, take the particularity of the application into account, and add some additional rules for probe selection.

## 11.3.2 Optimal Probes for Specified Applications

### 11.3.2.1 Fast Imaging

In a typical AFM setup, as shown in Figure 11.23a, probe response (normally, it is cantilever's deflection) is compared with a given setpoint, and the error signal is transmitted to a proportional-integral-differential (PID) feedback loop to generate correction signal used to regulate drive voltage to control scanner's movement in the $z$ direction, as such to make the AFM tip track the sample surface. Electronic responding rate, scanner, and probe's bandwidth together determine the response of feedback loop, and as shown in Figure 11.23b, bandwidth of scanner and probe are bottleneck. On certain AFM systems with higher bandwidth scanner, one can select the proper probes to increase imaging speed.

As we discussed in Section 11.3.1, the key to increase scan rate without sacrificing imaging quality is to increase imaging bandwidth, which would decrease time constant, as such to reduce time of obtaining data at each sampling point. Therefore, probe with high resonance frequency ($f_0$) and low quality factor $Q$ is optimal probe for fast scanning application. To increase resonance frequency, according to equation (11.11), we can increase spring constant $k$ or decrease the effective mass of the cantilever $m$. Since imaging force is proportional to $k$, decreasing m is a better choice to make high $f_0$ probe because the imaging force can still be controlled at low level to reach better imaging quality. Consequently, the probes for fast imaging usually are in small size to reduce their mass. Increasing damping can decrease $Q$; solid

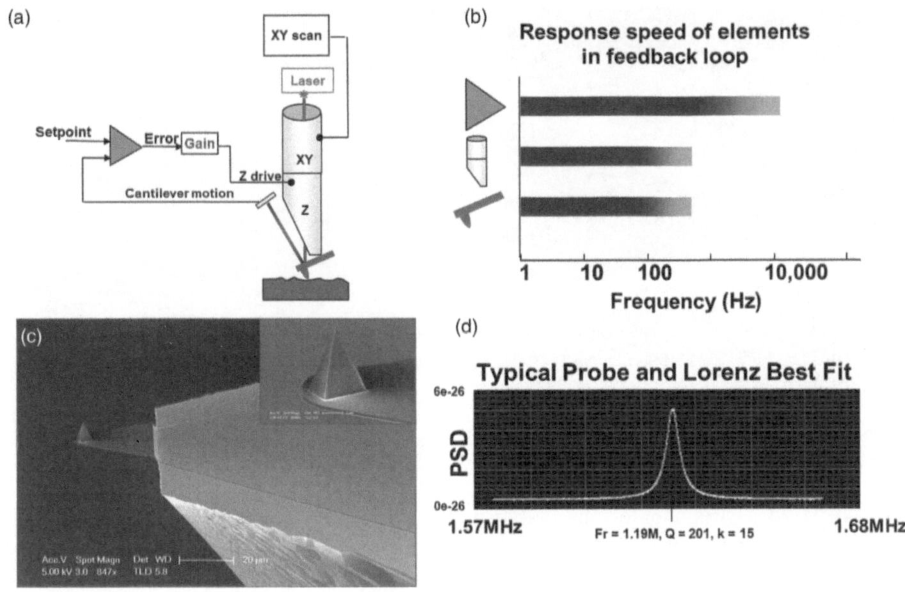


**FIGURE 11.23** (a) Schematic diagram of basic feedback loop of AFM system. (b) Schematic diagram of response speeds of electronic circuit, scanner, and probe. (c) Scanning electron microscopy (SEM) and (d) power spectral density (PSD) result of typical probe for fast scan purpose, FastScan-A from Bruker.

cantilever without hollow structure often is used to achieve this goal. Short, thin, solid triangular-shaped cantilever, as shown in Figure 11.23c and d, would balance $f_0$ and $Q$ to maximize probe's bandwidth so it can be qualified for fast imaging applications.

#### 11.3.2.2 Low Drift Fluid Imaging

Most of the cantilevers working in fluid have backside coating, to enhance the signal of reflected laser. The coating material generally is gold. The expansion coefficient difference between metal coating and cantilever itself causes the cantilever bending easily when the temperature changes and finally results in large thermal drift. A lot of work has been done to compensate the thermal drift to improve high-resolution imaging in fluid. Removing the backside coating can significantly reduce the thermal drift of cantilever,[99] although it leads to about 10-fold decrease of reflected laser single. Using cantilever without backside coating, some events happened on specific small areas of sample surface, which need high-precise positioning, like folding-unfolding transition of protein can be acquired successfully, which is failed to be done by normal cantilever.[100] One method for improving reflection signal is to add a reflection plate back to the position for laser reflection, shown in Figure 11.24a.[101] Another way is just keeping the gold coating where laser focuses on at the front of the cantilever and removing the coating in other areas.[101–105] The AFM probes with low-drift cantilever are now commercially available, like MLCT-Bio-DC(Bruker) or Uniqprobe (NanoSensors, NanoWorld).

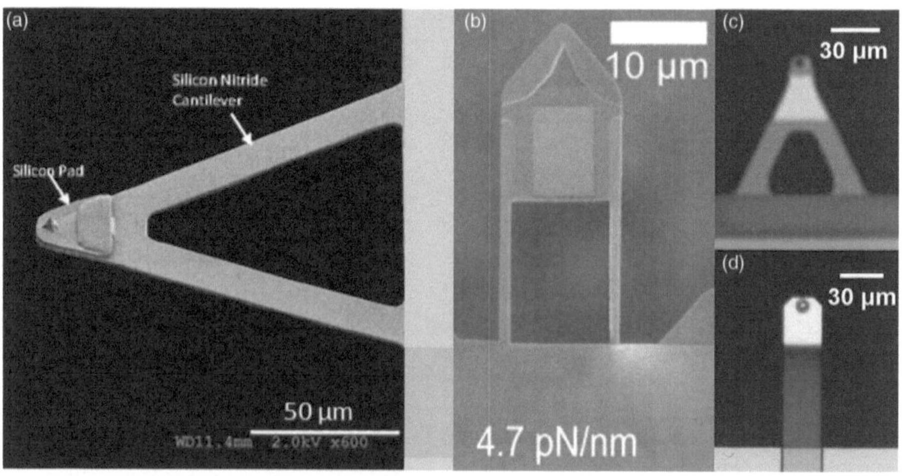


**FIGURE 11.24** Design for low thermal drift cantilevers. (a) A cantilever made by researcher. A silicon pad as a mirror for laser reflection. (b) A cantilever made by researcher. Modified cantilever to enhance reflection and decrease drift. (Reprinted with permission from Bull et al.[102] Copyright 2014 American Chemical Society.) (c) Optical image of commercial MLCT-Bio-DC probe used for low-drift fluid imaging from Bruker. Reflected gold layer is on front end of the cantilever. (The picture is from https://www.brukerafmprobes.com.) (d) Optical image of commercial Uniqprobe used for low-drift fluid imaging from NanoWorld. Reflected gold layer is on the front end of the cantilever. (The picture is from https://www.nanosensors.com.)

#### 11.3.2.3 Force Measurement

As we discussed above, if we do force measurement in force-curve based modes, modulus is calculated by the relationship between applied force and sample indentation; therefore, the key is to accurately measure the force $F$ and indentation $\delta$.

The deflection of cantilever is usually measured by optical lever, and AFM system use Hooker's Law, i.e., equation (11.9) to convert deflection to force. However, the deflection $D$ is measured in volts, as we discussed above, the DS is used to convert deflection in volts to nm, so equation (11.9) can be rewritten as

$$F\ (nN) = k\ \frac{N}{m}\ \times \text{DS}\ \frac{nm}{V}\ \times D \tag{11.17}$$

Equation (11.17) means under the same applied force, if the spring constant $k$ and DS are smaller, a larger deflection can be detected by AFM system; therefore, both smaller spring constant $k$ and DS result in better force sensitivity. According to equation (11.16), the shorter cantilever leads to smaller DS and thus better sensitivity to the deflection and finally to the force measurement. The conclusion is that to measure the force with better sensitivity, one should select the probe with smaller $k$ and shorter cantilever. In Section 11.2.2.1, we got the conclusion that the indentation depth error induced by defection sensitivity error could be amplified by a factor of $D/\delta$, based on this statement, seems smaller deflection is preferred, which is contradictory to the conclusion we just got. In force measurement, we need to consider both factors – the spring constant $k$ and DS must be small enough to make sure AFM can measure the accurate force with good sensitivity, and simultaneously, the spring constant $k$ also needs to be large enough to make the sample have enough indentation $\delta$ to reduce the indentation error. The empirical data is shown in Table 11.1. It is important to note that these data are only applied to the probe with several nm to tens nm end radius.

Equation (11.17) also explained that both the spring constant $k$ and DS should be calibrated before experiment for accurate force measurement. As the laser position on cantilever and cantilever position in probe holder may not be the same even using the same probe, DS should be calculated every time after remounting the probe or realigning the laser position. Generally, DS is calculated from the slope of force curve on a hard sample like sapphire. However, this operation may

**TABLE 11.1**
**Recommended Probes for Force Measurement**

| Young's Modulus | Commercial Probe Example | Suggested Spring Constant (N/m) |
|---|---|---|
| $1 < E < 20$ MPa | ScanAsyst-Air, SNL-A | 0.5 |
| $5 < E < 500$ MPa | RTESPA-150 | 5 |
| $200 < E < 2000$ MPa | RTESPA-300 | 40 |
| $1 < E < 20$ GPa | RTESPA-525 | 200 |
| $10 < E < 100$ GPa | DNISP-HS | 350 |

contaminate or break the tip. Spring constant could be calculated by thermal tune method for soft cantilever or Sader method for rectangular cantilever. These commonly used methods were discussed in other papers, and we will not cover them in this chapter.

A better way to get the spring constant and DS is to use laser Doppler vibrometer to get the accurate $k$ first, and then do "back calculation" to get DS. It is based on the following equations:

$$\frac{1}{2}kZ_1^2 = \beta\frac{1}{2}\,k_BT \tag{11.18}$$

$$Z_1 = \text{DS} \times \Delta\text{PSD} \tag{11.19}$$

where $Z_1$ is the deflection of first vibration mode due to thermal noise, $\beta$ is the correction factor, $k_B$ is the Boltzmann constant, and $T$ is the absolute temperature. Most of the modern AFM instruments have built-in feature for thermal noise calculation which could obtain spring constant from known DS and vice versa.

From equation (11.6), to measure the accurate Young's modulus with Hertz model (also applied to other models like DMT or JKR which need to consider tip radius), another key parameter tip radius needs to be taken into account. However, the definition of the radius should be clarified. Tip radius does not simply mean the nominal tip radius which usually can be found on probe package. A sphere tip has unique radius at any indentation. But for tip with other geometry, one probe may have different "radius". Figure 11.25 shows, for the cone tip, different indentation implies different effective tip radius. If the sample is soft, the same load force leads to larger indentation. Therefore, if one is using a non-sphere tip to do force measurement, effective tip radius should be calculated at sample indentation $\delta$.

To avoid fussy steps of probe calibration – spring constant, DS, and tip radius, pre-calibrated probes are preferred to do accurate force measurement. These probes usually have hemispheric end with controlled tip radius as well as calibrated spring constant. Bruker has some commercial pre-calibrated probes such as SAA-HPI-30, RTESPA-150-30, RTESPA-300-30, RTESPA-525-30. These individually pre-calibrated probes cover a range of spring constant from 0.25 N/s to 200 N/m with

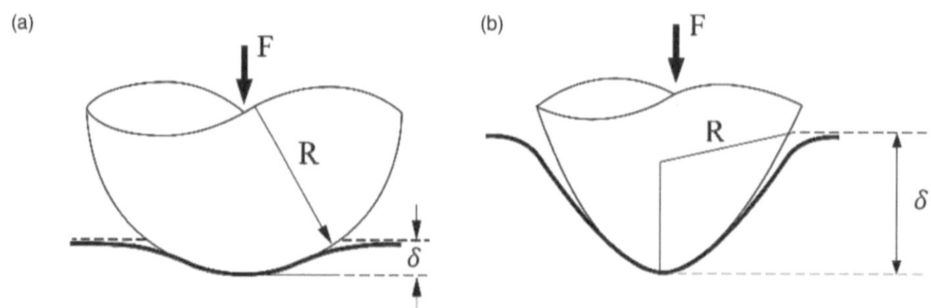


**FIGURE 11.25** The definition of the tip end radius. (a) A sphere tip has unique end radius $R$ at any indentation $\delta$. (b) A cone tip has different end radius at different indentation depth.

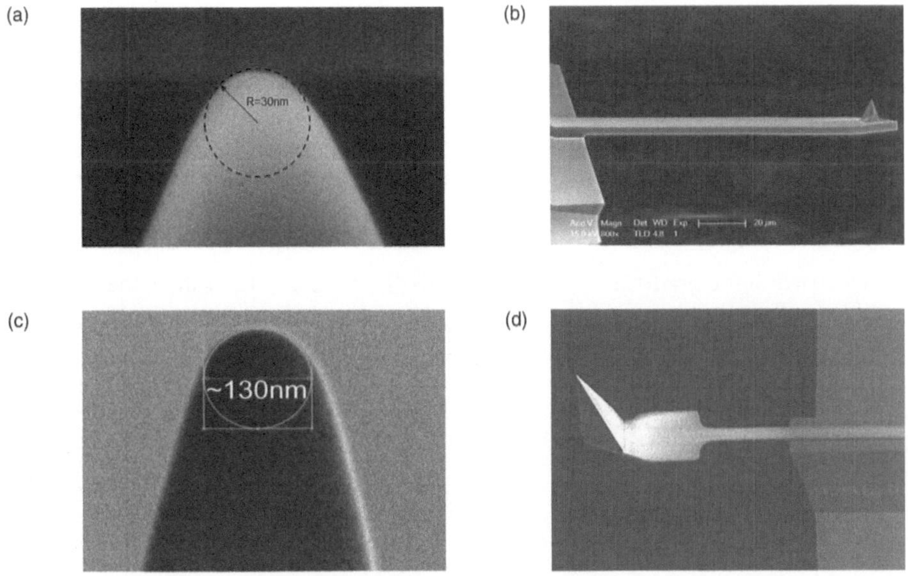


**FIGURE 11.26** SEM images of pre-calibrated probes. (a) and (b) RTESP-300-30, spring constant is 40 N/m and tip radius is 30 nm. (c) and (d) PFQNM-LC, spring constant is 0.1 N/m and tip radius is 65 nm. (There pictures are from www.brukerafmprobes.com.)

30 nm tip radius. There is also a pre-calibrated probe called PFQNM-LC which is optimized for living cell imaging and has 17 μm tip height and 65 nm tip radius. These probes are shown in Figure 11.26.

### 11.3.2.4 Electrical Measurement

To do electrical measurement, the basic requirement for the probe is that it should be conductive to allow bias to be applied and current to pass through. The most common conductive probe is silicon or silicon nitride probe with conductive coating, like Pt, Ir, and doped diamond. The coating increased conductivity, but the tradeoff is that the tip radius is also increased to usually tens of nm. As we discussed above, this will usually decrease the imaging resolution. Some special probes with highly conductive apex sharp boron-doped diamond coating tip can reach <5 nm tip radius by unique patented process, which can be used for high-resolution imaging in electrical modes. Another kind of conductive probe is solid metal probe made by Rocky Mountain Nanotechnology (RMN). RMN probes are constructed from pure platinum and placed on a standard AFM probe sized ceramic substrate. This kind of probe offers great conductivity and does not suffer thin-film adhesion problem which occurred with metal-coated probe. It can reach <10 nm tip radius for high-resolution applications.

EFM usually operates in lift mode. The main scanline is usually based on tapping mode, so the selected probe should first meet the requirements of tapping mode we discussed before. EFM measures electrical signal in interleave scanline, the probe lifts to tens to hundreds of nanometers, so it is difficult to reach <10 nm resolution as topography imaging in the EFM channel. Consequently, the effect of the tip radius on the spatial resolution is negligible compared with the effect of the lift height. Therefore, conductive probes with metal coating are often used for EFM measurement. In normal operation, the EFM phase signal is proportional to *Q/k*, so the probe with smaller spring constant has higher EFM resolution.

KPFM resolution has been analyzed before.[106] For amplitude modulation (AM) KPFM, the system tries to null electrical force $F(z)$ at driven frequency between the tip and the sample, while for frequency modulation (FM) KPFM, the system nulls the electrical force gradient $\frac{\partial F(z)}{\partial z}$. Equations (11.20) and (11.21) show the electrical force and force gradient between the tip with radius $R$ and an infinite flat sample, with a distance $z$.

$$F(z) = -\pi\epsilon_0 \frac{R^2}{z(z+R)} \Delta V^2 \tag{11.20}$$

$$\frac{\partial F(z)}{\partial z} = -\pi\epsilon_0 \left(\frac{1}{z} + \frac{1}{z+R}\right) \frac{R^2}{z(z+R)} \Delta V^2 \tag{11.21}$$

where $\epsilon_0$ is the dielectric constant of the free space and $\Delta V$ is the potential difference between the tip and the sample. In AM-KPFM, the KPFM signal is inversely proportional to the square of the distance between the probe and the sample, while in FM-KPFM it is inversely proportional to the third power of the distance between the probe and the sample. In Figure 11.27, a simple model is used to show that using the same probe with the same lift height, FM-KPFM has better resolution than AM-KPFM. In practice, it is more complex. Loppacher et al. mentioned that if the tip geometry matches the feature of the sample, high-resolution signal can be obtained.[107] Because EFM and KPFM only require potential applied, the highly doped silicon probe without coating also works well.

For electrical modes operating in contact mode, like cAFM, scanning capacitance microscopy (SCM), scanning spreading resistance microscopy (SSRM), and PFM, good electrical signal relies on solid contact between the tip and the sample. Wear resistance of the probe must be considered in this situation. Due to the wearing, the probe with metal coating will decrease the resolution because of the increase of tip radius and loss conductivity. Probes with solid boron-doped polycrystalline diamond coating or platinum silicide coating and RMN probes without coating are preferred for these applications.

#### 11.3.2.5 AFM-IR

Contact mode and tapping mode are widely used in AFM-IR application. Recently, the peak force tapping mode is increasingly being applied. Besides for choosing the appropriate resonance frequency ($f_0$) and quality factor ($Q$) according to the different modes, the key point to select the optimal probes for AFM-IR application is the gold coating with top-down method. The common AFM probes are composed of Si and $Si_xN_y$, and they will absorb IR light in specific wave number, which will affect the spectral test. Another reason for using gold coating is that when testing very thin

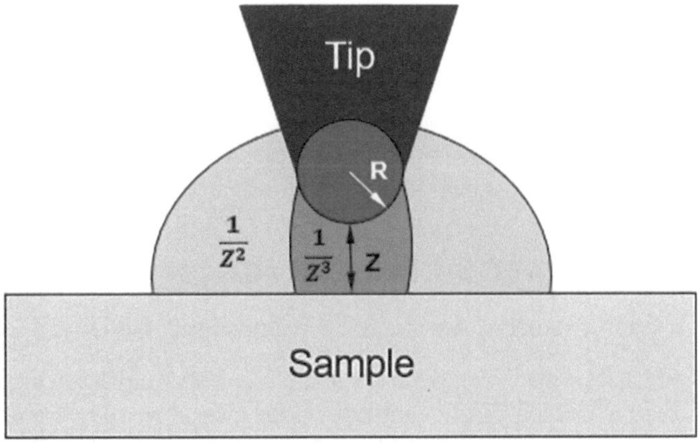


AM-KPFM $$F_{el}(z) = -\frac{1}{2}\frac{\partial C}{\partial Z}\Delta V^2 = -\pi\epsilon\left(\frac{R^2}{z(z+R)}\right)\Delta V^2 \propto \frac{1}{z^2}$$

FM-KPFM $$\frac{\partial F_{el}}{\partial z} = \left(\frac{1}{z} + \frac{1}{z+R}\right)F_{el}(z) \propto \frac{1}{z^3}$$

**FIGURE 11.27** A simple model to study spatial resolution of AM-KPFM and FM-KPFM. FM-KPFM signal is proportional to $1/z^3$ which is more localized than AM-KPFM $1/z^2$, so FM-KPFM has better resolution.

samples, the local electric field enhancement effect (lightning-rod effect) can be utilized to enhance the relatively weak IR signal on gold substrate. For the bottom-up method, the common probes without gold coating can be used because the probe and IR light are on both sides of the sample.

When researchers explore new probe types, they should also pay attention to observe whether the spectrum obtained by AFM-IR is consistent with that of Fourier Transform Infrared (FTIR). Some well-studied samples, such as polymethyl methacrylate (PMMA) or polystyrene (PS) films, can be used for confirmatory testing. Previous work proved that uneven spectral baselines may occur if probe is not selected properly. To avoid the above problem, the following probes are recommended for AFM-IR testing. PR-EX-nIR2-10/PR-EX-CNIR-B-10 probes with gold coating, spring constant ~0.2 N/m, resonance frequency ~13 kHz, and tip radius ~25 nm can be used in contact AFM-IR mode. PR-EX-TnIR-A-10 probes with gold coating, spring constant ~3 N/m, resonance frequency ~75 kHz, and tip radius ~25 nm and PR-EX-TnIR-C-10/PR-EX-TnIR-D-10 probes with gold coating, spring constant ~40 N/m, resonance frequency ~300 kHz, and tip radius ~25 nm can be used in tapping AFM-IR mode. The major difference is the spring constant which affects the imaging force. If the samples are made by cryo-microtomy, contact mode is generally used, and probes with low spring constant should be selected. For relatively hard samples, PR-EX-TnIR-A-10 probes also can be used, because ~3 N/m spring constant is not too high. If the sample is granular or a soft film, tapping mode is preferred to avoid sample damage during imaging, and probes with relative high spring constant should select to overcome the adhesion force during tapping. For the bottom-up method, probes without gold coating are allowed, because the probe and IR light are on two different sides of the sample. PR-EX-C450-10/PR-EX-CSIN-10 probes are commonly used, with no coating, spring constant ~0.2 N/m, resonance frequency ~13 kHz, and tip radius ~10 nm.

## 11.4 AFM ARTIFACTS RECOGNITION

### 11.4.1 Common Artifacts in Topography Imaging

AFM uses tip to scan over the sample surface and laser beam to detect the deflection or vibration of the cantilever. The laser, tip, and sample often introduce some artifacts. Researchers should recognize these artifacts and know how to avoid them.

The first common artifact in AFM imaging is the optical interference. The appearance is sinusoidal-like pattern on the image with a period typically 1.5–2.5 μm, shown in Figure 11.28a, while Figure 11.28b shows a normal image. The phenomenon is caused by the interference between the incident and reflected laser from sample surface. If the sample surface is highly reflective, the laser is at a bad position as shown in Figure 11.28c, or backside coating of the cantilever is none or poor, the interference artifact can happen. This artifact is more often seen in contact mode and PFT mode than tapping mode because tapping mode is operated in high frequency and the internal lock-in amplifier can improve the situation.

The optical interference artifact can be reduced by laser realignment to a good position as shown in Figure 11.28d or using cantilever with good reflective backside coating. It is also suggested to do interference check before AFM imaging. The steps are as follows: (1) Engage the tip on to the surface with normal settings; (2) Set feedback gains to zero to turn off the feedback; (3) Manually lift the tip for about 1 μm; (4) Use typical scan size to scan above the sample surface and monitor if there is

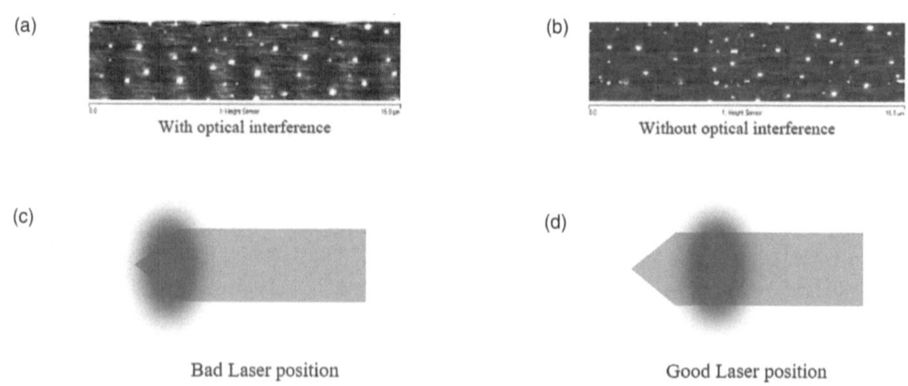


**FIGURE 11.28** (a) AFM image with optical interference. Sinusoidal-like pattern can be observed in the image. (b) AFM image without optical interference. (c) Bad laser position allows more laser to go to sample surface and leads to optical interference. (d) Good laser position. Most of the laser is reflected by cantilever.

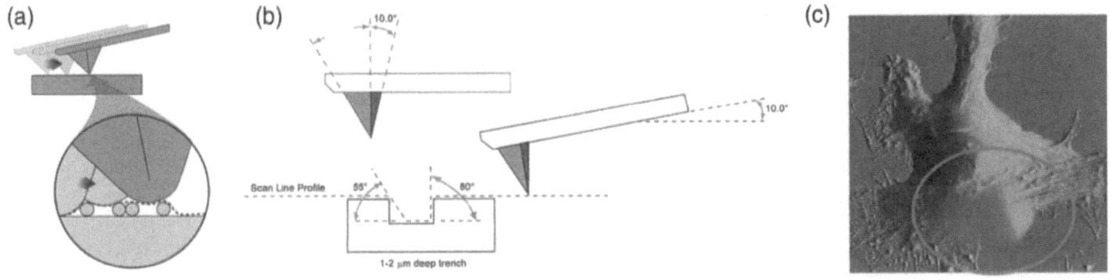


**FIGURE 11.29** (a) Effect of tip radius. Probe with large tip radius cannot resolve small features. Need to select probe with proper tip radius to make sure the sample features dominated in the final image. (b) When measuring deep trench, the sidewall angle of the tip should be considered. The convolution of the tip and the sample makes the slope of the trench wrong. Need to use high-aspect ratio probe for this application. (c) If the sample is very tall or the sample with higher slope features than the slope of tip sidewall, an almost linear ramp-like artifact will happen around the feature. This is because the sidewall of tip directly interacts with sample instead of the tip apex. The probe with proper tip sidewall slope needs to be considered for this application.

interference in image channels. If interference happens, realign the laser or change another probe. The criterion for a good laser position is that, for a short and wide cantilever, the laser should be focused on the free end of the cantilever and not protrude out the cantilever. For a long and thin cantilever, sometimes it is better to focus the laser 1/3 to 1/2 away from the free end of the cantilever to avoid interference.

The second common artifact is the tip-sample convolution. Every AFM image is the result of convolution of the tip and the sample. The geometry of the tip plays important role in image quality. The effects of tip radius and tip sidewall angles are shown in Figure 11.29. The radius issue has been discussed above, and operator needs to select probe with proper tip radius to make sure the sample features are dominated in the final image. For tip sidewall angle issue, which causes wrong slope when measuring deep trench and very tall or high slope samples, special probes like high-aspect ratio probes are recommended for this application.

If the tip goes bad because of tip contamination, tip breaking, or tip wearing, the tip shape and size also change, and it finally leads to some artifacts, shown in Figure 11.30. Figure 11.30a shows a typical AFM image captured using a dull or dirty tip with debris attached. The characteristic of the artifact is all the features in the

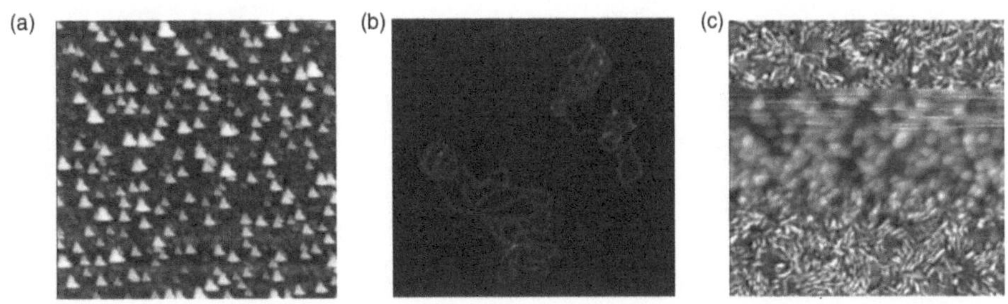


**FIGURE 11.30** (a) Effects of a dull or dirty tip. (b) DNA image with double tip artifact. (c) A typical artifact caused by tip contamination.

image have the same shape. Figure 11.30b shows double tip effect in a DNA image. Each DNA strand is doubled in this image, like a shadow. This effect is resulted from tip breaking or contamination. Unexpected, jagged points near the real tip also contact the sample and form an image. Figure 11.30c shows a typical artifact caused by tip contamination. The tip picked up loose debris during scanning so that the high-resolution image was gone, and instead, the small features are represented as larger rounded ones as well as streaking is produced. After the debris were swept out during the scanning, the clean, high-resolution image came back. Tip wearing often occurs when the probe is used for a long time, or the imaging force is not controlled well. As a result, tip radius becomes larger and leads to poor resolution. A sign of tip wearing is the roughness of the same area of sample surface which becomes smaller and smaller under the same imaging conditions.

To reduce or eliminate the artifacts caused by tip-sample convolution, the simplest and direct way is to replace a new probe with proper parameters. If the artifacts are caused by tip contamination, especially for organic contamination, wash the probes in 1%–5% sodium dodecyl sulfate (SDS) solution or use ultraviolet (UV) lamp to irradiate the probe and probe holder for ~2 minutes at full intensity to remove the contamination.

To distinguish the tip image and the "true" image, a simple way is to rotate the sample for 90° and image it again. The real features will also rotate with sample rotation, but artifacts do not.

The next common artifact is sample distortion caused by drifting, shown in Figure 11.1. We have discussed it in Section 11.2.1. The way to reduce or eliminate this artifact is to fix the sample firmly on the surface, make the surface clean, and control the thermal drift of system well.

Another common artifact is that the tip does not track the sample surface well, shown in Figure 11.31. This artifact is caused by bad tracking and can be easily recognized. To eliminate it, operator can adjust setpoint to increase the imaging force, reduce the scan rate, or increase gain to speed up the response of the z piezo, till the trace and retrace lines overlay.

Other artifacts in topography image such as tilt and bow introduced by scanner characteristics can be removed offline, which will be discussed in data processing section.

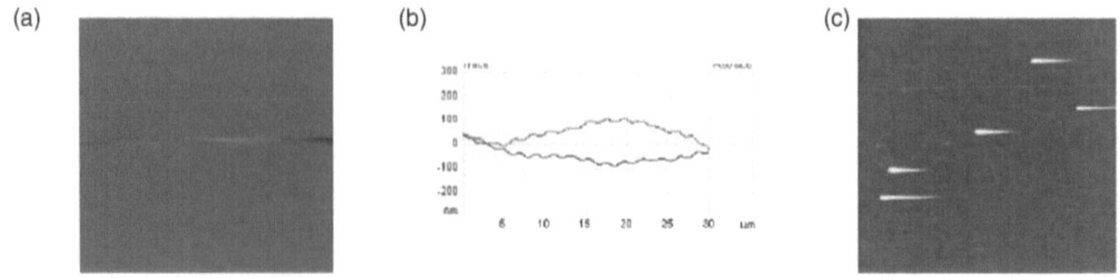


**FIGURE 11.31** (a) Height channel captured in contact mode, showed that tip does not track the surface and surface topography cannot be obtained. (b) The corresponding trace and retrace lines to the image in (a), trace and retrace lines do not overlay. (c) Tip does not track the surface well and streaks occur on the trailing edge of the features.

All above artifacts can happen in contact mode, tapping mode, and PFT mode. However, in different imaging modes there are some unique artifacts. A typical artifact in tapping mode is the "rings around the features".

Tapping mode uses cantilever amplitude as the feedback signal. In this mode, the cantilever is oscillating near its resonance frequency and intermittent touching with the sample surface. Tapping mode can work well in both attractive regime (low amplitude solution) and repulsive regime (high amplitude solution). The "bi-stable regime" results from the nonlinear characteristics of both attractive and repulsive force.[108–111] The "rings" artifacts are caused by "bi-stable regime". In Figure 11.32, it illustrates an ambiguity in the operation of tapping mode, because at a certain amplitude setpoint, there may be two different z values from two different regimes, leading to the artifacts. Figure 32a and b show the amplitude vs. z and phase vs. *z*, respectively. If the selected amplitude setpoint is in the bi-stable regime, from Figure 11.32a we can see the same amplitude is corresponding to two *z* positions. And from Figure 11.32b we can see the same z position corresponds to two different phase signals. In practice, one can get topography and phase images with artifacts like Figure 11.32c and d.

The regime is highly dependent on the operating frequency, and slightly adjusting the drive frequency can help cantilever work in only one regime and get stable image. Usually in repulsive regime, the system often gives more details and sharper image. It is because when operation in repulsive regime, the tip senses short-range repulsive interactions; on a rough sample surface, the tip-sample separation is the same at peaks and valleys; the tip can penetrate surface water layer, and image

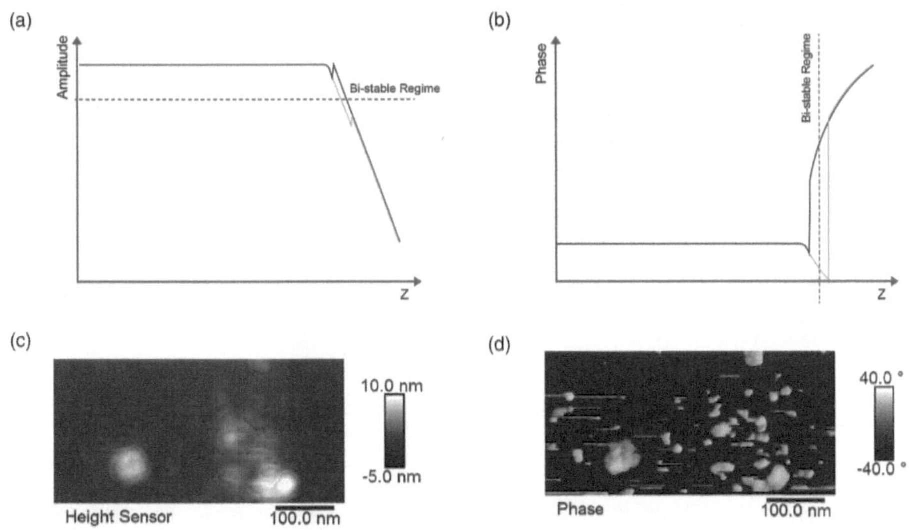


**FIGURE 11.32** Bi-stable regime in tapping mode. (a) Amplitude vs. *z* in tapping mode AFM. If selected amplitude setpoint (dash line) is in bi-stable regime, one amplitude corresponds two different *z* positions. (b) Phase vs. *z* in tapping mode. If operating in bi-stable regime, one *z* position corresponds two different phase signals. (c) Topography and (d) Phase images when operating in bi-stable regime.

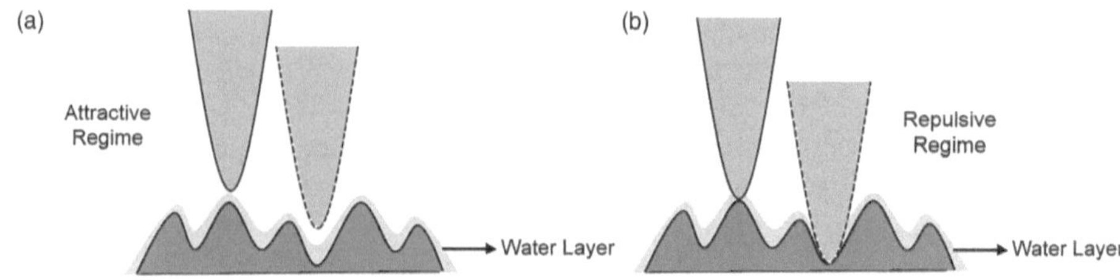


**FIGURE 11.33** Two operation regimes of tapping mode AFM. (a) Tip is operated in attractive regime. In this regime, tip senses the attractive force, which is long range interaction. Tip-sample distance at peaks is shorter than in valleys, so operation is less stable. Tip usually cannot penetrate surface water layer and phase signal does not relate to real surface properties. (b) Tip is operated in repulsive regime. In this regime, tip senses the repulsive force, which is short range interaction. Tip-sample distance is same at peaks and valleys, so operation is more stable. Tip can penetrate surface water layer, and image on solid surface. And phase signal relates to surface properties.

on solid surface, which makes the imaging more stable. If operating in attractive regime, because the tip cannot penetrate the surface water layer, the features often look larger, and the phase image often has less contrast. These two different operation regimes are shown in Figure 11.33.

If sample surface is charged, artifacts will also be introduced to topography image due to the additional electrical force between the tip and the sample, especially in tapping mode. In tapping mode AFM, the cantilever of the probe oscillates near its resonant frequency with the amplitude controlled by amplitude setpoint during imaging. If sample surface is electrically charged, the probe will oscillate in a new force field, and additional electrical force gradient leads to the change of spring constant of cantilever, thus changing the resonance frequency of the cantilever, according to equation (11.11). Figure 11.34a shows the amplitude change with the resonance frequency change. The brown line indicates the amplitude $A$ vs. drive frequency $\omega$ without electrical force, and the blue line indicates the amplitude $A$ vs. drive frequency $\omega$ affecting by additional repulsive electrical force. The repulsive force causes the

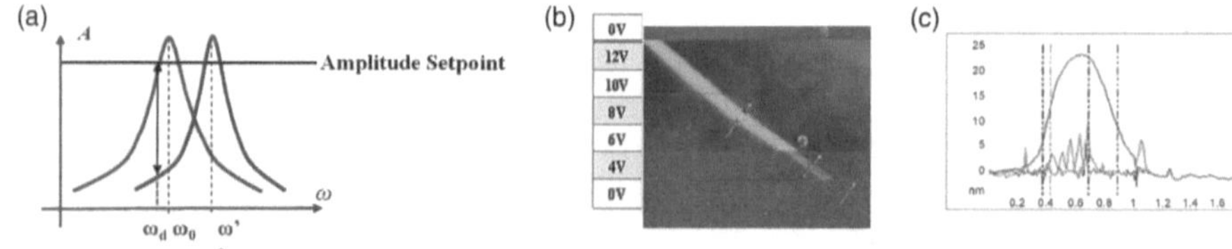


**FIGURE 11.34** (a) Tip-sample interaction will lead to resonance frequency change of the cantilever in tapping mode, at the same drive frequency, the amplitude also changes, and feedback system adjusts the $z$ piezo to recover the amplitude to setpoint, thus AFM records a pseudo height value. (b) AFM topography image captured at different tip bias under tapping mode. Different apparent heights of SWNTs on $SiO_x$/Si substrate were observed at different tip bias. The left table gives the tip bias value we used during scanning. (c) Cross-section profiles indicate the change of apparent height of SWNTs with tip bias.

resonance frequency increases from $\omega_0$ to $\omega_0'$, and if the system still drives the cantilever at $\omega_d$ near $\omega_0$, in this case, the cantilever amplitude decreases, the feedback system moves the z scanner to recover the amplitude to the given amplitude setpoint; therefore, artificial height is introduced to the topography image. Because the drive frequency is near the resonance frequency, no matter repulsive or attractive force is introduced, only in small frequency shift range the amplitude increases, in most cases, the amplitude decreases. It means, under common conditions, the measured height is usually higher than real height if the electrical force is involved in tapping mode. Figure 11.34b and c are AFM topography image and corresponding cross-section profiles. The image was captured at different tip bias in tapping mode, to simulate additional electrical force between the tip and the sample. The sample is single-walled carbon nanotubes (SWNTs) on $SiO_x$/Si substrate, which is typically with 1–2 nm height. The apparent height of SWNTs became higher than their real height and even reached to over 20 nm with high tip-sample electrical interaction. It indicates that the electrical force dramatically affects the measured height, and this artifact needs to be paid attention in tapping mode imaging.

This kind of artifact usually happens on conductive samples prepared on insulator substrates. Some electrical charges accumulate on the sample surface as time goes on, and the electrical charges cannot move freely. To avoid this artifact, static electricity eliminators, like static master and anti-static gun, are often used to remove the charges on sample surface before AFM imaging. In PFT mode, if the electrical force between the tip and the sample is not too high, the background subtraction mechanism can remove the electrical interaction, but if the interaction is too strong, the force curve will become unstable, which leads to challenges to get good AFM image.

### 11.4.2 Artifacts in Force Measurement

In force measurement, we usually focus on modulus and adhesion of sample. Adhesion force is highly dependent on the contact area between the tip and the sample. If one needs to compare adhesion force on two different areas of the sample, the contact areas must keep the same. However, it is very difficult to meet this requirement during adhesion mapping, because in most of the force measurement modes, for example, PFQNM, QI mode, Force Volume, system maintains a force setpoint or trigger force to control the z piezo movement to generate force curves at different imaging pixels as mechanical properties calculation. If sample surface is not homogeneous, the contact area will not maintain the same either. Thus, the artifacts will go to adhesion channel. Modulus calculation is highly dependent on the contact mechanism models; if indentation depth changes, effective radius and contact area may also change, and therefore, the artifacts will go to modulus channel.

In most of force measurement modes, modulus is calculated using certain contact mechanism model in real time for the whole modulus image. Therefore, heterogeneity of the sample may lead to artifacts in mechanical channels. Many composite materials have domains with large differences moduli. For example, as shown in Figure 11.35, the tip cannot indent hard matrix of the sample much, the indentation is typically a few nanometers, which matches the condition that indentation $\delta$ is far less than tip radius $R_T$ and Hertz or DMT model is usually used to calculate the modulus.

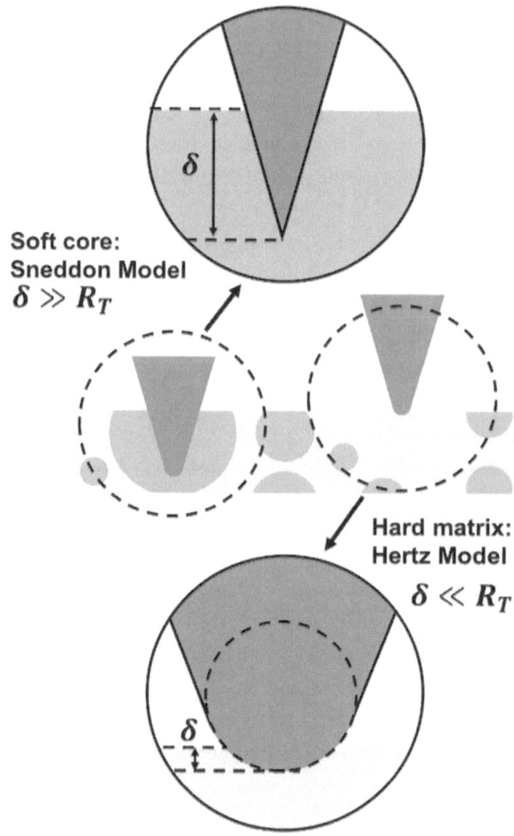


**FIGURE 11.35** Multi-phase sample may need different models to illustrate the modulus of different domains.

While the tip can indent deeply in soft core areas of the sample where the indentation can reach to tens or hundreds of nanometers, in this situation, Sneddon model is more suitable for modulus calculation. However, during imaging, usually only one pre-selected model can be used, which means accurate modulus cannot be obtained at two different areas simultaneously. Special offline data processing is needed to deal with this situation by setting a threshold of indentation depth. Indentation depth below and above this threshold will be fitted by different models. If threshold is hard to be defined because of the smaller differences of modulus, a cone-sphere combined model is recommended.[112] Different indentation depths also affect adhesion comparison because the contact areas are also different.

Surface fluctuation also leads to artifacts in mechanical channels. According to equations (11.1), (11.2) and (11.5), the indention depth $\delta$, contact area $a$, and Hertz/DMT modulus $E^*$ are all related to the effective radius $R^*$, which is dependent on both tip radius and sample surface radius. If Hertz/DMT model is chosen by modulus calculation, the measured fluctuation of modulus of sample might be artifact. An example is shown in Figure 11.36, imaging a mechanically homogenous sample is

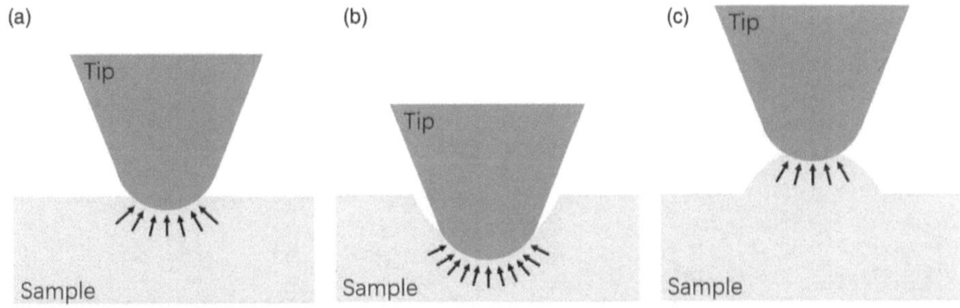


**FIGURE 11.36** AFM tip indent on (a) flat area, (b) concave area, and (c) convex area. The different surface radius and contact areas will lead to different measured modulus and adhesion.

under force imaging by AFM. Ideal situation is shown in Figure 11.36a, a sphere-like tip indents into a flatten surface. According to equation (11.3), in this area the surface radius is infinite, so effective radius $R^*$ equals to tip radius $R_T$. However, when the tip indents on a convex area, $R_S$ is positive, so $R^*$ is less than $R_T$ and measured modulus will be larger according to equation (11.5); even the convex area has the same modulus with flat area. On the other hand, at concave area the measured modulus will be smaller than the real modulus because surface radius is negative. Another explanation is that, in convex area the contact area will be smaller, but at the same imaging force, the pressure is higher so the indentation depth will be larger, which will make the area look softer than the flat area, while concave area looks stiffer than flat area. The measured adhesion is smaller in convex area and larger in concave area due to the different contact areas.

Similar situation also happens on the edge of measured materials. At the edge of the materials, due to the feature transitions and modulus change, the contact area is difficult to be defined. At these kinds of areas, the modulus image is usually pretty but is artifact, which is shown in Figure 11.37a. If the sample is tilted, the angle also causes problem for the contact areas and finally introduces artifacts, shown in Figure 11.37b.

Another critical artifact is caused by contamination. Sometimes sudden change of adhesion force or modulus can be observed in mechanical channels, as shown in

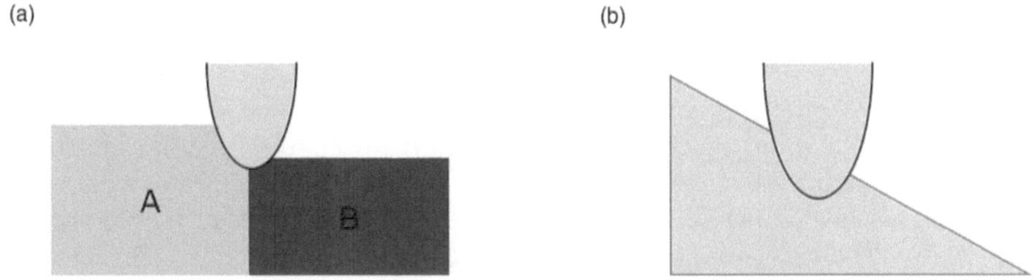


**FIGURE 11.37** (a) Force measurement at edge of materials. The contact area is difficult to be defined. (b) Force measurement on a tilt sample. The contact area also has problem.

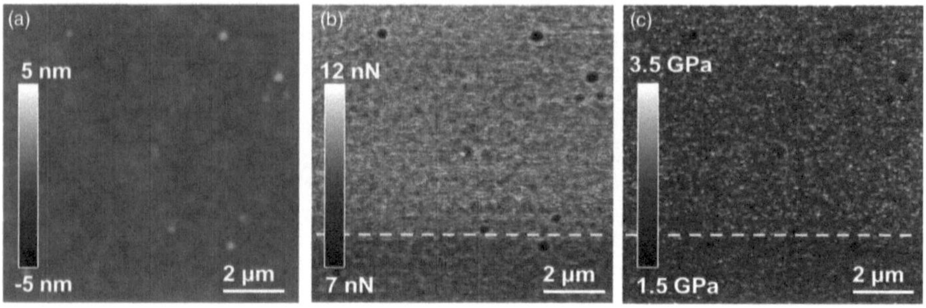


**FIGURE 11.38** Artifact caused by contamination. Images were acquired in PFQNM mode. (a) AFM topography image. (b) Adhesion image. (c) Modulus image.

Figure 11.38. The jumping of adhesion or modulus is because the contamination suddenly attaches onto the tip or just falls off. To avoid this kind of artifact, a clean tip on a clean surface for mechanical characterization is extremely important.

Other factors related to force imaging principle can also introduce artifacts. Here we take PFQNM mode as an example, which is based on PFT mode, to investigate the reason why there are artifacts in mechanical property channels. PFT mode is a breakthrough new AFM imaging mode introduced by Bruker in 2009,[113,114] which uses a peak force as the feedback signal. In this mode, a system uses a sinusoidal piezo motion to control the probe position, brings the probe intermittent contact with the sample surface at a frequency far away from the cantilever resonance frequency. In fact, the system performs a very fast force curve at every pixel in the image by modulating the z piezo, which can be used to calculate the mechanical properties of samples.

In a PFT cycle, not all deflection presents tip-sample interaction force; a system must have the ability to distinguish the deflection caused by real tip-sample interaction and parasitic deflections, such as free oscillation after cantilever snapping off the surface, the deflection triggered by harmonics of the piezo motion, deflection signal change caused by laser interference, and deflection caused by viscous forces, especially operating PFT mode in fluid. To decouple the real tip-sample interaction from the big background, the "background subtraction" technique is applied to the PFT cycle.[114] Multiple deflection data points are gathered when the tip is not interacting with the sample surface and used to generate an average baseline level, which includes the periodic parasitic deflections. Subtraction of the baseline is then applied to the raw deflection signal, to decrease the minimum controllable force from μN range to pN range. Figure 11.39 illustrates the background subtraction process in PFT mode.

In PFT mode, interference can happen in both topography and force channels; the latter will lead to wrong mechanical properties. Figure 11.40 shows PFT images of PS surface, which is quite uniform, but sinusoidal-like pattern appears in adhesion force image, shown in Figure 11.40b. The most probable reason is the interference of laser caused by a bad laser position we discussed in previous sections. This artifact also happens in other channels, like Young's modulus channel. Here we use adhesion image to understand why laser interference can cause artifact in mechanical channels.

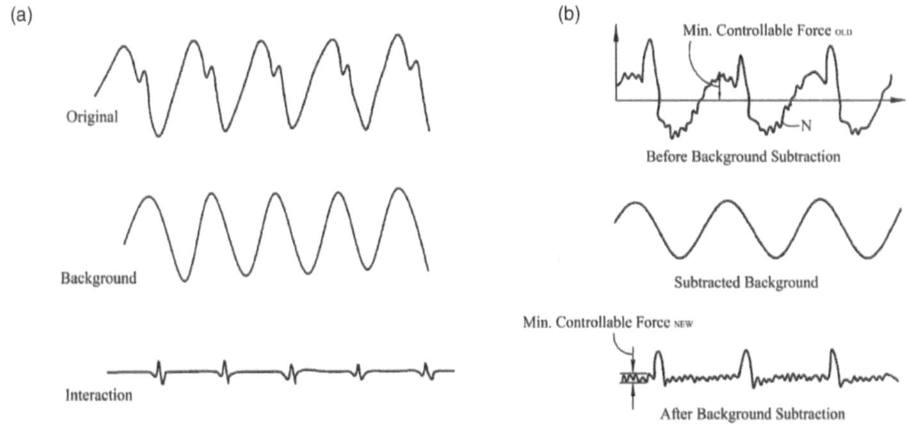


**FIGURE 11.39** Background subtraction process in PFT mode. (a) A theoretical model for background subtraction. The top is the original deflection signal with parasitic deflection. The real tip-sample interaction is buried within the background. The middle is the generated background. The bottom is the real interaction force which can be used for feedback control. (b) An experimental data applies background subtraction. Subtraction of the parasitic background from the deflection signal thereby lowering the minimum controllable force from μN range to pN range.

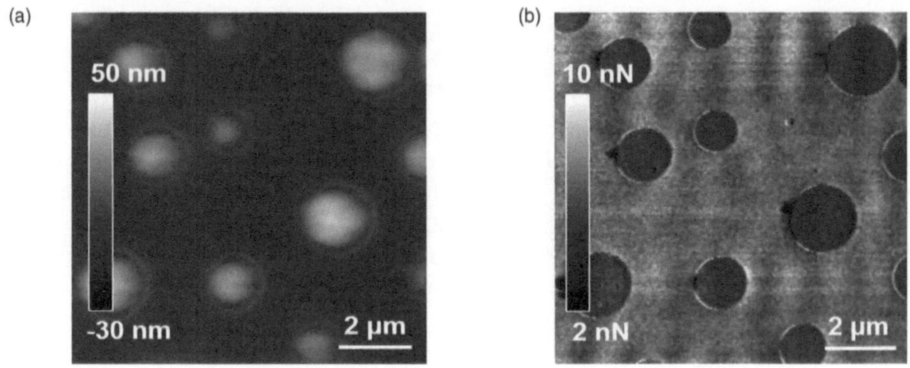


**FIGURE 11.40** (a) Topography image. (b) Adhesion force mapping shows obvious sinusoidal-like pattern caused by optical interference.

For normal force curve shown in Figure 11.41a, actual adhesion force is the force between minimum force and baseline. In normal operation, the periodic component in parasitic deflection only comes from cantilever ramping at driving frequency; after background subtraction, the linear fitted curve of baseline is just the real baseline, and the measured adhesion force is equal to the real adhesion force, as shown in Figure 11.41b. Optical interference introduces other frequencies, and the fitted baseline cannot be the real baseline because background subtraction cannot use a simple model to simulate the background. Therefore, the measured adhesion force may

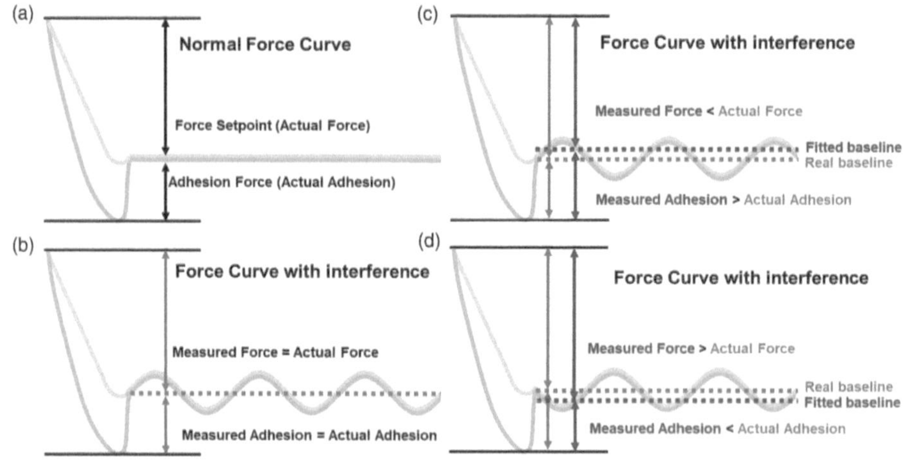


**FIGURE 11.41** Scheme of normal force curve and force curve with interference.

be lower or higher than real adhesion force, shown in Figure 11.41c and d, respectively. It also affects the imaging force because of the wrong baseline. As shown in Figure 11.41c, the measured peak force at setpoint is smaller than actual force, which will make z piezo to travel further distance downward, leading to lower height, larger indentation, and smaller modulus.

### 11.4.3 Artifacts in Electrical Measurement

Two commonly used electrical modes are EFM and KPFM. In a normal setup, EFM and KPFM work under lift mode. In lift mode, AFM system first performs a standard topography scan with the z feedback control enabled, and then the z feedback is turned off and the AFM tip is lifted off from the sample surface at user-defined height for electrical properties measurement. In this mode, a remarkable artifact is the crosstalk of topography and electrical signals; if the lift height is not large enough during the electrical mapping, the tip still has non-electrical interaction with sample surface. Figure 11.42 shows a typical topography-potential crosstalk artifact of a brush polymer sample on mica substrate measured in PeakForce KPFM (PFKPFM) mode. With the lift height increases by small steps, the potential contrast always changes, and finally an abrupt change of potential contrast is observed at ~27 nm lift height, then almost no change in potential with small lift height increases. It indicates that the potential contrast with chain pattern obtained below 27 nm lift height contains artifacts from topography of sample surface, caused by the non-electrical tip-sample interactions. To avoid this kind of artifacts, one can increase the lift height with small steps (~1–2 nm per step) till electrical signal has almost no change. However, do not lift tip too much, otherwise the electrical resolution will decrease a lot based on our discussion in previous session.

If electrical modes work on single-pass mode instead of dual-pass lift mode, for example, dual frequency KPFM and tapping mode-based frequency modulation KPFM (TP-KPFM-FM), the artifact from topography-electrical signal should be

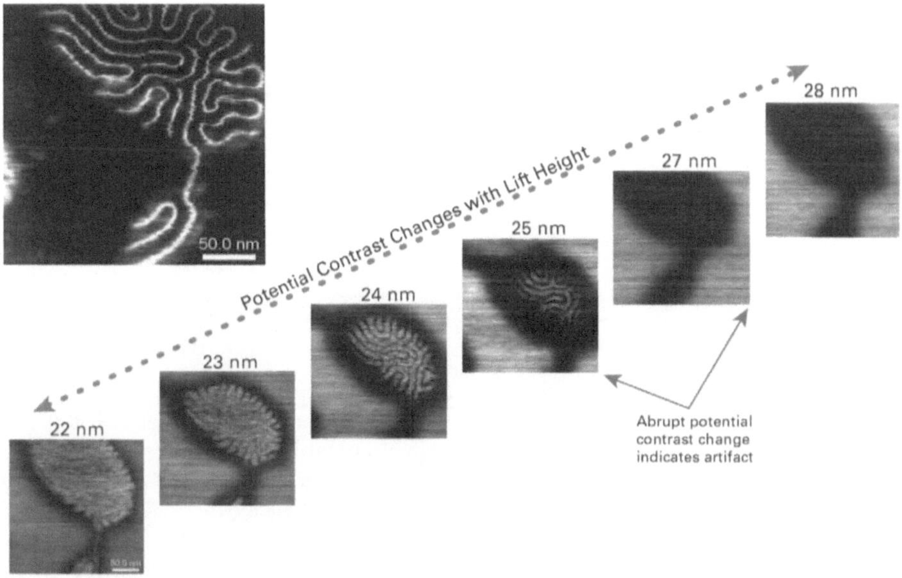


**FIGURE 11.42** PFKPFM images of a brush polymer sample on mica substrate. The inset at upper-left corner is the topography image. The bottom images show the potential contrast changes with lift height and an abrupt potential contrast change at ~27 nm lift height, which indicates below this lift height, the potential image contains artifacts from topography.

paid more attention. In dual frequency KPFM, an AC bias frequency for potential measurement is usually set to the value far below the cantilever resonance frequency and a low-pass filter is also applied, to reduce the crosstalk from topography, which is done by a vibrating AFM probe driven at near resonance frequency in tapping mode. In TP-KPFM-FM, the tapping frequency acts both for topography imaging and providing base frequency, which is modulated by AC signal, and the crosstalk cannot be avoided so this mode is only suitable for samples with small tapping phase contrast but large surface potential contrast.

Another typical artifact is false piezo response signal in PFM mode, which is caused by electrical force between the tip and the sample. In PFM measurement, the drive AC bias between the tip and the sample causes oscillating electrical force at the same frequency, such force would then cause oscillation of cantilever at this frequency and finally be detected by AFM system. Obviously, amplitude of this vertical signal is highly dependent on spring constant of cantilever, i.e., stiffer the probe, smaller the amplitude. The electrical force is dependent on surface charge distribution. However, in PFM, deflection of cantilever is used to detect sample deformation caused by applied external electric field, which characterizes converse piezoelectric effect of samples, while phase contrast between ferroelectric domains is either 0° or 180°. Therefore, in practical work, vertical signal detected in AFM is a sum of piezo effect and electrical force effect, and the phase contrast would deviate PFM's characteristics. To recognize this artifact, non-180° phase contrast in PFM phase image and obvious contrast in PFM amplitude image indicate the existence of electrostatic

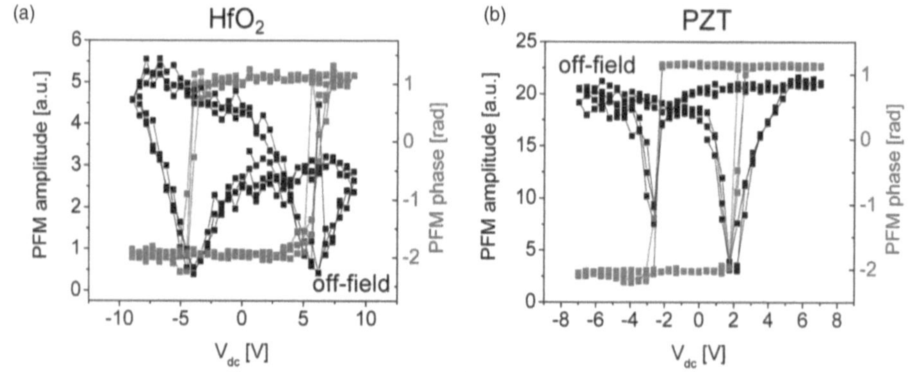


**FIGURE 11.43** Ferroelectric hysteresis loop measurement on (a) Non-ferroelectric thin film $HfO_2$. The amplitude has no saturated part at high electrical field, and (b) Ferroelectric thin film PZT, saturated amplitude is observed at high electrical field. (Reprinted with permission from Nina Balke et al.[115] Copyright 2015 American Chemical Society.)

contribution. In ferroelectric hysteresis loop measurement, the pure PFM signal has saturated part at high electrical field while the electrostatic-dominant loop hasn't.[115] Figure 11.43 shows PFM response of Non-ferroelectric thin film $HfO_2$ and Ferroelectric thin film PZT in ferroelectric hysteresis loop measurement. The different characteristics of amplitude response at high electrical field can be observed.

To minimize the artifact caused by electrical force, stiff probe is suggested. Sample pretreated by liquid bandage or ionized blower is also helpful for eliminating electrostatic contribution.

## 11.5 AFM DATA PROCESSING

The raw data of AFM images often contains waste information, like artifacts and noise, which must be removed before we do data analysis. There are many good AFM data processing software, like commercially available NanoScope Analysis from Bruker and MountainSPIP from Image Metrology, as well as freeware like WSXM[116] and Gwyddion.[117] These software provide powerful functions to process and analyze AFM data. Because most data processing functions are well described in the software manuals, it is not necessary to repeat how to use these functions here. In this section, we only focus on some important steps which are usually ignored by operators.

### 11.5.1 Remove Z Offset, Tilt, and Bow

Because AFM uses z piezo to generate height data, the raw data contains information of z piezo positions. Z scanner drift and unstable movement lead to z offset, the unavoidable sample tilt results in a tilt base plane in the image, and for tubular scanners, the "swing" in large scan size introduces bow to the image. The z offset, tilt, and bow are artifacts and should be removed in the topography images. Unfortunately, this critical action was ignored or done in a wrong way in many reported results.

Removing z offset, tilt, and bow is a basic function of every AFM data processing and analysis software. Here we take Bruker NanoScope Analysis software as an example; two commonly used commands to do this are "Flatten" and "Plane Fit". In other software, command names may be different, but what we discussed here also is also applicable.

"Flatten" uses all unmasked portions of scan lines to calculate individual least-square fit polynomials for each line and then subtract it from the raw data. Each line is fit individually to move the data to the center to eliminate the z offset (zeroth order flatten) or remove tilt and higher order bow from the image (first, second, or third order bow). A key step of this operation is to calculate the correct baseline for each data line and then subtracted out. Because software does not know where the substrate and the sample are, operator must tell software which data is used as baseline calculation by excluding the regions which are not substrates. Without this step, the image after "Flatten" will be introduced to other artifacts because wrong baseline is subtracted out from the raw data. Figure 11.44 shows an example, and the sample is some NPs on gold substrate. Figure 11.44a is the original data; we can see serious bow in vertical direction. Figure 11.44b is the result by directly applying "Flatten" command without any mask. In Figure 11.44b we still see black areas on both sides of the bright particles because we subtracted a wrong baseline. After excluding all particles from the background, the flattened image, Figure 11.44c, correctly shows the topography.

There are two ways to exclude the areas of the image from raw data which is used for polynomial fits. One way is to manually draw box cursor to exclude the features, shown in Figure 11.45a. This way is often used to process the sample with small number of discrete features. Another way is to adjust threshold setting and software will mark out the eligible areas automatically, shown in Figure 11.45b. This way is usually used to deal with large number of dense features. Operator can use "Flatten" command several times if the first-time operation cannot reach to good result. For example, in Figure 11.45b, in the first time "Flatten" one may not find a suitable threshold to select all areas that should be excluded. Just execute "Flatten" command once and then do it again. There are many good examples in publications[48,118–121] one can take as reference.

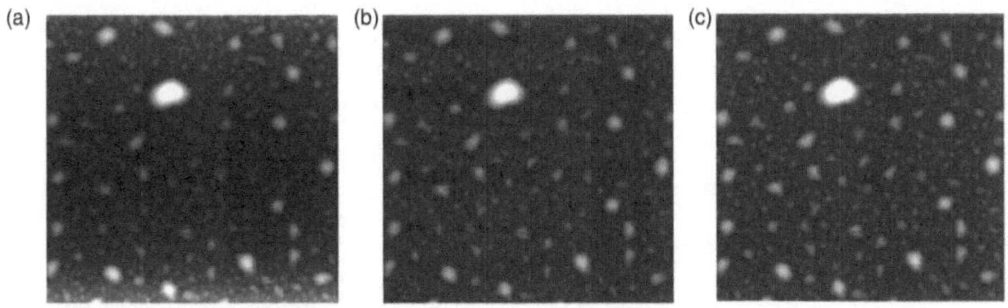


**FIGURE 11.44** NPs on gold substrate. (a) Before applying "Flatten". (b) Directly applying "Flatten" to the whole image. (c) Using threshold method to exclude features then apply "Flatten".

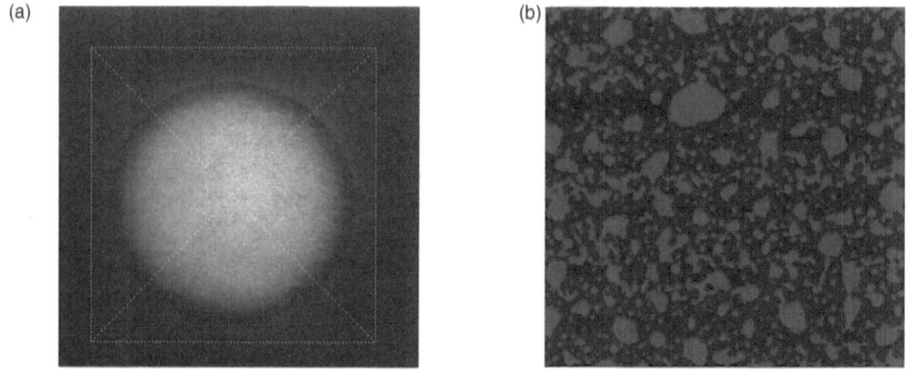


**FIGURE 11.45** Two methods to exclude non-baseline data when using "Flatten". (a) Manually draw box cursor to create a stopband. The data masked by the stopband will not involve baseline calculation. (b) Set a threshold and software select the excluded areas automatically. The areas marked by blue color are the excluded data selected by software based on threshold setting.

Another command which is used to remove z offset, tilt, and bow is "Plane Fit". "Plane Fit" is very similar to "Flatten", which computes a single polynomial of a selectable order for an image and subtracts it from the raw data. The two major differences are: first, "Plane Fit" is applied to XY plane by default, not as "Flatten" which is applied on each scan line; second, operator creates passbands not stopbands to select areas from the image to allow specific points used in calculation of the polynomial. By default, "Plane Fit" is more complex than "Flatten" because the calculation of the polynomial includes information from both X and Y directions. Usually "Flatten" is enough to deal with most of the raw topography data, and "Plane Fit" is only used in few cases which one can easily identify a large area substrate. The ways to identify the correct reference level are like the methods to identify correct baseline used in "Flatten". One way is to manually draw box cursor to select the areas for reference level calculation, another way is to use threshold to mark out the areas automatically.

However, in practical usages, such two methods are sometimes combined to achieve ideal results, especially in layered structures, which are widely observed in solar cell and lithium cell samples.[122,123] "Flatten" is usually used to eliminate dislocation between each scan lines, where "Plane Fit" could be used to define a reference level, which is critical to reveal relative height difference in layered structures. For example, AFM topographic image, as shown in Figure 11.46a, reveals surface of calcite. One could clearly observe dislocations during scanning and such artifacts could be removed using "Flatten". The result is shown in Figure 11.46b, in which the nice, layered structure of calcite crystal was reviewed. However, the image after "Flatten" still has artifacts because these layers are still tilt. To reach to a better view and measure correct height difference between layers, a steep step structure is preferred. Therefore, one could choose one layer in Figure 11.46b as reference level and do "Plane Fit". The final image shown in Figure 11.46c demonstrates an ideal flat and layered structure.

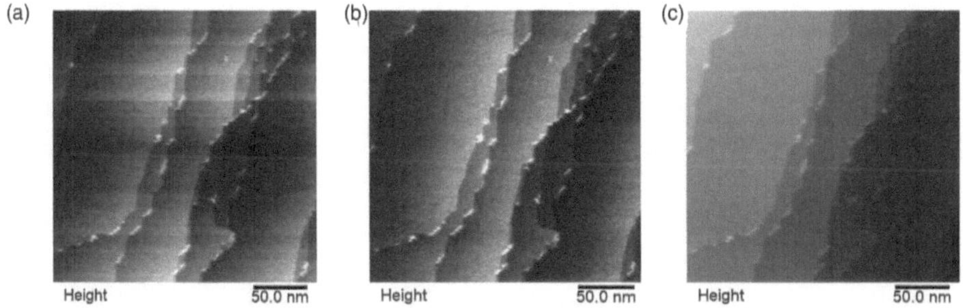


**FIGURE 11.46** Combined "Flatten" and "Plane Fit" together to get ideal flat and layered structure topography image. (a) Raw data of the layered structure. (b) Image only after "Flatten". (c) Selected one layer as reference level and applied "Plane Fit", then got the ideal image.

Only topography images use "Flatten" and "Plane Fit" to remove z offset, tilt, and bow. For physical properties channels, like phase, potential, modulus, adhesion, etc., the raw data is necessary to do future analysis and only in some special purposes we do this operation.

### 11.5.2 Remove Noise

A lot of factors contribute to the noise in the AFM image. To get a good image, one should remove the noise from the raw data as far as possible. Noise can come from system and the environment. There are low frequency noises that come from environmental vibration, loose probe, probe holder, and unstable piezo movement and acoustic waves, and high frequency noises come from electrical components and RF radiation.

In all AFM data processing software, there are a set of filter functions which can be used to remove the noise from raw data. Here we will still take Bruker NanoScope Analysis software as an example; "Low-pass" command can effectively suppress the high spatial frequency components by averaging a $3 \times 3$ pixel region centered on each pixel; "Gaussian" filter applies either a low-pass or a high-pass digital filter to an image along either X or Y direction to eliminate high frequency features or low frequency features; and "Spectrum 2D" function transforms image data from spatial domain into the frequency domain and back via a 2D fast Fourier transform (FFT) to remove or keep specific frequencies in the spectrum and then reconstruct the data. They may have different names in other software, but they are all basic AFM data processing functions. All of them can be used to remove noise and get an enhanced version of image data. The basic usage of them have been discussed in detail in software manual, here we focus on some typical examples.

Film material is very popular in green energy study, for example, in solar cell research, and surface analysis is a key part of the process development of solar cells as the roughness and textured surface contribute to high efficiency solar cell.[124] AFM topographic image of an organic film for solar cell study is shown in Figure 11.47a, which includes certain frequency noise. The FFT image, Figure 11.47b, shows the frequency distribution of this image, after one executed the "Spectrum 2D" function.

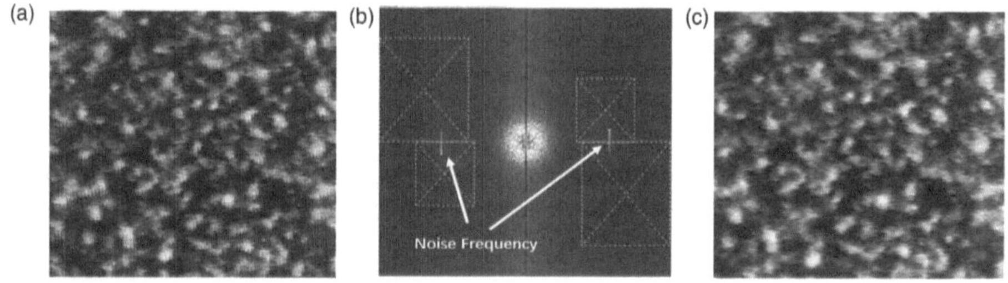


**FIGURE 11.47** Using "Spectrum 2D" function to remove noise at certain frequency. (a) AFM topographic image with noise. (b) "Spectrum 2D" function transfers (a) from spatial domain to frequency domain and the noise frequency is identified and blocked. (c) "Spectrum 2D" function transfers (b) from frequency domain back to spatial domain, the image with no noise is restructured.

In this image, the frequency of noise is shown clearly. Such noise could be removed by adding stopband. Then the image was transformed back to spatial domain, results in Figure 11.47c, in which the grain structure of organic surface is much clear. Karolien Saliou et al., in C-Si solar cell pyramid size distribution study, removed the noise and interferences by extracting selected frequencies obtained from FFT spectrum.[125] The resulted filtered topographic image was then obtained by the inverse FFT. The final image was further analyzed to obtain the structural, morphological, and geometrical characteristics of the surface. Tin (IV) phthalocyanine dichloride ($SnCl_2Pc$) thin film is also used in optical/electric study.[125–127] FFT analysis was used to study $SnCl_2Pc$ thin film deposition process.

Sometimes "Gaussian" and "Spectrum 2D" function would be employed together to obtain neat result. "Spectrum 2D" is often utilized to remove noise with several fixed frequency, while "Gaussian" filter is more general and can be used as final cleaning step to complete image processing. For example, AFM topographic image, as shown in Figure 11.48a, demonstrated surface features of polydimethylsiloxane (PDMS) thin film. One could distinguish periodic stripes on this image, which was clearly showed in FFT result. After eliminating these noises using FFT, the strips

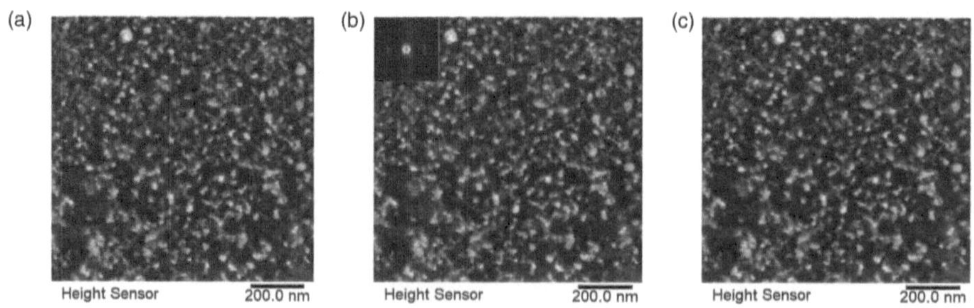


**FIGURE 11.48** "Gaussian" filters combined with "Spectrum 2D" applied on PDMS film sample. (a) The raw data. (b) Image restructured after using "Spectrum 2D" function. The up left inset is FFT image before removing noise. (c) Final result after applying low-pass "Gaussian" filter.

were significantly reduced, shown in Figure 11.48b. Followed by a low-pass Gaussian filter along horizontal direction, such noise was completely removed, and a neat image was shown in Figure 11.48c.

## Case Study

To make this chapter be with more practical significance, in this section, we would like to share some case studies in energy research. In these examples, basic rules discussed in previous sessions were followed, and the unique purpose of the experiments was also considered during practice.

### EXAMPLE I: PHOTOVOLTAIC MATERIALS RESEARCH

Solar-driven photocatalytic reactions have attracted significant notices due to its great potential in exploring sustainable fuels. Such processes are widely believed to be dependent on effective separation of photogenerated charges; therefore, it is vital to understand charge generation processes for improvement of photocatalysts' performance.[123,128–130] Dr. Can Li and his group developed a unique surface photovoltage microscopy (SPV) based on commercial AFM platform, Bruker Dimension Icon, to investigate charge separation behavior of photogenerated charges, as illustrated in diagram in Figure 11.49.

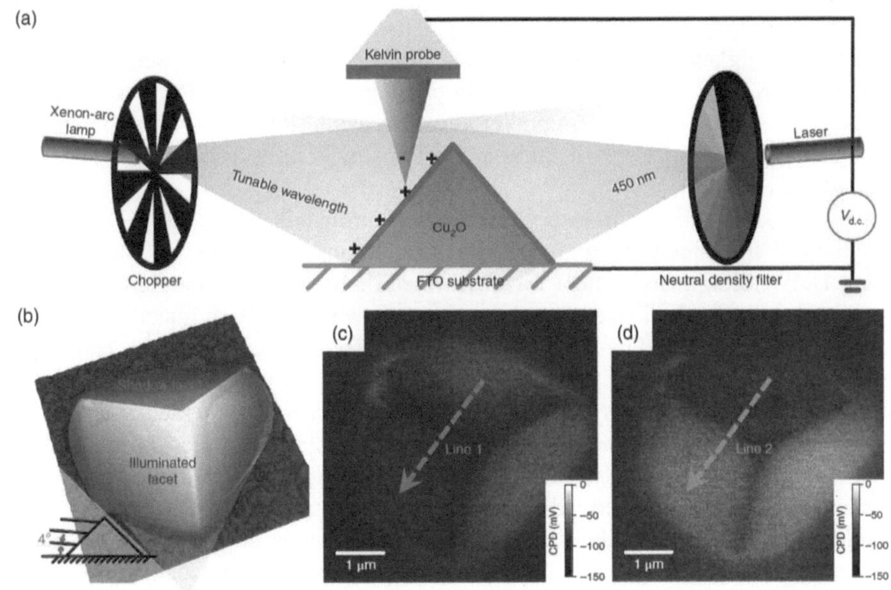


**FIGURE 11.49** (a) Schematic diagram of experimental setup for Dr. Can Li's SPV. (b) Illustration of asymmetric-irradiation-induced illuminated and shadow regions on a typical cubic $Cu_2O$ particle grown along the (111) orientation. (c) and (d) Corresponding surface potential images in the dark state (c) and under illumination (d) ($\lambda = 450$ nm). (Reprinted with permission from Chen et al.[123] Copyright 2018 Springer Nature.)

To conduct such photo induced KPFM imaging, AFM probe was scanned on desired facets while certain facets were illuminated; in other words, orientation was key for this study. Therefore, photo-active crystals should be deposited on substrate with desired facets, which suggested significance of firm contact between sample and substrate. In situ crystallization is preferred for such research, which would guarantee the generated crystal fixed on the substrate with designed orientation. For example, they used their SPV to study photocatalytic $Cu_2O$ particles' charge separation behavior.[123] To deposit $Cu_2O$ on fluorine-doped tin oxide (FTO) substrate, they utilized an electrodeposition method. The isolated cubic $Cu_2O$ crystals were deposited galvanostatically with designed current density from $Cu(NO_3)_2$ aqueous solution.

The surface potential images were captured using traditional AM-KPFM. Surface morphology was carried out in tapping mode while potential was measured in lift mode with a lift height around 100 nm. For tapping mode imaging under ambient conditions, a probe with spring constant of over 1 N/m was appropriate, while a conductive tip is preferred for surface potential measurement. Since charge difference on different crystal facets is goal of the study, spatial resolution around micrometer scale is enough. Therefore, a metal-coated silicon tip with radius around 20 nm is normally used in the SPV study.

In practical imaging, potential contrast in SPV study could be interfered by topography convolution, as described in Section 11.4.3. Moreover, one would image steep structures, such as crystals or particles, in SPV study, which would easily bring artifacts of topography into potential mapping. Therefore, optimizing probe tracing by using relative higher feedback gain, lower scanning speed, and larger setpoint in PFT mode-based KPFM or smaller setpoint in tapping mode-based KPFM is preferred. On the other hand, lift height in potential image would be another vital parameter, while too low lift height would cause artifact of topography in potential images, as detailed in Section 11.4.3.

Such prepared $Cu_2O$ single crystals were illuminated under two light sources with tunable light intensity and opposite illuminating directions, as illustrated in Figure 11.49a. SPV measurement was performed on the high-symmetric $Cu_2O$ under asymmetric photoexcitation, i.e., laser light was off, where the illuminated and shadow facets were illustrated in Figure 11.49b. After illuminating using xenon-arc lamp, SPV results showed difference of potential on shadow and illuminated facets, while they were similar when light was off, which demonstrated a photogenerated carrier concentration gradient. The holes and electrons were transferred to the illuminated and shadow facets, respectively. Further study suggested that intrinsic mobility difference of holes and electrons enabled a diffusion-controlled charge separation process. They exploited these findings to develop spatial separated redox co-catalysts on a single photocatalytic particle, which enhanced the performance of photocatalytic reaction by 300%.

Using such spatial resolved SPV setup, they have also obtained direct evidence of anisotropic photogenerated charge separation on $BiVO_4$.[122] Beyond that, this technique was also employed to study plasmonic photocatalysis and located plasmonic holes near gold-semiconductor interface.[130] It also revealed polarization dependence of plasmonic charge separation.[128] All these results implied great potential of AFM-based SPV technique in the investigation of photocatalysis and utilization of solar energy.

## EXAMPLE II: BIO-ENERGY MATERIALS RESEARCH

Microorganisms, like bacteria, alga, and fungus, which are good candidates to produce molecules as fuel, are very important in biofuel research. Microorganisms can also be used to develop bio-batteries or biocatalyst by combining them with special electrode. AFM is a powerful tool to characterize the topography and physical and chemical properties of microorganism samples. Young's modulus mapping can help understand the mechanical strength of biomaterials built by *Shewanella oneidensis*.[131] cAFM can reveal the electrical transmission characteristics of bacteria-metal complexes.[132,133] AFM-IR technique can be used to analyze the distribution of produced bio-fuel in *Streptomyces*.[134] Good topography imaging is the precondition of all these measurements. Many characterizations of live bacteria were done in fluid environment to enable the in situ behavior study.[135,136] Here we review a typical protocol of force measurement on live *E. coli* cell, based on force curve technology, like PFQNM and QI mode.[137–139]

Step I: Culturing bacteria

The protocol for bacteria culture can be found easily in molecular biology handbooks.[136] In general, a single colony is picked up from a well-cultured LB-agar plate and is inoculated into 3 mL LB broth. After incubation for ~3 hours at 37°C, LB broth is centrifuged under 5,000 rpm for 2 minutes. The supernatant is removed then, and bacteria is resuspended using 3 mL buffer solutions (PBS, HEPES are recommended).

Step II: Cleaning glass substrate

Glass substrate, typically a glass slide, is washed with pure water and pure ethanol and then immersed into freshly prepared piranha solution (98% $H_2SO_4$: 30% $H_2O_2$ = 7:3 v/v) for 30 minutes. After washing with pure water for 5 minutes, glass substrate is dried under pure $N_2$. This step is to make the substrate clean.

Step III: Glass substrate coating

A ~300 μL droplet of 0.01% poly-L-lysine (PLL) solution (Sigma) is added on to the clean glass substrate, stewing for 5 minutes to allow PLL absorption. Keep the substrate lying in a wet box to prevent PLL solution from drying. Then the substrate is washed with pure water and dried under pure $N_2$. Other polymer glues, like Cell-Tak™, Vectabond®, and poly-L-ornithine (PLO), can also be used to coat glass substrate. This step is to create a homogenous polymer layer to fix the bacteria.

Step IV: Bacteria absorption

A ~300 μL droplet of bacteria suspension is added onto PLL coated substrate. After ~5 minutes, use the suspension buffer to wash the substrate so that unbounded bacteria can be removed, and keep bounded bacteria on the substrate for further AFM imaging. Keep the liquid on the surface to prevent the bacteria from death. The volume of bacteria droplet and absorption time can be adjusted to optimize the surface density of absorbed bacteria. This step is to make fixed bacteria sample.

Step V: AFM probe selection
The cantilever should have Au backside coating to increase the reflectivity of the laser in liquid for signal noise ratio improvement. As discussed before, Al coating cannot be used in aqueous solution. Cantilever with high resonance frequency is recommended to allow fast mapping so thermal drift can be decreased. Soft and short cantilever is preferred for better force control and high sensitivity. A tip with large radius is relatively easier to handle the imaging if no high-resolution imaging is required, because a blunt tip cannot easily stab through cell and damage the fragile simple. MLCT-F, DNP-C, ScanAsyst-Fluid, AC40, and FasScan-D are good candidates for bacteria imaging. If higher spatial resolution is needed, a sharp tip like SNL-C, ScanAsyst-Fluid+ should be the better choice.

Step VI: Cantilever calibration
If only do topography imaging, this step is not necessary. However, cantilever calibration is critical step for the mechanical property's characterization. The DS of cantilever can be calibrated by performing several force curves on a clean and hard surface in the same buffer, because the medium also affects DS. The DS calculation is shown in Figure 11.50a. The hard surface can be glass slide, silicon plate, or sapphire substrate. Then use "thermal tune" to calculate the spring constant of cantilever, which is a good way for spring constant calibration for soft cantilevers. The thermal tune method can be accessed on most of the commercial AFM systems.

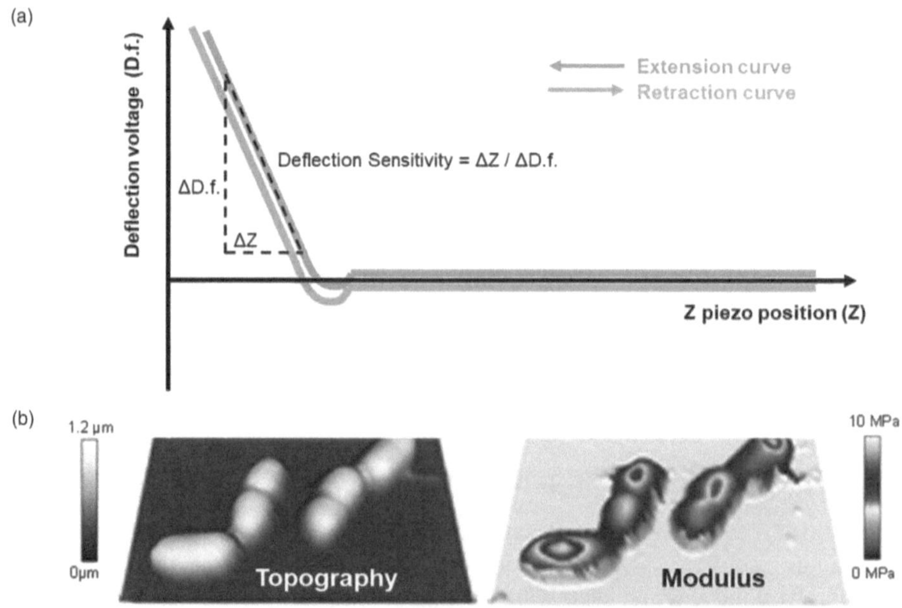


**FIGURE 11.50** (a) DS calibration of cantilever. (b) Live bacteria topography and modulus imaging.

Step VII: Imaging bacteria

Here we take QI mode for example. The largest force applied on sample is called trigger force. The trigger force can be set from 0.3 to 0.8 nN. Adjust the trigger force to optimize the indention depth to meet the contact mechanism model which is selected for analysis. The traveling distance of z piezo can be from 0.5 to 1 µm. Optimize the z traveling distance to allow at least one-third of the force curve is obtained on the sample surface. It will make sure there are enough data points for fitting. The z speed can be 20/100 µm/s. Increase the z speed to save the time till the force curve is no longer stable or with different background. Figure 11.50b is an example of mechanical measurement results of live *E. coli* bacterial.[135] Topography image can be overlaid with mechanical channel to show the mechanical properties distribution of the sample.

## EXAMPLE III: PIEZOELECTRIC MATERIALS RESEARCH

Bismuth ferrite ($BiFeO_3$, BFO) is one of the most popular piezoelectric materials, which shows both (anti-)ferromagnetism and ferroelectricity at room temperature.[140] Due to large piezoelectric coefficient, unique magnetic-electric coupling, visible wavelength region, and bulk photovoltaic effect, it is promising for solar cells,[141] hydrogen production,[142] and energy storage devices.[143] With the trend of device miniaturization, it is necessary to characterize BFO at nanoscale with AFM. In this section, we take BFO as an example to review some practical guidance of AFM experiments.

BFO samples for AFM measurement could be particle, nanowire, thin film, and bulk ceramic. The BFO particle and nanowire can be prepared by hydrothermal method.[144] Using homemade or commercial BFO ceramic as target, thin film could be grown by pulsed laser epitaxy.[145] Phase pure BFO ceramic can be synthesized by traditional solid-state reaction method using $Bi_2O_3$ and $Fe_2O_3$ powders as precursors.[145]

In BFO study, cAFM can be used to measure the photocurrent and domain wall current to investigate the potential application of BFO in optical and data storage devices.[146–149] PFM is used to visualize the domain structure and quantizes the built-in electric field to investigate the ferroelectric photovoltaic effect.[150–152] EFM and KPFM[149] are used to examine the charge accumulation, surface potential, and photovoltage of BFO to promote its applications in photovoltaic and photocatalytic field. For those AFM electrical measurements, the BFO samples need electrical contact with the sample stage. The details were discussed in Section 11.2.2.2. For bulk sample, to obtain a uniform electric field, top electrode can be made by coating a Pt or Au layer.[153]

Liu et al. observed the ferroelectric photovoltaic effect of tetragonal-phase BFO (T-BFO) directly with cAFM and demonstrated a controllable photovoltaic process, which provides a possible way to design high-performance electric–optical integrating devices.[154] In cAFM measurement, with the tip at virtual ground, a DC bias is applied to T-BFO through bottom electrode $La_{0.67}Sr_{0.33}MnO_3$ (LSMO). While scanning in contact mode, a linear amplifier senses the current passing through the sample and the conductive Pt/Ir-coated tip. The conductive coating guarantees the closed current pathway through controller, sample stage, sample, probe, probe holder, electrical module, and controller. The current strength strongly depends on the contact between the tip and the sample. Large setpoint leads to large imaging force in both cAFM and PFTUNA modes, which results in larger current, because the contact area is large, so resistance becomes smaller. In PFTUNA mode, low peak force amplitude has the same effect because the contact time becomes longer. When it comes to unknown sample, it is a good choice to start with low DC bias and large current sensitivity. Moreover, slow scan rate will also result in more stable current signal.

PFM is based on contact mode, which can obtain piezo signal through inverse piezoelectric effect. Liou et al. used PFM to explore the dynamic evolution of ferroelectric domain and the phase transition of BFO under optical stimulus.[155] Accompanied with the $d_{33}$ loop, the mechanism of light control was revealed, and possible developments for photovoltaic devices and data storage devices were suggested. In this work, $LaAlO_3$ substrates were chosen as bottom electrode to guarantee relative uniform electric field. Pt/Ir-coated conductive probe with spring constant ~7 N/m was used to apply an AC bias with 1 V amplitude at ~7 kHz. The conductive Pt/Ir coating makes it possible to apply bias. Probe with other conductive coating also works. Short cantilever with better DS is preferred for small domain characterization. The moderate spring constant (2~10 N/m) guarantees enough imaging force, which is crucial for signal tracing. To minimize the artifacts caused by electrostatic force, stiff probe is suggested; typical conductive probes for electrical measurement usually have spring constant ~2–3 N/m, but in this work, 7 N/m probe was selected to get purer piezo signal. The resonance frequency of such tip usually ranges from 60 to 150 kHz, which is flexible to operate the PFM far away from or near its contact resonance frequency. In this work, the frequency of the AC bias is 7 kHz and far below the contact resonance frequency, and real piezo response is dominated in the signal. But for most of the piezoelectric materials, the piezo response is so small, usually ranges from several picometer to hundreds of picometer, that it is not easy to be detected. To optimize the signal noise ratio, PFM can be operated near the contact resonance frequency to amplify the signal by ~$Q$ times; therefore, for quantitative measurement, the mechanical amplification effect must be considered. However, it is difficult to get accurate $Q$ so it is almost impossible to do accurate quantitative measurement when operating near contact resonance frequency.

EFM and KPFM usually work under dual-pass mode, and the first scan detects the topography in tapping mode or PFT mode, while the second scan obtains electrostatic effect or surface potential at a constant height above the sample surface. Chen et al. provided direct evidence of the hydrogenation-induced oxygen vacancy on the BFO surface by KPFM study and proved that high oxygen vacancy concentration benefits an enhanced light absorption capability in photocatalytic process.[156] In this work, a conductive Co/Cr-coated tip with a resonant frequency of 75 kHz and a spring constant of 2.8 N/m (MESP cantilever, Bruker) was used. The sample powder was dispersed on the ITO conductive glass, to make sure bias can be applied between the tip and the sample. In EFM and KPFM measurement, electrically conductive probes will be a good choice for most case, while uncoated highly doped Si probes will give contrast either. The sensitivity of EFM and KPFM is proportional to $\frac{Q}{k}$, and small spring constant and large $Q$ factor will result in higher sensitivity. Lift height is critical for EFM and KPFM. Too low lift height will result in topography crosstalk, while too high lift height leads to resolution loss of electrical signal. Typically, the lift height ranges from 50 to 150 nm. To verify whether topography crosstalk happens, one can adjust the lift height. The topography will fade obviously with the increase of lift height, while electrostatic and potential contrast change slightly with small lift height change.

## EXAMPLE IV: PEROVSKITES MATERIALS RESEARCH

In recent years, perovskite materials, especially the hybrid organic–inorganic perovskites (OIP), are becoming the new stars in the field of solar cells because of their excellent photoelectric conversion efficiency. One advantage of hybrid perovskite materials is that it can regulate the composition at the molecular level, to effectively regulate its bandgap and change the material properties. In general, PFM can be used to measure the piezoelectric properties of the perovskite materials. To further characterize the domains of perovskite materials at the nanoscale and investigate in situ whether they are susceptible to electrical bias, AFM-IR technique can be used. AFM-IR is a novel method that combines the lateral resolution of AFM with the specificity of absorption spectroscopy.[157] It was initially developed in the mid-IR and has attracted much interest for enabling label-free composition mapping, material identification, and conformational analysis at the nanoscale. AFM-IR data obtained direct evidence of $MA^+$ electromigration in solar cells with OIP lateral structure.[158] Recently, this technique has been extended to the visible and near-IR spectral ranges, an advance that has enabled the determination of the local bandgap in particular kind of perovskite films.[159–160]

In the setup, a pulsed wavelength-tunable laser is used to illuminate the sample via total internal reflection, as shown in Figure 11.51a. As it is a bottom-up method, PR-EX-C450-10 probe with ~450 μm long silicon cantilever, no coating, and a nominal spring constant ~0.2 N/m was used in this setup. To enable AFM-IR characterization at both mid-IR and visible ranges, a polycrystalline OIP lateral device was fabricated on a ZnS prism surface by spin-coating.

The amplitude of the cantilever oscillation (AFM-IR signal) is proportional to the sample absorption and yields nanoscale absorption spectra when sweeping the laser wavelength while holding the tip at a given location. And AFM-IR absorption images are obtained by illuminating the sample at a fixed wavelength. Probe with proper spring constant should be used, and during imaging, proper imaging force is controlled by setpoint to prevent sample from damage with large imaging force or poor tracking due to small imaging force. In both cases it results in the instability of the collected spectra, so that ghost peaks will appear. In the data processing, "flatten" is needed in AFM topography images to remove the z offset, tilt, and bow, as we discussed in previous session; but raw data is needed for AFM-IR images.

Figure 11.51b shows the topography of the OIP layer in the area between the device electrodes. Figure 11.51c shows the AFM-IR images obtained by illuminating the sample at 1468 $cm^{-1}$, corresponding to the -$CH_3$ asymmetric deformation of the methylammonium ion. Figure 11.51d shows the AFM-IR images obtained by illuminating the sample at 13,250 $cm^{-1}$, corresponding to electronic excitation just above the bandgap. Figures 11.51c and d show stripe domains on most grains, and these

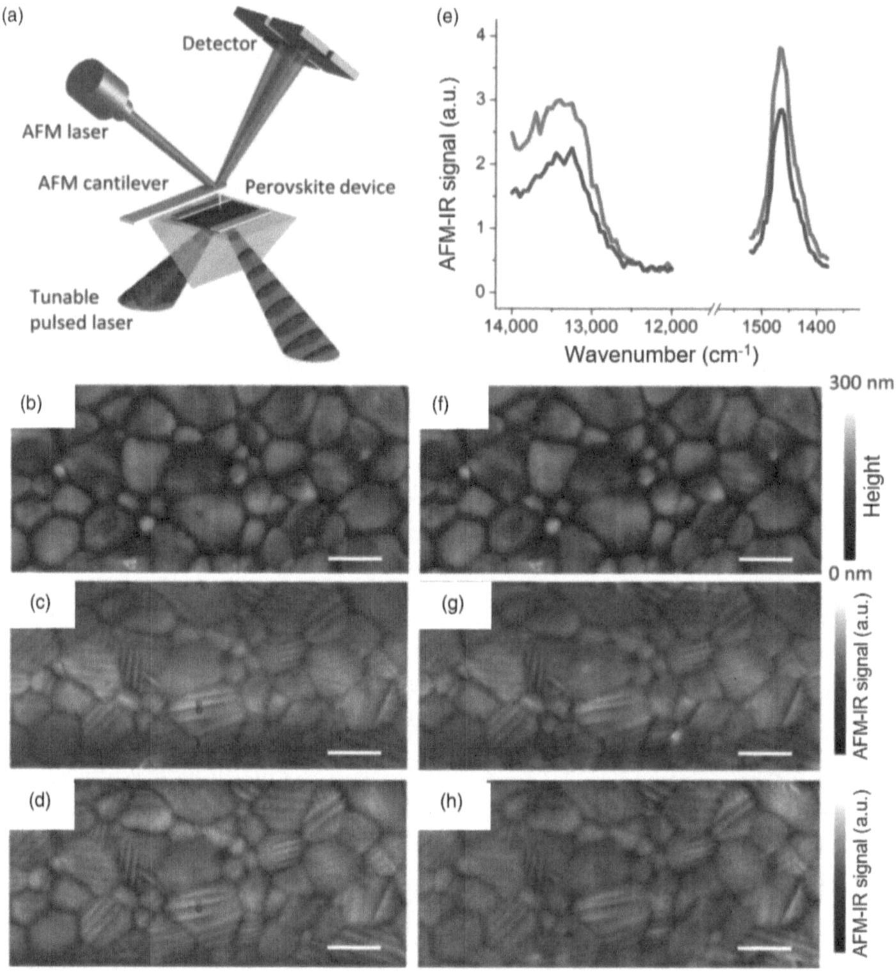


**FIGURE 11.51** Observation of ferro-elastic domains by AFM-IR and their insensitivity to the applied electric field. (a) Schematic diagram of the AFM-IR setup. (b) AFM topography image of the sample area between electrodes, and (c), (d) are corresponding AFM-IR images. (c) $-CH_3$ asymmetric deformation of the methylammonium ion (1,468 cm$^{-1}$) and (d) electronic transition above the bandgap (13,250 cm$^{-1}$ and 1.64 eV) of the as-prepared sample. (e) Representative electronic (left) and vibrational (right) absorption spectra obtained from contiguous bright (red) and dark (blue) striations visible in AFM-IR images. (f) AFM topography image and corresponding AFM-IR images obtained at 1,468 cm$^{-1}$ (g) and 13,250 cm$^{-1}$ (h) after applying a bias of 0.86 V/μm for 1 minutes (in plane electric field). Scale bars are 2 μm. (Reprinted with permission from Strelcov et al.[60])

striations should be ascribed to the presence of ferro-elastic domains. Figure 11.51e shows representative nanoscale spectra measured from two adjacent striations. To test the effect of electric field, the lateral device structure in sample was leveraged to apply an electric field in the sample plane. The AFM topography images and the AFM-IR images, as shown in Figure 11.51f, g and h, obtained after applying a lateral electric field of 0.86 V/μm for 1 minute, did not show any differences. In summary, no change in topography and in the AFM-IR images were observed with an applied electric field.[60]

Through AFM-IR technology, the chemical image of a specific area on the sample surface with a fixed wavenumber or the IR spectrum of a specific point can be obtained by applying different electrical bias to the sample. It is of great reference significance to determine the grain size and grain boundary of the perovskite materials. Because grain size and grain boundaries are important factors that influence the long-term stability of OIP devices, which is one of the most important issues hindering their widespread adoption, the results presented here could have important implications for their practical application.[161]

## 11.6 SUMMARY

AFM is a powerful tool to investigate energy materials. It is a multi-disciplinary technology that involved chemistry, physics, mechanics, and electronics. Thus, it is important that AFM users must first be well trained and have expertise in the field. Careful sample preparation and knowledge-guided selection of AFM probe are the first steps towards successful experiment. They are as important as parameter optimizations during AFM operation. Scientific and reasonable processing of raw data in the later stage ensures that we can obtain meaningful data that match the sample properties.

## REFERENCES


1. Huang, Z.; Jiang, J.; Hua, Y.; Li, C.; Wagner, M.; Lewerenz, H. J.; Soriaga, M. P.; Anfuso, C., Atomic force microscopy for solar fuels research: An introductory review. *Energy and Environment Focus* 2015, *4* (4), 260–277.
2. Sun, H.; Ye, M.; Sun, W., *High Resolution AFM and its Applications*. Singapore: Springer Nature Singapore Pte Ltd. 2018, 179–235.
3. Bibi, I.; Hussain, S.; Majid, F.; Kamal, S.; Ata, S.; Sultan, M.; Din, M. I.; Iqbal, M.; Nazir, A., Structural, Dielectric and magnetic studies of perovskite Gd1-xMxCrO3 (M = La, Co, Bi) nanoparticles: Photocatalytic degradation of dyes. *Zeitschrift Fur Physikalische Chemie-International Journal of Research in Physical Chemistry & Chemical Physics* 2019, *233* (10), 1431–1445.
4. Byranvand, M. M.; Song, S.; Pyeon, L.; Kang, G.; Lee, G.-Y.; Park, T., Simple post annealing-free method for fabricating uniform, large grain-sized, and highly crystalline perovskite films. *Nano Energy* 2017, *34*, 181–187.
5. Ajmal, S.; Bibi, I.; Majid, F.; Ata, S.; Kamran, K.; Jilani, K.; Nouren, S.; Kamal, S.; Ali, A.; Iqbal, M., Effect of Fe and Bi doping on $LaCoO_3$ structural, magnetic, electric and catalytic properties. *Journal of Materials Research and Technology-JMR&T* 2019, *8* (5), 4831–4842.
6. Chang, C.-Y.; Wang, C.-P.; Raja, R.; Wang, L.; Tsao, C.-S.; Su, W.-F., High-efficiency bulk heterojunction perovskite solar cell fabricated by one-step solution process using single solvent: Synthesis and characterization of material and film formation mechanism. *Journal of Materials Chemistry A* 2018, *6* (9), 4179–4188.
7. Almadori, Y.; Moerman, D.; Martinez, J. L.; Leclere, P.; Grevin, B., Multimodal noncontact atomic force microscopy and Kelvin probe force microscopy investigations of organolead tribromide perovskite single crystals. *Beilstein Journal of Nanotechnology* 2018, *9*, 1695–1704.

8. Benetti, D.; Jokar, E.; Yu, C.-H.; Fathi, A.; Zhao, H.; Vomiero, A.; Diau, E. W.-G.; Rosei, F., Hole-extraction and photostability enhancement in highly efficient inverted perovskite solar cells through carbon dot-based hybrid material. *Nano Energy* 2019, *62*, 781–790.
9. Cai, Y.; Zhang, Z.; Zhou, Y.; Liu, H.; Qin, Q.; Lu, X.; Gao, X.; Shui, L.; Wu, S.; Liu, J., Enhancing the efficiency of low-temperature planar perovskite solar cells by modifying the interface between perovskite and hole transport layer with polymers. *Electrochimica Acta* 2018, *261*, 445–453.
10. Adhikari, T.; Shahiduzzaman, M.; Yamamoto, K.; Lebel, O.; Nunzi, J.-M., Interfacial modification of the electron collecting layer of low-temperature solution-processed organometallic halide photovoltaic cells using an amorphous perylenediimide. *Solar Energy Materials and Solar Cells* 2017, *160*, 294–300.
11. Abdel-Aal, S. K.; Abdel-Rahman, A. S., Graphene influence on the structure, magnetic, and optical properties of rare-earth perovskite. *Journal of Nanoparticle Research* 2020, *22* (9), 1–10.
12. Zheng, P.; Xiang, L.; Chang, J.; Lin, Q.; Xie, L.; Lan, T.; Liu, J.; Gong, Z.; Tang, T.; Shuai, L.; Luo, X.; Chen, N.; Zeng, H., Nanomechanics of Lignin–Cellulase interactions in aqueous solutions. *Biomacromolecules* 2021, *22* (5), 2033–2042.
13. Yang, K.; Jia, L.; Liu, X.; Wang, Z.; Wang, Y.; Li, Y.; Chen, H.; Wu, B.; Yang, L.; Pan, F., Revealing the anion intercalation behavior and surface evolution of graphite in dual-ion batteries via in situ AFM. *Nano Research* 2020, *13* (2), 412–418.
14. Laskowski, F. A. L.; Oener, S. Z.; Nellist, M. R.; Gordon, A. M.; Bain, D. C.; Fehrs, J. L.; Boettcher, S. W., Nanoscale semiconductor/catalyst interfaces in photoelectrochemistry. *Nature Materials* 2020, *19* (1), 69–76.
15. Wan, J.; Hao, Y.; Shi, Y.; Song, Y.-X.; Yan, H.-J.; Zheng, J.; Wen, R.; Wan, L.-J., Ultra-thin solid electrolyte interphase evolution and wrinkling processes in molybdenum disulfide-based lithium-ion batteries. *Nature Communications* 2019, *10* (1), 3265.
16. Farahi, R. H.; Charrier, A. M.; Tolbert, A.; Lereu, A. L.; Ragauskas, A.; Davison, B. H.; Passian, A., Plasticity, elasticity, and adhesion energy of plant cell walls: Nanometrology of lignin loss using atomic force microscopy. *Scientific Reports* 2017, *7* (1), 152.
17. Shen, X.; Li, Y.; Qian, T.; Liu, J.; Zhou, J.; Yan, C.; Goodenough, J. B., Lithium anode stable in air for low-cost fabrication of a dendrite-free lithium battery. *Nature Communications* 2019, *10* (1), 900.
18. Shen, C.; Hu, G.; Cheong, L.-Z.; Huang, S.; Zhang, J.-G.; Wang, D., Direct observation of the growth of lithium dendrites on graphite anodes by operando EC-AFM. *Small Methods* 2018, *2* (2), 1700298.
19. Peng, Z.; Zhao, N.; Zhang, Z.; Wan, H.; Lin, H.; Liu, M.; Shen, C.; He, H.; Guo, X.; Zhang, J.-G.; Wang, D., Stabilizing Li/electrolyte interface with a transplantable protective layer based on nanoscale LiF domains. *Nano Energy* 2017, *39*, 662–672.
20. Gu, Y.; Wang, W.-W.; Li, Y.-J.; Wu, Q.-H.; Tang, S.; Yan, J.-W.; Zheng, M.-S.; Wu, D.-Y.; Fan, C.-H.; Hu, W.-Q.; Chen, Z.-B.; Fang, Y.; Zhang, Q.-H.; Dong, Q.-F.; Mao, B.-W., Designable ultra-smooth ultra-thin solid-electrolyte interphases of three alkali metal anodes. *Nature Communications* 2018, *9* (1), 1339.
21. Zheng, J.; Zheng, H.; Wang, R.; Ben, L.; Lu, W.; Chen, L.; Chen, L.; Li, H., 3D visualization of inhomogeneous multi-layered structure and Young's modulus of the solid electrolyte interphase (SEI) on silicon anodes for lithium ion batteries. *Physical Chemistry Chemical Physics* 2014, *16* (26), 13229–13238.
22. Zhang, J.; Yang, X.; Wang, R.; Dong, W.; Lu, W.; Wu, X.; Wang, X.; Li, H.; Chen, L., Influences of additives on the formation of a solid electrolyte interphase on MnO electrode studied by atomic force microscopy and force spectroscopy. *The Journal of Physical Chemistry C* 2014, *118* (36), 20756–20762.

23. Haro, M.; Kumar, P.; Zhao, J.; Koutsogiannis, P.; Porkovich, A. J.; Ziadi, Z.; Bouloumis, T.; Singh, V.; Juarez-Perez, E. J.; Toulkeridou, E.; Nordlund, K.; Djurabekova, F.; Sowwan, M.; Grammatikopoulos, P., Nano-vault architecture mitigates stress in silicon-based anodes for lithium-ion batteries. *Communications Materials* 2021, *2* (1), 16.
24. Terreblanche, J. S.; Thompson, D. L.; Aldous, I. M.; Hartley, J.; Abbott, A. P.; Ryder, K. S., Experimental visualization of commercial lithium ion battery cathodes: Distinguishing between the microstructure components using atomic force microscopy. *The Journal of Physical Chemistry C* 2020, *124* (27), 14622–14631.
25. Zhang, H.; Wang, D.; Shen, C., In-situ EC-AFM and ex-situ XPS characterization to investigate the mechanism of SEI formation in highly concentrated aqueous electrolyte for Li-ion batteries. *Applied Surface Science* 2020, *507*, 145059.
26. Huang, S.; Wang, S.; Hu, G.; Cheong, L.-Z.; Shen, C., Modulation of solid electrolyte interphase of lithium-ion batteries by LiDFOB and LiBOB electrolyte additives. *Applied Surface Science* 2018, *441*, 265–271.
27. Kang, H.; Chen, Y.; Xu, L.; Lin, Y.; Feng, Q.; Yao, H.; Zheng, Y., Top-down strategy synthesis of fluorinated graphdiyne for lithium ion battery. *RSC Advances* 2019, *9* (54), 31406–31412.
28. Haro, M.; Singh, V.; Steinhauer, S.; Toulkeridou, E.; Grammatikopoulos, P.; Sowwan, M., Nanoscale heterogeneity of multilayered Si anodes with embedded nanoparticle scaffolds for Li-ion batteries. *Advanced Science* 2017, *4* (10), 1700180.
29. Zhang, H.; Shen, C.; Huang, Y.; Liu, Z., Spontaneously formation of SEI layers on lithium metal from LiFSI/DME and LiTFSI/DME electrolytes. *Applied Surface Science* 2021, *537*, 147983.
30. Huang, S.; Cheong, L.-Z.; Wang, D.; Shen, C., Thermal stability of solid electrolyte interphase of lithium-ion batteries. *Applied Surface Science* 2018, *454*, 61–67.
31. Pittenger, B.; Osechinskiy, S.; Yablon, D.; Mueller, T., Nanoscale DMA with the atomic force microscope: A new method for measuring viscoelastic properties of nanostructured polymer materials. *JOM* 2019, *71* (10), 3390–3398.
32. Aljarrah, M. F.; Masad, E., Nanoscale viscoelastic characterization of asphalt binders using the AFM-nDMA test. *Materials and Structures* 2020, *53* (4), 110.
33. Hertz, H. *On the Contact of Rigid Elastic Solids and on Hardness*. London: Macmillan, 1896.
34. Niu, Y.-F.; Yang, Y.; Gao, S.; Yao, J.-W., Mechanical mapping of the interphase in carbon fiber reinforced poly(ether-ether-ketone) composites using peak force atomic force microscopy: Interphase shrinkage under coupled ultraviolet and hydro-thermal exposure. *Polymer Testing* 2016, *55*, 257–260.
35. Niu, Y.-F.; Yang, Y.; Wang, X.-R., Investigation of the interphase structures and properties of carbon fiber reinforced polymer composites exposed to hydrothermal treatments using peak force quantitative nanomechanics technique. *Polymer Composites* 2018, *39* (S2), E791–E796.
36. Eliyahu, M.; Emmanuel, S.; Day-Stirrat, R. J.; Macaulay, C. I., Mechanical properties of organic matter in shales mapped at the nanometer scale. *Marine and Petroleum Geology* 2015, *59*, 294–304.
37. Coq Germanicus, R.; Mercier, D.; Agrebi, F.; FÈBvre, M.; Mariolle, D.; Descamps, P.; LeclÈRe, P., Quantitative mapping of high modulus materials at the nanoscale: Comparative study between atomic force microscopy and nanoindentation. *Journal of Microscopy* 2020, *280* (1), 51–62.
38. Morales-Rivas, L.; González-Orive, A.; Garcia-Mateo, C.; Hernández-Creus, A.; Caballero, F. G.; Vázquez, L., Nanomechanical characterization of nanostructured bainitic steel: Peak Force Microscopy and Nanoindentation with AFM. *Scientific Reports* 2015, *5* (1), 17164.

39. Álvarez-Asencio, R.; Sababi, M.; Pan, J.; Ejnermark, S.; Ekman, L.; Rutland, M. W., Role of microstructure on corrosion initiation of an experimental tool alloy: A quantitative nanomechanical property mapping study. *Corrosion Science* 2014, *89*, 236–241.
40. Li, Y.; Yang, J.; Pan, Z.; Tong, W., Nanoscale pore structure and mechanical property analysis of coal: An insight combining AFM and SEM images. *Fuel* 2020, *260*, 116352.
41. Trtik, P.; Kaufmann, J.; Volz, U., On the use of peak-force tapping atomic force microscopy for quantification of the local elastic modulus in hardened cement paste. *Cement and Concrete Research* 2012, *42* (1), 215–221.
42. Megevand, B.; Pruvost, S.; Lins, L. C.; Livi, S.; Gérard, J.-F.; Duchet-Rumeau, J., Probing nanomechanical properties with AFM to understand the structure and behavior of polymer blends compatibilized with ionic liquids. *RSC Advances* 2016, *6* (98), 96421–96430.
43. Igarashi, T.; Fujinami, S.; Nishi, T.; Asao, N.; Nakajima, A. K., Nanorheological mapping of rubbers by atomic force microscopy. *Macromolecules* 2013, *46* (5), 1916–1922.
44. Nieswandt, K.; Georgopanos, P.; Abetz, C.; Filiz, V.; Abetz, V., Synthesis of poly(3-vinylpyridine)-block-polystyrene Diblock copolymers via surfactant-free RAFT emulsion polymerization. *Materials* 2019, *12* (19), 3145.
45. Graham, H. K.; Hodson, N. W.; Hoyland, J. A.; Millward-Sadler, S. J.; Garrod, D.; Scothern, A.; Griffiths, C. E. M.; Watson, R. E. B.; Cox, T. R.; Erler, J. T.; Trafford, A. W.; Sherratt, M. J., Tissue section AFM: In situ ultrastructural imaging of native biomolecules. *Matrix Biology* 2010, *29* (4), 254–260.
46. Matsko, N.; Mueller, M., AFM of biological material embedded in epoxy resin. *Journal of Structural Biology* 2004, *146* (3), 334–343.
47. Kelchtermans, M.; Lo, M.; Dillon, E.; Kjoller, K.; Marcott, C., Characterization of a polyethylene–polyamide multilayer film using nanoscale infrared spectroscopy and imaging. *Vibrational Spectroscopy* 2016, *82*, 10–15.
48. Hieulle, J.; Stecker, C.; Ohmann, R.; Ono, L. K.; Qi, Y., Scanning probe microscopy applied to organic-inorganic halide perovskite materials and solar cells. *Small Methods* 2018, *2* (1), 1700295.
49. Lang, S.-Y.; Shi, Y.; Hu, X.-C.; Yan, H.-J.; Wen, R.; Wan, L.-J., Recent progress in the application of in situ atomic force microscopy for rechargeable batteries. *Current Opinion in Electrochemistry* 2019, *17*, 134–142.
50. Jarzembski, A.; Shaskey, C.; Park, K., Review: Tip-based vibrational spectroscopy for nanoscale analysis of emerging energy materials. *Frontiers in Energy* 2018, *12* (1), 43–71.
51. Limani, N.; Boudet, A.; Blanchard, N.; Jousselme, B.; Cornut, R., Local probe investigation of electrocatalytic activity. *Chemical Science* 2020, *12* (1), 71–98.
52. Luo, D.; Sun, H.; Li, Y., Kelvin probe force microscopy in nanoscience and nanotechnology. In C. S. S. R. Kumar (Ed.) *Surface Science Tools for Nanomaterials Characterization*. Berlin: Springer, 2015, pp. 117–158.
53. Choi, T.; Lee, S.; Choi, Y. J.; Kiryukhin, V.; Cheong, S. W., Switchable ferroelectric diode and photovoltaic effect in $BiFeO_3$. *Science* 2009, *324* (5923), 63–66.
54. Röhm, H.; Leonhard, T.; Hoffmann, M. J.; Colsmann, A., Ferroelectric poling of methylammonium lead iodide thin films. *Advanced Functional Materials* 2019, *30* (5), 1908657.
55. Li, T.; Zeng, K., Probing of local multifield coupling phenomena of advanced materials by scanning probe microscopy techniques. *Advanced Materials* 2018, *30* (47), e1803064.
56. Qian, Y.; Wang, P.; Rao, L.; Song, C.; Yin, H.; Wang, X.; Zhou, G.; Notzel, R., Electric dipole of InN/InGaN quantum dots and holes and giant surface photovoltage directly measured by Kelvin probe force microscopy. *Scientific Reports* 2020, *10* (1), 5930.

57. Dappe, Y. J.; Almadori, Y.; Dau, M. T.; Vergnaud, C.; Jamet, M.; Paillet, C.; Journot, T.; Hyot, B.; Pochet, P., Grevin, B., Charge transfers and charged defects in WSe2/graphene-SiC interfaces. *Nanotechnology* 2020, *31* (25), 255709.
58. Kim, J. H., Kim, Y. K.; Lee, J. S., Perovskite tandems advance solar hydrogen production. *Joule* 2019, *3* (13), 2892–2894.
59. Li, J., et al., Touching is believing: Interrogating halide perovskite solar cells at the nanoscale via scanning probe microscopy. *NPJ Quantum Materials* 2017, *2* (1), 1–7.
60. Strelcov, E., et al., $CH_3NH_3PbI_3$ perovskites: Ferroelasticity revealed. *Science Advances* 2017, *3* (4), e1602165.
61. Xing, J., et al., Ultrafast ion migration in hybrid perovskite polycrystalline thin films under light and suppression in single crystals. *Physical Chemistry Chemical Physics* 2016, *18* (44), 30484–30490.
62. Levratovsky, Y.; Gross, E., High spatial resolution mapping of chemically-active self-assembled N-heterocyclic carbenes on Pt nanoparticles. *Faraday Discussions* 2016, *188*, 345–353.
63. Wu, C. Y., et al., High-spatial-resolution mapping of catalytic reactions on single particles. *Nature* 2017, *541* (7638), 511–515.
64. Weiss, E. A., et al., Si/$SiO_2$-templated formation of ultraflat metal surfaces on glass, polymer, and solder supports: Their use as substrates for self-assembled monolayers. *Langmuir* 2007, *23* (19), 9686–9694.
65. Morsch, S., et al., Insights into epoxy network nanostructural heterogeneity using AFM-IR. *ACS Applied Materials & Interfaces* 2016, *8* (1), 959–966.
66. Tang, F.; Bao, P.; Su, Z., Analysis of nanodomain composition in high-impact polypropylene by atomic force microscopy-infrared. *Analytical Chemistry* 2016, *88* (9), 4926–4930.
67. Tang, F., et al., In-situ spectroscopic and thermal analyses of phase domains in high-impact polypropylene. *Polymer* 2018, *142*, 155–163.
68. Guo, M., et al., Flexible robust and high-density FeRAM from array of organic ferroelectric nano-lamellae by self-assembly. *Advanced Science* 2019, *6* (6), 1801931.
69. Zhang, M.; Wang, Z. H.; Zhou, B.; Jia, X. G.; Ma, Q. S.; Yuan, N. Y.; Zheng, X. J.; Ding, J. N.; Zhang, W. H., Green anti-solvent processed planar perovskite solar cells with efficiency beyond 19%. *Solar RRL* 2018, *2* (2), 1700213.
70. Kurouski, D.; Dazzi, A.; Zenobi, R.; Centrone, A., Infrared and Raman chemical imaging and spectroscopy at the nanoscale. *Chemical Society Reviews* 2020, *49* (11), 3315–3347.
71. Waeytens, J., Doneux, T.; Napolitano, S., Evaluating mechanical properties of polymers at the nanoscale level via atomic force microscopy–infrared spectroscopy. *ACS Applied Polymer Materials* 2018, *1* (1), 3–7.
72. Wang, Z., et al., Molecular orientation in individual electrospun nanofibers studied by polarized AFM–IR. *Macromolecules* 2019, *52* (24), 9639–9645.
73. Manne, S.; Massie, J.; Elings, V. B.; Hansma, P. K.; Gewirth, A. A., Electrochemistry on a gold surface observed with the atomic force microscope. *Journal of Vacuum Science & Technology B: Microelectronics and Nanometer Structures Processing, Measurement, and Phenomena* 1991, *9* (2), 950–954.
74. Reggente, M.; Passeri, D.; Rossi, M.; Tamburri, E.; Terranova, M. L., Electrochemical atomic force microscopy: In situ monitoring of electrochemical processes. *AIP Conference Proceedings* 2017, *1873* (1), 020009.
75. Lang, S.-Y.; Shen, Z.-Z.; Hu, X.-C.; Shi, Y.; Guo, Y.-G.; Jia, F.-F.; Wang, F.-Y.; Wen, R.; Wan, L.-J., Tunable structure and dynamics of solid electrolyte interphase at lithium metal anode. *Nano Energy* 2020, *75*, 104967.

76. Zhang, Z.; Smith, K.; Jervis, R.; Shearing, P. R.; Miller, T. S.; Brett, D. J. L., Operando electrochemical atomic force microscopy of solid–electrolyte interphase formation on graphite anodes: The evolution of SEI morphology and mechanical properties. *ACS Applied Materials & Interfaces* 2020, *12* (31), 35132–35141.
77. Luchkin, S. Y.; Lipovskikh, S. A.; Katorova, N. S.; Savina, A. A.; Abakumov, A. M.; Stevenson, K. J., Solid-electrolyte interphase nucleation and growth on carbonaceous negative electrodes for Li-ion batteries visualized with in situ atomic force microscopy. *Scientific Reports* 2020, *10* (1), 1–10.
78. Wan, J.; Hao, Y.; Shi, Y.; Song, Y.-X.; Yan, H.-J.; Zheng, J.; Wen, R.; Wan, L.-J., Ultra-thin solid electrolyte interphase evolution and wrinkling processes in molybdenum disulfide-based lithium-ion batteries. *Nature Communications* 2019, *10* (1), 1–10.
79. Nault, L.; Taofifenua, C.; Anne, A.; Chovin, A.; Demaille, C.; Besong-Ndika, J.; Cardinale, D.; Carette, N.; Michon, T.; Walter, J., Electrochemical atomic force microscopy imaging of redox-immunomarked proteins on native potyviruses: From subparticle to single-protein resolution. *ACS Nano* 2015, *9* (5), 4911–4924.
80. Deng, J.; Nellist, M. R.; Stevens, M. B.; Dette, C.; Wang, Y.; Boettcher, S. W., Morphology dynamics of single-layered Ni(OH)(2)/NiOOH nanosheets and subsequent Fe incorporation studied by in situ electrochemical atomic force microscopy. *Nano Letters* 2017, *17* (11), 6922–6926.
81. Valtiner, M.; Ankah, G. N.; Bashir, A.; Renner, F. U., Atomic force microscope imaging and force measurements at electrified and actively corroding interfaces: Challenges and novel cell design. *Review of Scientific Instruments* 2011, *82* (2), 023703.
82. Shen, C.; Wang, S.; Jin, Y.; Han, W. Q., In situ AFM imaging of solid electrolyte interfaces on HOPG with ethylene carbonate and fluoroethylene carbonate-based electrolytes. *ACS Applied Materials and Interfaces* 2015, *7* (45), 25441–25447.
83. Wang, W.-W.; Gu, Y.; Yan, H.; Li, S.; He, J.-W.; Xu, H.-Y.; Wu, Q.-H.; Yan, J.-W.; Mao, B.-W., Evaluating solid-electrolyte interphases for lithium and lithium-free anodes from nanoindentation features. *Chem* 2020, 6 (10), 2728–2745.
84. Smith, C. I.; Harrison, P.; Farrell, T.; Weightman, P., The nature and stability of the Au(110)/electrochemical interface produced by flame annealing. *Journal of Physics: Condensed Matter* 2012, *24* (48), 482002.
85. Kreta, A.; Rodošek, M.; Perše, L. S.; Orel, B.; Gaberšček, M.; Vuk, A. Š., In situ electrochemical AFM, ex situ IR reflection–absorption and confocal Raman studies of corrosion processes of AA 2024-T3. *Corrosion Science* 2016, *104*, 290–309.
86. Chen, Y.; Niu, Y.; Lin, C.; Li, J.; Lin, Y.; Xu, G.; Palmer, R. E.; Huang, Z., Insight into the intrinsic mechanism of improving electrochemical performance via constructing the preferred crystal orientation in lithium cobalt dioxide. *Chemical Engineering Journal* 2020, *399*, 125708.
87. Liu, Z. H.; Brown, N. M. D., Studies using AFM and STM of the correlated effects of the deposition parameters on the topography of gold on mica. *Thin Solid Films* 1997, *300* (1), 84–94.
88. Shen, Z.-Z.; Zhou, C.; Wen, R.; Wan, L.-J., Surface mechanism of catalytic electrodes in lithium-oxygen batteries: How nanostructures mediate the interfacial reactions. *Journal of the American Chemical Society* 2020, *142* (37), 16007–16015.
89. Domi, Y.; Ochida, M.; Tsubouchi, S.; Nakagawa, H.; Yamanaka, T.; Doi, T.; Abe, T.; Ogumi, Z., In situ AFM study of surface film formation on the edge plane of HOPG for lithium-ion batteries. *The Journal of Physical Chemistry C* 2011, *115* (51), 25484–25489.
90. Herrera, S. E.; Tesio, A. Y.; Clarenc, R.; Calvo, E. J., AFM study of oxygen reduction products on HOPG in the LiPF 6–DMSO electrolyte. *Physical Chemistry Chemical Physics* 2014, *16* (21), 9925–9929.

91. Chen, H.; Qin, Z.; He, M.; Liu, Y.; Wu, Z., Application of electrochemical atomic force microscopy (EC-AFM) in the corrosion study of metallic materials. *Materials (Basel)* 2020, *13* (3), 668.
92. Wanless, E. J.; Senden, T. J.; Hyde, A. M.; Sawkins, T. J.; Heath, G. A., A new electrochemical cell for atomic force microscopy. *Review of Scientific Instruments* 1994, *65* (4), 1019–1020.
93. Nellist, M. R.; Laskowski, F. A.; Qiu, J.; Hajibabaei, H.; Sivula, K.; Hamann, T. W.; Boettcher, S. W., Potential-sensing electrochemical atomic force microscopy for in operando analysis of water-splitting catalysts and interfaces. *Nature Energy* 2018, *3* (1), 46–52.
94. Garcia, R.; Perez, R., Dynamic atomic force microscopy methods. *Surface Science Reports* 2002, *47* (6–8), 197–301.
95. Schroeter, K.; Petzold, A.; Henze, T.; Thurn-Albrecht, T., Quantitative analysis of scanning force microscopy data using harmonic models. *Macromolecules* 2009, *42* (4), 1114–1124.
96. Giessibl, F. J., Forces and frequency shifts in atomic-resolution dynamic-force microscopy. *Physical Review B* 1997, *56* (24), 16010–16015.
97. Butt, H. J.; Cappella, B.; Kappl, M., Force measurements with the atomic force microscope: Technique, interpretation and applications. *Surface Science Reports* 2005, *59* (1–6), 1–152.
98. Putman, C. A. J.; Degrooth, B. G.; Vanhulst, N. F.; Greve, J., A detailed analysis of the optical beam deflection technique for use in atomic force microscopy. *Journal of Applied Physics* 1992, *72* (1), 6–12.
99. Churnside, A. B.; Sullan, R. M. A.; Nguyen, D. M.; Case, S. O.; Bull, M. S.; King, G. M.; Perkins, T. T., Routine and timely sub-picoNewton force stability and precision for biological applications of atomic force microscopy. *Nano Letters* 2012, *12* (7), 3557–3561.
100. He, C.; Hu, C.; Hu, X.; Hu, X.; Xiao, A.; Perkins, T. T.; Li, H., Direct observation of the reversible two-state unfolding and refolding of an $\alpha/\beta$ protein by single-molecule atomic force microscopy. *Angewandte Chemie International Edition* 2015, *54* (34), 9921–9925.
101. Schumacher, Z.; Miyahara, Y.; Aeschimann, L.; Grütter, P., Improved atomic force microscopy cantilever performance by partial reflective coating. *Beilstein Journal of Nanotechnology* 2015, *6*, 1450–1456.
102. Bull, M. S.; Sullan, R. M. A.; Li, H.; Perkins, T. T., Improved single molecule force spectroscopy using micromachined cantilevers. *ACS Nano* 2014, *8* (5), 4984–4995.
103. Edwards, D. T.; Faulk, J. K.; Sanders, A. W.; Bull, M. S.; Walder, R.; LeBlanc, M.-A.; Sousa, M. C.; Perkins, T. T., Optimizing 1-µs-resolution single-molecule force spectroscopy on a commercial atomic force microscope. *Nano Letters* 2015, *15* (10), 7091–7098.
104. Edwards, D. T.; Faulk, J. K.; LeBlanc, M.-A.; Perkins, T. T., Force spectroscopy with 9-µs resolution and sub-pN stability by tailoring AFM cantilever geometry. *Biophysical Journal* 2017, *113* (12), 2595–2600.
105. Walder, R.; Van Patten, W. J.; Ritchie, D. B.; Montange, R. K.; Miller, T. W.; Woodside, M. T.; Perkins, T. T., High-precision single-molecule characterization of the folding of an HIV RNA hairpin by atomic force microscopy. *Nano Letters* 2018, *18* (10), 6318–6325.
106. Luo, D; Sun, H.; Li, Y, *Kelvin Probe Force Microscopy in Nanoscience and Nanotechnology*. Berlin Heidelberg: Springer-Verlag, 2015, pp. 117–158.
107. Zerweck, U.; Loppacher, C.; Otto, T.; Grafström, S.; Eng, L. M., Accuracy and resolution limits of Kelvin probe force microscopy. *Physical Review B* 2005, *71* (12),125424.

108. Anczykowski, B.; Kruger, D.; Babcock, K. L.; Fuchs, H., Basic properties of dynamic force spectroscopy with the scanning force microscope in experiment and simulation. *Ultramicroscopy* 1996, *66* (3–4), 251–259.
109. Anczykowski, B.; Kruger, D.; Fuchs, H., Cantilever dynamics in quasinoncontact force microscopy: Spectroscopic aspects. *Physical Review B* 1996, *53* (23), 15485–15488.
110. Garcia, R.; San Paulo, A., Dynamics of a vibrating tip near or in intermittent contact with a surface. *Physical Review B* 2000, *61* (20), 13381–13384.
111. San Paulo, A.; Garcia, R., Tip-surface forces, amplitude, and energy dissipation in amplitude-modulation (tapping mode) force microscopy. *Physical Review B* 2001, *64* (19).
112. Briscoe, B. J.; Sebastian, K. S.; Adams, M. J., The effect of indenter geometry on the elastic response to indentation. *Journal of Physics D: Applied Physics* 1994, *27* (6), 1156–1162.
113. Su, C.; Lombrozo, P. M., Method and apparatus of high speed property mapping. United States Patent, 2010, (Patent No. US 7,658,097 B2).
114. Hu, Y.; Hu, S.; Su, C., Method and apparatus of operating a scanning probe microscope. United States Patent Application Publication, 2010, (Pub. No. US 2010/0122385 A1).
115. Nina Balke, P. M., et al., Differentiating ferroelectric and nonferroelectric electromechanical effects with scanning probe microscopy. *ACS Nano* 2015, *9* (6), 6484–6492.
116. Horcas, I.; Fernandez, R.; Gomez-Rodriguez, J. M.; Colchero, J.; Gomez-Herrero, J.; Baro, A. M., WSXM: A software for scanning probe microscopy and a tool for nanotechnology. *Review of Scientific Instruments* 2007, *78* (1), 013705.
117. Klapetek, P., *Quantitative Data Processing in Scanning Probe Microscopy*, 2nd edition. Amsterdam, Netherlands: Elsevier, 2018.
118. Chen, W.; Li, K.; Wang, Y.; Feng, X.; Liao, Z.; Su, Q.; Lin, X.; He, Z., Black phosphorus quantum dots for hole extraction of typical planar hybrid perovskite solar cells. *Journal of Physical Chemistry Letters* 2017, *8* (3), 591–598.
119. Ding, J.; Zhao, Y.; Du, S.; Sun, Y.; Cui, H.; Zhan, X.; Cheng, X.; Jing, L., Controlled growth of MAPbBr(3) single crystal: Understanding the growth morphologies of vicinal hillocks on (100) facet to form perfect cubes. *Journal of Materials Science* 2017, *52* (13), 7907–7916.
120. Duan, L.; Guli, M.; Zhang, Y.; Yi, H.; Haque, F.; Uddin, A., The air effect in the burn-in thermal degradation of nonfullerene organic solar cells. *Energy Technology* 2020, *8* (5), 1901401.
121. Fairfield, D. J.; Sai, H.; Narayanan, A.; Passarelli, J. V.; Chen, M.; Palasz, J.; Palmer, L. C.; Wasielewski, M. R.; Stupp, S. I., Structure and chemical stability in perovskite-polymer hybrid photovoltaic materials. *Journal of Materials Chemistry A* 2019, *7* (4), 1687–1699.
122. Zhu, J.; Fan, F.; Chen, R.; An, H.; Feng, Z.; Li, C., Direct imaging of highly anisotropic photogenerated charge separations on different facets of a single $BiVO_4$ photocatalyst. *Angewandte Chemie International Edition* 2015, *54* (31), 9111–9114.
123. Chen, R.; Pang, S.; An, H.; Zhu, J.; Ye, S.; Gao, Y.; Fan, F.; Li, C., Charge separation via asymmetric illumination in photocatalytic $Cu_2O$ particles. *Nature Energy* 2018, *3* (8), 655–663.
124. Green, M., Thin-film solar cells: Review of materials, technologies and commercial status. *Journal of Materials Science: Materials in Electronics* 2007, *18*, 15–19.
125. Saliou, K.; Hilt, F.; Fischer, G.; Hildebrandt, T.; Grand, P.-P.; Drahi, E., Powerful topographic analyzing method using fast Fourier transform for c-Si solar cells and advanced technologies. *AIP Conference Proceedings* 2019, *2147* (1), 020013.
126. Obaidulla, S. M., et al., Surface roughening and scaling behavior of vacuum-deposited SnCl2Pc organic thin films on different substrates. *Applied Physics Letters* 2015, *107*, 221910.

127. Karan, S.; Mallik, B., Power spectral density analysis and photoconducting behavior in copper(ii) phthalocyanine nanostructured thin films. *Physical Chemistry Chemical Physics* 2008, *10* (45), 6751–6761.
128. Gao, Y.; Nie, W.; Zhu, Q.; Wang, X.; Wang, S.; Fan, F.; Li, C., The polarization effect in surface-plasmon-induced photocatalysis on Au/$TiO_2$ nanoparticles. *Angewandte Chemie International Edition* 2020, *59* (41), 18218–18223.
129. Mu, L.; Zhao, Y.; Li, A.; Wang, S.; Wang, Z.; Yang, J.; Wang, Y.; Liu, T.; Chen, R.; Zhu, J.; Fan, F.; Li, R.; Li, C., Enhancing charge separation on high symmetry $SrTiO_3$ exposed with anisotropic facets for photocatalytic water splitting. *Energy & Environmental Science* 2016, *9* (7), 2463–2469.
130. Wang, S.; Gao, Y.; Miao, S.; Liu, T.; Mu, L.; Li, R.; Fan, F.; Li, C., Positioning the water oxidation reaction sites in plasmonic photocatalysts. *Journal of the American Chemical Society* 2017, *139* (34), 11771–11778.
131. Leung, K. M.; Wanger, G.; Guo, Q.; Gorby, Y.; Southam, G.; Lau, W. M.; Yang, J., Bacterial nanowires: Conductive as silicon, soft as polymer. *Soft Matter* 2011, *7* (14), 6617–6621.
132. Wu, R.; Tian, X.; Xiao, Y.; Ulstrup, J.; Mølager Christensen, H. E.; Zhao, F.; Zhang, J., Selective electrocatalysis of biofuel molecular oxidation using palladium nanoparticles generated on *Shewanella oneidensis* MR-1. *Journal of Materials Chemistry A* 2018, *6* (23), 10655–10662.
133. Leung, K. M.; Wanger, G.; El-Naggar, M. Y.; Gorby, Y.; Southam, G.; Lau, W. M.; Yang, J., *Shewanella oneidensis* MR-1 bacterial nanowires exhibit p-type, tunable electronic behavior. *Nano Letters* 2013, *13* (6), 2407–2411.
134. Deniset-Besseau, A.; Prater, C. B.; Virolle, M.-J.; Dazzi, A., Monitoring TriAcylGlycerols accumulation by atomic force microscopy based infrared spectroscopy in streptomyces species for biodiesel applications. *The Journal of Physical Chemistry Letters* 2014, *5* (4), 654–658.
135. Bhat, S. V.; Sultana, T.; Körnig, A.; McGrath, S.; Shahina, Z.; Dahms, T. E. S., Correlative atomic force microscopy quantitative imaging-laser scanning confocal microscopy quantifies the impact of stressors on live cells in real-time. *Scientific Reports* 2018, *8* (1), 8305.
136. Benn, G.; Pyne, A. L. B.; Ryadnov, M. G.; Hoogenboom, B. W., Imaging live bacteria at the nanoscale: Comparison of immobilisation strategies. *Analyst* 2019, *144* (23), 6944–6952.
137. Dufrêne, Y. F.; Martínez-Martín, D.; Medalsy, I.; Alsteens, D.; Müller, D. J., Multiparametric imaging of biological systems by force-distance curve–based AFM. *Nature Methods* 2013, *10* (9), 847–854.
138. Smolyakov, G.; Formosa-Dague, C.; Severac, C.; Duval, R. E.; Dague, E., High speed indentation measures by FV, QI and QNM introduce a new understanding of bionanomechanical experiments. *Micron* 2016, *85*, 8–14.
139. Chopinet, L.; Formosa, C.; Rols, M. P.; Duval, R. E.; Dague, E., Imaging living cells surface and quantifying its properties at high resolution using AFM in QI™ mode. *Micron* 2013, *48*, 26–33.
140. Huang, Y. L.; Nikonov, D.; Addiego, C.; Chopdekar, R. V.; Prasad, B.; Zhang, L.; Chatterjee, J.; Liu, H. J.; Farhan, A.; Chu, Y. H.; Yang, M.; Ramesh, M.; Qiu, Z. Q.; Huey, B. D.; Lin, C. C.; Gosavi, T.; Iniguez, J.; Bokor, J.; Pan, X.; Young, I.; Martin, L. W.; Ramesh, R., Manipulating magnetoelectric energy landscape in multiferroics. *Nature Communications* 2020, *11* (1), 2836.
141. Nakashima, S.; Higuchi, T.; Yasui, A.; Kinoshita, T.; Shimizu, M.; Fujisawa, H., Enhancement of photovoltage by electronic structure evolution in multiferroic Mn-doped $BiFeO_3$ thin films. *Scientific Reports* 2020, *10* (1), 15108.

142. You, H.; Wu, Z.; Zhang, L.; Ying, Y.; Liu, Y.; Fei, L.; Chen, X.; Jia, Y.; Wang, Y.; Wang, F.; Ju, S.; Qiao, J.; Lam, C. H.; Huang, H., Harvesting the vibration energy of $BiFeO_3$ nanosheets for hydrogen evolution. *Angewandte Chemie International Edition in English* 2019, *58* (34), 11779–11784.
143. Mathurin, J.; Pancani, E.; Deniset-Besseau, A.; Kjoller, K.; Prater, C. B.; Gref, R.; Dazzi, A., How to unravel the chemical structure and component localization of individual drug-loaded polymeric nanoparticles by using tapping AFM-IR. *Analyst* 2018, *143* (24), 5940–5949.
144. Wu, S.; Zhang, J.; Liu, X.; Lv, S.; Gao, R.; Cai, W.; Wang, F.; Fu, C., Micro-area ferroelectric, piezoelectric and conductive properties of single BiFeO(3) nanowire by scanning probe microscopy. *Nanomaterials (Basel)* 2019, *9* (2), 190.
145. Sharma, Y.; Agarwal, R.; Collins, L.; Zheng, Q.; Ievlev, A. V.; Hermann, R. P.; Cooper, V. R.; Kc, S.; Ivanov, I. N.; Katiyar, R. S.; Kalinin, S. V.; Lee, H. N.; Hong, S.; Ward, T. Z., Self-assembled room temperature multiferroic $BiFeO_3$-$LiFe_5O_8$ nanocomposites. *Advanced Functional Materials* 2019, *30* (3), 1906849.
146. Jiang, J.; Bai, Z. L.; Chen, Z. H.; He, L.; Zhang, D. W.; Zhang, Q. H.; Shi, J. A.; Park, M. H.; Scott, J. F.; Hwang, C. S.; Jiang, A. Q., Temporary formation of highly conducting domain walls for non-destructive read-out of ferroelectric domain-wall resistance switching memories. *Nature Materials* 2018, *17* (1), 49–56.
147. Zhang, Y.; Lu, H.; Xie, L.; Yan, X.; Paudel, T. R.; Kim, J.; Cheng, X.; Wang, H.; Heikes, C.; Li, L.; Xu, M.; Schlom, D. G.; Chen, L. Q.; Wu, R.; Tsymbal, E. Y.; Gruverman, A.; Pan, X., Anisotropic polarization-induced conductance at a ferroelectric-insulator interface. *Nature Nanotechnology* 2018, *13* (12), 1132–1136.
148. Zang, Y.; Xie, D.; Chen, Y.; Wu, X.; Ren, T.; Wei, J.; Zhu, H.; Plant, D., Electrical and thermal properties of a carbon nanotube/polycrystalline BiFeO3/Pt photovoltaic heterojunction with CdSe quantum dots sensitization. *Nanoscale* 2012, *4* (9), 2926–2930.
149. Fan, H.; Fan, Z.; Li, P.; Zhang, F.; Tian, G.; Yao, J.; Li, Z.; Song, X.; Chen, D.; Han, B.; Zeng, M.; Wu, S.; Zhang, Z.; Qin, M.; Lu, X.; Gao, J.; Lu, Z.; Zhang, Z.; Dai, J.; Gao, X.; Liu, J.-M., Large electroresistance and tunable photovoltaic properties of ferroelectric nanoscale capacitors based on ultrathin super-tetragonal $BiFeO_3$ films. *Journal of Materials Chemistry C* 2017, *5* (13), 3323–3329.
150. Guo, R.; You, L.; Lin, W.; Abdelsamie, A.; Shu, X.; Zhou, G.; Chen, S.; Liu, L.; Yan, X.; Wang, J.; Chen, J., Continuously controllable photoconductance in freestanding $BiFeO_3$ by the macroscopic flexoelectric effect. *Nature Communications* 2020, *11* (1), 2571.
151. Knoche, D. S.; Steimecke, M.; Yun, Y.; Muhlenbein, L.; Bhatnagar, A., Anomalous circular bulk photovoltaic effect in $BiFeO_3$ thin films with stripe-domain pattern. *Nature Communications* 2021, *12* (1), 282.
152. Knoche, D. S.; Yun, Y.; Ramakrishnegowda, N.; Muhlenbein, L.; Li, X.; Bhatnagar, A., Domain and switching control of the bulk photovoltaic effect in epitaxial $BiFeO_3$ thin films. *Scientific Reports* 2019, *9* (1), 13979.
153. Soergel, E., Piezoresponse force microscopy (PFM). *Journal of Physics D: Applied Physics* 2011, *44* (46), 464003.
154. Lu, Z.; Li, P.; Wan, J. G.; Huang, Z.; Tian, G.; Pan, D.; Fan, Z.; Gao, X.; Liu, J. M., Controllable photovoltaic effect of microarray derived from epitaxial tetragonal $BiFeO_3$ films. *ACS Applied Materials & Interfaces* 2017, *9* (32), 27284–27289.
155. Liou, Y. D.; Chiu, Y. Y.; Hart, R. T.; Kuo, C. Y.; Huang, Y. L.; Wu, Y. C.; Chopdekar, R. V.; Liu, H. J.; Tanaka, A.; Chen, C. T.; Chang, C. F.; Tjeng, L. H.; Cao, Y.; Nagarajan, V.; Chu, Y. H.; Chen, Y. C.; Yang, J. C., Deterministic optical control of room temperature multiferroicity in $BiFeO_3$ thin films. *Nature Materials* 2019, *18* (6), 580–587.
156. Chen, D.; Niu, F.; Qin, L.; Wang, S.; Zhang, N.; Huang, Y., Defective $BiFeO_3$ with surface oxygen vacancies: Facile synthesis and mechanism insight into photocatalytic performance. *Solar Energy Materials and Solar Cells* 2017, *171*, 24–32.

157. Dazzi, A., et al., Local infrared microspectroscopy with subwavelength spatial resolution with an atomic force microscope tip used as a photothermal sensor. *Optics Letters* 2005, *30* (18), 2388–2390.
158. Yuan, Y., et al., Photovoltaic switching mechanism in lateral structure hybrid perovskite solar cells. *Advanced Energy Materials* 2015, *5* (15), 1500615.
159. Katzenmeyer, A. M., et al., Absorption spectroscopy and imaging from the visible through mid-infrared with 20nm resolution. *Analytical Chemistry* 2015, *87* (6), 3154–3159.
160. Chae, J., et al., Chloride incorporation process in $CH_3NH_3PbI_3$–x Clx perovskites via nanoscale bandgap maps. *Nano Letters* 2015, *15* (12), 8114–8121.
161. Liu, Y., et al., Light-ferroic interaction in hybrid organic–inorganic perovskites. *Advanced Optical Materials* 2019, *7* (23), 1901451.

# Index

Note: **Bold** page numbers refer to tables; *italic* page numbers refer to figures.

## M



## N

## Q



## R

## S

## T